Understanding Society

An Introductory Reader

FIFTH EDITION

MARGARET L. ANDERSEN
University of Delaware

KIM A. LOGIO
Saint Joseph's University

HOWARD F. TAYLOR
Princeton University

Australia • Brazil • Japan • Korea • Mexico • Singapore • Spain • United Kingdom • United States

Understanding Society: An Introductory Reader, Fifth Edition
Margaret L. Andersen, Kim A. Logio, Howard F. Taylor

Product Director: Marta Lee-Perriard

Product Manager: Jenny Harrison

Content Developer: Lori Bradshaw, S4Carlisle

Media Developer: John Chell

Product Assistant: Chelsea Meredith

Marketing Manager: Kara Kindstrom

Content Project Manager: Rita Jaramillo

Art Director: Michelle Klunker

Manufacturing Planner: Judy Inouye

Production Service: Lynn Lustberg, MPS Limited

Text Researcher: Pinky Subi

Copy Editor: Heather McElwain

Cover Designer: Lou Ann Thesing

Cover Image: © Mike Theiss/National Geographic/Getty Images, Inc.

Compositor: MPS Limited

Library of Congress Control Number: 2014942473

Student Edition:
ISBN: 978-1-305-09370-6

Cengage Learning
20 Channel Center Street
Boston, MA 02210
USA

Cengage Learning is a leading provider of customized learning solutions with office locations around the globe, including Singapore, the United Kingdom, Australia, Mexico, Brazil, and Japan. Locate your local office at **www.cengage.com/global**.

Cengage Learning products are represented in Canada by Nelson Education, Ltd.

To learn more about Cengage Learning Solutions, visit **www.cengage.com**.

Purchase any of our products at your local college store or at our preferred online store **www.cengagebrain.com**.

Printed in the United States of America
Print Number: 01 Print Year: 2014

Contents

Preface

This anthology is intended to introduce students to the value of a sociological perspective. Most of those who will read the articles in this collection are first- or second-year students—students who are likely taking their first course in sociology, and most of whom are not sociology majors. What we hope to do is excite students about the value of a sociological perspective, showing them how sociologists explain and interpret the patterns, structures, and changes in contemporary society.

With this in mind, several themes guided our decisions about the articles to be included. One major criterion is that the articles will be accessible to undergraduate readers, even while grounded in sound sociological concepts, research, and theory. We include articles with a mix of methodological styles, including some that may not have been written by sociologists but that engage a sociological perspective. With each edition, we revise what is included to reflect topics of contemporary relevance. The anthology has a strong focus on diversity, both in the sections on class, race, gender, and age, and in other selections throughout the book. We also include articles that reflect the global forces that are so affecting current social changes. And, to introduce students to the basic theoretical perspectives in sociology, we also include short pieces from some of the classic thinkers of our discipline.

The fifth edition thus keeps the same themes as the first four editions, but we have included thirty-four new articles that examine some of the latest developments in society, such as the economic downturn, socialization in the military, migration and immigration, and multigenerational families, to name a few. More detail about these and other topics follows.

The table of contents is modeled on the companion text, *Sociology: The Essentials* by Andersen, Taylor, and Logio, but it can easily be adapted for use with other introductory books or as a stand-alone text. In keeping with what reviewers told us about how many articles students could read in a week, we have somewhat shortened this edition, but we also wanted to include enough articles to give instructors flexibility in what they assigned.

Thus, the major themes of this anthology are:

- **Contemporary research:** We wanted students to see examples of strong contemporary research, presented in a fashion that is accessible to beginning undergraduates. The articles included here feature different styles of sociological research. In fact, we now include an overview of sociological research methods with Earl Babbie's selection, "Human Inquiry and Science." We do so to give students a basic introduction to the value and logic of sociological research. But, in addition, different articles utilize different methodological approaches. For example, Rosanna Hertz's article, "'Why Can't I Have What I Want?' Timing Employment, Marriage, and Motherhood," is based on extensive interviews with mothers who are not married. Peter Edelman's "How Has the Picture of Poverty Changed?" shows students how secondary data can be used to articulate an argument. These are but two examples of the diversity of sociological research found in this anthology, but they show students how research designs and methods differ, depending on the nature of the research question being asked.
- **Current events in society:** We want students to see how a sociological perspective can help them understand events in the contemporary world. Students need to understand that the issues they see in society have sociological dimensions. Thus, we include an article by Ronald Brownstein, "Children of the Great Recession: A Tour of the Generational Landscape, from Struggles to Successes, Coast to Coast," which will help students understand some of their fears about entering the job market during the economic downturn. Brian Halweil's "The Rise of Food Democracy" analyzes the global forces shaping the distribution and consumption of food—a subject of increasing interest to students. Even the current topic of football head concussions is explored in the context of changing norms of masculinity in "Examining Media Contestation of Masculinity and Head Trauma in the National Football League" by Eric Anderson and Edward M. Kian.
- **Diversity:** In keeping with our understanding that society is increasingly diverse, we selected articles that show the range of experiences people have by virtue of differences in race, gender, class, sexual orientation, and other characteristics (such as age and religion). Many of the selections bring a comprehensive analysis of race, class, and gender to the subject at hand, thus adding to students' understanding of how diverse groups experience the social structure of society. We also add here a discussion of diversity in the context of education with the inclusion of Dan Goodley's "Inclusive Disability Studies." Additionally, David Naguib Pellow and Robert J. Brulle in "Poisoning the Planet: The Struggle for Environmental Justice" show the racial dimensions to the broad concern of environmental degradation.
- **Global perspective:** We completely revised the section on global stratification, anchoring it with a very good introduction by Maxine Baca Zinn and D. Stanley Eitzen ("Globalization: An Introduction") that lays out the

different dimensions of globalization. This section now also includes an analysis of how social class shapes the experience of workers in the global economy (Katie Quan, "Global Strategies for Workers"), as well as a piece on food production that will interest students who are now more aware of the global patterns in food production and distribution. We also added a piece on global migration (Khalid Koser, "Why Migration Matters") to help students understand the flow of labor in a global economy. Throughout this section, students can learn how their own consumption practices are shaped by globalization in ways they might not have recognized.

- **Applying sociological knowledge:** Our students commonly ask, "What can you do with a sociological perspective?" We think this is an important question and one with many different answers. Sociologists use their knowledge in a variety of ways: to influence social policy formation, to interpret current events, and to educate people about common misconceptions and stereotypes, to name a few. Many of the new readings have a particularly contemporary appeal because the topics relate directly to changes in students' lives. For example, Danielle M. Currier (in "Strategic Ambiguity: Protecting Emphasized Femininity and Hegemonic Masculinity in the Hookup Culture") provides a contemporary analysis of gender dynamics in the hookup culture on college campuses. Also, Jen'nan Ghazal Read's article, "Muslims in America," provides an important perspective on a group that is widely misunderstood. We hope that students can see themselves and the issues in their lives throughout the anthology.
- **Classical theory:** We think it is important that introductory students learn about the contributions of classical sociological theorists. Thus, we have kept articles by Weber, Marx and Engels, DuBois, and Goffman that will showcase some of the most important classics. As with all of the selections, we include discussion questions at the end of each reading to help students think about how such classic pieces are reflected in contemporary issues. For example, Max Weber's argument about the Protestant ethic and the spirit of capitalism is fascinating to think about in the contemporary context of increased consumerism and increased class inequality. Students might ask whether contemporary patterns of wealth and consumption no longer reflect the asceticism and moral calling about which Weber wrote. W. E. B. DuBois's reflections on double consciousness also continue to be relevant in discussions of race and group perceptions.

NEW TO THE FIFTH EDITION

The fifth edition of *Understanding Society* is organized to follow the outline of most introductory courses. We replaced many of the articles in the thematic sections to reflect more current research and current events, particularly the sections on globalization, gender, families, sexuality, and education. We have continued the pedagogical features that we introduced in the earlier editions. Thus, at the conclusion of each article, we include a list of **key concepts** that are reflected in the article. These concepts are then defined in the **glossary** at the end of the book. We also include **student exercises** at the end of each major section. Through these exercises, students can—on their own or in group projects—apply what they have learned to their own observations of social behavior. The exercises also enhance the theme of applying sociological knowledge to everyday life because they engage students in hands-on activities that will enrich their understanding of the course material.

Brief introductions before each article place the article in context and help frame students' understanding of the selection. **Discussion questions** at the end of each reading help students think about the implications of what they have read.

The **thirty-four new readings** in the book were selected to engage student interest, to reflect the richness of sociological thought, and to add articles that address issues that have emerged since the publication of the last edition (such as the Affordable Care Act, the influence of the Tea Party, and the legalization of same-sex marriage, to name a few). In fact, we are always amazed at how, in a short period of time, important new topics develop that influence our decisions about the content of this anthology.

The book includes several new articles that explore popular culture—an increasingly important subject in the experiences of student readers. Thus, Susan Jane Gilman's "Klaus Barbie, and Other Dolls I'd Like to See" takes a humorous approach to illustrating the import of popular Barbie dolls on young women's gender identities. In a different way, George Ritzer's "The McDonaldization of Society" shows students how this common organizational form now penetrates so many dimensions of daily life.

Other new articles include those that examine the new politics of the Tea Party ("Cultures of the Tea Party" by Andrew J. Perrin, Steven J. Tepper, Neal Caren, and Sally Morris); the ongoing problem of health care disparities ("Beyond the Affordable Care Act: Achieving Real Improvements in Americans' Health" by David R. Williams, Mark B. McClellan, and Alice M. Rivlin); and, among other topics, the growing diversity of the U.S. population (William Frey's "Zooming In on Diversity").

In sum, with this anthology we hope to capture student interest in sociology, provide interesting research and theory, incorporate the analysis of diversity into the core of the sociological perspective, analyze the increasingly global dimensions of society, and show students how what they learn about sociology can be applied to real issues and problems.

PEDAGOGICAL FEATURES

In addition to the sociological content of this reader, a number of pedagogical features enrich student learning and help instructors teach with the book. Each essay has a **brief introductory paragraph** that identifies the major themes and questions being raised in the article. A short list of **key concepts** at the end of each reading helps students understand how the reading is related to basic ideas in the field of sociology. All of the core concepts are defined in the **glossary** at the end of the book. This will especially help instructors who use this anthology as a stand-alone text. We also follow each article with **discussion questions** that students can use to improve their critical thinking and to reinforce their understanding of the article's major points. Many of these questions could also be used as the basis for class discussion, student papers, or research exercises and projects. And new to this edition are the **student exercises** included at the end of each major part. Finally, a **subject/name index** helps students and faculty locate specific topics and authors in the book.

ACKNOWLEDGMENTS

Many people helped in a variety of ways as we revised this anthology. We especially thank Dana Alvaré for her assistance in locating articles as well as tolerating many of the seemingly mindless, but essential, details of copying, making PDF files, and locating materials for the new edition. We also appreciate the support of Dana Brittingham whose work makes a lot of things possible. Thank you, Patrick Harker (President, University of Delaware), Scott Douglass (Executive Vice President of the University of Delaware), and Patricia Wilson (Vice President at the University of Delaware) for providing funding that supported the research for this book.

We sincerely appreciate the enthusiasm and support provided by Cengage's editorial team. Thank you Seth Dobrin for your support for this work, as well as Lori Bradshaw for consulting on the development and overseeing the book's production. And, most especially, we thank Richard, Jim, Nolan, Owen, and Isabel for their love and support and for reminding us every day of the most important things in life.

About the Editors

Margaret L. Andersen is the Edward F. and Elizabeth Goodman Rosenberg Professor of Sociology at the University of Delaware, where she also holds joint appointments in Women's Studies and Black American Studies and serves as the Vice Provost for Faculty Affairs and Diversity. She is the author of *On Land and On Sea: A Century of Women in the Rosenfeld Collection; Thinking about Women: Sociological Perspectives on Sex and Gender; Living Art: The Life of Paul R. Jones, African American Art Collector; Race, Class, and Gender* (with Patricia Hill Collins); *Sociology: Understanding a Diverse Society* (with Howard F. Taylor); and *Sociology: The Essentials* (co-authored with Howard Taylor and Kim A. Logio). She is a recipient of the American Sociological Association's Jessie Bernard Award, Merit Award from the Eastern Sociological Society, and the Sociologists for Women in Society's Feminist Lecturer Award. She has served as Vice President of the American Sociological Association and as President of the Eastern Sociological Society. She has won two teaching awards at the University of Delaware.

Kim A. Logio is Associate Professor and Chair of the Department of Sociology at Saint Joseph's University in Philadelphia, where she teaches courses in research methods, childhood obesity, and women and health. Her research on adolescent body image has appeared in *Violence against Women* and has been presented at the American Sociological Association meetings. Her research on childhood obesity and adolescent health focuses on race, class, and gender differences among young people's access to healthy food and knowledge of nutrition. She is also the coauthor of *Adventures in Criminal Justice Research* with George Dowdall, Earl Babbie, and

Fred Halley. She has been heard on National Public Radio discussing childhood obesity. She lives in Pennsylvania with her husband and three children.

Courtesy of Howard F. Taylor

Howard F. Taylor is Professor of Sociology at Princeton University. He is the author of *Balance in Small Groups; The IQ Game; Sociology: Understanding a Diverse Society* (with Margaret L. Andersen); and *Sociology: The Essentials* (with Margaret L. Andersen). He is the winner of the DuBois-Johnson-Frazier Award, given by the American Sociological Association for distinguished research in race and ethnic relations, and has received the Princeton University President's Award for Distinguished Teaching. He is past president of the Eastern Sociological Society and is currently writing a book titled *The Triple Whammy: Race, Class, and Gender Bias in the SAT.*

PART I

The Sociological Perspective

1

The Sociological Imagination

C. WRIGHT MILLS

First published in 1959, C. Wright Mills's essay, taken from his book The Sociological Imagination, *is a classic statement about the sociological perspective. Mills was a man of his times, and his sexist language intrudes on his argument, but the questions he poses about the connection between history, social structure, and peoples' biographies (or lived experiences) still resonate today. His central theme is that the task of sociology is to understand how social and historical structures impinge on the lives of different people in society.*

Nowadays men often feel that their private lives are a series of traps. They sense that within their everyday worlds, they cannot overcome their troubles, and in this feeling, they are often quite correct: What ordinary men are directly aware of and what they try to do are bounded by the private orbits in which they live; their visions and their powers are limited to the close-up scenes of job, family, neighborhood; in other milieux, they move vicariously and remain spectators. And the more aware they become, however vaguely, of ambitions and of threats which transcend their immediate locales, the more trapped they seem to feel.

Underlying this sense of being trapped are seemingly impersonal changes in the very structure of continent-wide societies. The facts of contemporary history are also facts about the success and the failure of individual men and women. When a society is industrialized, a peasant becomes a worker; a feudal lord is liquidated or becomes a businessman. When classes rise or fall, a man is employed or unemployed; when the rate of investment goes up or down, a man takes new heart or goes broke. When wars happen, an insurance salesman becomes a rocket launcher; a store clerk, a radar man; a wife lives alone; a child grows up without a father. Neither the life of an individual nor the history of a society can be understood without understanding both.

Yet men do not usually define the troubles they endure in terms of historical change and institutional contradiction. The well-being they enjoy, they do not usually impute to the big ups and downs of the societies in which they live.

Seldom aware of the intricate connection between the patterns of their own lives and the course of world history, ordinary men do not usually know what this connection means for the kinds of men they are becoming and for the kinds of history-making in which they might take part. They do not possess the quality of mind essential to grasp the interplay of man and society, of biography and history, of self and world. They cannot cope with their personal troubles in such ways as to control the structural transformations that usually lie behind them....

The sociological imagination enables its possessor to understand the larger historical scene in terms of its meaning for the inner life and the external career of a variety of individuals. It enables him to take into account how individuals, in the welter of their daily experience, often become falsely conscious of their social positions. Within that welter, the framework of modern society is sought, and within that framework the psychologies of a variety of men and women are formulated. By such means the personal uneasiness of individuals is focused upon explicit troubles and the indifference of publics is transformed into involvement with public issues.

The first fruit of this imagination—and the first lesson of the social science that embodies it—is the idea that the individual can understand his own experience and gauge his own fate only by locating himself within his period, that he can know his own chances in life only by becoming aware of those of all individuals in his circumstances. In many ways it is a terrible lesson; in many ways a magnificent one. We do not know the limits of man's capacities for supreme effort or willing degradation, for agony or glee, for pleasurable brutality or the sweetness of reason. But in our time we have come to know that the limits of "human nature" are frighteningly broad. We have come to know that every individual lives, from one generation to the next, in some society; that he lives out a biography, and that he lives it out within some historical sequence. By the fact of his living he contributes, however minutely, to the shaping of this society and to the course of its history, even as he is made by society and by its historical push and shove.

The sociological imagination enables us to grasp history and biography and the relations between the two within society. That is its task and its promise. To recognize this task and this promise is the mark of the classic social analyst....

No social study that does not come back to the problems of biography, of history and of their intersections within a society has completed its intellectual journey. Whatever the specific problems of the classic social analysts, however limited or however broad the features of social reality they have examined, those who have been imaginatively aware of the promise of their work have consistently asked three sorts of questions:

1. What is the structure of this particular society as a whole? What are its essential components, and how are they related to one another? How does it differ from other varieties of social order? Within it, what is the meaning of any particular feature for its continuance and for its change?
2. Where does this society stand in human history? What are the mechanics by which it is changing? What is its place within and its meaning for the development of humanity as a whole? How does any particular feature

we are examining affect, and how is it affected by, the historical period in which it moves? And this period—what are its essential features? How does it differ from other periods? What are its characteristic ways of history-making?

3. What varieties of men and women now prevail in this society and in this period? And what varieties are coming to prevail? In what ways are they selected and formed, liberated and repressed, made sensitive and blunted? What kinds of "human nature" are revealed in the conduct and character we observe in this society in this period? And what is the meaning for "human nature" of each and every feature of the society we are examining?

Whether the point of interest is a great power state or a minor literary mood, a family, a prison, a creed—these are the kinds of questions the best social analysts have asked. They are the intellectual pivots of classic studies of man in society—and they are the questions inevitably raised by any mind possessing the sociological imagination. For that imagination is the capacity to shift from one perspective to another—from the political to the psychological; from examination of a single family to comparative assessment of the national budgets of the world; from the theological school to the military establishment; from considerations of an oil industry to studies of contemporary poetry. It is the capacity to range from the most impersonal and remote transformations to the most intimate features of the human self—and to see the relations between the two. Back of its use there is always the urge to know the social and historical meaning of the individual in the society and in the period in which he has his quality and his being.

That, in brief, is why it is by means of the sociological imagination that men now hope to grasp what is going on in the world, and to understand what is happening in themselves as minute points of the intersections of biography and history within society. In large part, contemporary man's self-conscious view of himself as at least an outsider, if not a permanent stranger, rests upon an absorbed realization of social relativity and of the transformative power of history. The sociological imagination is the most fruitful form of this self-consciousness. By its use men whose mentalities have swept only a series of limited orbits often come to feel as if suddenly awakened in a house with which they had only supposed themselves to be familiar. Correctly or incorrectly, they often come to feel that they can now provide themselves with adequate summations, cohesive assessments, comprehensive orientations. Older decisions that once appeared sound now seem to them products of a mind unaccountably dense. Their capacity for astonishment is made lively again. They acquire a new way of thinking, they experience a transvaluation of values: in a word, by their reflection and by their sensibility, they realize the cultural meaning of the social sciences.

Perhaps the most fruitful distinction with which the sociological imagination works is between "the personal troubles of milieu" and "the public issues of social structure." This distinction is an essential tool of the sociological imagination and a feature of all classic work in social science.

Troubles occur within the character of the individual and within the range of his immediate relations with others; they have to do with his self and with those limited areas of social life of which he is directly and personally aware.

Accordingly, the statement and the resolution of troubles properly lie within the individual as a biographical entity and within the scope of his immediate milieu—the social setting that is directly open to his personal experience and to some extent his willful activity. A trouble is a private matter: values cherished by an individual are felt by him to be threatened.

Issues have to do with matters that transcend these local environments of the individual and the range of his inner life. They have to do with the organization of many such milieux into the institutions of an historical society as a whole, with the ways in which various milieux overlap and interpenetrate to form the larger structure of social and historical life. An issue is a public matter: some value cherished by publics is felt to be threatened. Often there is a debate about what that value really is and about what it is that really threatens it. This debate is often without focus if only because it is the very nature of an issue, unlike even widespread trouble, that it cannot very well be defined in terms of the immediate and everyday environments of ordinary men. An issue, in fact, often involves a crisis in institutional arrangements, and often too it involves what Marxists call "contradictions" or "antagonisms."

In these terms, consider unemployment. When, in a city of 100,000, only one man is unemployed, that is his personal trouble, and for its relief we properly look to the character of the man, his skills, and his immediate opportunities. But when in a nation of 50 million employees, 15 million men are unemployed, that is an issue, and we may not hope to find its solution within the range of opportunities open to any one individual. The very structure of opportunities has collapsed. Both the correct statement of the problem and the range of possible solutions require us to consider the economic and political institutions of the society, and not merely the personal situation and character of a scatter of individuals.

Consider war. The personal problem of war, when it occurs, may be how to survive it or how to die in it with honor; how to make money out of it; how to climb into the higher safety of the military apparatus; or how to contribute to the war's termination. In short, according to one's values, to find a set of milieux and within it to survive the war or make one's death in it meaningful. But the structural issues of war have to do with its causes; with what types of men it throws up into command; with its effects upon economic and political, family and religious institutions, with the unorganized irresponsibility of a world of nation-states.

Consider marriage. Inside a marriage a man and a woman may experience personal troubles, but when the divorce rate during the first four years of marriage is 250 out of every 1,000 attempts, this is an indication of a structural issue having to do with the institutions of marriage and the family and other institutions that bear upon them.

Or consider the metropolis—the horrible, beautiful, ugly, magnificent sprawl of the great city. For many upper-class people, the personal solution to "the problem of the city" is to have an apartment with private garage under it in the heart of the city, and forty miles out, a house by Henry Hill, garden by Garrett Eckbo, on a hundred acres of private land. In these two controlled environments—with a small staff at each end and a private helicopter connection—most people could solve many of the problems of personal milieux caused by the facts of the city. But all this, however splendid, does not solve the public issues that the structural fact of the city poses. What should be done with this wonderful monstrosity? Break it all up into scattered units, combining residence and work? Refurbish it as it stands? Or, after evacuation, dynamite it and build new cities according to new plans in new places? What should those plans be? And who is to decide and to accomplish whatever choice is made? These are structural issues; to confront them and to solve them requires us to consider political and economic issues that affect innumerable milieux.

In so far as an economy is so arranged that slumps occur, the problem of unemployment becomes incapable of personal solution. In so far as war is inherent in the nation-state system and in the uneven industrialization of the world, the ordinary individual in his restricted milieu will be powerless—with or without psychiatric aid—to solve the troubles this system or lack of system imposes upon him. In so far as the family as an institution turns women into darling little slaves and men into their chief providers and unweaned dependents, the problem of a satisfactory marriage remains incapable of purely private solution. In so far as the overdeveloped megalopolis and the overdeveloped automobile are built-in features of the overdeveloped society, the issues of urban living will not be solved by personal ingenuity and private wealth.

What we experience in various and specific milieux, I have noted, is often caused by structural changes. Accordingly, to understand the changes of many personal milieux we are required to look beyond them. And the number and variety of such structural changes increase as the institutions within which we live become more embracing and more intricately connected with one another. To be aware of the idea of social structure and to use it with sensibility is to be capable of tracing such linkages among a great variety of milieux. To be able to do that is to possess the sociological imagination....

KEY CONCEPTS

issues

sociological imagination

sociology

troubles

DISCUSSION QUESTIONS

1. Using either today's newspaper or some other source of news, identify one example of what C. Wright Mills would call an issue. How is this issue reflected in the personal troubles of people it affects? Why would Mills call it a social issue?
2. What are the major historical events that have influenced the biographies of people in your generation? In your parents' generation? What does this tell you about the influence of society and history on biography?

2

Invitation to Sociology: A Humanistic Perspective

PETER BERGER

Peter Berger's classic book, Invitation to Sociology, *introduces the idea of debunking—that is, the process that sociologists use to see behind the taken-for-granted ways of thinking about social reality. As you read his essay, think about what it means to see behind the facades of commonsense explanations of society.*

... It can be said that the first wisdom of sociology is this—things are not what they seem. This too is a deceptively simple statement. It ceases to be simple after a while. Social reality turns out to have many layers of meaning. The discovery of each new layer changes the perception of the whole....

People who like to avoid shocking discoveries, who prefer to believe that society is just what they were taught in Sunday School, who like the safety of the rules and the maxims of what Alfred Schuetz has called the "world-taken-for-granted," should stay away from sociology. People who feel no temptation before closed doors, who have no curiosity about human beings, who are content to admire scenery without wondering about the people who live in those houses on the other side of that river, should probably also stay away from sociology. They will find it unpleasant or, at any rate, unrewarding. People who are interested in human beings only if they can change, convert or reform them should also be warned, for they will find sociology much less useful than they hoped. And people whose interest is mainly in their own conceptual constructions will do just as well to turn to the study of little white mice. Sociology will be satisfying, in the long run, only to those who can think of nothing more entrancing than to watch men and to understand things human.

... The sociologist uses the term in a more precise sense, though, of course, there are differences in usage within the discipline itself. The sociologist thinks of "society" as denoting a large complex of human relationships, or to put it in more technical language, as referring to a system of interaction. The word "large" is difficult to specify quantitatively in this context. The sociologist may speak of a "society" including millions of human beings (say, "American society"), but he may also use the term to refer to a numerically much smaller collectivity (say, "the society of sophomores on this campus"). Two people chatting on a street corner will hardly constitute a "society," but three people stranded on an island certainly will. The applicability of the concept, then, cannot be decided on quantitative grounds alone. It rather applies when a complex of relationships is sufficiently succinct to be analyzed by itself, understood as an autonomous entity, set against others of the same kind....

To ask sociological questions, then, presupposes that one is interested in looking some distance beyond the commonly accepted or officially defined goals of human actions. It presupposes a certain awareness that human events have different levels of meaning, some of which are hidden from the consciousness of everyday life. It may even presuppose a measure of suspicion about the way in which human events are officially interpreted by the authorities, be they political, juridical or religious in character....

The sociological perspective involves a process of "seeing through" the facades of social structures. We could think of this in terms of a common experience of people living in large cities. One of the fascinations of a large city is the immense variety of human activities taking place behind the seemingly anonymous and endlessly undifferentiated rows of houses. A person who lives in such a city will time and again experience surprise or even shock as he discovers the strange pursuits that some men engage in quite unobtrusively in houses that, from the outside, look like all the others on a certain street. Having had this experience once or twice, one will repeatedly find oneself walking down a street, perhaps late in the evening, and wondering what may be going on under the bright lights showing through a line of drawn curtains. An ordinary family engaged in pleasant talk with guests? A scene of desperation amid illness or death? Or a scene of debauched pleasures? Perhaps a strange cult or a dangerous conspiracy? The facades of the houses cannot tell us, proclaiming nothing but an architectural conformity to the tastes of some group or class that may not even inhabit the street any longer. The social mysteries lie behind the facades. The wish to penetrate to these mysteries is an [analogy] to sociological curiosity. In some cities that are suddenly struck by calamity, this wish may be abruptly realized. Those who have experienced wartime bombings know of the sudden encounters with unsuspected (and sometimes unimaginable) fellow tenants in the air-raid shelter of one's apartment building. Or they can recollect the startling morning sight of a house hit by a bomb during the night, neatly sliced in half, the facade torn away and the previously hidden interior mercilessly revealed in the daylight. But in most cities that one may normally live in, the facades must be penetrated by one's own inquisitive intrusions. Similarly, there are historical situations in which the facades of society are violently torn apart and all but the most incurious are forced to see that there was a reality behind the facades all along. Usually this does not happen and the facades continue to confront us with seemingly rocklike permanence. The perception of the reality behind the facades then demands a considerable intellectual effort....

... We would contend, then, that there is a debunking motif inherent in sociological consciousness. The sociologist will be driven time and again, by the very logic of his discipline, to debunk the social systems he is studying. This unmasking tendency need not necessarily be due to the sociologist's temperament or inclinations. Indeed, it may happen that the sociologist, who as an individual may be of a conciliatory disposition and quite disinclined to disturb the comfortable assumptions on which he rests his own social existence, is nevertheless compelled by what he is doing to fly in the face of what those around him take for granted. In other words, we would contend that the roots of the debunking motif in sociology are not psychological but methodological. The sociological frame of reference, with its built-in procedure of looking for levels of reality other than those given in the official interpretations of society, carries with it a logical imperative to unmask the pretensions and the propaganda by which men cloak their actions with each other. This unmasking imperative is one of the characteristics of sociology particularly at home in the temper of the modern era.

KEY CONCEPTS

debunking | social structure | society

DISCUSSION QUESTIONS

1. What does Berger mean by the "unmasking tendency" of sociology?
2. Pay attention to how a particular social issue is portrayed by a common news source (for example, job loss or a violent crime). What is the common explanation given for this phenomenon? How might a sociological explanation differ, and how does this illustrate Berger's concept of *debunking?*

Applying Sociological Knowledge: An Exercise for Students

One of the points in this section is how a sociological perspective differs from an individualistic or even psychological perspective. Pick one of the following topics: teen pregnancy, unemployment, or child abuse. Compare and contrast what factors might be important to consider when explaining this social issue using a sociological perspective versus a psychological perspective.

PART II

Sociological Research

3

Human Inquiry and Science

EARL BABBIE

In this piece, Babbie summarizes how everyday curiosity develops into scientific inquiry. We can distinguish between knowledge we just accept and knowledge we obtain through direct observation and experience. Science relies on empirical evidence to support new conclusions. Sociology, in particular, looks at patterns of behaviors among groups of people rather than individual actions that lead to individual outcomes. Understanding the scientific approach to studying the world around us is an important part of doing sociology.

… Let's start by examining a few things you probably know already.

You know the world is round. You probably also know it's cold on the dark side of the moon (the side facing away from the sun), and you know people speak Chinese in China. You know that vitamin C can prevent colds and that unprotected sex can result in AIDS.

How do you know? Unless you've been to the dark side of the moon lately or done experimental research on the virtues of vitamin C, you know these things because somebody told them to you, and you believed what you were told. You may have read in *National Geographic* that people speak Chinese languages in China, and because that made sense to you, you didn't question it. Perhaps your physics or astronomy instructor told you it was cold on the dark side of the moon, or maybe you heard it on the news.

Some of the things you know seem absolutely obvious to you. If someone asked you how you know the world is round, you'd probably say, "Everybody knows that." There are a lot of things everybody knows. Of course, everyone used to "know" that the world was flat.

Most of what you and I know is a matter of agreement and belief. Little of it is based on personal experience and discovery. A big part of growing up in any society, in fact, is the process of learning to accept what everybody around us "knows" is so. If you don't know those same things, you can't really be a part of the group. If you were to question seriously whether the world is really round, you'd quickly find yourself set apart from other people. You might be sent to live in a hospital with other people who question things like that.

SOURCE: From Babbie, THE PRACTICE OF SOCIAL RESEARCH, 13E.

Although most of what we know is a matter of believing what we've been told, there's nothing wrong with us in that respect. It's simply the way human societies are structured, and it's a quite useful quality. The basis of knowledge is agreement. Because we can't learn all we need to know by means of personal experience and discovery alone, things are set up so we can simply believe what others tell us. We know some things through tradition and some things from "experts." I'm not saying you should never question this received knowledge: I'm just drawing your attention to the way you and society normally get along regarding what's so.

There are other ways of knowing things, however. In contrast to knowing things through agreement, we can know them through direct experience—through observation. If you dive into a glacial stream flowing through the Canadian Rockies, you don't need anyone to tell you it's cold. The first time you stepped on a thorn, you knew it hurt before anyone told you.

When our experience conflicts with what everyone else knows, though, there's a good chance we'll surrender our experience in favor of the agreement.

Let's take an example. Imagine you've come to a party at my house. It's a high-class affair, and the drinks and food are excellent. In particular, you're taken by one of the appetizers I bring around on a tray: a breaded, deep-fried appetizer that's especially zesty. You have a couple—they're so delicious! You have more. Soon you're subtly moving around the room to be wherever I am when I arrive with a tray of these nibblies.

Finally, you can't contain yourself any more. "What are they?" you ask. "How can I get the recipe?" And I let you in on the secret: "You've been eating breaded, deep-fried worms!" Your response is dramatic: Your stomach rebels, and you throw up all over the living-room rug. Argh! What a terrible thing to serve guests!

The point of the story is that both of your feelings about the appetizer were quite real. Your initial liking for them, based on your own direct experience, was certainly real. But so was your feeling of disgust when you found out that you'd been eating worms. It should be evident, however, that this feeling of disgust was strictly a product of the agreements you have with those around you that worms aren't fit to eat. That's an agreement you entered into the first time your parents found you sitting in a pile of dirt with hall of a wriggling worm dangling from your lips. When they pried your mouth open and reached down your throat in search of the other half of the worm, you learned that worms are not acceptable food in our society.

Aside from these agreements, what's wrong with worms? They are probably high in protein and low in calories. Bite-sized and easily packaged, they are a distributor's dream. They are also a delicacy for some people who live in societies that lack our agreement that worms are disgusting. Some people might love the worms but be turned off by the deep-fried breading.

Here's another question to consider: "Are worms 'really' good or 'really' bad to eat?" And here's a more interesting question: "How could you know which was really so?"...

LOOKING FOR REALITY

Reality is a tricky business. You probably already suspect that some of the things you "know" may not be true, but how can you really know what's real? People have grappled with this question for thousands of years.

KNOWLEDGE FROM AGREEMENT REALITY

One answer that has arisen out of that grappling is science, which offers an approach to both agreement reality and experiential reality. Scientists have certain criteria that must be met before they will accept the reality of something they have not personally experienced. In general, a scientific assertion must have both logical and empirical support: It must make sense, and it must not contradict actual observation. Why do earthbound scientists accept the assertion that the dark side of the moon is cold? First, it makes sense, because the moon's surface heat comes from the sun's rays, and the dark side of the moon is dark because it's always turned away from the sun. Second, scientific measurements made on the moon's dark side confirm this logical expectation. So, scientists accept the reality of things they don't personally experience—they accept an agreement reality—but they have special standards for doing so.

… [S]cience offers a special approach to the discovery of reality through personal experience. In other words, it offers a special approach to the business of inquiry. **Epistemology** is the science of knowing; **methodology** (a subfield of epistemology) might be called the science of finding out.…

THE FOUNDATIONS OF SOCIAL SCIENCE

Science is sometimes characterized as logico-empirical. This ungainly term carries an important message: As we noted earlier, the two pillars of science are logic and observation. That is, a scientific understanding of the world must both make sense and correspond to what we observe. Both elements are essential to science and relate to the three major aspects of the enterprise of social science: theory, data collection, and data analysis.

To oversimplify just a bit, scientific **theory** deals with the logical aspect of science—providing systematic explanations—whereas data collection deals with the observational aspect. Data analysis looks for patterns in observations and, where appropriate, compares what is logically expected with what is actually observed.…

THEORY, NOT PHILOSOPHY OR BELIEF

Today, social theory has to do with what is, not with what should be. For many centuries, however, social theory did not distinguish between these two orientations. Social philosophers liberally mixed their observations of what happened around them, their speculations about why, and their ideas about how things ought to be. Although modern social researchers may do the same from time to time, as scientists they focus on how things actually are and why.

This means that scientific theory—and, more broadly, science itself—cannot settle debates about values. Science cannot determine whether capitalism is better or worse than socialism. What it can do is determine how these systems perform, but only in terms of some set of agreed-on criteria. For example, we could determine scientifically whether capitalism or socialism most supports human dignity and freedom only if we first agreed on some measurable definitions of dignity and freedom. Our conclusions would then be limited to the meanings specified in our definitions. They would have no general meaning beyond that.

By the same token, if we could agree that suicide rates, say, or giving to charity were good measures of the quality of a religion, then we could determine scientifically whether Buddhism or Christianity is the better religion. Again, our conclusion would be inextricably tied to our chosen criteria. As a practical matter, people seldom agree on precise criteria for determining issues or value, so science is seldom useful in settling such debates. In fact, questions like these are so much a matter of opinion and belief that scientific inquiry is often viewed as a threat to what is "already known."...

Social science, then, can help us know only what is and why. We can use it to determine what ought to be, but only when people agree on the criteria for deciding what outcomes are better than others—an agreement that seldom occurs....

SOCIAL REGULARITIES

In large part, social research aims to find patterns of regularity in social life. Certainly at first glance the subject matter of the physical sciences seems to be more governed by regularities than does that of the social sciences. A heavy object falls to earth every time we drop it, but a person may vote for a particular candidate in one election and against that same candidate in the next. Similarly, ice always melts when heated enough, but habitually honest people sometimes steal. Despite such examples, however, social affairs do exhibit a high degree of regularity that research can reveal and theory can explain.

To begin with, the tremendous number of formal norms in society create a considerable degree of regularity. For example, traffic laws in the United States induce the vast majority of people to drive on the right side of the street rather than the left. Registration requirements for voters lead to some predictable patterns in which classes of people vote in national elections. Labor laws create a high degree of uniformity in the minimum age of paid workers as well as the minimum amount they are paid. Such formal prescriptions regulate, or regularize, social behavior.

Aside from formal prescriptions, we can observe other social norms that create more regularities. Among registered voters, Republicans are more likely than Democrats to vote for Republican candidates. University professors tend to earn more money than unskilled laborers do. Men tend to earn more than women.... The list of regularities could go on and on.

Three objections are sometimes raised in regard to such social regularities. First, some of the regularities may seem trivial. For example, Republicans vote for Republicans; everyone knows that. Second, contradictory cases may be cited, indicating that the "regularity" isn't totally regular. Some laborers make more money than some professors do. Third, it may be argued that, unlike the heavy objects that cannot decide not to fall when dropped, the people involved in the regularity could upset the whole thing if they wanted to....

AGGREGATES, NOT INDIVIDUALS

The regularities of social life that social scientists study generally reflect the collective behavior of many individuals. Although social scientists often study motivations that affect individuals, the individual as such is seldom the subject of social science. Instead, social scientists create theories about the nature of group, rather than individual, life. The term, *aggregate,* includes, groups, organizations, collectives, and so forth. Whereas psychologists focus on what happens *inside* individuals, social scientists study what goes on *between* them: examining everything from couples to small groups and organizations, and on up to whole societies and even interactions between societies.

Sometimes the collective regularities are amazing. Consider the birthrate, for example. People have babies for a wide variety of personal reasons. Some do it because their own parents want grandchildren. Some feel it's a way of completing their womanhood or manhood. Others want to hold their marriages together, enjoy the experience of raising children, perpetuate the family name, or achieve a kind of immortality. Still others have babies by accident.

If you have fathered or given birth to a baby, you could probably tell a much more detailed, idiosyncratic story. Why did you have the baby when you did, rather than a year earlier or later? Maybe you lost your job and had to delay a year before you could afford to have the baby. Maybe you only felt the urge to become a parent after someone close to you had a baby. Everyone who had a baby last year had his or her own reasons for doing so. Yet, despite this vast diversity, and despite the idiosyncrasy of each individual's reasons, the overall birthrate in a society—the number of live births per 1,000 population—is remarkably consistent from year to year....

Social science theories, then, typically deal with aggregated, not individual, behavior. Their purpose is to explain why aggregate patterns of behavior are so regular even when the individuals participating in them may change over time. We could even say that social scientists don't seek to explain people at all. They try to understand the systems in which people operate, the systems that explain why people do what they do. The elements in such a system are not people but *variables*.

CONCEPTS AND VARIABLES

Our most natural attempts at understanding usually take place at the level of the concrete and idiosyncratic. That's just the way we think.

Imagine that someone says to you, "Women ought to get back into the kitchen where they belong." You're likely to hear that comment in terms of what you know about the speaker. If it's your old uncle Harry who is also strongly opposed to daylight saving time, zip codes, and personal computers, you're likely to think his latest pronouncement simply fits into his rather dated point of view about things in general. If, on the other hand, the statement is muttered by an incumbent politician trailing a female challenger in an electoral race, you'll probably explain his comment in a completely different way.

In both examples, you're trying to understand the behavior of a particular individual. Social research seeks insights into classes or types of individuals. Social researchers would want to find out about the kind of people who share that view of women's "proper" role. Do those people have other characteristics in common that may help explain their views?

Even when researchers focus their attention on a single case study—such as a community or a juvenile gang—their aim is to gain insights that would help people understand other communities and other juvenile gangs. Similarly, the attempt to fully understand one individual carries the broader purpose of understanding people or types of people in general.

When this venture into understanding and explanation ends, social researchers will be able to make sense out of more than one person. In understanding what makes a group of people hostile to women who are active outside the home, they gain insight into all the individuals who share that hostility. This is possible because, in an important sense, they have not been studying antifeminists as much as they have been studying antifeminism. It might then turn out that Uncle Harry and the politician have more in common than first appeared.

Antifeminism is spoken of as a **variable** because it varies. Some people display the attitude more than others do. Social researchers are interested in understanding the system of variables that causes a particular attitude to be strong in one instance and weak in another.

The idea of a system composed of variables may seem rather strange, so let's look at an analogy. The subject of a physician's attention is the patient. If the patient is ill, the physician's purpose is to help the patient get well. By contrast, a medical researcher's subject matter is different—the variables that cause a disease, for example. The medical researcher may study the physician's patient, but for the researcher, that patient is relevant only as a carrier of the disease.

That is not to say that medical researchers don't care about real people. They certainly do. Their ultimate purpose in studying diseases is to protect people from them. But in their research, they are less interested in individual patients than they are in the patterns governing the appearance of the disease. In fact, when they can study a disease meaningfully without involving actual patients, they do so.

Social research, then, involves the study of variables and their relationships. Social theories are written in a language of variables, and people get involved only as the "carriers" of those variables....

KEY CONCEPTS

epistemology methodology theory variable

DISCUSSION QUESTIONS

1. Give an example of a "commonly accepted" belief that might actually be challenged if we did research? How much research is needed before we accept it as "common knowledge?"
2. If this class required a research project, what research would you want to do? What question might you ask? How might you go about researching this question? What observations could you make?

4

Promoting Bad Statistics

JOEL BEST

In this article, Joel Best points out how numbers publicly used to describe social problems can be misleading. He shows how advocacy about a given problem can distort accurate, empirical observations. In addition, he emphasizes that statistical information is produced in a social context.

In contemporary society, social problems must compete for attention. To the degree that one problem gains media coverage, moves to the top of politicians' agendas, or becomes the subject of public concern, others will be neglected. Advocates find it necessary to make compelling cases for the importance of

SOURCE: "Promoting Bad Statistics" by Joel Best from SOCIETY, March 2001, pp. 10–15. Reprinted by permission of author.

particular social problems. They choose persuasive wording and point to disturbing examples, and they usually bolster their case with dramatic statistics.

Statistics have a fetish-like power in contemporary discussions about social problems. We pride ourselves on rational policy making, and expertise and evidence guide our rationality. Statistics become central to the process: numbers evoke science and precision; they seem to be nodules of truth, facts that distill the simple essence of apparently complex social processes. In a culture that treats facts and opinions as dichotomous terms, numbers signify truth—what we call "hard facts." In virtually every debate about social problems, statistics trump "mere opinion."

Yet social problems statistics often involve dubious data. While critics occasionally call some number into question, it generally is not necessary for a statistic to be accurate—or even plausible—in order to achieve widespread acceptance. Advocates seeking to promote social problems often worry more about the processes by which policy makers, the press, and the public come to focus on particular problems, than about the quality of their figures. I seek here to identify some principles that govern this process. They are, if you will, guidelines for creating and disseminating dubious social problems statistics.

Although we talk about facts as though they exist independently of people, patiently awaiting discovery, someone has to produce—or construct—all that we know. Every social statistic reflects the choices that go into producing it. The key choices involve definition and methodology: Whenever we count something, we must first define what it is we hope to count, and then choose the methods by which we will go about counting. In general, the press regards statistics as facts, little bits of truth. The human choices behind every number are forgotten; the very presentation of a number gives each claim credibility. In this sense, statistics are like fetishes.

ANY NUMBER IS BETTER THAN NO NUMBER

By this generous standard, a number need not bear close inspection, or even be remotely plausible. To choose an example first brought to light by Christina Hoff Sommers, a number of recent books, both popular and scholarly, have repeated the garbled claim that anorexia kills 150,000 women annually. (The figure seems to have originated from an estimate for the total number of women who are anorexic; only about 70 die each year from the disease.) It should have been obvious that something was wrong with this figure. Anorexia typically affects *young* women. Each year, roughly 8,500 females aged 15–24 die from all causes; another 47,000 women aged 25–44 also die. What are the chances, then, that there could be 150,000 deaths from anorexia each year? But, of course, most of us have no idea how many young women die each year—("It must be a lot...."). When we hear that anorexia kills 150,000 young women per year, we assume that whoever cites the number must know that it is true. It is, after all, a number and therefore presumably factual.

Oftentimes, social problems statistics exist in splendid isolation. When there is only one number, that number has the weight of authority. It is accepted and repeated. People treat the statistic as authoritative because it is a statistic. Often, these lone numbers come from activists seeking to draw attention to neglected social phenomena. One symptom of societal neglect is that no one has bothered to do much research or compile careful records; there often are no official statistics or other sources for more accurate numbers. When reporters cover the story, they want to report facts. When activists have the only available figures, their numbers look like facts, so, in the absence of other numbers, the media simply report the activists' statistics.

Once a number appears in one news report, that story becomes a potential source for everyone seeking information about the social problem; officials, experts, activists, and other reporters routinely repeat figures that appear in press reports.

NUMBERS TAKE ON LIVES OF THEIR OWN

David Luckenbill has referred to this as "number laundering." A statistic's origin—perhaps simply as someone's best guess—is soon forgotten, and through repetition, the figure comes to be treated as a straightforward fact—accurate and authoritative. The trail becomes muddy, and people lose track of the estimate's original source, but they become confident that the number must be correct because it appears everywhere.

It barely matters if critics challenge a number, and expose it as erroneous. Once a number is in circulation, it can live on, regardless of how thoroughly it may have been discredited. Today's improved methods of information retrieval—electronic indexes, full-text databases, and the Internet—make it easier than ever to locate statistics. Anyone who locates a number can, and quite possibly will, repeat it. That annual toll of 150,000 anorexia deaths has been thoroughly debunked, yet the figure continues to appear in occasional newspaper stories. Electronic storage has given us astonishing, unprecedented access to information, but many people have terrible difficulty sorting through what's available and distinguishing good information from bad. Standards for comparing and evaluating claims seem to be wanting. This is particularly true for statistics that are, after all, numbers and therefore factual, requiring no critical evaluation. Why not believe and repeat a number that everyone else uses? Still, some numbers do have advantages.

BIG NUMBERS ARE BETTER THAN LITTLE NUMBERS

Remember: social problems claims must compete for attention; there are many causes and a limited amount of space on the front page of the *New York Times*. Advocates must find ways to make their claims compelling; they favor melodrama—terrible villains, sympathetic, vulnerable victims, and big numbers.

Big numbers suggest that there is a big problem, and big problems demand attention, concern, action. They must not be ignored.

Advocates seeking to attract attention to a social problem soon find themselves pressed for numbers. Press and policy makers demand facts ("You say it's a problem? Well, how big a problem is it?"). Activists believe in the problem's seriousness, and they often spend much of their time talking to others who share that belief. They know that the problem is much more serious, much more common than generally recognized ("The cases we know about are only the tip of the iceberg."). When asked for figures, they thus offer their best estimates, educated guesses, guesstimates, ballpark figures, or stabs in the dark. Mitch Snyder, the most visible spokesperson for the homeless in the early 1980s, explained on ABC's "Nightline" how activists arrived at the figure of three million homeless: "Everybody demanded it. Everybody said we want a number.... We got on the phone, we made a lot of calls, we talked to a lot of people, and we said, 'Okay, here are some numbers.' They have no meaning, no value." Because activists sincerely believe that the new problem is big and important, and because they suspect that there is a very large dark figure of unreported or unrecorded cases, activists' estimates tend to be high, and to err on the side of exaggeration.

This helps explain the tendency to estimate the scope of social problems in large, suspiciously round figures. There are, we are told, one million victims of elder abuse each year, two million missing children, three million homeless, 60 million functionally illiterate Americans; child pornography may be, depending on your source, a $1 billion or $46 billion industry, and so on. Often, these estimates are the only available numbers.

The mathematician John Allen Paulos argues that innumeracy—the mathematical counterpart to illiteracy—is widespread and consequential. He suggests that innumeracy particularly shapes the way we deal with large numbers. Most of us understand hundreds, even thousands, but soon the orders of magnitude blur into a single category: "It's a lot." Even the most implausible figures can gain widespread acceptance. When missing-children advocates charged that nearly two million children are missing each year, anyone might have done the basic math: there are about 60 million children under 18; if two million are missing, that would be one in 30; that is, every year, the equivalent of one child in every American schoolroom would be missing. A 900-student school would have 30 children missing from its student body each year. To be sure, the press debunked this statistic in 1985, but only four years after missing children became a highly publicized issue and the two-million estimate gained wide circulation. And, of course, having been discredited, the number survives and can still be encountered on occasion.

It is remarkable how often contemporary discussions of social problems make no effort to define what is at issue. Often, we're given a dramatic, compelling example, perhaps a tortured, murdered child, then told that this terrible case is an example of a social problem—in this case, child abuse—and finally given a statistic: "There are more than three million reports of child abuse each year." The example, coupled with the problem's name, seems sufficient to make the definition self-evident. However, definitions cannot always be avoided.

DEFINITIONS: BETTER BROAD THAN NARROW

Because broad definitions encompass more kinds of cases, they justify bigger numbers, and we have already noted the advantages of big numbers. No definition is perfect; there are two principal ways definitions of social problems can be flawed. On the one hand, a definition might be too broad and encompass more than it ought to include. That is, broad definitions tend to identify what methodologists call false positives; they include some cases that arguably ought not to be included as part of the problem. On the other hand, a definition that is too narrow may exclude false negatives, cases that perhaps ought to be included as part of the problem.

In general, activists trying to promote a new social problem view false negatives as more troubling than false positives. Activists often feel frustrated trying to get people concerned about some social condition that has been ignored. The general failure to recognize and acknowledge that something is wrong is part of what the activists want to correct; therefore, they may be especially careful not to make things worse by defining the problem too narrowly. A definition that is too narrow fails to recognize a problem's full extent; in doing so, it helps perpetuate the history of neglecting the problem. Some activists favor definitions broad enough to encompass every case that ought to be included; that is, they promote broad definitions in hopes of eliminating all false negatives.

However, broad definitions may invite criticism. They include cases that not everyone considers instances of social problems; that is, while they minimize false negatives, they do so at the cost of maximizing cases that critics may see as false positives. The rejoinder to this critique returns us to the idea of neglect and the harm it causes. Perhaps, advocates acknowledge, their definitions may seem to be too broad, to encompass cases that seem too trivial to be counted as instances of the social problem. But how can we make that judgment? Here advocates are fond of pointing to terrible examples, to the victim whose one, brief, comparatively mild experience had terrible personal consequences; to the child who, having been exposed to a flasher, suffers a lifetime of devastating psychological consequences. Perhaps, advocates say, other victims with similar experiences suffer less or at least seem to suffer less. But is it fair to define a problem too narrowly to include everyone who suffers? Shouldn't our statistics measure the problem's full extent? While social problems statistics often go unchallenged, critics occasionally suggest that some number is implausibly large, or that a definition is too broad.

DEFENDING NUMBERS BY ATTACKING CRITICS

When activists have generated a statistic as part of a campaign to arouse concern about some social problem, there is a tendency for them to conflate the number with the cause. Therefore, anyone who questions a statistic can be suspected of being unsympathetic to the larger claims, indifferent to the victims' suffering, and so on. *Ad hominem* attack on the motives of individuals challenging numbers is a standard response to statistical confrontations. These attacks allow advocates to refuse to budge; making *ad hominem* arguments lets them imply that their opponents don't want to acknowledge the truth, that their statistics are derived from ideology, rather than methodology. If the advocates' campaign has been reasonably successful, they can argue that there is now widespread appreciation that this is a big, serious problem; after all, the advocates' number has been widely accepted and repeated, surely it must be correct. A fallback stance—useful in those rare cases where public scrutiny leaves one's own numbers completely discredited—is to treat the challenge as meaningless nitpicking. Perhaps our statistics were flawed, the advocates acknowledge, but the precise number hardly makes a difference ("After all, even one victim is too many.").

Similarly, criticizing definitions for being too broad can provoke angry reactions. For advocates, such criticisms seem to deny victim's suffering, minimize the extent of the problem, and by extension endorse the status quo. If broader definitions reflect progress, more sensitive appreciation of the true scope of social problems, then calls for narrowing definitions are retrograde, insensitive refusals to confront society's flaws.

Of course, definitions must be operationalized if they are to lead to statistics. It is necessary to specify how the problem will be measured and the statistic produced. If there is to be a survey, who will be sampled? And how will the questions be worded? In what order will they be asked? How will the responses be coded? Most of what we call social-scientific methodology requires choosing how to measure social phenomena. Every statistic depends upon these choices. Just as advocates' preference for large numbers leads them to favor broad definitions, the desirability of broad definitions shapes measurement choices.

MEASURES: BETTER INCLUSIVE THAN EXCLUSIVE

Most contemporary advocates have enough sociological sophistication to allude to the dark figure—that share of a social problem that goes unreported and unrecorded. Official statistics, they warn, inevitably underestimate the size of social problems. This undercounting helps justify advocates' generous estimates (recall all those references to "the tip of the iceberg"). Awareness of the dark figure also justifies measurement decisions that maximize researchers' prospects for discovering and counting as many cases as possible.

Consider the first federally sponsored National Incidence Studies of Missing, Abducted, Runaway, and Thrownaway Children (NISMART). This was an attempt to produce an accurate estimate for the numbers of missing children. To estimate family abductions (in which a family member kidnaps a child) researchers conducted a telephone survey of households. The researchers made a variety of inclusive measurement decisions: an abduction could involve moving a child as little as 20 feet; it could involve the child's complete cooperation; there was no minimum time that the abduction had to last; those involved may not have considered what happened an abduction; and there was no need that the child's whereabouts be unknown (in most family abductions identified by

NISMART, the child was not with someone who had legal custody, but everyone knew where the child was). Using these methods of measurement, a non-custodial parent who took a child for an unauthorized visit, or who extended an authorized visit for an extra night, was counted as having committed a "family abduction." If the same parent tried to conceal the taking or to prevent the custodial parent's contact with the child, the abduction was classified in the most serious ("policy-focal") category. The NISMART researchers concluded that there were 163,200 of these more serious family abductions each year, although evidence from states with the most thorough missing-children reporting systems suggests that only about 9,000 cases per year come to police attention. In other words, the researchers' inclusive measurement choices led to a remarkably high estimate. Media coverage of the family-abduction problem coupled this high figure with horrible examples—cases of abductions lasting years, involving long-term sexual abuse, ending in homicide, and so on. Although most of the episodes identified by NISMART's methods were relatively minor, the press implied that very serious cases were very common ("It's a big number!").

There is nothing atypical about the NISMART example. Advocacy research has become an important source of social problems statistics. Advocates hope research will produce large numbers, and they tend to believe that broad definitions are justified. They deliberately adopt inclusive research measurements that promise to minimize false negatives and generate large numbers. These measurement decisions almost always occur outside public scrutiny and only rarely attract attention. When the media report numbers, percentages, and rates, they almost never explain the definitions and measurements used to produce those statistics.

While many statistics seem to stand alone, occasions do arise when there are competing numbers or contradictory statistical answers to what seems to be the same question. In general, the media tend to treat such competing numbers with a sort of even-handedness.

COMPETING NUMBERS ARE EQUALLY GOOD

Because the media tend to treat numbers as factual, and to ignore definitions and measurement choices, inconsistent numbers pose a problem. Clearly, both numbers cannot be correct. Where a methodologist might try to ask how different advocates arrived at different numbers (in hopes of showing that one figure is more accurate than another, or at least of understanding how the different numbers might be products of different methods), the press is more likely to account for any difference in terms of the competitors' conflicting ideologies or agendas.

Consider the case of the estimates for the crowd size at the 1995 Million Man March. The event's very name set a standard for its success: as the date for the March approached, its organizers insisted that it would attract a million people, while their critics predicted that the crowd would never reach that size. On the day of the March, the organizers announced success: there were, they said, 1.5 to 2 million people present. Alas, the National Park Service Park Police, charged by Congress with estimating the size of demonstrations on the Capitol Mall, calculated that the March drew only 400,000 people (still more than any previous civil rights demonstration). The Park Police knew the Mall's dimensions, took aerial photos, and multiplied the area covered by the crowd by a multiplier based on typical crowd densities. The organizers, like the organizers of many previous demonstrations on the Mall, insisted that the Park Police estimate was far too low. Enter a team of aerial photo analysts from Boston University who eventually calculated that the crowd numbered 837,000 plus or minus 25 percent (i.e., they suggested there might have been a million people in the crowd).

The press covered these competing estimates in standard "he said–she said" style. Few reporters bothered to ask why the two estimates were different. The answer was simple: the BU researchers used a different multiplier. Where the Park Police estimated that there was one demonstrator per 3.6 square feet (actually a fairly densely-packed crowd), the BU researchers calculated that there was a person for every 1.8 square feet (the equivalent of being packed in a crowded elevator). But rather than trying to compare or evaluate the processes by which people arrived at the different estimates, most press reports treated the numbers as equally valid, and implied that the explanation for the difference lay in the motives of those making the estimates.

The March organizers (who wanted to argue that the demonstration had been successful) produced a high number; the Park Police (who, the March organizers insisted, were biased against the March) produced a low one; and the BU scientists (presumably impartial and authoritative) found something in between. The BU estimate quickly found favor in the media: it let the organizers save face (because the BU team conceded the crowd might have reached one million); it seemed to split the difference between the high and low estimates; and it apparently came from experts. There was no effort to judge the competing methods and assumptions behind the different numbers, for example, to ask whether it was likely that hundreds of thousands of men stood packed as close together as the BU researchers imagined for the hours the demonstration lasted.

This example, like those discussed earlier, reveals that public discussions of social statistics are remarkably unsophisticated. Social scientists advance their careers by using arcane inferential statistics to interpret data. The standard introductory undergraduate statistics textbook tends to zip through descriptive statistics on the way to inferential statistics. But it is descriptive statistics—simple counts, averages, percentages, rates, and the like—that play the key role in public discussions over social problems and social policy. And the level of those discussions is not terribly advanced. There is too little critical thinking about social statistics. People manufacture, and other people repeat, dubious figures. While this can involve deliberate attempts to deceive and manipulate, this need not be the case. Often, the people who create the numbers—who, as it were, make all those millions—believe in them. Neither the advocates who create statistics, nor the reporters who repeat them, nor the larger public questions the figures.

What Paulos calls innumeracy is partly to blame—many people aren't comfortable with basic ideas of numbers and calculations. But there is an even more fundamental issue: many of us do not appreciate that every number is a social construction, produced by particular people using particular methods. The naïve, but widespread, tendency is to treat statistics as fetishes, that is, as almost magical nuggets of fact, rather than as someone's efforts to summarize, to simplify complexity. If we accept the statistic as a fetish, then several of the guidelines I have outlined make perfect sense. Any number is better than no number, because the number represents truth. Numbers take on lives of their own because they are true, and their truth justifies their survival. The best way to defend a number is to attack its critics' motives, because anyone who questions a presumably true number must have dubious reasons for doing so. And, when we are confronted with competing numbers, those numbers are equally good, because, after all, they are somehow equivalent bits of truth. At the same time, the guidelines offer those who must produce numbers justifications for favoring big numbers, broad definitions, and inclusive methods. Again, this need not be cynical. Often, advocates are confident that they know the truth, and they approach collecting statistics as a straightforward effort to generate the numbers needed to document what they, after all, know to be true.

Any effort to improve the quality of public discussion of social statistics needs to begin with the understanding that numbers are socially constructed. Statistics are not nuggets of objective fact that we discover; rather, they are people's creations. Every statistic reflects people's decisions to count, their choices of what to count and how to go about counting it, and so on. These choices inevitably shape the resulting numbers.

Public discussions of social statistics need to chart a middle path between naivete (the assumption that numbers are simply true) and cynicism (the suspicion that figures are outright lies told by people with bad motives). This middle path needs to be critical. It needs to recognize that every statistic has to be created, to acknowledge that every statistic is imperfect, yet to appreciate that statistics still offer an essential way of summarizing complex information. Social scientists have a responsibility to promote this critical stance in the public, within the press, and among advocates.

KEY CONCEPTS

false negative

innumeracy

social construction

DISCUSSION QUESTIONS

1. What does Best mean by claiming that numbers are social constructions?
2. Find an example in the media in which someone is using numbers to promote concern about a particular social problem. Based on Best's article, what questions would you need to ask to find out if the numbers are accurate?

Applying Sociological Knowledge: An Exercise for Students

Were you doing research on a topic of interest to you, what would your major question be? Would you need to do qualitative or quantitative research to answer your question? What pitfalls would you need to avoid so as not to misuse statistics and numbers in reporting your research?

PART III

Culture

5

Body Ritual among the Nacirema

HORACE MINER

Horace Miner's classic piece takes a look at a culture that is very much taken for granted by its members. In doing so, he shows the rich detail that outsiders might bring to describing a culture, even when its own members may not see it quite the same way because culture tends to be assumed and not critically questioned.

...[T]he magical beliefs and practices of the Nacirema present such unusual aspects that it seems desirable to describe them as an example of the extremes to which human behavior can go.

Professor Linton first brought the ritual of the Nacirema to the attention of anthropologists twenty years ago (1936:326), but the culture of this people is still very poorly understood. They are a North American group living in the territory between the Canadian Cree, the Yaqui and Tarahumare of Mexico, and the Carib and Arawak of the Antilles. Little is known of their origin, although tradition states that they came from the east. According to Nacirema mythology, their nation was originated by a culture hero, Notgnihsaw, who is otherwise known for two great feats of strength—the throwing of a piece of wampum across the river Pa-To-Mac and the chopping down of a cherry tree in which the Spirit of Truth resided.

Nacirema culture is characterized by a highly developed market economy which has evolved in a rich natural habitat. While much of the people's time is devoted to economic pursuits, a large part of the fruits of these labors and a considerable portion of the day are spent in ritual activity. The focus of this activity is the human body, the appearance and health of which loom as a dominant concern in the ethos of the people. While such a concern is certainly not unusual, its ceremonial aspects and associated philosophy are unique.

The fundamental belief underlying the whole system appears to be that the human body is ugly and that its natural tendency is to debility and disease. Incarcerated in such a body, man's only hope is to avert these characteristics through the use of the powerful influences of ritual and ceremony. Every household has one or more shrines devoted to this purpose. The more powerful individuals in

SOURCE: Horace Miner, "Body Ritual among the Nacirema," *American Anthropologist* 58 (June 1956): 503–507.

the society have several shrines in their houses and, in fact, the opulence of a house is often referred to in terms of the number of such ritual centers it possesses. Most houses are of wattle and daub construction, but the shrine rooms of the more wealthy are walled with stone. Poorer families imitate the rich by applying pottery plaques to their shrine walls.

While each family has at least one such shrine, the rituals associated with it are not family ceremonies but are private and secret. The rites are normally only discussed with children, and then only during the period when they are being initiated into these mysteries. I was able, however, to establish sufficient rapport with the natives to examine these shrines and to have the rituals described to me.

The focal point of the shrine is a box or chest which is built into the wall. In this chest are kept the many charms and magical potions without which no native believes he could live. These preparations are secured from a variety of specialized practitioners. The most powerful of these are the medicine men, whose assistance must be rewarded with substantial gifts. However, the medicine men do not provide the curative potions for their clients, but decide what the ingredients should be and then write them down in an ancient and secret language. This writing is understood only by the medicine men and by the herbalists who, for another gift, provide the required charm.

The charm is not disposed of after it has served its purpose, but is placed in the charm-box of the household shrine. As these magical materials are specific for certain ills, and the real or imagined maladies of the people are many, the charm-box is usually full to overflowing. The magical packets are so numerous that people forget what their purposes were and fear to use them again. While the natives are very vague on this point, we can only assume that the idea in retaining all the old magical materials is that their presence in the charm-box, before which the body rituals are conducted, will in some way protect the worshipper.

Beneath the charm-box is a small font. Each day every member of the family, in succession, enters the shrine room, bows his head before the charm-box, mingles different sorts of holy water in the font, and proceeds with a brief rite of ablution. The holy waters are secured from the Water Temple of the community, where the priests conduct elaborate ceremonies to make the liquid ritually pure.

In the hierarchy of magical practitioners, and below the medicine men in prestige, are specialists whose designation is best translated "holy-mouth-men." The Nacirema have an almost pathological horror of and fascination with the mouth, the condition of which is believed to have a supernatural influence on all social relationships. Were it not for the rituals of the mouth, they believe that their teeth would fall out, their gums bleed, their jaws shrink, their friends desert them, and their lovers reject them. They also believe that a strong relationship exists between oral and moral characteristics. For example, there is a ritual ablution of the mouth for children which is supposed to improve their moral fiber.

The daily body ritual performed by everyone includes a mouth-rite. Despite the fact that these people are so punctilious about care of the mouth, this rite involves a practice which strikes the uninitiated stranger as revolting. It was reported to me that the ritual consists of inserting a small bundle of hog hairs into the mouth, along with certain magical powders, and then moving the bundle in a highly formalized series of gestures.

In addition to the private mouth-rite, the people seek out a holy-mouth-man once or twice a year. These practitioners have an impressive set of paraphernalia, consisting of a variety of augers, awls, probes, and prods. The use of these objects in the exorcism of the evils of the mouth involves almost unbelievable ritual torture of the client. The holy-mouth-man opens the client's mouth and, using the above mentioned tools, enlarges any holes which decay may have created in the teeth. Magical materials are put into these holes. If there are no naturally occurring holes in the teeth, large sections of one or more teeth are gouged out so that the supernatural substance can be applied. In the client's view, the purpose of these ministrations is to arrest decay and to draw friends. The extremely sacred and traditional character of the rite is evident in the fact that the natives return to the holy-mouth-men year after year, despite the fact that their teeth continue to decay.

It is to be hoped that, when a thorough study of the Nacirema is made, there will be careful inquiry into the personality structure of these people. One has but to watch the gleam in the eye of a holy-mouth-man, as he jabs an awl into an exposed nerve, to suspect that a certain amount of sadism is involved. If this can be established, a very interesting pattern emerges, for most of the population shows definite masochistic tendencies. It was to these that Professor Linton referred in discussing a distinctive part of the daily body ritual which is performed only by men. This part of the rite involves scraping and lacerating the surface of the face with a sharp instrument....

The medicine men have an imposing temple, or *latipso,* in every community of any size. The more elaborate ceremonies required to treat very sick patients can only be performed at this temple. These ceremonies involve not only the thaumaturge[1] but a permanent group of vestal maidens who move sedately about the temple chambers in distinctive costume and headdress.

The *latipso* ceremonies are so harsh that it is phenomenal that a fair proportion of the really sick natives who enter the temple ever recover. Small children whose indoctrination is still incomplete have been known to resist attempts to take them to the temple because "that is where you go to die." Despite this fact, sick adults are not only willing but eager to undergo the protracted ritual purification, if they can afford to do so. No matter how ill the supplicant or how grave the emergency, the guardians of many temples will not admit a client if he cannot give a rich gift to the custodian. Even after one has gained admission and survived the ceremonies, the guardians will not permit the neophyte to leave until he makes still another gift.

The supplicant entering the temple is first stripped of all his or her clothes. In every-day life the Nacirema avoids exposure of his body and its natural functions. Bathing and excretory acts are performed only in the secrecy of the household shrine, where they are ritualized as part of the body-rites. Psychological shock results from the fact that body secrecy is suddenly lost upon entry into the *latipso*. A man, whose own wife has never seen him in an excretory act, suddenly finds himself naked and assisted by a vestal maiden while he performs his natural functions into a sacred vessel. This sort of ceremonial treatment is necessitated by the fact that the excreta are used by a diviner to ascertain the course and nature of the client's sickness. Female clients, on the other hand, find their naked bodies are subjected to the scrutiny, manipulation and prodding of the medicine men.

Few supplicants in the temple are well enough to do anything but lie on their hard beds. The daily ceremonies, like the rites of the holy-mouth-men, involve discomfort and torture. With ritual precision, the vestals awaken their miserable charges each dawn and roll them about on their beds of pain while performing ablutions, in the formal movements of which the maidens are highly trained. At other times they insert magic wands in the supplicant's mouth or force him to eat substances which are supposed to be healing. From time to time the medicine men come to their clients and jab magically treated needles into their flesh. The fact that these temple ceremonies may not cure, and may even kill the neophyte, in no way decreases the people's faith in the medicine men.

There remains one other kind of practitioner, known as a "listener." This witch-doctor has the power to exorcise the devils that lodge in the heads of people who have been bewitched. The Nacirema believe that parents bewitch their own children. Mothers are particularly suspected of putting a curse on children while teaching them the secret body rituals. The counter-magic of the witch-doctor is unusual in its lack of ritual. The patient simply tells the "listener" all his troubles and fears, beginning with the earliest difficulties he can remember. The memory displayed by the Nacirema in these exorcism sessions is truly remarkable. It is not uncommon for the patient to bemoan the rejection he felt upon being weaned as a babe, and a few individuals even see their troubles going back to the traumatic effects of their own birth.

In conclusion, mention must be made of certain practices which have their base in native esthetics but which depend upon the pervasive aversion to the natural body and its functions. There are ritual fasts to make fat people thin and ceremonial feasts to make thin people fat. Still other rites are used to make women's breasts larger if they are small, and smaller if they are large. General dissatisfaction with breast shape is symbolized in the fact that the ideal form is virtually outside the range of human variation. A few women afflicted with almost inhuman hypermammary development are so idolized that they make a handsome living by simply going from village to village and permitting the natives to stare at them for a fee.

Reference has already been made to the fact that excretory functions are ritualized, routinized, and relegated to secrecy. Natural reproductive functions are similarly distorted. Intercourse is taboo as a topic and scheduled as an act. Efforts are made to avoid pregnancy by the use of magical materials or by limiting intercourse to certain phases of the moon. Conception is actually very infrequent. When pregnant, women dress so as to hide their condition. Parturition takes place in secret, without friends or relatives to assist, and the majority of women do not nurse their infants.

Our review of the ritual life of the Nacirema has certainly shown them to be a magic-ridden people. It is hard to understand how they have managed to exist so long under the burdens which they have imposed upon themselves. But even such exotic customs as these take on real meaning when they are viewed with the insight provided by Malinowski when he wrote (1948:70):

> Looking from far and above, from our high places of safety in the developed civilization, it is easy to see all the crudity and irrelevance of magic. But without its power and guidance early man could not have mastered his practical difficulties as he has done, nor could man have advanced to the higher stages of civilization.

NOTE

1. Miracle worker or magician. —ED.

REFERENCES

Linton, Ralph. 1936. *The Study of Man*. New York: D. Appleton-Century Co.

Malinowski, Bronislaw. 1948. *Magic, Science, and Religion*. Glencoe: The Free Press.

KEY CONCEPTS

culture

ritual

DISCUSSION QUESTIONS

1. Why is it so difficult to see one's own culture as an outsider might?
2. Culture becomes a taken-for-granted reality. What experiences might lead one to question (or debunk) one's usual understanding of one's own culture?

6

Gamers, Hackers, and Facebook—Computer Cultures, Virtual Community, and Postmodern Identity

ROSS HAENFLER

Ross Haenfler uses cybercultures to explore basic sociological concepts pertaining to cultures and subcultures. Among other things, he shows how cyberspace interactions shape different cultural forms.

If you are of typical college-age people, you might have difficulty recalling a time before video games, downloaded music, and DVDs, let alone a time before personal computers. From cell phones and personal digital assistants to video games and the Internet, computers play a role in nearly every aspect of our lives. New technologies have spawned new subcultures and given established subcultures a new arena in which to interact. Immersive fantasy games such as *World of Warcraft* allow players to become heroes in fanciful realms, and *Second Life* enables residents to own virtual land, run virtual businesses, and even attend virtual concerts performed by real-life musicians. Message boards and listservs connect music fans help social activists network, and bring together people of every possible interest, from bird watching to sports. Blogs enable amateur journalists a forum to write (or rant) about politics, religion, and pop culture, and chat rooms and multi-user domains, or MUDs, serve as virtual cafes where people can socialize or cruise for a date. Auction sites, such as eBay, make buying and selling nearly anything a mere mouse click away. Whether it's Xbox Online, iTunes, Facebook, or MySpace, many of us spend an increasing amount of our lives online, forging meaningful communities and online identities....

SOURCE: GOTHS, GAMERS, & GRRRLS: DEVIANCE AND YOUTH SUBCULTURES by Haenfler (2009) 3884w from Chp. 8 "Gamers, Hackers, and Facebook - Computer Culturesa, Virtual Community, and Postmodern Identity" pp. 83–94. By permission of Oxford University Press, USA.

VIRTUAL SUBCULTURES AND SCENES

Hundreds of millions of people around the world use the Net, many of them for the pleasure and emotional support of socializing online (Hornsby 2005). Civic organizations, activists, political junkies, and a myriad of other groups form subcultures on the Net. Here we'll discuss one of the original online subcultures, hackers, and one of the fastest growing, players of online video games....

Massive Multiplayer Online Games

Since the first video game, *Spacewar!* in 1961, the video game industry has grown exponentially, taking in billions of dollars a year in the United States and rivaling the film industry in profitability. Whether you grew up with Atari, Nintendo, PlayStation, or Xbox, video games have become ubiquitous in contemporary youth culture. Some players have even gone professional, securing sponsorships and competing in tournaments as part of a pro circuit. The latest surge in video game popularity has been Massively Multiplayer Online Role Playing Games [MMORPG] such as *Everquest, Star Wars Galaxies, World of Warcraft, Ultima Online,* and *Dungeons and Dragons Online. World of Warcraft* alone has over 11.5 million subscriptions worldwide, including players in the United States, Korea, New Zealand, China, Australia, United Kingdom, Singapore, France, Germany, and Spain. In each of these games, players create a virtual persona called an avatar, character, or "toon." Unlike more conventional video games in which players assume a role created by the game designers (for example, you play Lara Croft in *Tomb Raider,* the Master Chief in *Halo*, or Marcus Fenix in *Gears of War*), MMORPG enthusiasts create and customize their own characters, including abilities, skills, appearance, profession, possessions or weapons, and names. Most games charge a subscription fee of around $15, making the most popular games big business.

... Players and game master use their imaginations to create a "shared fantasy" in which almost anything can happen, rolling dice to determine if they successfully accomplish tasks such as attacking a monster with a sword or sneaking past a guard undetected.... While MMORPGs require no dice, and replace the game master with faceless game developers, the notion of a party of adventurers questing in a fantasy world populated by mythical creatures remains.

After creating a character, players enter a vast virtual world of spectacular geographies and fantastic opponents. They explore the terrain, meet non-player (computer controlled) characters (NPCs), and undertake perilous quests. While players can play "solo," most band together forming groups in which each member performs a specialized role.

... Despite the lack of face-to-face contact, norms, values, and a sort of social order emerge in every MMORPG. Some players "role play" their characters, creating an in-game personality and speaking and acting accordingly. They craft a virtual self that may or may not reflect their own presentation of self. Just as in the non-virtual world, players exist in a status hierarchy and in organizational structures.

... VIRTUAL COMMUNITY—FACEBOOK, MYSPACE, AND MORE

If I were to ask you to tell me about your community, you would most likely describe the physical space and people in your neighborhood, town, or city. We tend to think of community in narrow terms and almost always tie community to geography, a physical place. In the information age, however, we need a broader definition. A *community* is a social network of people who somehow interact and have something in common such as geographic place (e.g., a university community), common interests (e.g., the poker community), distinct identity (e.g., the Latino and gay communities), or shared values (e.g., the Baptist community). Though we typically think of it in terms of place, a community is not explicitly tied to one location but is instead another way of identifying people who claim an identity, such as "the lesbian community," "the African American community," or "the Pagan community." Community can bring people with similar interests together, as in "the mountain biking community" or "the peace community." The Internet, especially, calls into question the idea that community is necessarily connected to a physical location.

Chances are you are one of the millions of people who have created a personal online profile page using MySpace, Facebook, or a similar social networking site. Facebook has over 200 million users worldwide. Each of these sites encourages users to post personal information such as favorite music, movies, and activities as well as pictures, blogs, videos, and songs. Members can customize their site and form and join groups based upon similar interests, anything from horror films to indie rock music. Many users are young adults in the United States and Europe, though the sites have spread to many countries and attracted people of all age groups. Users appreciate the opportunity to reconnect with old friends and to make new ones. In fact, given that you can immediately "screen" users' age, interests, and motivation (for example to make friends or find a date), profile sites are an *efficient* way to make friends.

In 1983 novelist William Gibson coined the term "*cyberspace*" which has since come to describe the virtual, computerized realm of Web pages, chat rooms, emails, video games, and blogs.[1] *Virtual communities* are communities of people who regularly interact and form ongoing relationships primarily via the Internet (Rheingold 2000). People who interact online are not automatically part of a community—virtual community entails more than surfing Web sites, making a Facebook page, reading emails, or engaging in brief chats on bulletin boards. Just because someone plays an MMORPG does not necessarily mean they are part of a virtual community; after all, you could theoretically play the game and never chat with another human player. On the other hand, someone who regularly plays an MMORPG with the same people (as part of a guild, for example), gets to know them a bit beyond playing the game, and develops personal relationships in the game is part of a community.

In postindustrial society we have more freedom to choose our communities. Before the advent of advanced communication technologies, affordable travel opportunities, and job mobility, people were more or less restricted to their local or regional communities—their hometown, with its churches, schools, and civic organizations. Now, to a certain extent, we can select the communities we are drawn to.... Our loyalties may be divided among many different communities, and we can leave and join communities relatively easily. Think about the myriad interest groups and subcultures that connect people online. You like goldfish? Chat with other enthusiasts at www.koivet.com. Enjoy bird watching? If you can't locate members of the Audubon Society (or if you can find them, but don't *like* them), join a bird-watching listserv. The main idea is this: community, now more than ever, is flexible and less tied to *geography*.

... The Web changes the nature of subcultures, potentially expanding community, but in a different form. If you can buy your favorite underground music for less money online, maybe you'll frequent your local independent record store less often, and eventually what was once perhaps a hub of a local scene might fade away.

Subcultures have typically relied upon physical spaces in which members can get together: clubs, record stores, skate parks, street corners, pubs, alternative fashion boutiques, and so on. You can think of these spaces as part of the *subcultural geography,* the terrain in which youth congregate and live the subculture day to day. Virtual subcultures have their own geographies, digital hangouts that bring together participants from all over the world. Kendall (2002) likens chat rooms/MUDs to virtual pubs, "neighborhood" hangouts where regulars meet and gossip (Kendall 2002). Correll (1995) claims that members of the Lesbian Café BBS (bulletin board system) talked about their virtual space as if it were a physical place in which they interacted. Instead of dropping by the neighborhood pub after work to have a laugh and catch up on news, many of us are logging in to virtual communities, often several times a day (and often *during* work!).

... Many people are skeptical of virtual communities, sometimes called "*computer-mediated communities*" to emphasize how interaction takes place via, or through, computers. In addition to concerns about online sexual predators, identity thieves, and other criminals they worry that computers will make us more isolated—the more we're "plugged in" the less we're interacting face to face (Nie and Erbring 2000). We all know the stereotype of the isolated computer nerd who substitutes virtual friendships for "real" ones, implying that virtual communities are less "authentic" than face-to-face relationships (Miller and Slater 2000). While the differences between face-to-face and virtual communities pose meaningful sociological questions, we should be cautious in assuming that new technologies automatically undermine community and wary of moral panics about "gaming addiction." After all, people initially had the same worries that the telephone would impede rather than help build genuine community.

ELEMENTS OF THE VIRTUAL SCENE

All of the standard elements of nonvirtual subcultures, from style and status hierarchy to jargon and gender ideology, have their equivalents in the virtual world. Although we do not have enough space to cover them all, a few are especially interesting in the way they transcend the virtual/nonvirtual divide.

... Geek and Gamer Language—Netspeak, 133t sp33k, txtspk

Just like any subculture, virtual cultures have produced their own languages. If I were to show you "/ooc 24 wiz lfg AQ3 pst," would you understand that I am communicating an "out of character" message, playing a "24th level wizard," am "looking for a group" to complete "armor quest 3," and would like you to "please send a tell" to me if I can join your group? Do the acronyms "ROFL," "brb," "afk," "lol," "mt," "pwn," or "lmao" mean anything to you? To online gamers and members of other virtual communities they mean "rolling on the floor laughing," "be right back," "away from keyboard," "laughing out loud," "mistell" (or "main tank," depending on context), "owned," and "laughing my ass off," respectively. Just as nonvirtual subcultures create and use their own vocabulary, virtual subcultures like hackers employ their own dialect consisting of shorthand words, acronyms, and computer jargon, sometimes called "133t" speak, short for elite where the numeral "1" substitutes for the letter "l" and "3" for "e." More commonly known as netspeak or text speak, abbreviated words and acronyms are now interwoven into daily language and text messaging. Words are symbols that convey meaning, and text becomes part of symbolic interaction. Knowledge of gaming-specific acronyms and ability to decipher text speak separate insiders from outsiders, contributing to a sense of community for those "in the know."

In real life, we use much more than words to communicate our intended meaning. Normal communication cues such as voice tone, facial expressions, gestures, and posture help us convey the meanings we want to accompany our words. Online talking with text alone leaves a lot of room for misinterpretation and misunderstanding, requiring other ways of conveying emotion and meaning. Many of you have probably used "emoticons" ("CONventions for expressing EMOTIons") to add feeling and emphasis to email messages (Hornsby 2005). Thus, as you surely know, :) and : (become smiley faces that can indicate a whole range of feelings, depending on the context: excitement, happiness, contentment. Likewise ;) is a winking smiley and connotes a shared joke, sarcasm, teasing, flirting, or similar meaning. In the MMORPG world, players can enact emotion and a presentation of self through their avatars. Toons dance, bow, smirk, threaten, flirt, scowl, cheer, clap, and blow kisses at the direction of their player-puppet masters.

Online language is no longer confined to virtual settings, as email shorthand and leet speak have moved to the nonvirtual world. In the last several years I have noticed (often with dismay) an increasing number of student papers including text message writing "u" substituted for "you" and "b4" for "before." The sheer numbers of people playing online games ensures the blending of texting and talking, with gamers exclaiming "Woot!!" (typed as w00t! in online gaming) to express joy in real life, "gee gee" (for gg, or good game) to congratulate, and calling each other "noobs" (for newbie, or newcomer) as a joking insult. Language, both virtual and nonvirtual, is fluid and will continue to evolve as the virtual and nonvirtual worlds overlap.

Virtual Gender and Nerd Masculinity

Long considered the domain of adolescent boys, video games often represent women in very sexualized ways. The hit series of *Tomb Raider* games is one of the few with a female protagonist; yet with her tiny waist and enormous breasts she hardly represents the typical female form. Another extremely popular series, *Grand Theft Auto,* has been maligned for the way players can direct the thuggish main character to commit violence against women. To regain health, players direct the main character to pay for sex with a female prostitute—afterwards players can beat and rob her to regain the money they just spent. To claim that video games are the source of real-life degradation of women is a simplistic attempt by moral entrepreneurs to create a moral panic or engage in symbolic politics; video games make an easy scapegoat for larger social problems, including sexism and violence against women. Nevertheless, in the context of a sexist culture, games that often depict violent men as tough and admirable heroes and women as seminaked sexual objects do perpetuate stereotypical gender representations and roles.

Despite the sexist depictions of women in games and the male-dominated tech world, computer culture has long been associated with geeks and nerds—the "computer geek" is a cultural icon. The "nerd" status serves a purpose in youth culture as one of the identities that other groups define themselves against. Jocks, for example, are almost always defined in part against the stereotypical nerd—jocks are popular, strong, self-confident, and attractive to the opposite sex, while nerds are unpopular, weak, shy, and asexual. In a sense, nerd "connotes a lack of masculinity," particularly dating/sexual incompetence and little athletic ability (Kendall 2002, 80).

... Some self-described nerds proudly claim (or reclaim) their deviant status, embracing the nerd identity (Wright 1996). As with all deviance, meaning depends upon context—calling someone a nerd can be a demeaning slur or a show of affection and solidarity (among nerds). Technical expertise becomes a mark of superiority to more popular kids, particularly for male nerds. Female nerds face a more difficult situation; being stereotyped as unattractive has more negative consequences for women than for men. Computer programming has traditionally been thoroughly male dominated, making fitting into the "boys' club" a challenge for many young women. In addition, women who do manage to break into the boys' club risk being viewed as somehow less feminine or perhaps even intimidating to men because of their perceived intelligence and expertise (Seymour and Hewitt 1999).

Nerd masculinity encompasses both a critique and reinforcement of hegemonic masculinity. Like straight edgers, self-described nerds and geeks are sharply critical of the stereotypical young male bent on sexual conquest of women and domination of other men. Yet nerds, like skinheads, punks, and others who question what it means to be a man, do not fully resist hegemonic masculinity. For example, in MUDs and MMORPGs young men regularly talk trash as if they were on the basketball court or football field. They also talk about women as sexual objects even as they tease one another about their lack of experience with women and refuse to adopt the "asshole" persona necessary, in their minds, to be attractive to women (Kendall 2002). In virtual competitive games, players often express dominance and power, claiming to have "owned" (often typed pwned in-game) the other team and hurling insults at opposing players or less-skilled teammates. Many games reward players for high kill counts and set up online rankings—a kid who could never hold his own in gym class or on the football field could be the king of *Halo* rankings—virtual status, to be sure, but appealing nonetheless.

POSTMODERN IDENTITY AND THE VIRTUAL SELF

Most of us have, at one time or another, dreamed of being someone we are not—maybe a movie star, professional athlete, revolutionary, or supermodel. Virtual worlds may help us live out these dreams as, in a sense, we can all be rich, powerful, and beautiful, even achieving a measure of fame. We have the opportunity to recreate and remake, to an extent, who we are. In chat rooms we can express a different personality, in online games we can gender-bend by playing a toon of the opposite sex, and in the *Simms* we can pick up people in clubs and take them back to our virtual mansions. All of these possibilities raise questions about how we think of our "self" and our identities. Online games, especially, offer an opportunity to construct a self relatively free from some of the constraints of the material world.... Players choose their appearance, associations, professions, and so on. Yet it is important not to overemphasize the freedom offered in online forums. Participants bring with them their knowledge and experience from the material world. While opportunities to experiment with identity abound, status hierarchies emerge nonetheless. Players value some identities and expressions of self more than others. Having the most rare, expensive, or difficult-to-acquire armor becomes a status symbol in game, much as a sports car might in the material world—the difference is a players' assumption that anyone can have the armor with considerable effort.

We tend to think of our personal identity as coherent, ongoing, and stable—we might periodically abandon an identity (such as student) or adopt a new identity (such as becoming a parent), but ultimately we are who we are. Symbolic interactionists take a much more fluid and social view of the self. The *self* is *process* including one's thoughts, feelings, and choices as well as being something we *do* rather than simply *are*.... Our self emerges in interaction with others. Rather than packing a coherent self with us from time to time and place to place, we express our self (or many selves) depending upon the context of a particular interaction. Sociologists disagree to what extent the self continually changes, and virtual interaction has added another complex piece to the puzzle.

In diverse, mobile, technologically advanced, rapidly changing societies is it possible to construct a stable and coherent self? Rather than having one relatively stable identity, we have multiple identities, some only briefly. Think of your own life. You might have many different selves, one you express at work, one for your family, and another for your sorority. *Postmodern identity* is temporary, fragmented, unstable, and fluid. Rather than being deeply personal and unified, the self is relational and fragmented (Gergen 1991). People assume a variety of seemingly contradictory identities, such as an athlete who is also a band geek, a religious preacher who loves gory horror films, or a porn star happily married with a family. Rather than being tied to a consistent, stable "self," we bring a different, flexible self, so to speak, to each context (Zurcher 1977).

Changing computer and communication technologies are central to many theories of the self (see Agger 2004). Think about the ways you can communicate that differ from when your grandparents were your age: fax machines, email, cell phones, video conferencing, and answering machines. Now consider the technologies you may take for granted that people two generations ago could only dream of: satellite and cable TV, personal digital assistants, laptop computers. These technologies enable hackers, MMORPG players, and personal profile users to literally construct virtual selves unfettered by the same rules that apply to face-to-face relationships. The *virtual self* is "the person connected to the world and to others through electronic means such as the Internet, television, and cell phones" (Agger 2004, 1). It is a state of being, created and experienced through technology. Perhaps people feel more comfortable exploring and enacting taboo identities in the anonymity of cyberspace? Thus a shy, reserved, even socially awkward person can be outgoing, boisterous, and charming in the online world. You can never be sure when someone's online identity matches their offline characteristics and when they are *masquerading,* or pretending to be something they are not (Turkle 1997; Kendall 2002). How do you determine someone's authenticity if you can't even reliably determine their age, sex, race, or real-life actions (Williams 2003)?

CONCLUSIONS

We are in a state of profound ambivalence about technology. We love the comforts and conveniences it affords us, but we are wary of its dangers, as films like *2001: A Space Odyssey, The Matrix,* and *Terminator* demonstrate. As more and more people "plug in" to the web, new moral panics arise: Internet stalkers, porn and video game addiction, and identity thieves. Yet for all the panic, many of us cannot wait to upgrade our cell phones, iPods, and Facebook pages, injecting a bit more of ourselves into the virtual universe.

The latest communication technologies are still so new that we continue to make a false distinction between "virtual" and "real" life, as if online experience is somehow secondary, less meaningful, and less real than face-to-face interaction.... Instead, we should be asking how (and *if*) virtual scenes differ from nonvirtual and, more importantly, how they overlap. Hackers, MMORPGs, chat rooms, blogs, and personal profile/networking sites force us to ask how the Internet might change our very conceptualization of subcultures/scenes. They show us how the boundaries between what we see as virtual and real are blurry.... People meet online and then agree to meet offline, but they also meet offline and subsequently get together online. Our communities, and our identities, transcend the virtual-real divide.

NOTE

1. William Gibson, *Neuromancer* (New York: Ace Books, 1983).

REFERENCES

Agger, Ben. 2004. *The Virtual Self: A Contemporary Sociology*. Oxford: Blackwell Publishing.

Correll, Shelley. 1995. "The Ethnography of an Electronic Bar: The Lesbian Café." *Journal of Contemporary Ethnography* 24 (3): 270–298.

Gergen, Kenneth. 1991. *The Saturated Self: Dilemmas of Identity in Contemporary Life*. New York: Basic Books.

Hornsby, Anne M. 2005. "Surfing the Net for Community: A Durkheimian Analysis of Electronic Gatherings." In *Illuminating Social Life: Classical and Contemporary Theory Revisited*, ed. Peter Kivisto, 59–91. Thousand Oaks, CA: Pine Forge Press.

Kendall, Lori. 2002. *Hanging Out in the Virtual Pub: Masculinities and Relationships Online*. Berkeley: University of California Press.

Miller, Daniel, and Don Slater. 2000. *The Internet: An Ethnographic Approach*. New York: Berg.

Nie, Norman H., and Lutz Erbring. 2000. "Our Shrinking Social Universe." *Public Perspective* 11 (3): 44–45.

Rheingold, Howard. 2000. *The Virtual Community: Homesteading on the Electronic Frontier*, rev. ed., Cambridge, MA: MIT Press.

Seymour, Elaine, and Nancy M. Hewitt. 1999. *Talking About Leaving: Why Undergraduates Leave the Sciences*. Boulder, CO: Westview Press.

Turkle, Sherry. 1997. *Life on the Screen: Identity in the Age of the Internet*. New York: Simon & Schuster.

Williams, J. Patrick. 2003. "The Straightedge Subculture on the Internet: A Case Study of Style-display Online." *Media International Australia Incorporating Culture and Policy* 107: 61–74.

Wright, R. 1996. The Occupational Masculinity of Computing." In *Masculinities in Organizations*, ed. C. Cheng, 77–96. Thousand Oaks, CA: Sage.

Zurcher, Louis. 1977. *The Mutable Self*. Beverly Hills, CA: Sage.

KEY CONCEPTS

community
culture
self
social interaction
subculture
virtual communities

DISCUSSION QUESTIONS

1. What does Haenfler mean by "computer-mediated communities?" How is social interaction in such communities different from and similar to social interaction in face-to-face communities?
2. Do you participate in any online communities? If so, which one(s)? How would you describe the cultural characteristics of this community?

7

Global Culture

Sameness or Difference?

MANFRED B. STEGER

Steger enters a debate here about the impact of globalization on local cultures. He shows some of the impact of global capitalism on local cultures, but also argues that local contexts have a role in whether or not cultures become the same or remain different.

Does globalization make people around the world more alike or more different? This is the question most frequently raised in discussions on the subject of cultural globalization. A group of commentators we all might call "pessimistic hyperglobalizers" argue in favour of the former. They suggest that we are not moving towards a cultural rainbow that reflects the diversity of the world's existing cultures. Rather, we are witnessing the rise of an increasingly homogenized popular culture underwritten by a Western "culture industry" based in New York, Hollywood, London, and Milan. As evidence for their interpretation, these commentators point to Amazonian Indians wearing Nike training shoes, denizens of the Southern Sahara purchasing Texaco baseball caps, and Palestinian youths proudly displaying their Chicago Bulls sweatshirts in downtown Ramallah. Referring to the diffusion of Anglo-American values and consumer goods as the "Americanization of the world," the proponents of this cultural

SOURCE: GLOBALIZATION: A VERY SHORT INTRODUCTION by Steger (2003) 1308w from pp. 70–75.

homogenization thesis argue that Western norms and lifestyles are overwhelming more vulnerable cultures. Although there have been serious attempts by some countries to resist these forces of "cultural imperialism"—for example, a ban on satellite dishes in Iran, and the French imposition of tariffs and quotas on imported film and television—the spread of American popular culture seems to be unstoppable.

But these manifestations of sameness are also evident inside the dominant countries of the global North. American sociologist George Ritzer coined the term "McDonaldization" to describe the wide-ranging sociocultural processes by which the principles of the fast-food restaurant are coming to dominate more and more sectors of American society as well as the rest of the world. On the surface, these principles appear to be rational in their attempts to offer efficient and predictable ways of serving people's needs. However, looking behind the façade of repetitive TV commercials that claim to "love to see you smile," we can identify a number of serious problems. For one, the generally low nutritional value of fast-food meals—and particularly their high fat content—has been implicated in the rise of serious health problems such as heart disease, diabetes, cancer, and juvenile obesity. Moreover, the impersonal, routine operations of "rational" fast-service establishments actually undermine expressions of forms of cultural diversity. In the long run, the McDonaldization of the world amounts to the imposition of uniform standards that eclipse human creativity and dehumanize social relations.

Perhaps the most thoughtful analyst in this group of pessimistic hyperglobalizers is American political theorist Benjamin Barber. In his popular book on the subject, he warns his readers against the cultural imperialism of what he calls "McWorld"—a soulless consumer capitalism that is rapidly transforming the world's diverse populations into a blandly uniform market. For Barber, McWorld is a product of a superficial American popular culture assembled in the 1950s and 1960s, driven by expansionist commercial interests. Music, video, theatre, books, and theme parks are all constructed as American image exports that create common tastes around common logos, advertising slogans, stars, songs, brand names, jingles, and trademarks.

Barber's insightful account of cultural globalization also contains the important recognition that the colonizing tendencies of McWorld provoke cultural and political resistance in the form of "Jihad"—the parochial impulse to reject and repel the homogenizing forces of the West wherever they can be found.... Jihad draws on the furies of religious fundamentalism and ethnonationalism which constitute the dark side of cultural particularism. Fuelled by opposing universal aspirations, Jihad and McWorld are locked in a bitter cultural struggle for popular allegiance. Barber asserts that both forces ultimately work against a participatory form of democracy, for they are equally prone to undermine civil liberties and thus thwart the possibility of a global democratic future.

Optimistic hyperglobalizers agree with their pessimistic colleagues that cultural globalization generates more sameness, but they consider this outcome to be a good thing. For example, American social theorist Francis Fukuyama explicitly welcomes the global spread of Anglo-American values and lifestyles, equating the Americanization of the world with the expansion of democracy and free markets. But optimistic hyperglobalizers do not just come in the form of American chauvinists who apply the old theme of manifest destiny to the global arena. Some representatives of this camp consider themselves staunch cosmopolitans who celebrate the Internet as the harbinger of a homogenized "techno-culture." Others are free-market enthusiasts who embrace the values of global consumer capitalism.

It is one thing to acknowledge the existence of powerful homogenizing tendencies in the world, but it is quite another to assert that the cultural diversity existing on our planet is destined to vanish. In fact, several influential commentators offer a contrary assessment that links globalization to new forms of cultural expression. Sociologist Roland Robertson, for example, contends that global cultural flows often reinvigorate local cultural niches. Hence, rather than being totally obliterated by the Western consumerist forces of sameness, local difference and particularity still play an important role in creating unique local contexts. Robertson rejects the cultural homogenization thesis and speaks instead of "glocalization"—a complex interaction of the global and local characterized by cultural borrowing. The resulting expressions of cultural "hybridity" cannot be reduced to clear-cut manifestations of "sameness" or "difference." … [S]uch

The American Way of Life

Number of types of packaged bread available at a Safeway in Lake Ridge, Virginia	104
Number of those breads containing no hydrogenated fat or diglycerides	0
Amount of money spent by the fast-food industry on television advertising per year	$3 billion
Amount of money spent promoting the National Cancer Institute's "Five A Day" programme, which encourages the consumption of fruits and vegetables to prevent cancer and other diseases	$1 million
Number of "coffee drinks" available at Starbucks, whose stores accommodate a stream of over 5 million customers per week, most of whom hurry in and out	26
Number of "coffee drinks" in the 1950s coffee houses of Greenwich Village, New York City	2
Number of new models of cars available to suburban residents in 2001	197
Number of convenient alternatives to the car available to most such residents	0
Number of U.S. daily newspapers in 2000	1,483
Number of companies that control the majority of those newspapers	6
Number of leisure hours the average American has per week	35
Number of hours the average American spends watching television per week	28

Sources: Eric Schossier, *Fast Food Nation* (Houghton & Mifflin, 2001), p. 47; *Consumer Reports Buying Guide 2001* (Consumers Union, 2001), pp. 147–163; Laurie Garrett, *Betrayal of Trust* (Hyperion, 2000), p. 353; *The World Almanac and Book of Facts 2001* (World Almanac Books, 2001), p. 315; www.starbucks.com.

processes of hybridization have become most visible in fashion, music, dance, film, food, and language.

In my view, the respective arguments of hyperglobalizers and sceptics are not necessarily incompatible. The contemporary experience of living and acting across cultural borders means both the loss of traditional meanings and the creation of new symbolic expressions. Reconstructed feelings of belonging coexist in uneasy tension with a sense of placelessness. Cultural globalization has contributed to a remarkable shift in people's consciousness. In fact, it appears that the old structures of modernity are slowly giving way to a new "postmodern" framework characterized by a less stable sense of identity and knowledge.

Given the complexity of global cultural flows, one would actually expect to see uneven and contradictory effects. In certain contexts, these flows might change traditional manifestations of national identity in the direction of a popular culture characterized by sameness; in others they might foster new expressions of cultural particularism; in still others they might encourage forms of cultural hybridity. Those commentators who summarily denounce the homogenizing effects of Americanization must not forget that hardly any society in the world today possesses an "authentic," self-contained culture. Those who despair at the flourishing of cultural hybridity ought to listen to exciting Indian rock songs, admire the intricacy of Hawaiian pidgin, or enjoy the culinary delights of Cuban-Chinese cuisine. Finally, those who applaud the spread of consumerist capitalism need to pay attention to its negative consequences, such as the dramatic decline of communal sentiments as well as the commodification of society and nature....

KEY CONCEPTS

cultural imperialism dominant culture global culture

DISCUSSION QUESTIONS

1. How would you answer Steger's central question, "Does globalization make people around the world more alike?"
2. What impacts does global capitalism have on diversity in world cultures? How do you see this in your particular environment?

Applying Sociological Knowledge: An Exercise for Students

Take a look around your campus and make note of any differences in the clothing that people are wearing that you would associate with cross-cultural differences. What issues would you face in your family, community, or friendship network were you to adopt that style of dress? What do these issues teach you about how cultural expectations shape social norms and group conformity?

PART IV

Socialization and the Life Course

8

Barbie Girls versus Sea Monsters

Children Constructing Gender

MICHAEL A. MESSNER

In this article, Messner analyzes the gender differences among preschool soccer teams. His analysis uncovers how young boys and girls "do gender" when they interact and play with and among one another. Messner also discusses how the youth soccer league is structured in gendered ways. Finally, the research shows that popular culture icons, like Barbie dolls, provide symbols of gendered expectations for children.

In the past decade, studies of children and gender have moved toward greater levels of depth and sophistication (e.g., Jordan and Cowan 1995; McGuffy and Rich 1999; Thorne 1993). In her groundbreaking work on children and gender, Thorne (1993) argued that previous theoretical frameworks, although helpful, were limited: The top-down (adult-to-child) approach of socialization theories tended to ignore the extent to which children are active agents in the creation of their worlds—often in direct or partial opposition to values or "roles" to which adult teachers or parents are attempting to socialize them. Developmental theories also had their limits due to their tendency to ignore group and contextual factors while overemphasizing "the constitution and unfolding of *individuals* as boys or girls" (Thorne 1993, 4). In her study of grade school children, Thorne demonstrated a dynamic approach that examined the ways in which children actively construct gender in specific social contexts of the classroom and the playground. Working from emergent theories of performativity, Thorne developed the concept of "gender play" to analyze the social processes through which children construct gender. Her level of analysis was not the individual but "*group life*—with social relations, the organization and meanings of social situations, the collective practices through which children and adults create and recreate gender in their daily interactions" (Thorne 1993, 4).

SOURCE: "Barbie Girls vs. Sea Monsters" by Michael Messner from GENDER & SOCIETY, Vol. 14, No. 6, pp. 765–784. Copyright © 2000. Reprinted by permission of Sage Publications.

A key insight from Thorne's research is the extent to which gender varies in salience from situation to situation. Sometimes, children engage in "relaxed, cross sex play"; other times—for instance, on the playground during boys' ritual invasions of girls' spaces and games—gender boundaries between boys and girls are activated in ways that variously threaten or (more often) reinforce and clarify these boundaries. However, these varying moments of gender salience are not free-floating; they occur in social contexts such as schools and in which gender is formally and informally built into the division of labor, power structure, rules, and values (Connell 1987).

The purpose of this article is to use an observation of a highly salient gendered moment of group life among four- and five-year-old children as a point of departure for exploring the conditions under which gender boundaries become activated and enforced. I was privy to this moment as I observed my five-year-old son's first season (including weekly games and practices) in organized soccer. Unlike the long-term, systematic ethnographic studies of children conducted by Thorne (1993) or Adler and Adler (1998), this article takes one moment as its point of departure. I do not present this moment as somehow "representative" of what happened throughout the season; instead, I examine this as an example of what Hochschild (1994, 4) calls "magnified moments," which are "episodes of heightened importance, either epiphanies, moments of intense glee or unusual insight, or moments in which things go intensely but meaningfully wrong. In either case, the moment stands out; it is metaphorically rich, unusually elaborate and often echoes [later]." A magnified moment in daily life offers a window into the social construction of reality. It presents researchers with an opportunity to excavate gendered meanings and processes through an analysis of institutional and cultural contexts. The single empirical observation that serves as the point of departure for this article was made during a morning. Immediately after the event, I recorded my observations with detailed notes. I later slightly revised the notes after developing the photographs that I took at the event.

I will first describe the observation—an incident that occurred as a boys' four- and five-year-old soccer team waited next to a girls' four- and five-year-old soccer team for the beginning of the community's American Youth Soccer League (AYSO) season's opening ceremony. I will then examine this moment using three levels of analysis.

The interactional level: How do children "do gender," and what are the contributions and limits of theories of performativity in understanding these interactions?

The level of structural context: How does the gender regime, particularly the larger organizational level of formal sex segregation of AYSO, and the concrete, momentary situation of the opening ceremony provide a context that variously constrains and enables the children's interactions?

The level of cultural symbol: How does the children's shared immersion in popular culture (and their differently gendered locations in this immersion) provide symbolic resources for the creation, in this situation, of apparently categorical differences between the boys and the girls?

Although I will discuss these three levels of analysis separately, I hope to demonstrate that interaction, structural context, and culture are simultaneous and mutually intertwined processes, none of which supersedes the others.

BARBIE GIRLS VERSUS SEA MONSTERS

It is a warm, sunny Saturday morning. Summer is coming to a close, and schools will soon reopen. As in many communities, this time of year in this small, middle- and professional-class suburb of Los Angeles is marked by the beginning of another soccer season. This morning, 156 teams, with approximately 1,850 players ranging from 4 to 17 years old, along with another 2,000 to 3,000 parents, siblings, friends, and community dignitaries have gathered at the local high school football and track facility for the annual AYSO opening ceremonies. Parents and children wander around the perimeter of the track to find the assigned station for their respective teams. The coaches muster their teams and chat with parents. Eventually, each team will march around the track, behind their new team banner, as they are announced over the loudspeaker system and are applauded by the crowd. For now though, and for the next 45 minutes to an hour, the kids, coaches, and parents must stand, mill around, talk, and kill time as they await the beginning of the ceremony.

The Sea Monsters is a team of four- and five-year-old boys. Later this day, they will play their first-ever soccer game. A few of the boys already know each other from preschool, but most are still getting acquainted. They are wearing their new uniforms for the first time. Like other teams, they were assigned team colors—in this case, green and blue—and asked to choose their team name at their first team meeting, which occurred a week ago. Although they preferred "Blue Sharks," they found that the name was already taken by another team and settled on "Sea Monsters." A grandmother of one of the boys created the spiffy team banner, which was awarded a prize this morning. As they wait for the ceremony to begin, the boys inspect and then proudly pose for pictures in front of their new award-winning team banner. The parents stand a few feet away—some taking pictures, some just watching. The parents are also getting to know each other, and the common currency of topics is just how darned cute our kids look, and will they start these ceremonies soon before another boy has to be escorted to the bathroom?

Queued up one group away from the Sea Monsters is a team of four- and five-year-old girls in green and white uniforms. They too will play their first game later today, but for now, they are awaiting the beginning of the opening ceremony. They have chosen the name "Barbie Girls," and they also have a spiffy new team banner. But the girls are pretty much ignoring their banner, for they have created another, more powerful symbol around which to rally. In fact, they are the only team among the 156 marching today with a team float—a red Radio Flyer wagon base, on which sits a Sony boom box playing music, and a 3-foot-plus-tall Barbie doll on a rotating pedestal. Barbie is dressed in the team colors—indeed, she sports a custom-made green-and-white cheerleader-style outfit, with the Barbie Girls' names written on the skirt. Her normally all-blonde hair has been streaked with Barbie Girl green and features a green bow, with white polka dots. Several of the girls on the team also have supplemented their uniforms with green bows in their hair.

The volume on the boom box nudges up and four or five girls begin to sing a Barbie song. Barbie is now slowly rotating on her pedestal, and as the girls sing more gleefully and more loudly, some of them begin to hold hands and walk around the float, in sync with Barbie's rotation. Other same-aged girls from other teams are drawn to the celebration and, eventually, perhaps a dozen girls are singing the Barbie song. The girls are intensely focused on Barbie, on the music, and on their mutual pleasure.

As the Sea Monsters mill around their banner, some of them begin to notice, and then begin to watch and listen as the Barbie Girls rally around their float. At first, the boys are watching as individuals, seemingly unaware of each other's shared interest. Some of them stand with arms at their sides, slack-jawed, as though passively watching a television show. I notice slight smiles on a couple of their faces, as though they are drawn to the Barbie Girls' celebratory fun. Then, with side glances, some of the boys begin to notice each other's attention on the Barbie Girls. Their faces begin to show signs of distaste. One of them yells out, "NO BARBIE!" Suddenly, they all begin to move—jumping up and down, nudging and bumping one other—and join into a group chant: "NO BARBIE! NO BARBIE! NO BARBIE!" They now appear to be every bit as gleeful as the girls, as they laugh, yell, and chant against the Barbie Girls.

The parents watch the whole scene with rapt attention. Smiles light up the faces of the adults, as our glances sweep back and forth, from the sweetly celebrating Barbie Girls to the aggressively protesting Sea Monsters. "They are SO different!" exclaims one smiling mother approvingly. A male coach offers a more in-depth analysis: "When I was in college," he says, "I took these classes from professors who showed us research that showed that boys and girls are the same. I believed it, until I had my own kids and saw how different they are." "Yeah," another dad responds, "Just look at them! They are so different!"

The girls, meanwhile, show no evidence that they hear, see, or are even aware of the presence of the boys who are now so loudly proclaiming their opposition to the Barbie Girls' songs and totem. They continue to sing, dance, laugh, and rally around the Barbie for a few more minutes, before they are called to reassemble in their groups for the beginning of the parade.

After the parade, the teams reassemble on the infield of the track but now in a less organized manner. The Sea Monsters once again find themselves in the general vicinity of the Barbie Girls and take up the "NO BARBIE!" chant again. Perhaps put out by the lack of response to their chant, they begin to dash, in twos and threes, invading the girls' space, and yelling menacingly. With this, the Barbie Girls have little choice but to recognize the presence of the boys—some look puzzled and shrink back, some engage the boys and chase them off. The chasing seems only to incite more excitement among the boys. Finally, parents intervene and defuse the situation, leading their children off to their cars, homes, and eventually to their soccer games.

THE PERFORMANCE OF GENDER

In the past decade, especially since the publication of Judith Butler's highly influential *Gender Trouble* (1990), it has become increasingly fashionable among academic feminists to think of gender not as some "thing" that one "has" (or not) but rather as situationally constructed through the performances of active agents. The idea of gender as performance analytically foregrounds the agency of individuals in the construction of gender, thus highlighting the situational fluidity of gender: here, conservative and reproductive, there, transgressive and disruptive. Surely, the Barbie Girls versus Sea Monsters scene described above can be fruitfully analyzed as a moment of crosscutting and mutually constitutive gender performances: The girls—at least at first glance—appear to be performing (for each other?) a conventional four- to five-year-old version of emphasized femininity. At least on the surface, there appears to be nothing terribly transgressive here. They are just "being girls," together. The boys initially are unwittingly constituted as an audience for the girls' performance but quickly begin to perform (for each other?—for the girls, too?) a masculinity that constructs itself in opposition to Barbie, and to the girls, as not feminine. They aggressively confront—first through loud verbal chanting, eventually through bodily invasions—the girls' ritual space of emphasized femininity, apparently with the intention of disrupting its upsetting influence. The adults are simultaneously constituted as an adoring audience for their children's performances and as parents who perform for each other by sharing and mutually affirming their experience-based narratives concerning the natural differences between boys and girls.

In this scene, we see children performing gender in ways that constitute themselves as two separate, opposed groups (boys vs. girls) and parents performing gender in ways that give the stamp of adult approval to the children's performances of difference, while constructing their own ideological narrative that naturalizes this categorical difference. In other words, the parents do not seem to read the children's performances of gender as social constructions of gender. Instead, they interpret them as the inevitable unfolding of natural, internal differences between the sexes....

The parents' response to the Barbie Girls versus Sea Monsters performance suggests one of the main limits and dangers of theories of performativity. Lacking an analysis of structural and cultural context, performances of gender can all too easily be interpreted as free agents' acting out the inevitable surface manifestations of a natural inner essence of sex difference. An examination of structural and cultural contexts, though, reveals that there was nothing inevitable about the girls' choice of Barbie as their totem, nor in the boys' response to it.

THE STRUCTURE OF GENDER

In the entire subsequent season of weekly games and practices, I never once saw adults point to a moment in which boy and girl soccer players were doing the *same* thing and exclaim to each other, "Look at them! They are *so similar!*" The actual similarity of the boys and the girls, evidenced by nearly all of the kids' routine actions throughout a soccer season—playing the game, crying over a skinned knee, scrambling enthusiastically for their snacks after the games, spacing out on a bird or a flower instead of listening to the coach at practice—is a key to understanding the salience of the Barbie Girls versus Sea Monsters moment for gender relations. In the face of a multitude of moments that speak to similarity, it was this anomalous Barbie Girls versus Sea Monsters moment—where the boundaries of gender were so clearly enacted—that the adults seized to affirm their commitment to difference. It is the kind of moment—to use Lorber's (1994, 37) phrase—where "believing is seeing," where we selectively "see" aspects of social reality that tell us a truth that we prefer to believe, such as the belief in categorical sex difference. No matter that our eyes do not see evidence of this truth most of the rest of the time.

In fact, it was not so easy for adults to actually "see" the empirical reality of sex similarity in everyday observations of soccer throughout the season. That is due to one overdetermining factor: an institutional context that is characterized by informally structured sex segregation among the parent coaches and team managers, and by formally structured sex segregation among the children. The structural analysis developed here is indebted to Acker's (1990) observation that organizations, even while appearing "gender neutral," tend to reflect, re-create, and naturalize a hierarchical ordering of gender....

Adult Divisions of Labor and Power

There was a clear—although not absolute—sexual division of labor and power among the adult volunteers in the AYSO organization. The Board of Directors consisted of 21 men and 9 women, with the top two positions—commissioner and assistant commissioner—held by men. Among the league's head coaches, 133 were men and 23 women. The division among the league's assistant coaches was similarly skewed. Each team also had a team manager who was responsible for organizing snacks, making reminder calls about games and practices, organizing team parties and the end-of-the-year present for the coach. The vast majority of team managers were women. A common slippage in the language of coaches and parents revealed the ideological assumptions underlying this position: I often noticed people describe a team manager as the "team mom." In short, as Table 1 shows, the vast majority of the time, the formal authority of the head coach and assistant coach was in the hands of a man, while the backup, support role of team manager was in the hands of a woman.

TABLE 1 Adult Volunteers as Coaches and Team Managers, by Gender (in percentages) (*N* = 156 teams)

	Head Coaches	Assistant Coaches	Team Managers
Women	15	21	86
Men	85	79	14

These data illustrate Connell's (1987, 97) assertion that sexual divisions of labor are interwoven with, and mutually supportive of, divisions of power and authority among women and men. They also suggest how people's choices to volunteer for certain positions are shaped and constrained by previous institutional practices. There is no formal AYSO rule that men must be the leaders, women the supportive followers. And there are, after all, *some* women coaches and *some* men team managers. So, it may appear that the division of labor among adult volunteers simply manifests an accumulation of individual choices and preferences. When analyzed structurally, though, individual men's apparently free choices to volunteer disproportionately for coaching jobs, alongside individual women's apparently free choices to volunteer disproportionately for team manager jobs, can be seen as a logical collective result of the ways that the institutional structure of sport has differentially constrained and enabled women's and men's previous options and experiences (Messner 1992). Since boys and men have had far more opportunities to play organized sports and thus to gain skills and knowledge, it subsequently appears rational for adult men to serve in positions of knowledgeable authority, with women serving in a support capacity (Boyle and McKay 1995). Structure—in this case, the historically constituted division of labor and power in sport—constrains current practice. In turn, structure becomes an object of practice, as the choices and actions of today's parents re-create divisions of labor and power similar to those that they experienced in their youth.

The Children: Formal Sex Segregation

As adult authority patterns are informally structured along gendered lines, the children's leagues are formally segregated by AYSO along lines of age and sex. In each age-group, there are separate boys' and girls' leagues. The AYSO in this community included 87 boys' teams and 69 girls' teams. Although the four- to five-year-old boys often played their games on a field that was contiguous with games being played by four- to five-year-old girls, there was never a formal opportunity for cross-sex play. Thus, both the girls' and the boys' teams could conceivably proceed through an entire season of games and practices in entirely homosocial contexts. In the all-male contexts that I observed throughout the season, gender never appeared to be overtly salient among the children, coaches, or parents. It is against this backdrop that I might suggest a working hypothesis about structure and the variable salience of gender: The formal sex segregation of children does not, in and of itself, make gender overtly salient. In fact, when children are absolutely segregated, with no opportunity for cross-sex interactions, gender may appear to disappear as an overtly salient organizing principle. However, when formally sex-segregated children are placed into immediately contiguous locations, such as during the opening ceremony, highly charged gendered interactions between the groups (including invasions and other kinds of border work) become more possible.

Although it might appear to some that formal sex segregation in children's sports is a natural fact, it has not always been so for the youngest age-groups in

AYSO. As recently as 1995, when my older son signed up to play as a five-year-old, I had been told that he would play in a coed league. But when he arrived to his first practice and I saw that he was on an all-boys team, I was told by the coach that AYSO had decided this year to begin sex segregating all age-groups, because "during halftimes and practices, the boys and girls tend to separate into separate groups. So the league thought it would be better for team unity if we split the boys and girls into separate leagues." I suggested to some coaches that a similar dynamic among racial ethnic groups (say, Latino kids and white kids clustering as separate groups during halftimes) would not similarly result in a decision to create racially segregated leagues. That this comment appeared to fall on deaf ears illustrates the extent to which many adults' belief in the need for sex segregation—at least in the context of sport—is grounded in a mutually agreed-upon notion of boys' and girls' "separate worlds," perhaps based in ideologies of natural sex difference.

The gender regime of AYSO, then, is structured by formal and informal sexual divisions of labor and power. This social structure sets ranges, limits, and possibilities for the children's and parents' interactions and performances of gender, but it does not determine them. Put another way, the formal and informal gender regime of AYSO made the Barbie Girls versus Sea Monsters moment possible, but it did not make it inevitable. It was the agency of the children and the parents within that structure that made the moment happen. But why did this moment take on the symbolic forms that it did? How and why do the girls, boys, and parents construct and derive meanings from this moment, and how can we interpret these meanings? These questions are best grappled within in the realm of cultural analysis.

THE CULTURE OF GENDER

The difference between what is "structural" and what is "cultural" is not clear-cut. For instance, the AYSO assignment of team colors and choice of team names (cultural symbols) seem to follow logically from, and in turn reinforce, the sex segregation of the leagues (social structure). These cultural symbols such as team colors, uniforms, songs, team names, and banners often carried encoded gendered meanings that were then available to be taken up by the children in ways that constructed (or potentially contested) gender divisions and boundaries.

Team Names

Each team was issued two team colors. It is notable that across the various age-groups, several girls' teams were issued pink uniforms—a color commonly recognized as encoding feminine meanings—while no boys' teams were issued pink uniforms. Children, in consultation with their coaches, were asked to choose their own team names and were encouraged to use their assigned team colors as cues to theme of the team name (e.g., among the boys, the "Red Flashes," the "Green Pythons," and the blue-and-green "Sea Monsters"). When I analyzed the team names of the 156 teams by age-group and by sex, three categories emerged:

1. Sweet names: These are cutesy team names that communicate small stature, cuteness, and/or vulnerability. These kinds of names would most likely be widely read as encoded with feminine meanings (e.g., "Blue Butterflies," "Beanie Babes," "Sunflowers," "Pink Flamingos," and "Barbie Girls").
2. Neutral or paradoxical names: Neutral names are team names that carry no obvious gendered meaning (e.g., "Blue and Green Lizards," "Team Flubber," "Galaxy," "Blue Ice"). Paradoxical names are girls' team names that carry mixed (simultaneously vulnerable *and* powerful) messages (e.g., "Pink Panthers," "Flower Power," "Little Tigers").
3. Power names: These are team names that invoke images of unambiguous strength, aggression, and raw power (e.g., "Shooting Stars," "Killer Whales," "Shark Attack," "Raptor Attack," and "Sea Monsters").

... [A]cross all age-groups of boys, there was only one team name coded as a sweet name—"The Smurfs," in the 10- to 11-year-old league. Across all age categories, the boys were far more likely to choose a power name than anything else, and this was nowhere more true than in the youngest age-groups, where 35 of 40 (87 percent) of boys' teams in the four-to-five and six-to-seven age-groups took on power names. A different pattern appears in the girls' team name choices, especially among the youngest girls. Only 2 of the 12 four- to five-year-old girls' teams chose power names, while 5 chose sweet names and 5 chose neutral/paradoxical names. At age six to seven, the numbers begin to tip toward the boys' numbers but still remain different, with half of the girls' teams now choosing power names. In the middle and older girls' groups, the sweet names all but disappear, with power names dominating, but still a higher proportion of neutral/paradoxical names than among boys in those age-groups.

Barbie Narrative versus Warrior Narrative

How do we make sense of the obviously powerful spark that Barbie provided in the opening ceremony scene described above? Barbie is likely one of the most immediately identifiable symbols of femininity in the world. More conservatively oriented parents tend to happily buy Barbie dolls for their daughters, while perhaps deflecting their sons' interest in Barbie toward more sex-appropriate "action toys." Feminist parents, on the other hand, have often expressed open contempt—or at least uncomfortable ambivalence—toward Barbie. This is because both conservative and feminist parents see dominant cultural meanings of emphasized femininity as condensed in Barbie and assume that these meanings will be imitated by their daughters. Recent developments in cultural studies, though, should warn us against simplistic readings of Barbie as simply conveying hegemonic messages about gender to unwitting children (Attfield 1996; Seiter 1995). In addition to critically analyzing the cultural values (or "preferred meanings") that may be encoded in Barbie or other children's toys, feminist scholars of cultural studies point to the necessity of examining "reception, pleasure, and agency," and especially "the fullness of reception contexts" (Walters 1999, 246). The Barbie Girls versus Sea Monsters moment can be analyzed as a "reception context," in which differently situated boys, girls, and parents variously used Barbie to construct pleasurable intergroup bonds, as well as boundaries between groups.

... Indeed, as the Barbie Girls rallied around Barbie, their obvious pleasure did not appear to be based on a celebration of quiet passivity (as feminist parents might fear). Rather, it was a statement that they—the Barbie Girls—were here in this public space. They were not silenced by the boys' oppositional chanting. To the contrary, they ignored the boys, who seemed irrelevant to their celebration. And, when the boys later physically invaded their space, some of the girls responded by chasing the boys off. In short, when I pay attention to what the girls *did* (rather than imposing on the situation what I *think* Barbie "should" mean to the girls), I see a public moment of celebratory "girl power."

And this may give us better basis from which to analyze the boys' oppositional response. First, the boys may have been responding to the threat of displacement they may have felt while viewing the girls' moment of celebratory girl power. Second, the boys may simultaneously have been responding to the fears of feminine pollution that Barbie had come to symbolize to them. But why might Barbie symbolize feminine pollution to little boys? A brief example from my older son is instructive. When he was about three, following a fun day of play with the five-year-old girl next door, he enthusiastically asked me to buy him a Barbie like hers. He was gleeful when I took him to the store and bought him one. When we arrived home, his feet had barely hit the pavement getting out of the car before an eight-year-old neighbor boy laughed at and ridiculed him: "A *Barbie?* Don't you know that Barbie is a *girl's toy?*" No amount of parental intervention could counter this devastating peer-induced injunction against boys' playing with Barbie. My son's pleasurable desire for Barbie appeared almost overnight to transform itself into shame and rejection. The doll ended up at the bottom of a heap of toys in the closet, and my son soon became infatuated, along with other boys in his preschool, with Ninja Turtles and Power Rangers....

By kindergarten, most boys appear to have learned—either through experiences similar to my son's, where other boys police the boundaries of gender-appropriate play and fantasy and/or by watching the clearly gendered messages of television advertising—that Barbie dolls are not appropriate toys for boys (Rogers 1999, 30). To avoid ridicule, they learn to hide their desire for Barbie, either through denial and oppositional/pollution discourse and/or through sublimation of their desire for Barbie into play with male-appropriate "action figures" (Pope et al. 1999). In their study of a kindergarten classroom, Jordan and Cowan (1995, 728) identified "warrior narratives ... that assume that violence is legitimate and justified when it occurs within a struggle between good and evil" to be the most commonly agreed-upon currency for boys' fantasy play. They observe that the boys seem commonly to adapt story lines that they have seen on television. Popular culture—film, video, computer games, television, and comic books—provides boys with a seemingly endless stream of Good Guys versus Bad Guys characters and stories—from cowboy movies, Superman and Spiderman to Ninja Turtles, Star Wars, and Pokémon—that are available for the boys to appropriate as the raw materials for the construction of their own warrior play....

A cultural analysis suggests that the boys' and the girls' previous immersion in differently gendered cultural experiences shaped the likelihood that they would derive and construct different meanings from Barbie—the girls through pleasurable and symbolically empowering identification with "girl power" narratives; the boys through oppositional fears of feminine pollution (and fears of displacement by girl power?) and with aggressively verbal, and eventually physical, invasions of the girls' ritual space. The boys' collective response thus constituted them differently, *as boys*, in opposition to the girls' constitution of themselves *as girls*. An individual girl or boy, in this moment, who may have felt an inclination to dissent from the dominant feelings of the group (say, the Latina Barbie Girl who, her mother later told me, did not want the group to be identified with Barbie, or a boy whose immediate inner response to the Barbie Girls' joyful celebration might be to join in) is most likely silenced into complicity in this powerful moment of border work.

What meanings did this highly gendered moment carry for the boys' and girls' teams in the ensuing soccer season? Although I did not observe the Barbie Girls after the opening ceremony, I did continue to observe the Sea Monsters' weekly practices and games. During the boys' ensuing season, gender never reached this "magnified" level of salience again—indeed, gender was rarely raised verbally or performed overtly by the boys. On two occasions, though, I observed the coach jokingly chiding the boys during practice that "if you don't watch out, I'm going to get the Barbie Girls here to play against you!" This warning was followed by gleeful screams of agony and fear, and nervous hopping around and hugging by some of the boys. Normally, though, in this sex-segregated, all-male context, if boundaries were invoked, they were not boundaries between boys and girls but boundaries between the Sea Monsters and other boys' teams, or sometimes age boundaries between the Sea Monsters and a small group of dads and older brothers who would engage them in a mock scrimmage during practice. But it was also evident that when the coach was having trouble getting the boys to act together, as a group, his strategic and humorous invocation of the dreaded Barbie Girls once again served symbolically to affirm their group status. They were a team. They were the boys.

CONCLUSION

The overarching goal of this article has been to take one empirical observation from everyday life and demonstrate how a multilevel (interactioinst, structural, cultural) analysis might reveal various layers of meaning that give insight into the everyday social construction of gender. This article builds on observations made by Thorne (1993) concerning ways to approach sociological analyses of children's worlds. The most fruitful approach is not to ask why boys and girls are so different but rather to ask how and under what conditions boys and girls constitute themselves as separate, oppositional groups. Sociologists need not debate whether gender is "there"—clearly, gender is always already there, built as it is into the structures, situations, culture, and consciousness of children and adults. The key issue is under what conditions gender is activated as a salient organizing principle in social life and under what conditions it may be less salient. These are important questions, especially since the social organization of categorical gender difference has always been so clearly tied to gender hierarchy (Acker 1990; Lorber 1994). In the Barbie Girls versus Sea Monsters moment, the performance of gendered boundaries and the construction of boys' and girls' groups as categorically different occurred in the context of a situation systematically structured by sex segregation, sparked by the imposing presence of a shared cultural symbol that is saturated with gendered meanings, and actively supported and applauded by adults who basked in the pleasure of difference, reaffirmed.

I have suggested that a useful approach to the study of such "how" and "under what conditions" questions is to employ multiple levels of analysis. At the most general level, this project supports the following working propositions.

Interactionist theoretical frameworks that emphasize the ways that social agents "perform" or "do" gender are most useful in describing how groups of people actively create (or at times disrupt) the boundaries that delineate seemingly categorical differences between male persons and female persons. In this case, we saw how the children and the parents interactively performed gender in a way that constructed an apparently natural boundary between the two separate worlds of the girls and the boys.

Structural theoretical frameworks that emphasize the ways that gender is built into institutions through hierarchical sexual divisions of labor are most useful in explaining under what conditions social agents mobilize variously to disrupt or to affirm gender differences and inequalities. In this case, we saw how the sexual division of labor among parent volunteers (grounded in their own histories in the gender regime of sport), the formal sex segregation of the children's leagues, and the structured context of the opening ceremony created conditions for possible interactions between girls' teams and boys' teams.

Cultural theoretical perspectives that examine how popular symbols that are injected into circulation by the culture industry are variously taken up by differently situated people are most useful in analyzing how the meanings of cultural symbols, in a given institutional context, might trigger or be taken up by social agents and used as resources to reproduce, disrupt, or contest binary conceptions of sex difference and gendered relations of power. In this case, we saw how a girls' team appropriated a large Barbie around which to construct a pleasurable and empowering sense of group identity and how the boys' team responded with aggressive denunciations of Barbie and invasions....

... The eventual interactions between the boys and the girls were made possible—although by no means fully determined—by the structure of the gender regime and by the cultural resources that the children variously drew on.

On the other hand, the gendered division of labor in youth soccer is not seamless, static, or immune to resistance. One of the few woman head coaches, a very active athlete in her own right, told me that she is "challenging the sexism" in AYSO by becoming the head of her son's league. As post–Title IX women increasingly become mothers and as media images of competent, heroic female athletes become more a part of the cultural landscape for children, the gender regimes of children's sports may be increasingly challenged (Dworkin and Messner 1999). Put another way, the dramatically shifting opportunity structure and cultural imagery of post–Title IX sports have created opportunities for new kinds of interactions, which will inevitably challenge and further shift institutional structures. Social structures simultaneously constrain and enable, while agency is simultaneously reproductive and resistant.

REFERENCES

Acker, Joan. 1990. Hierarchies, jobs, bodies: A theory of gendered organizations. *Gender & Society* 4:139–58.

Adler, Patricia A., and Peter Adler. 1998. *Peer power: Preadolescent culture and identity.* New Brunswick, NJ: Rutgers University Press.

Attfield, Judy. 1996. Barbie and Action Man: Adult toys for girls and boys, 1959–93. In *The gendered object*, edited by Pat Kirkham, 80–89. Manchester, UK, and New York: Manchester University Press.

Boyle, Maree, and Jim McKay. 1995. "You leave your troubles at the gate": A case study of the exploitation of older women's labor and "leisure" in sport. *Gender & Society* 9:556–76.

Butler, Judith. 1990. *Gender trouble: Feminism and the subversion of identity.* New York and London: Routledge.

Connell, R.W. 1987. *Gender and power.* Stanford, CA: Stanford University Press.

Dworkin, Shari L., and Michael A. Messner. 1999. Just do … what?: Sport, bodies, gender. In *Revisioning gender*, edited by Myra Marx Ferree, Judith Lorber, and Beth B. Hess, 341–61. Thousand Oaks, CA: Sage.

Hochschild, Arlie Russell. 1994. The commercial spirit of intimate life and the abduction of feminism: Signs from women's advice books. *Theory, Culture & Society* 11:1–24.

hooks, bell. 1993. "Keeping Close to Home: Class and Education." Pp. 99–11 in *Working Class Women in the Academy*, edited by Michelle Tokarczyk and Elizabeth Fay. Amherst, MA: University of Massachusetts Press.

Jordan, Ellen, and Angela Cowan. 1995. Warrior narratives in the kindergarten classroom: Renegotiating the social contract? *Gender & Society* 9:727–43.

Lorber, Judith. 1994. *Paradoxes of gender.* New Haven, CT, and London: Yale University Press.

McGuffy, C. Shawn, and B. Lindsay Rich. 1999. Playing in the gender transgression zone: Race, class and hegemonic masculinity in middle childhood. *Gender & Society* 13:608–27.

Messner, Michael A. 1992. *Power at play: Sports and the problem of masculinity.* Boston: Beacon.

Pope, Harrison G., Jr., Roberto Olivarda, Amanda Gruber, and John Borowiecki. 1999. Evolving ideals of male body image as seen through action toys. *International Journal of Eating Disorders* 26:65–72.

Rogers, Mary F. 1999. *Barbie culture.* Thousand Oaks, CA: Sage.

Seiter, Ellen. 1995. *Sold separately: Parents and children in consumer culture.* New Brunswick, NJ: Rutgers University Press.

Thorne, Barrie. 1993. *Gender play: Girls and boys in school.* New Brunswick, NJ: Rutgers University Press.

Walters, Suzanna Danuta. 1999. Sex, text, and context: (In) between feminism and cultural studies. In *Revisioning gender*, edited by Myra Marx Ferree, Judith Lorber, and Beth B. Hess, 222–57. Thousand Oaks, CA: Sage.

KEY CONCEPTS

doing gender | gender segregation | gender socialization

DISCUSSION QUESTIONS

1. Think back to a time when you may have played in organized groups (sports, camps, or some other activity). How is the example of the soccer league described in this article similar to what you may have experienced? How is it different?
2. What are some popular children's toys today? How do they socialize children into gendered roles? Are there toys today that construct gender differently than when you were a child?

9

Klaus Barbie, and Other Dolls I'd Like to See

SUSAN JANE GILMAN

In this humorous account, Susan Jane Gilman asks you to imagine the impact that having more empowering role models might have for the socialization of young girls. As you read it, you might think about how your childhood play influenced your socialization.

For decades, Barbie has remained torpedo-titted, open-mouthed, tippy-toed and vagina-less in her cellophane coffin—and, ever since I was little, she has threatened me.

Most women I know are nostalgic for Barbie. "Oh," they coo wistfully, "I used to *loooove* my Barbies. My girlfriends would come over, and we'd play for hours...."

Not me. As a child, I disliked the doll on impulse; as an adult, my feelings have actually fermented into a heady, full-blown hatred.

My friends and I never owned Barbies. When I was young, little girls in my New York City neighborhood collected "Dawns." Only seven inches high, Dawns were, in retrospect, the underdog of fashion dolls. There were four in the collection: Dawn, dirty-blond and appropriately smug; Angie, whose name and black hair allowed her to pass for Italian or Hispanic; Gloria, a redhead with bangs and green eyes (Irish, perhaps, or a Russian, Jew?); and Dale, a black doll with a real afro.

Oh, they had their share of glitzy frocks—the tiny wedding dress, the gold lamé ball gown that shredded at the hem. And they had holes punctured in the bottoms of their feet so you could impale them on the model's stand of the "Dawn Fashion Stage" (sold separately), press a button and watch them revolve jerkily around the catwalk. But they also had "mod" clothes like white go-go boots and a multicolored dashiki outfit called "Sock It to Me" with rose-colored sunglasses. Their hair came in different lengths and—although probably only a six-year-old doll fanatic could discern this—their facial expressions and features were indeed different. They were as diverse as fashion dolls could be in 1972, and in this way, I realize now, they were slightly subversive.

Of course, at that age, my friends and I couldn't spell subversive, let alone wrap our minds around the concept. But we sensed intuitively that Dawns were more democratic than Barbies. With their different colors and equal sizes, they were closer to what we looked like. We did not find this consoling—for we hadn't yet learned that our looks were something that required consolation. Rather, our love of Dawns was an offshoot of our own healthy egocentrism. We were still at that stage in our childhood when little girls want to be everything special, glamorous and wonderful—and believe they can be.

As a six-year-old, I remember gushing, "I want to be a ballerina, and a bride, and a movie star, and a model, and a queen...." To be sure, I was a disgustingly girly girl. I twirled. I skipped. I actually wore a tutu to school. (I am not kidding.) For a year, I refused to wear blue. Whenever the opportunity presented itself, I dressed up in my grandmother's pink chiffon nightgowns and rhinestone necklaces and paraded around the apartment like the princess of the universe. I dressed like my Dawn dolls—and dressed my Dawn dolls like me. It was a silly, fabulous narcissism—but one that sprang from a crucial self-love. These dolls were part of my fantasy life and an extension of my ambitions. Tellingly, my favorite doll was Angie, who had dark brown hair, like mine.

But at some point, most of us prima ballerinas experienced a terrible turning point. I know I did. I have an achingly clear memory of myself, standing before a mirror in all my finery and jewels, feeling suddenly ridiculous and miserable. *Look at yourself,* I remember thinking acidly. *Nobody will ever like you.* I could not have been older than eight. And then later, another memory: my friend Allison confiding in me, "The kids at my school, they all hate my red hair." Somewhere, somehow, a message seeped into our consciousness telling us that we weren't good enough to be a bride or a model or a queen or anything because we weren't pretty enough. And this translated into not smart enough or likable enough, either.

Looks, girls learn early, collapse into a metaphor for everything else. They quickly become the defining criteria for our status and our worth. And somewhere along the line, we stop believing in our own beauty and its dominion. Subsequently, we also stop believing in the power of our minds and our bodies.

Barbie takes over.

Barbie dolls had been around long before I was born, but it was precisely around the time my friends and I began being evaluated on our "looks" that we became aware of the role Barbie played in our culture.

Initially, my friends and I regarded Barbies with a sort of vague disdain. With their white-blond hair, burnt orange "Malibu" skin, unblinking turquoise eyes and hot-pink convertibles, Barbie dolls represented a world utterly alien to us. They struck us as clumsy, stupid, overly obvious. They were clearly somebody else's idea of a doll—and a doll meant for vapid girls in the suburbs. Dawns, my friend Julie and I once agreed during a sleepover, were far more hip.

But eventually, the message of Barbie sunk in. Literally and metaphorically, Barbies were bigger than Dawns. They were a foot high. They merited more plastic! More height! More visibility! And unlike Dawns, which were pulled off the market in the mid-'70s, Barbies were ubiquitous and perpetual bestsellers.

We urban, Jewish, black, Asian and Latina girls began to realize slowly and painfully that if you didn't look like Barbie, you didn't fit in. Your status was diminished. You were less beautiful, less valuable, less worthy. *If you didn't look like Barbie, companies would discontinue you.* You simply couldn't compete.

I'd like to think that, two decades later, my anger about this would have cooled off—not heated up. (I mean, it's a *doll* for chrissake. Get over it.) The problem, however, is that despite all the flag-waving about multiculturalism and girls' self-esteem these days, I see a new generation of little girls receiving the same message I did twenty-five years ago, courtesy of Mattel. I'm currently a "big sister" to a little girl who recently moved here from Mexico. When I first began spending time with her, she drew pictures of herself as she is: a beautiful seven-year-old with café au lait skin and short black hair. Then she began playing with Barbies. Now she draws pictures of both herself and her mother with long, blond hair. "I want long hair," she sighs, looking woefully at her drawing.

A coincidence? Maybe, but Barbie is the only toy in the Western world that human beings actively try to mimic. Barbie is not just a children's doll; it's an adult cult and an aesthetic obsession. We've all seen the evidence. During Barbie's thirty-fifth anniversary, a fashion magazine ran a "tribute to Barbie," using live models posing as dolls. A New York museum held a "Barbie retrospective," enshrining Barbie as a pop artifact—at a time when most human female pop artists continue to work in obscurity. Then there's Pamela Lee. The Barbie Halls of Fame. The websites, the newsletters, the collectors clubs. The woman whose goal is to transform herself, via plastic surgery, into a real Barbie. Is it any wonder then that little girls have been longing for generations to "look like Barbie"—and that the irony of this goes unchallenged?

For this reason, I've started calling Barbie dolls "Klaus Barbie dolls" after the infamous Gestapo commander. For I now clearly recognize what I only sensed as a child. This "pop artifact" is an icon of Aryanism. Introduced after the second world war, in the conservatism of the Eisenhower era (and rumored to be modeled after a German prostitute by a man who designed nuclear warheads), Barbies, in their "innocent," "apolitical" cutesiness, propagate the ideals of the Third Reich. They ultimately succeed where Hitler failed: They instill in legions of little girls a preference for whiteness, for blond hair, blue eyes and delicate features, for an impossible *über*figure, perched eternally and submissively in high heels. In the Cult of the Blond, Barbies are a cornerstone. They reach the young, and they reach them quickly. *Barbie, Barbie!* The Aqua song throbs. *I'm a Barbie girl!*

It's true that, in the past few years, Mattel has made an effort to create a few slightly more p.c. versions of its best-selling blond. Walk down the aisle at Toys-R-Us (and they wonder why kids today can't spell), and you can see a few boxes of American Indian Barbie, Jamaican Barbie, Cowgirl Barbie. Their skin tone is darker and their outfits ethnicized, but they have the same Aryan features and the same "tell-me-any-thing-and-I'll-believe-it" expressions on their plastic faces. Ultimately, their packaging reinforces their status as "Other." These are "special" and "limited" edition Barbies, the labels announce: clearly *not* the standard.

And, Barbie's head still pops off with ease. Granted, this makes life a little sweeter for the sadists on the playground (there's always one girl who gets more pleasure out of destroying Barbie than dressing her), but the real purpose is to make it easier to swap your Barbies' Lilliputian ball gowns. Look at the literal message of this: Hey, girls, a head is simply a neck plug, easily disposed of in the name of fashion. Lest anyone think I'm nit-picking here, a few years ago, a "new, improved" Talking Barbie hit the shelves and created a brouhaha because one of the phrases it parroted was *Math is hard.* Once again, the cerebrum took a backseat to "style." Similarly, the latest "new, improved" Barbie simply trades in one impossible aesthetic for another: The bombshell has now become the waif. Why? According to a Mattel spokesperson, a Kate Moss figure is better suited for today's fashions. Ah, such an improvement.

Now, I am not, as a rule, anti-doll. Remember, I once wore a tutu and collected the entire Dawn family myself. I know better than to claim that dolls are nothing but sexist gender propaganda. Dolls can be a lightning rod for the imagination, for companionship, for learning. And they're *fun*—something that must never be undervalued.

But dolls often give children their first lessons in what a society considers valuable—and beautiful. And so I'd like to see dolls that teach little girls something more than fashion-consciousness and self-consciousness. I'd like to see dolls that expand girls' ideas about what is beautiful instead of constricting them. And how about a few role models instead of runway models as playmates? If you can make a Talking Barbie, surely you can make a Working Barbie. If you can have a Barbie Townhouse, surely you can have a Barbie business. And if you can construct an entire Barbie world out of pink and purple plastic, surely you can construct some "regular" Barbies that are more than white and blond. And remember, Barbie's only a doll! So give it a little more inspired goofiness, some real *pizzazz!*

Along with Barbies of all shapes and colors, here are some Barbies I'd personally like to see:

Dinner Roll Barbie. A Barbie with multiple love handles, double chin, a real, curvy belly, generous tits and ass and voluminous thighs to show girls that voluptuousness is also beautiful. Comes with miniature basket of dinner rolls, bucket o'fried chicken, tiny Entenmann's walnut ring, a brick of Sealtest ice cream, three packs of potato chips, a T-shirt reading "Only the Weak Don't Eat" and, of course, an appetite.

Birkenstock Barbie. Finally, a doll made with horizontal feet and comfortable sandals. Made from recycled materials.

Bisexual Barbie. Comes in a package with Skipper and Ken.

Butch Barbie. Comes with short hair, leather jacket, "Silence=Death" T-shirt, pink triangle buttons, Doc Martens, pool cue and dental dams. Packaged in cardboard closet with doors flung wide open. Barbie Carpentry Business sold separately.

Our Barbies, Ourselves. Anatomically correct Barbie, both inside and out, comes with spreadable legs, her own speculum, magnifying glass and detailed diagrams of female anatomy so that little girls can learn about their bodies in a friendly, nonthreatening way. Also included: tiny Kotex, booklets on sexual responsibility. Accessories such as contraceptives, sex toys, expanding uterus with fetus at various stages of development and breast pump are all optional, underscoring that each young women has the right to choose what she does with her own Barbie.

Harley Barbie. Equipped with motorcycle, helmet, shades. Tattoos are nontoxic and can be removed with baby oil.

Body Piercings Barbie. Why should Earring Ken have all the fun? Body Piercings Barbie comes with changeable multiple earrings, nose ring, nipple rings, lip ring, navel ring and tiny piercing gun. Enables girls to rebel, express alienation and gross out elders without actually having to puncture themselves.

Blue Collar Barbie. Comes with overalls, protective goggles, lunch pail, UAW membership, pamphlet on union organizing and pay scales for women as compared to men. Waitressing outfits and cashier's register may be purchased separately for Barbies who are holding down second jobs to make ends meet.

Rebbe Barbie. So why not? Women rabbis are on the cutting edge in Judaism. Rebbe Barbie comes with tiny satin *yarmulke*, prayer shawl, *tefillin*, silver *kaddish* cup, Torah scrolls. Optional: tiny *mezuzah* for doorway of Barbie Dreamhouse.

B-Girl Barbie. Truly fly Barbie in midriff-baring shirt and baggy jeans. Comes with skateboard, hip hop accessories and plenty of attitude. Pull her cord, and she says things like, "I don't *think* so," "Dang, get outta my face" and "You go, girl." Teaches girls not to take shit from men and condescending white people.

The Barbie Dream Team. Featuring Quadratic Equation Barbie (a Nobel Prize–winning mathematician with her own tiny books and calculator), Microbiologist Barbie (comes with petri dishes, computer and Barbie Laboratory) and Bite-the-Bullet Barbie, an anthropologist with pith helmet, camera, detachable limbs, fake blood and kit for performing surgery on herself in the outback.

Transgender Barbie. Formerly known as G.I. Joe.

KEY CONCEPTS

gender identity · role model · socialization

DISCUSSION QUESTIONS

1. What influence did the toys you played with as a child have on the gender identity you have developed?
2. What are the consequences of behaving outside of the expectations others have of you? What does this tell you about socialization as a form of social control?

10

Leaving Home for College: Expectations for Selective Reconstruction of Self

DAVID KARP, LYNDA LYTLE HOLMSTROM, AND PAUL S. GRAY

Many young adults leave home for the first time when they go away to college. This article addresses the changes young students go through when they leave high school and the family home. The authors discuss how personal changes in identity and perceptions of self are more significant than the geographical move to college.

In their important and much discussed critique of American culture, *Habits of the Heart* (1985), Robert Bellah and his colleagues remark that American parents are of two minds about the prospect of their children leaving home. The thought that their children will leave is difficult, but perhaps more troublesome is the thought that they might not. In contrast to many cultures, American parents place great emphasis on their children establishing independence at a relatively early age. Still, as Bellah's wry comment suggests, they are deeply ambivalent about their children leaving home. The data presented in this paper, part of a larger project on family dynamics during the year that a child applies for admission to college, show that such ambivalence is shared by the children. Our goal here is to document some of the social psychological complexities of achieving independence in America by analyzing the perspectives of 23 primarily upper-middle-class high school seniors as they moved through the college application process and contemplated leaving home.[1]

Of course, a great deal has been written about the internal conflict that surrounds any significant personal change (most obviously, Erik Erikson 1963, 1968, 1974, 1980; see also Manaster 1977; O'Mally 1995). Although researchers have attended to the phenomenon of "incompletely launched young adults" (Heer, Hodge, and Felson 1985; Grigsby and McGowan 1986; Schnaiberg and Goldenberg 1989), little has been written about how relatively sheltered, middle- to upper-middle-class children think about "leaving the nest." Leaving home for college is perhaps among the greatest changes that the economically comfortable students we interviewed have thus far encountered in their lives. For them, going

SOURCE: Reprinted from Symbolic Interaction. Vol. 2, No. 3, pp. 253-276.

to college carries great significance as a coming-of-age moment, in part because it has been long anticipated and not to do so would be unacceptable from a normative stand point. Literature on students who "beat the odds" by going to college suggests that this is also an important transition for them, but one carrying fundamentally different meanings. Unlike the middle- or upper-middle-class students we interviewed, who are trying, at the least, to retain their class position, students arriving at college from less privileged backgrounds must confront wholly new cultural worlds (Rodriguez 1982; Smith 1993; hooks 1993).

The 23 students with whom we were able to complete interviews simply assumed, as did their parents, that they would go to college.[2] Among the 30 sets of parents, all but four individuals had attended college (and two received some different training beyond high school). All of the adults, however, felt strongly about the necessity of college attendance for their children. One of the four who did not go to college, a self-made and extraordinarily successful entrepreneur, did offer some reservation about the utility of an education in the rough and tumble "real world." Even so, both he and his wife were highly invested in getting their son into a prestigious college. While all of the children knew their parents' expectations and fully expected to meet them, we did speak with two students who had some misgivings about whether they really wanted to go to college. Like their counterparts in our sample, these students knew they would go, but still entertained private doubts about their interest in and motivation for college work.[3]

What does it mean to become independent of one's parents, family, and high school friendship groups? As Anna Freud noted, "few situations in life are more difficult to cope with than the attempts of adolescent children to liberate themselves" (Bassoff 1988, p. xi). Young people are ambivalent regarding independence; it is hard to break away. Their ambivalence embodies both symbolic and pragmatic dimensions. Symbolically, independence is the desired outcome of a necessary process of differentiation (Blos 1962). The task for adolescents is "to find their own way in the world and develop confidence that they are strong enough to survive outside the protective family circle" (Bassoff 1988, p. 3). To establish their own identity and sense of purpose, "... they need to wrench themselves away from those who threaten their developing selfhood" (Bassoff 1988, p. 3; see also Campbell, Adams, and Dobson 1984; Katchadourian and Boli 1994). However, independence also has a pragmatic side. In college, young people can "start over"; they can make new friends, establish intimate relationships, and develop the skills and knowledge to help them become self-supporting adults. "But the truth is that they are not sure they can take care of themselves or that they want to be left alone" (Bassoff 1988, p. 3)....

IDENTITY AFFIRMATION, IDENTITY RECONSTRUCTION, AND IDENTITY DISCOVERY

While the students in this study anticipated college as a time during which they would maintain, refine, build upon, and elaborate certain of their identities, they also anticipated negotiating some fundamental identity changes. The students saw college as the time for discovering who they *really* were. They anticipated finding wholly new and permanent life identities during the college years. In addition, they believed that going to college provides a unique opportunity to consciously establish some new identities. Repeatedly, students described the importance of going away to college in terms of an opportunity to discard disliked identities while making a variety of "fresh starts." Their words suggest that college-bound students look forward to re-creating themselves in a context far removed (often geographically, but always symbolically) from their family, high school, and community. The immediately following sections attend, in turn, to how upper-middle-class high school seniors (1) anticipate change, (2) strategize about solidifying certain identities, [and] (3) evaluate identities they wish to escape....

Anticipating Change

Along with such turning points as marriage, having children, and making an occupational commitment, it is plain that leaving for college is self-consciously understood as a dramatic moment of personal transformation. The students with whom we spoke all saw leaving home as a critical juncture in their lives. One measure of consensus in the way our 23 respondents interpreted the meaning of leaving home is the similarity of their words. Students used nearly identical phrases in describing the transition to college as the time to "move on," "discover who I really am," to "start over," to "become an adult," to "become independent," to "begin a new life." The students, moreover, explicitly saw going to college as the "next stage" of their lives....

While all the students interviewed recognized the need for change and were looking forward to it, their certainty about the appropriateness of moving on did not prevent them from feeling anxiety and ambivalence about the transition to college. Theirs is an anticipation composed of optimism, excitement, anxiety, and sometimes fear.

> [I'm] starting the rest of my life. I mean, deciding what I'm going to do and figuring out my future. I mean, that's one thing I'm looking forward to, but it's also one thing I'm not looking forward to. I have mixed feelings about that. It's exciting to figure out your future. In another sense it's scary to have all of the responsibility. (White male attending a public school)

These comments suggest that the prospect of leaving home generates an anticipatory socialization process characterized by multiple and sometimes contradictory feelings and emotions. Students long for independence, anticipate the excitement that accompanies all fresh starts, but worry about their ability to fully meet the challenge....

Affirming Who I Really Am

The one concrete and critical choice that college-bound students must make is which school, in fact, to attend. This decision is often an agonizing one for both students and their parents and involves very high levels of "emotion work" (Hochschild 1983). The significance of making the college choice and the anxiety that it occasions go well beyond questions of money, course curricula, or the physical amenities of the institutions themselves. What makes the decision so difficult is that the students know they are choosing the context in which their new identities will be established.... The fateful issue in the minds of the students is whether people with their identity characteristics and aspirations will be able to flourish. Consequently, it is not surprising that the most consistent and universal pattern in our data is the effort expended by students to find a school where "a person like me" will feel comfortable....

In the most global way, prospective students were searching for a place where the students seemed friendly. On several occasions, students remarked that they were turned on or off to a school because their "tour guide" was either really nice or not friendly enough....

In contrast to the students-like-me theme, an interesting sub-set of seniors expressed a strong interest in diversity. These students not only wanted to meet new people, but different kinds of new people. Students who wanted diversity were excited at the prospect of meeting people different from themselves as a critical learning experience. It is important to note that it was primarily the minority students we interviewed who looked for diversity as they contemplated colleges. An Asian student put it this way:

> The more mixed the better. I think interaction with other ethnic and racial groups is very healthy. If possible, I would not mind having, you know, like an Afro-American roommate. I'd love to. (Asian male attending a public school) ...

While the statements immediately above illustrate that students make careful assessments about the goodness of fit between certain aspects of themselves and the character of different colleges, a dominant theme in the interviews concerned change. Students repeatedly commented that, during their college years, they expected their identities to shift in two fundamental ways. First, they anticipated discovering "who I am" in the broadest sense. Second, they saw college as providing a fresh start because they could discard some of their disliked, sticky identities, often acquired as early as grade school.

CREATING THE PERSON I WANT TO BE

... Seen in terms of Erving Goffman's (1959) dramaturgical model of interaction, going to college provides a new stage and audience, together allowing for new identity performances. Goffman notes (1959, p. 6) that "When an individual appears before others his actions will influence the definition of the situation which they come to have. Sometimes the individual will act in a thoroughly calculating manner, expressing himself in a given way solely in order to give the kind of impression to others that is likely to evoke from them a specific response he is concerned to obtain." To the extent that such impression-management is most centrally dependent upon information control, leaving home provides an unparalleled opportunity to abandon labels that have most contributed to disliked and unshakable identities. When students speak of college as providing a fresh start, they have in mind the possibility of fashioning new roles and identities. Going to college promises the chance to edit, to revise, to re-write certain parts of their biographies.

> It's sort of like starting a new life. I'll have connections to the past, but I'm obviously starting with a clean slate.... Because no one cares how you did in your high school after you're in college. So everyone's equal now. (White male attending a public high school) ...

As students described their hopes about college, the theme of "fresh starts" was almost universally voiced.... Leaving home, friends, and community offers students the possibility to jettison identities which are the product of others' consistent definitions of them over many years. Going to college provides a unique opportunity to display new identities consistent with the person they wish to become.

The data presented thus far are meant to convey the symbolic weightiness of the transition from high school to college. Every student with whom we spoke saw leaving home as a critical biographical moment. They see it as a definitive life stage when their capacity for independence will be fully tested for the first time. Some have had a taste of independence at summer camps and the like, but the transition to college is viewed as the "real thing." Their words, we have been suggesting, indicate that they see strong connections among leaving home, gaining independence, achieving adult status, and transforming their identities. Students carefully attempt to pick a college where they will fit in, thus indicating the importance of retaining and consolidating certain parts of their identities (see Schreier 1991). In addition, they believe that they will discover, in a holistic sense, who they "really" are during the college years.

WILL THEY MISS ME?

... The family is a social system in which roles are interconnected and interdependent. When a child goes off to college, the system is disturbed and the family will try to adapt to the new circumstances. College-bound seniors worry about this process of adaptation. They speculate that their remaining siblings will miss them, or will be left to face the unremitting attentiveness and concern of parents. They also wonder about prospective changes in their parents' marital relationship. In particular, they are concerned for their mothers, whom they identify as being more invested than their fathers in keeping the family system *status quo ante*. Finally, and most significantly, these late adolescents manifest insecurity about their place in the family, especially now that they are leaving. Several of them remarked ruefully, "I should hope they feel some grief [laughter]." "I think they'll be lonelier. I hope they will." "They'll miss me, I hope.... I hope they feel my presence being gone.... They don't have to be, like, mourning my departure, but just a little bit would be nice." It's not that they actually want their parents and siblings to suffer, but missing them would be proof positive that their membership in the family was valued, and that their future place in the family system is assured, in spite of their changing addresses....

In many of our conversations, it appeared that the worst thing about going away to college was that the young people would no longer be able to participate in many aspects of family life. However, perhaps no issue symbolizes the worry associated with leaving home as powerfully as pending decisions over space in the household. How quickly one's bedroom is claimed by other members of the family is, for many of these students, a commentary on the fragility of their position. Although Silver (1996) points out that both the home room and college dorm room are used to symbolically affirm family relations, our conversations with students were more focused on the meanings they attached to their bedrooms at home. One senior said, "They always joke around and they say, 'Oh, we're going to make your room into a den.'"...

Some of the seniors are beginning to understand that the nature of relations with their parents will be altered forever. They will have much more discretion concerning what to reveal about themselves, and therefore much more control over the impression they choose to give their parents. As one young woman put it, "I will experience a lot of things without them there, so that they won't know that they've happened ... [unless] I tell them or if they can see a difference in me." Others expressed shared anxieties about personal transformations and the consequent stability of their place in the family constellation....

What are we to make of these worries, speculations, and musings? College-bound young adults genuinely want to remain attached to their families, even as they are yearning for true independence. Getting into college is understood as a point of departure which has the potential to alter fundamentally their relationship with their family. However, in spite of their worries, most students see the transition to college as a good thing—a positive transformation with life-long consequences. They cannot predict precisely how their relations with parents and siblings will change, but they know for sure that they have initiated a process that will alter the character of these primary relationships. Such knowledge is plainly implicated in the calculus of ambivalence they feel about leaving home:

> It's like, if you want to be treated like an adult, you have to act like an adult. If you want to be treated like a child, act like a child. If you want to be treated like an adult the rest of your life, you've got to start sometime. (White male attending a public high school)

"You've got to start sometime." That, of course, is exactly what they are doing as they embark on their great adventure of self-discovery, into college first and hopefully, thereby, toward full adulthood.

NOTES

1. We used father's occupation as a proxy for social class. We characterized our sample as predominantly upper-middle class. A sampling of the types of father's occupations that warrant this description includes physician, lawyer, professor, administrator, and architect. A few occupations were either higher or lower in status.
2. Either because we could not reach them or because they declined to be interviewed, we did not speak to eight of the 31 students originally included in our sample. The number is 31 because one of the 30 families had twins.
3. One student, "who declined to be interviewed, did not complete the college application process during his senior year in high school. He was the only student in our sample who did not anticipate attending college in the year following high school graduation.

REFERENCES

Bassoff, Evelyn. 1988. *Mothers and Daughters: Loving and Letting Go*. New York: Penguin Books.

Bellah, Robert, Richard Madsen, William Sullivan, Ann Swidler, and Steven Tipton. 1985. *Habits of the Heart: Individualism and Commitment in American Life*. Berkeley: University of California Press.

Blos, Peter. 1962. *On Adolescence: A Psychoanalytic Interpretation*. New York: Free Press.

Campbell, Eugene, Gerald Adams, and William Dobson. 1984. "Familial Correlates of Identity Formation in Late Adolescence: A Study of the Predictive Utility of Connectedness and Individuality in Family Relations." *Journal of Youth and Adolescence* 13: 509–525.

Erikson, Erik. 1963. *Childhood and Society*, 2nd ed. New York: W. W. Norton.

———. 1968. *Identity: Youth and Crisis*. New York: W. W. Norton.

———. 1974. *Dimensions of a New Identity*. New York: W. W. Norton.

———. 1980. *Identity and the Life Cycle*. New York: W. W. Norton.

Goffman, Erving. 1959. *The Presentation of Self in Everyday Life*. Garden City, NY: Doubleday Anchor.

Grigsby, Jill, and Jill McGowan. 1986. "Still in the Nest: Adult Children Living with Their Parents." *Sociology and Social Research* 70: 146–148.

Heer, David, Robert Hodge, and Marcus Felson. 1985. "The Cluttered Nest: Evidence That Young Adults Are More Likely to Live at Home Now Than in the Recent Past." *Sociology and Social Research* (69): 436–441.

Hochschild, Arlie. 1983. *The Managed Heart: Commercialization of Human Feeling*. Berkeley: University of California Press.

hooks, bell. 1993. "Keeping Close to Home: Class and Education." Pp. 99–111 in *Working-Class Women in the Academy*, edited by Michelle Tokarczyk and Elizabeth Fay. Amherst, MA: The University of Massachusetts Press.

Katchadourian, Herant, and John Boli. 1994. *Cream of the Crop: The Impact of Elite Education in the Decade After College*. New York: Basic Books.

Manaster, Guy. 1977. *Adolescent Development and the Life Tasks*. Boston: Allyn and Bacon.

O'Mally, Dawn. 1995. *Adolescent Development: Striking a Balance Between Attachment and Autonomy*. Ph.D. dissertation, Department of Psychology, Harvard University, Cambridge, MA.

Rodriguez, Richard. 1982. *Hunger of Memory: The Education of Richard Rodriguez*. Boston: David R. Godine.

Schnaiberg, Allan, and Sheldon Goldenberg. 1998. "From Empty Nest to Crowded Nest: The Dynamics of Incompletely-Launched Young Adults." *Social Problems* 36: 251–269.

Schreier, Barbara. 1991. *Fitting In: Four Generations of College Life*. Chicago: Chicago Historical Society.

Silver, Ira. 1996. "Role Transitions, Objects, and Identity." *Symbolic Interaction* 19: 1–20.

Smith, Patricia. 1993. "Grandma Went to Smith, All Right, But She Went from Nine to Five: A Memoir." Pp. 126–139 in *Working-Class Women in the Academy,* edited by Michelle Tokarczyk and Elizabeth Fay. Amherst, MA: The University of Massachusetts Press.

KEY CONCEPTS

anticipatory socialization   identity   rite of passage

DISCUSSION QUESTIONS

1. What changes did you go through (or are you going through) during your first year of college? How do these changes influence your self-identity and how others perceive you?
2. When you go home for vacations and visits, how does home feel differently now that you have lived away? What feels the same?

11

Anybody's Son Will Do

GWYNNE DYER

Resocialization is a process by which existing social roles are radically altered or replaced. Entry to the military is a good illustration of this process, as described here. Part of the process involves shaping identities to become part of a group, thus in some sense losing one's individual identity.

All soldiers belong to the same profession, no matter what country they serve, and it makes them different from everybody else. They have to be different, for their job is ultimately about killing and dying, and those things are not a natural vocation for any human being. Yet all soldiers are born civilians. The method for turning young men into soldiers—people who kill other people and expose themselves to death—is basic training. It's essentially the same all over the world, and it always has been, because young men everywhere are pretty much alike.

Human beings are fairly malleable, especially when they are young, and in every young man there are attitudes for any army to work with: the inherited values and postures, more or less dimly recalled, of the tribal warriors who were once the model for every young boy to emulate. Civilization did not involve a sudden clean break in the way people behave, but merely the progressive distortion and redirection of all the ways in which people in the old tribal societies used to behave, and modern definitions of maleness still contain a great deal of the old warrior ethic. The anarchic machismo of the primitive warrior is not what modern armies really need in their soldiers, but it does provide them with promising raw material for the transformation they must work in their recruits.

Just how this transformation is wrought varies from time to time and from country to country. In totally militarized societies—ancient Sparta, the samurai class of medieval Japan, the areas controlled by organizations like the Eritrean People's Liberation Front today—it begins at puberty or before, when the young boy is immersed in a disciplined society in which only the military values are allowed to penetrate. In more sophisticated modern societies, the process is briefer and more concentrated, and the way it works is much more visible. It is, essentially, a conversion process in an almost religious sense—and as in all conversion phenomena, the emotions are far more important than the specific ideas....

... Soldiers are not just robots; they are ordinary human beings with national and personal loyalties, and many of them do feel the need for some patriotic or ideological justification for what they do. But which nation, which ideology, does not matter: men will fight as well and die as bravely for the Khmer Rouge as for "God, King, and Country."...

... Armies know this. It is their business to get men to fight, and they have had a long time to work out the best way of doing it....

The way armies produce this sense of brotherhood in a peacetime environment is basic training: a feat of psychological manipulation on the grand scale which has been so consistently successful and so universal that we fail to notice it as remarkable. In countries where the army must extract its recruits in their late teens, whether voluntarily or by conscription, from a civilian environment that does not share the military values, basic training involves a brief but intense period of indoctrination whose purpose is not really to teach the recruits basic military skills, but rather to change their values and their loyalties. "I guess you could say we brainwash them a little bit," admitted a U.S. Marine drill instructor, "but you know they're good people."

It's easier if you catch them young. You can train older men to be soldiers; it's done in every major war. But you can never get them to believe that they like it, which is the major reason armies try to get their recruits before they are twenty. There are other reasons too, of course, like the physical fitness, lack of dependents, and economic dispensability of teenagers, that make armies prefer them, but the most important qualities teenagers bring to basic training are enthusiasm and naiveté. Many of them actively want the discipline and the closely structured environment that the armed forces will provide, so there is no need for the recruiters to deceive the kids about what will happen to them after they join....

Young civilians who have volunteered and have been accepted by the Marine Corps arrive at Parris Island, the Corps's East Coast facility for basic training, in a state of considerable excitement and apprehension: most are aware that they are about to undergo an extraordinary and very difficult experience. But they do not make their own way to the base; rather, they trickle in to Charleston airport on various flights throughout the day on which their training platoon is due to form, and are held there, in a state of suppressed but mounting nervous tension, until late in the evening. When the buses finally come to carry them the seventy-six miles to Parris Island, it is often after midnight—and this is not an administrative oversight. The shock treatment they are about to receive will work most efficiently if they are worn out and somewhat disoriented when they arrive.

The basic training organization is a machine, processing several thousand young men every month, and every facet and gear of it has been designed with the sole purpose of turning civilians into Marines as efficiently as possible. Provided it can have total control over their bodies and their environment for approximately three months, it can practically guarantee converts. Parris Island provides that controlled environment, and the recruits do not set foot outside it again until they graduate as Marine privates eleven weeks later....

For the young recruits, basic training is the closest thing their society can offer to a formal rite of passage, and the institution probably stands in an unbroken line of descent from the lengthy ordeals by which young males in pre-civilized groups were initiated into the adult community of warriors. But in civilized societies it is a highly functional institution whose product is not anarchic warriors, but trained soldiers.

Basic training is not really about teaching people skills; it's about changing them, so that they can do things they wouldn't have dreamt of otherwise. It works by applying enormous physical and mental pressure to men who have been isolated from their normal civilian environment and placed in one where the only right way to think and behave is the way the Marine Corps wants them to....

The first three days the raw recruits spend at Parris Island are actually relatively easy, though they are hustled and shouted at continuously. It is during this time that they are documented and inoculated, receive uniforms, and learn the basic orders of drill that will enable young Americans (who are not very accustomed to this aspect of life) to do everything simultaneously in large groups. But the most important thing that happens in "forming" is the surrender of the recruits' own clothes, their hair—all the physical evidence of their individual civilian identities....

> Forming Day One makes me nervous. You've got a whole new mob of recruits, you know, sixty or seventy depending, and they don't know anything. You don't know what kind of a reaction you're going to get from the stress you're going to lay on them, and it just worries me the first day.
>
> Things could happen, I'm not going to lie to you. Something might happen. A recruit might decide he doesn't want any part of this stuff and maybe take a poke at you or something like that. In a situation like that it's going to be a spur-of-the-moment thing and that worries me.
>
> —*USMC drill instructor*

But it rarely happens. The frantic bustle of forming is designed to give the recruit no time to think about resisting what is happening to him. And so the recruits emerge from their initiation into the system, stripped of their civilian clothes, shorn of their hair, and deprived of whatever confidence in their own identity they may previously have had as eighteen-year-olds, like so many blanks ready to have the Marine identity impressed upon them.

The first stage in any conversion process is the destruction of an individual's former beliefs and confidence, and his reduction to a position of helplessness and need. It isn't really as drastic as all that, of course, for three days cannot cancel out eighteen years; the inner thoughts and the basic character are not erased. But the recruits have already learned that the only acceptable behaviour is to repress any unorthodox thoughts and to mimic the character the Marine Corps wants. Nor are they, on the whole, reluctant to do so, for they want to be Marines. From the moment they arrive at Parris Island, the vague notion that has been passed down for a thousand generations that masculinity means being a warrior becomes an explicit article of faith, relentlessly preached: to be a man means to be a Marine....

Even the seeming inanity of close-order drill has a practical role in the conversion process. It has been over a century since mass formations of men were of any use on the battlefield, but every army in the world still drills its troops, especially during basic training, because marching in formation, with every man moving his body in the same way at the same moment, is a direct physical way of learning two things a soldier must believe: that orders have to be obeyed automatically and instantly, and that you are no longer an individual, but part of a group....

KEY CONCEPTS

conversion group resocialization

DISCUSSION QUESTIONS

1. In situations other than the military, might resocialization take place? What does that process look like?
2. Why is it important in military socialization for the recruit to identify so strongly with the group that she or he relinquishes an individual identity?

Applying Sociological Knowledge: An Exercise for Students

Socialization occurs in many different contexts. Pick one of the following situations that involve entry to a new role: becoming a college student, becoming a parent, or getting your first job. What changes do you experience in this new role? How are the expectations associated with your new status communicated to you? Are these formally or informally communicated? What does this teach you about the socialization process?

PART V

Society and Social Interaction

12

The Presentation of Self in Everyday Life

ERVING GOFFMAN

Erving Goffman likens social interaction to a "con game," in which we are consistently trying to put forward a certain impression or "self" in order to get something from others. Although many will not see human behavior so cynically, Goffman's analysis sheds light on how people try to manage the impression that others have of them.

When an individual plays a part, he implicitly requests his observers to take seriously the impression that is fostered before them. They are asked to believe that the character they see actually possesses the attributes he appears to possess, that the task he performs will have the consequences that are implicitly claimed for it, and that, in general, matters are what they appear to be. In line with this, there is the popular view that the individual offers his performance and puts on his show "for the benefit of other people." It will be convenient to begin a consideration of performances by turning the question around and looking at the individual's own belief in the impression of reality that he attempts to engender in those among whom he finds himself.

At one extreme, one finds that the performer can be fully taken in by his own act; he can be sincerely convinced that the impression of reality which he stages is the real reality. When his audience is also convinced in this way about the show he puts on—and this seems to be the typical case—then for the moment at least, only the sociologist or the socially disgruntled will have any doubts about the "realness" of what is presented.

At the other extreme, we find that the performer may not be taken in at all by his own routine. This possibility is understandable, since no one is in quite as good an observational position to see through the act as the person who puts it on. Coupled with this, the performer may be moved to guide the conviction of his audience only as a means to other ends, having no ultimate concern in the conception that they have of him or of the situation. When the individual has no belief in his own act and no ultimate concern with the beliefs of his audience, we may call him cynical, reserving the term "sincere" for individuals who believe in the impression fostered by their own performance. It should be understood that the cynic, with all his professional disinvolvement, may obtain unprofessional pleasures from his masquerade, experiencing a kind of gleeful spiritual aggression from the fact that he can toy at will with something his audience must take seriously.

It is not assumed, of course, that all cynical performers are interested in deluding their audiences for purposes of what is called "self-interest" or private gain. A cynical individual may delude his audience for what he considers to be their own good, or for the good of the community, etc. For illustrations of this we need not appeal to sadly enlightened showmen such as Marcus Aurelius or Hsun Tzu. We know that in service occupations practitioners who may otherwise be sincere are sometimes forced to delude their customers because their customers show such a heartfelt demand for it. Doctors who are led into giving placebos, filling station attendants who resignedly check and recheck tire pressures for anxious women motorists, shoe clerks who sell a shoe that fits but tell the customer it is the size she wants to hear—these are cynical performers whose audiences will not allow them to be sincere. Similarly, it seems that sympathetic patients in mental wards will sometimes feign bizarre symptoms so that student nurses will not be subjected to a disappointingly sane performance. So also, when inferiors extend their most lavish reception for visiting superiors, the selfish desire to win favor may not be the chief motive; the inferior may be tactfully attempting to put the superior at ease by simulating the kind of world the superior is thought to take for granted.

I have suggested two extremes: an individual may be taken in by his own act or be cynical about it. These extremes are something a little more than just the ends of a continuum. Each provides the individual with a position which has its own particular securities and defenses, so there will be a tendency for those who have traveled close to one of these poles to complete the voyage. Starting with lack of inward belief in one's role, the individual may follow the natural movement described by Park:

> It is probably no mere historical accident that the word person, in its first meaning, is a mask. It is rather a recognition of the fact that everyone is always and everywhere, more or less consciously, playing a role.... It is in these roles that we know each other; it is in these roles that we know ourselves.[1]

In a sense, and in so far as this mask represents the conception we have formed of ourselves—the role we are striving to live up to—this mask is our truer self, the self we would like to be. In the end, our conception of our role becomes second nature and an integral part of our personality. We come into the world as individuals, achieve character, and become persons.[2]...

Front, then, is the expressive equipment of a standard kind intentionally or unwittingly employed by the individual during his performance. For preliminary purposes, it will be convenient to distinguish and label what seem to be the standard parts of front.

First, there is the "setting," involving furniture, decor, physical layout, and other background items which supply the scenery and stage props for the spate of human action played out before, within, or upon it....

If we take the term "setting" to refer to the scenic parts of expressive equipment, one may take the term "personal front" to refer to the other items of expressive equipment, the items that we most intimately identify with the performer himself and that we naturally expect will follow the performer wherever he goes. As part of personal front we may include: insignia of office or rank; clothing; sex, age, and racial characteristics; size and looks; posture; speech patterns; facial expressions; bodily gestures; and the like. Some of these vehicles for conveying signs, such as racial characteristics, are relatively fixed and over a span of time do not vary for the individual from one situation to another. On the other hand, some of these sign vehicles are relatively mobile or transitory, such as facial expression, and can vary during a performance from one moment to the next....

In addition to the fact that different routines may employ the same front, it is to be noted that a given social front tends to become institutionalized in terms of the abstract stereotyped expectations to which it gives rise, and tends to take on a meaning and stability apart from the specific tasks which happen at the time to be performed in its name. The front becomes a "collective representation" and a fact in its own right.

When an actor takes on an established social role, usually he finds that a particular front has already been established for it. Whether his acquisition of the role was primarily motivated by a desire to perform the given task or by a desire to maintain the corresponding front, the actor will find that he must do both.

Further, if the individual takes on a task that is not only new to him but also unestablished in the society, or if he attempts to change the light in which his task is viewed, he is likely to find that there are already several well-established fronts among which he must choose. Thus, when a task is given a new front we seldom find that the front it is given is itself new....

NOTES

1. Robert Ezra Park, *Race and Culture* (Glencoe, IL: The Free Press, 1950), p. 249.
2. Ibid., p. 250.

KEY CONCEPTS

dramaturgical model | impression management | presentation of self

DISCUSSION QUESTIONS

1. How many "selves" do you think you could "play" or "do" in order to accomplish something with another person? Discuss two such selves, and try to get them to be quite different from each other.
2. "All the world's a stage," wrote William Shakespeare. So might Goffman have said this. How does his analysis of the presentation of self in everyday life suggest that life is a drama where we all play our parts?

13

The Impact of Internet Communications on Social Interaction

THOMAS WELLS BRIGNALL III AND THOMAS VAN VALEY

The increased use of the Internet has altered the way people interact. This article applies Goffman's theory about the presentation of self to social interaction that takes place on the Internet. The "rules" for social interaction and the skill that develops when you engage in face-to-face interaction are changed when communicating online. The authors suggest that young people today are cyberkids and experience social interaction and education differently because of the Internet.

INTRODUCTION

The Internet is clearly on the way to becoming an integral tool of business, communication, and popular culture in the United States and in other parts of the world. The Internet is also being presented as a pedagogical tool for much of public education. However, its extraordinary growth is not without concern. Of particular relevance is the issue of the potential impact of the Internet and especially computer-mediated communications on the nature and quality of social interaction among cyberkids.

SOURCE: "The Impact of Internet Communications on Social Interaction." by Thomas Wells Brignall III and Thomas Van Valey. 2005. Sociological Spectrum 25: 335–348. Reprinted with permission of Taylor & Francis Group.

According to NetValue, people who chat online are among the heaviest users of the Internet (2002, p. 1). Furthermore, 20% of female chatters are teenage girls. NetValue, Nielsen, and eMarketer (Ramsey 2000) all conclude that while teenagers are not yet the dominant demographic of Internet users, their usage is growing very rapidly. Regardless of country of origin, recent data on Internet usage indicates that young people are becoming some of the heaviest users of the Internet (Nielsen 2002, p. 1). According to Nielsen (2002), 74% of United States residents between the ages of 12 and 18 are now using the Internet.

Among Internet users between 12 and 18, 35% spend 31–60 minutes per day online, and 44% spend more than an hour per day online. Indeed, almost 4% of these cyberkids spend four or more hours online each day. America Online reported in their national survey of more than 6,700 teens and parents of teens that "fifty-six percent of teens (aged 18 to 19) prefer Internet to the telephone" (Pastore 2002, p. 1). The survey also reported that in order to keep up their communications with friends, more than 81% of teens use e-mail, while 70% use instant messaging.

Certainly, one can argue that online communication is not yet the dominant form of communication among young people. However, there is little question that the phenomenon of online communication among teens and children is growing rapidly. Moreover, it is not hard to imagine a time in the near future when elementary and secondary school students may spend several hours a day doing schoolwork, communicating with friends, teachers, and family, and seeking entertainment, all via the computer and the Internet. It also follows, therefore, that substantial portions of students' experiences with interpersonal communications (especially with persons not already known to them) are likely to be computer-mediated. If children and teenagers are already using computers as a significant form of education, communication, and entertainment, it may well be that less time is being spent having face-to-face interactions with peers.

If the amount of time spent in face-to-face interactions among youth is shrinking, there may be significant consequences for their development of social skills and their presentation of self. Several decades ago, Goffman (1959, 1967) suggested that individuals who lack the normative communication, cultural, and civility skills in a society would find it difficult to interact with others successfully. At the time, Goffman was referring to the variety of visual and auditory cues that occur in face-to-face communications. Today, with computer-mediated communications, it is possible that none of the cues that Goffman wrote about may be present during online communication.

Some authors have already suggested that online behavior is different from offline behavior (Rheingold 1993; Postman 1992; Jones 1995; Miller 1996). Examples of such differences in behavior include individuals who are willing to misrepresent themselves by feigning a different gender, skin color, sexual orientation, physical condition, or age. Other differences in observed behaviors include the open display of group norm violations such as aggressive behavior, racism, sexism, homophobia, personal attacks, harassment, and a tendency for individuals to quickly abandon groups and conversations, refusing to deal with issues they find difficult to immediately resolve. If these authors are correct, it is critical that the specific elements of interpersonal communication involved in computer-mediated interactions be identified and compared with face-to-face interactions. This article examines the nature of Internet communications that take place among young people (particularly the elements of the presentation of self that do not occur, or occur with limited frequency), and suggests some potential consequences for the education and socialization of our youth.

THEORIES OF SOCIAL INTERACTION

In *Asylums* (1961), Goffman described how small rituals have replaced the big rituals that occurred in traditional interpersonal relations. "What remains are brief rituals one individual performs for and to another, attesting to civility and good will on the performer's part and to the recipient's possession of a small patrimony of sacredness" (Goffman 1961, p. 63). However, the choice of the particular set of rules an individual chooses to follow derives from requirements established in social encounters. Therefore, an individual's concept of self is shaped by the sum of the social interactions in which that individual engages.

Goffman (1967) further argues that the self is the actor in an ongoing play that responds to the judgments of others. While each individual has more than one role, he or she is saved from role strain by "audience segregation," that is, by employing multiple roles in their interactions with others. Because an individual can play a unique role in each social situation, that individual can effectively be a different person in each situation without contradiction. However, audience segregation regularly breaks down, and an individual can present a role incompatible with ones presented on other occasions. In these situations, role strain occurs.

Children learn how to cope with such role strains through their own social experience and by watching others navigate social interactions that involve contradictory or competing roles. Indeed, Goffman argues that coping with role strain and developing impression management are necessary skills for the success of individuals in everyday social interactions. Children also must learn how to manage "front stage" and "back stage" behavior. The front stage is open to judgment by an audience, where the back stage is a place where actors can discuss, polish, or refine their performances without revealing themselves to the same audience. However, because there are multiple layers of front stages and back stages, individuals learn how much they can reveal to other characters.

All front stage roles contain a number of elements that are visual or auditory in nature including physical appearance (e.g., demographic characteristics, and physical features, such as size, make up, hairstyle, posture), manner of speaking (e.g., the use of standard vs. slang dialect, accents, regional vocabulary choices, and voice inflections), and the use of various props (e.g., clothing, car, and food preferences). Together, these elements help to create the role that is presented to others. However, individuals in face-to-face interactions not only present these more obvious indicators of their roles, they also give out more subtle cues such as posture, hand gestures, tone of voice, movement in a conversation, eye contact, and levels of social formality. Moreover, many of these indicators of role (both the obvious and the subtle) vary widely across groups.

However, in order to maintain a positive on-going relationship in any difficult face-to-face circumstance, an individual must learn the appropriate socialization rituals. Knowing these rituals and being able to play a proper front stage role is crucial in order for an individual to get along with others. Indeed, the appearance of getting along with others is sometimes far more important than whether individuals actually like one another. Once again, it is difficult for individuals to succeed if they lack the proper social skills of the various groups with which they interact. Only with practice, will individuals develop and learn to improve their interaction skills, and develop a better presentation when communicating to individuals via their front and back stages....

ONLINE INTERACTION

From the beginning of the Internet, it was clear that the interaction taking place online was a new form of social interaction. What was not clear, and still is to be determined, are the consequences of this new form of social interaction. The Internet itself is neither negative nor positive. It is inanimate, an object or a tool that can be used in various ways. To reify the Internet and suggest that it is somehow inherently liberating or enslaving is misleading. Nevertheless, it is important to look at how interaction on the Internet differs from other forms of interaction. These differences may play havoc with traditional social interaction rituals....

Because online interaction is different, it is entirely possible that children with high levels of online interactions would adopt different techniques of social interaction. Several authors such as LaRose, Eastin, and Gregg (2001) and Schmitz (1997) cultivate the notion that for many individuals, online communication helps facilitate face-to-face communications with current relationships and sometimes with new relationships. Boyd and Walther (2002) even argue that Internet social support is superior to face-to-face social support. According to them, online social support offers benefits that face-to-face social networks cannot: anonymity, constant access to better quality expertise, and enhanced modes of expression, with less chance of embarrassment and without incurring an obligation to the support provider. However, Spears and Lea (1992) argue that the absence of social and contextual cues undermine the perception of leadership, status, and power, and leads to reduced impact of social norms and therefore to deregulated, anti-normative behavior.

THE IMPACT OF INTERNET COMMUNICATIONS

Our fundamental position is that online social interaction is one form of roleplay, and thus an element in the development of the self. Moreover, if a substantial amount of communication is accomplished online, either at home, at school, or elsewhere, children and cyberkids are likely to develop the skills necessary for online interaction, but they are also likely to lack some of the skills that are involved in face-to-face interaction.

Tapscott (1998) claims that the children of the Internet generation are already way ahead of their parents and many other adults (who do not yet understand what the Internet is or how it works). However, Tapscott and others may have made a crucial mistake in their logic. They have forgotten that political and economic power are in the hands of adults. Therefore, the members of the cyberkid generation must understand, adapt, and modify their interactions in order to get along in society.

Moreover, if students develop unique interaction rituals based on online communication without enough experience or understanding of traditional face-to-face interaction rituals, the likelihood for friction and/or conflict to occur undoubtedly increases. Such youths may be perceived as rude, insolent, disconnected, spoiled, or apathetic. It can be argued that new cultural and social phenomena have typically produced tensions between the generations. However, not having the skills or the knowledge of how to communicate with people who have different values and attitudes complicates the traditional struggles between youth and their elders. Furthermore, such rifts will not only be between the young and the old, but between any groups that are different from one another....

The skills and lessons of socialization that students need to learn in order to cope in everyday life, however, cannot be manufactured by computer simulations or video games (at least not yet). Classrooms with computers hooked up to the Internet predispose students to work as individuals rather than as members of any social group. Although students can interact with others when they are on the Internet, they are often able to choose with whom they wish to interact and how they want to manage it. Even if the students are interacting with fellow classmates while online, is it not reasonable that the use of a computer in mediating those interactions will alter the interaction rituals in which they engage?...

DISCUSSION

Given that we know so little about the nature of the social interactions that take place over the Internet, not only but especially by youth, it is clear that much research is needed. One obvious arena would include studies of young children followed over time to determine if there are recognizable online interaction rituals and what consequences they may have on the social development of the children. There is a growing body of research on this general issue—whether computer-mediated communications "displace" more traditional face-to-face forms of communication. A special issue of *IT & Society* recently focused on it, reporting the results of a number of time-diary studies in the United States and elsewhere. However, the conclusions are mixed. Some studies have found reduced levels of time spent with friends and family (Nie and Hillygus 2002; Pronovost 2002; Nie and Erbring 2002). Others have either reported no difference (Robinson et al. 2002;

Gershuny 2002) or increased levels of sociability among Internet users (Neustadl and Robinson 2002; Cummings, Sproull, and Kiesler 2002).

We agree with Wellman and Gulia (1999) that "the internet … is not a separate reality. People bring to their online interactions such baggage as their gender, stage in the life cycle, cultural milieu, socioeconomic status, and offline connections with others" (p. 3). Online and face-to-face social interactions share some elements, and many individuals interact using both forms of communication.

This discussion is not about the members of the current generations who have grown up with the Internet. It is about future generations where the Internet is used as a primary source of communication, the focal center of schoolwork and research, the main source of entertainment, and the primary medium for the development of contemporary issues. If the strength of the Internet of the future is the fact that individuals can choose with whom they want to interact, then it may also be one of the Internet's weaknesses when it comes to the development of social interaction skills. The demands of learning to get along with others are likely to become drowned out by self-interested pursuits. The possibility of a narrow world perspective seems certain for those individuals who choose to isolate themselves from people and ideas with whom they feel uncomfortable. If the easiest solution to avoid dissonance is to avoid situations that produce it, then the potential for an unrealistic social process is high. We are not opposing or supporting computer-mediated communications or the Internet in schools. We are simply suggesting that it is important to assess the social impacts of the Internet while the frequency of use is still relatively low. Once they become ubiquitous, the possibility of such research has been forever lost.

REFERENCES

Boyd, S. and Joseph, B. Walther. 2002. "Attraction to Computer-Mediated Social Support." Pp. 153–188 in *Communication Technology and Society: Audience Adoption and Uses*, edited by C. A. Lin and D. Atkin. Cresskill, NJ: Hampton Press.

Cummings, J., L. Sproull, and S. Kiesler. 2002. "Beyond Hearing: Where Real World and Online Support Meet." *Group Dynamics: Theory, Research, and Practices*, 6(1):78–88.

Gershuny, J. 2002. "Mass Media, Leisure, and Home IT: A Panel Time-Diary Approach." Pp. 53–66. Retrieved June 8, 2002 from http://www.IT and Society.org.

Goffman, E. 1959. *The Presentation of Self in Everyday Life.* New York: Anchor.

Goffman, E. 1961. *Asylums: Essays on the Social Situation of Mental Patients and Other Inmates.* New York: Anchor.

Goffman, E. 1967. *Interaction Ritual: Essays on Face to Face Behavior.* New York: Pantheon Books.

Jones, G. S. 1995. *CyberSociety: Computer-Mediated Communication and Community.* Thousand Oaks, CA: Sage Publications.

LaRose, R., M. S. Eastin, and J. Gregg. 2001. "Reformulating the Internet Paradox: Social Cognitive Explanations of Internet Use and Depression." Retrieved January 30, 2003 from www.behavior.net/JOB/v1n1/paradox.html.

Miller, E. S. 1996. *Civilizing Cyberspace: Policy, Power, and the Information Superhighway.* New York: ACM Press.

NetValue. (2002). "Internet Use Patterns." Edited by R. A. Cole. Retrieved December 5, 2002 from www.netvalue.com/corp/actionnaires/index.htm.

Neustadl, A. and J. P. Robinson. 2002. "Media Use Differences Between Internet Users and Nonusers in the General Social Survey." Pp. 100–120. Retrieved June 8, 2002 from http://www.IT and Society.org.

Nie, N. H. and L. Erbring. (2002). "Internet and Society: A Preliminary Report." *IT and Society* 1(1):275–283.

Nie, N. H. and D. S. Hillygus. (2002). "The Impact of Internet Use on Sociability: Time-diary Findings." *IT and Society* 1(1):1–20.

Nielsen Net Ratings. (2002). "Internet User Growth." Retrieved September 19, 2002 from http://www.nielsen-netratings.com/news.jsp.

Pastore, M. (2002). "Internet Key to Communication Among Youth." Retrieved January 30, 2003 from http://cyberatlas.internet.com/big_picture/demographics/article/0,5901_961881,00.html.

Postman, N. (1992). *Technopoly*. New York: Vintage Books.

Pronovost, G. (2002). "The Internet and Time Displacement: A Canadian Perspective." *IT and Society* 1(1):44–53.

Ramsey, G. (2000). *The E-demographics and Usage Patterns Report: September 2000.* Retrieved October 23, 2000 from http://www.emarketer.com/ereports/ecommerce_b2b/.

Rheingold, H. (1993). *The Virtual Community: Homesteading on the Electronic Frontier.* Cambridge, Massachusetts: Addison-Wesley Publishing Company Reading.

Robinson, J. P., M. Kestnbaum, A. Neustadl, and A. Alvarez. 2002. "The Internet and Other Uses of Time." Pp. 244–262 in *The Internet in Everyday Life*, edited by B. Wellman and C. Haythornthwaite. Malden, MA: Blackwell Publishing.

Schmitz, J. (1997). "Structural Relations, Electronic Media, and Social Change: The Public Electronic Network and the Homeless." Pp. 80–101 in *Virtual Culture: Identity and Communication in Cybersociety*, edited by S. G. Jones. London: Sage.

Spears, R. and M. Lea. (1992). "Social Influence and the Influence of the "Social" in Computer-mediated Communication." Pp. 30–65 in *Contexts of Computer-mediated Communication*, edited by M. Lea. London: Harvester-Wheatsheaf.

Tapscott, D. (1998). *Growing Up Digital: The Rise of the Net Generation*. New York: McGraw-Hill Trade.

Wellman, B. and M. Gulia. 1999. "Net Surfers Don't Ride Alone: Virtual Community as Community." Pp. 331–367 in *Networks in the Global Village*, edited by Barry Wellman. Boulder, Co: Westview Press.

KEY CONCEPTS

ritual | role strain | social interaction

DISCUSSION QUESTIONS

1. How have Facebook, MySpace, and other online communication sites changed the way you interact with friends? In your opinion, is the nature of the friendship changed when communicating online as opposed to in person?
2. What "rules" do you think exist when e-mailing? For example, is there a difference in the way you address a professor in an e-mail compared to the way you address a friend or classmate? Consider e-mail etiquette. Is there such a thing? Should there be?

14

Code of the Street

ELIJAH ANDERSON

Elijah Anderson's study of interaction on the street shows the vast array of implicit "codes" of behavior or rules that guide street interaction. His analysis helps explain the complexity of street interaction and provides a sociological explanation of street violence.

In some of the most economically depressed and drug- and crime-ridden pockets of the city, the rules of civil law have been severely weakened, and in their stead a "code of the street" often holds sway. At the heart of this code is a set of prescriptions and proscriptions, or informal rules, of behavior organized around a desperate search for respect that governs public social relations, especially violence, among so many residents, particularly young men and women. Possession of respect—and the credible threat of vengeance—is highly valued for shielding the ordinary person from the interpersonal violence of the street. In this social context of persistent poverty and deprivation, alienation from broader society's institutions, notably that of criminal justice, is widespread. The code of the street emerges where the influence of the police ends and personal responsibility for one's safety is felt to begin, resulting in a kind of "people's law," based on "street justice." This code involves a quite primitive form of social exchange that holds would-be perpetrators accountable by promising an "eye for an eye," or a certain "payback" for transgressions. In service to this ethic, repeated displays of "nerve" and "heart" build or reinforce a credible reputation for vengeance that works to deter aggression and disrespect, which are sources of great anxiety on the inner-city street....

In approaching the goal of painting an ethnographic picture of these phenomena, I engaged in participant-observation, including direct observation, and conducted in-depth interviews. Impressionistic materials were drawn from various social settings around the city, from some of the wealthiest to some of the most economically depressed, including carryouts, "stop and go" establishments, Laundromats, taverns, playgrounds, public schools, the Center City indoor mall known as the Gallery, jails, and public street corners. In these settings I encountered a wide variety of people—adolescent boys and young women (some incarcerated, some not), older men, teenage mothers, grandmothers, and male and female schoolteachers, black and white, drug dealers, and common criminals. To protect the privacy and confidentiality of my subjects, names and certain details have been disguised....

Of all the problems besetting the poor inner-city black community, none is more pressing than that of interpersonal violence and aggression. This phenomenon wreaks havoc daily on the lives of community residents and increasingly spills over into downtown and residential middle-class areas. Muggings, burglaries, carjackings, and drug-related shootings, all of which may leave their victims or innocent bystanders dead, are now common enough to concern all urban and many suburban residents.

The inclination to violence springs from the circumstances of life among the ghetto poor—the lack of jobs that pay a living wage, limited basic public services (police response in emergencies, building maintenance, trash pickup, lighting, and other services that middle-class neighborhoods take for granted), the stigma of race, the fallout from rampant drug use and drug trafficking, and the resulting alienation and absence of hope for the future. Simply living in such an environment places young people at special risk of falling victim to aggressive behavior. Although there are often forces in the community that can counteract the negative influences—by far the most powerful is a strong, loving, "decent" (as inner-city residents put it) family that is committed to middle-class values—the despair is pervasive enough to have spawned an oppositional culture, that of "the street," whose norms are often consciously opposed to those of mainstream society. These two orientations—decent and street—organize the community socially, and the way they coexist and interact has important consequences for its residents, particularly for children growing up in the inner city. Above all, this environment means that even youngsters whose home lives reflect mainstream values—and most of the homes in the community do—must be able to handle themselves in a street-oriented environment.

This is because the street culture has evolved a "code of the street," which amounts to a set of informal rules governing interpersonal public behavior, particularly violence. The rules prescribe both proper comportment and the proper way to respond if challenged. They regulate the use of violence and so supply a rationale allowing those who are inclined to aggression to precipitate violent encounters in an approved way. The rules have been established and are enforced mainly by the street-oriented; but on the streets the distinction between street and decent is often irrelevant. Everybody knows that if the rules are violated, there are penalties. Knowledge of the code is thus largely defensive, and it is literally necessary for operating in public. Therefore, though families with a decency orientation are usually opposed to the values of the code, they often reluctantly encourage their children's familiarity with it in order to enable them to negotiate the inner-city environment.

At the heart of the code is the issue of respect—loosely defined as being treated "right" or being granted one's "props" (or proper due) or the deference one deserves. However, in the troublesome public environment of the inner city, as people increasingly feel buffeted by forces beyond their control, what one deserves in the way of respect becomes ever more problematic and uncertain. This situation in turn further opens up the issue of respect to sometimes intense interpersonal negotiation, at times resulting in altercations. In the street culture, especially among young people, respect is viewed as almost an external entity, one that is hard-won but easily lost—and so must constantly be guarded. The rules of the code in fact provide a framework for negotiating respect. With the right amount of respect, individuals can avoid being bothered in public. This security is important, for if they *are* bothered, not only may they face physical danger, but they will have been disgraced or "dissed" (disrespected). Many of the forms dissing can take may seem petty to middle-class people (maintaining eye contact for too long, for example), but to those invested in the street code, these actions, a virtual slap in the face, become serious indications of the other person's intentions. Consequently, such people become very sensitive to advances and slights, which could well serve as a warning of imminent physical attack or confrontation.

The hard reality of the world of the street can be traced to the profound sense of alienation from mainstream society and its institutions felt by many poor inner-city black people, particularly the young. The code of the street is actually a cultural adaptation to a profound lack of faith in the police and the judicial system—and in others who would champion one's personal security. The police, for instance, are most often viewed as representing the dominant white society and as not caring to protect inner-city residents. When called, they may not respond, which is one reason many residents feel they must be prepared to take extraordinary measures to defend themselves and their loved ones against those who are inclined to aggression. Lack of police accountability has in fact been incorporated into the local status system: the person who is believed capable of "taking care of himself" is accorded a certain deference and regard, which translates into a sense of physical and psychological control. The code of the street thus emerges where the influence of the police ends and where personal responsibility for one's safety is felt to begin. Exacerbated by the proliferation of drugs and easy access to guns, this volatile situation results in the ability of the street-oriented minority (or those who effectively "go for bad") to dominate the public spaces....

The attitudes and actions of the wider society are deeply implicated in the code of the street. Most people residing in inner-city communities are not totally invested in the code; it is the significant minority of hard-core street youth who maintain the code in order to establish reputations that are integral to the extant social order. Because of the grinding poverty of the communities these people inhabit, many have—or feel they have—few other options for expressing themselves. For them the standards and rules of the street code are the only game in town.

And as was indicated above, the decent people may find themselves caught up in problematic situations simply by being at the wrong place at the wrong time, which is why a primary survival strategy of residents here is to "see but don't see." The extent to which some children—particularly those who through upbringing have become most alienated and those who lack strong and conventional social support—experience, feel, and internalize racist rejection and contempt from mainstream society may strongly encourage them to express contempt for the society in turn. In dealing with this contempt and rejection, some youngsters consciously invest themselves and their considerable mental resources in what amounts to an oppositional culture, a part of which is the code of the street. They do so to preserve themselves and their own self-respect. Once they do, any respect they might be able to garner in the wider system pales in comparison with the respect available in the local system; thus they often lose interest in even attempting to negotiate the mainstream system.

At the same time, many less alienated young people have assumed a street-oriented demeanor as way of expressing their blackness while really embracing a much more moderate way of life; they, too, want a nonviolent setting in which to live and one day possibly raise a family. These decent people are trying hard to be part of the mainstream culture, but the racism, real and perceived, that they encounter helps legitimize the oppositional culture and, by extension, the code of the street. On occasion they adopt street behavior; in fact, depending on the demands of the situation, many people attempt to code switch, moving back and forth between decent and street behavior....

In addition, the community is composed of working-class and very poor people since those with the means to move away have done so, and there has also been a proliferation of single-parent households in which increasing numbers of kids are being raised on welfare. The result of all this is that the inner-city community has become a kind of urban village, apart from the wider society and limited in terms of resources and human capital. Young people growing up here often receive only the truncated version of mainstream society that comes from television and the perceptions of their peers....

According to the code, the white man is a mysterious entity, a part of an enormous monolithic mass of arbitrary power, in whose view black people are insignificant. In this system and in the local social context, the black man has very little clout; to salvage something of value, he must outwit, deceive, oppose, and ultimately "end-run" the system.

Moreover, he cannot rely on this system to protect him; the responsibility is his, and he is on his own. If someone rolls on him, he has to put his body, and often his life, on the line. The physicality of manhood thus becomes extremely important. And urban brinksmanship is observed and learned as a matter of course....

Urban areas have experienced profound structural economic changes, as deindustrialization—the movement from manufacturing to service and high-tech—and the growth of the global economy have created new economic conditions. Job opportunities increasingly go abroad to Singapore, Taiwan, India, and Mexico, and to nonmetropolitan America, to satellite cities like King of Prussia, Pennsylvania. Over the last fifteen years, for example, Philadelphia has lost 102,500 jobs, and its manufacturing employment has declined by 53 percent. Large numbers of inner-city people, in particular, are not adjusting effectively to the new economic reality. Whereas low-wage jobs—especially unskilled and low-skill factory jobs—used to exist simultaneously with poverty and there was hope for the future, now jobs simply do not exist, the present economic boom notwithstanding. These dislocations have left many inner-city people unable to earn a decent living. More must be done by both government and business to connect inner-city people with jobs.

The condition of these communities was produced not by moral turpitude but by economic forces that have undermined black, urban, working-class life and by a neglect of their consequences on the part of the public. Although it is true that persistent welfare dependency, teenage pregnancy, drug abuse, drug dealing, violence, and crime reinforce economic marginality, many of these behavioral problems originated in frustrations and the inability to thrive under conditions of economic dislocation. This in turn leads to a weakening of social and family structure, so children are increasingly not being socialized into mainstream values and behavior. In this context, people develop profound alienation and may not know what to do about an opportunity even when it presents itself. In other words, the social ills that the companies moving out of these neighborhoods today sometimes use to justify their exodus are the same ones that their corporate predecessors, by leaving, helped to create.

Any effort to place the blame solely on individuals in urban ghettos is seriously misguided. The focus should be on the socioeconomic structure, because it was structural change that caused jobs to decline and joblessness to increase in many of these communities. But the focus also belongs on the public policy that has radically threatened the well-being of many citizens. Moreover, residents of these communities lack good education, job training, and job networks, or connections with those who could help them get jobs. They need enlightened employers able to understand their predicament and willing to give them a chance. Government, which should be assisting people to adjust to the changed economy, is instead cutting what little help it does provide....

The emergence of an underclass isolated in urban ghettos with high rates of joblessness can be traced to the interaction of race prejudice, discrimination, and the effects of the global economy. These factors have contributed to the profound social isolation and impoverishment of broad segments of the inner-city black population. Even though the wider society and economy have been experiencing accelerated prosperity for almost a decade, the fruits of it often miss the truly disadvantaged isolated in urban poverty pockets.

In their social isolation an oppositional culture, a subset of which is the code of the street, has been allowed to emerge, grow, and develop. This culture is essentially one of accommodation with the wider society, but different from past efforts to accommodate the system. A larger segment of people are now not simply isolated but ever more profoundly alienated from the wider society and its institutions. For instance, in conducting the fieldwork for this book, I visited numerous inner-city schools, including elementary, middle, and high schools, located in areas of concentrated poverty. In every one, the so-called oppositional culture was well entrenched. In one elementary school, I learned from interviewing kindergarten, first-grade, second-grade, and fourth-grade teachers that through the first grade, about a fifth of the students were invested in the code of the street; the rest are interested in the subject matter and eager to take instruction from the teachers—in effect, well disciplined. By the fourth grade, though, about three-quarters of the students have bought into the code of the street or the oppositional culture.

As I have indicated throughout this work, the code emerges from the school's impoverished neighborhood, including overwhelming numbers of single-parent homes, where the fathers, uncles, and older brothers are frequently incarcerated—so frequently, in fact, that the word "incarcerated" is a prominent part of the young child's spoken vocabulary. In such communities there is not only a high rate of crime but also a generalized diminution of respect for law. As the residents go about meeting the exigencies of public life, a kind of people's law results.... Typically, the local streets are, as we saw, tough and dangerous places where people often feel very much on their own, where they themselves must be personally responsible for their own security, and where in order to be safe and to travel the public spaces unmolested, they must be able to show others that they are familiar with the code—that physical transgressions will be met in kind.

In these circumstances the dominant legal codes are not the first thing on one's mind; rather, personal security for self, family, and loved ones is. Adults, dividing themselves into categories of street and decent, often encourage their children in this adaptation to their situation, but at what price to the children and at what price to wider values of civility and decency? As the fortunes of the inner city continue to decline, the situation becomes ever more dismal and intractable....

KEY CONCEPTS

deindustrialization · norms · urban underclass

DISCUSSION QUESTIONS

1. List several ways that subtle or nonverbal behavior becomes important "on the street."
2. What specific ways does Anderson see street behavior as stemming from social structural conditions for African Americans?

Applying Sociological Knowledge: An Exercise for Students

For a 48-hour period, keep a detailed log of every form of technology that you use for social interaction (including various forms of communication, networking, scheduling, and so on). How is this different from or similar to face-to-face interaction? Try to imagine life without some of these technologies. How do you think they have changed the character of social interaction? Can you also imagine what technologies might shape social interaction in the future?

PART VI

Groups and Organizations

15

The McDonaldization of Society

GEORGE RITZER

Using the concept of McDonaldization, Ritzer shows how this particular organization now permeates our culture. He shows how this organizational model is both affected by and, in turn, affects, social change throughout society.

Ray Kroc (1902–1984), the genius behind the franchising of McDonald's restaurants, was a man with big ideas and grand ambitions. But even Kroc could not have anticipated the astounding impact of his creation. McDonald's is the basis of one of the most influential developments in contemporary society. Its reverberations extend far beyond its point of origin in the United States and in the fast-food business. It has influenced a wide range of undertakings, indeed the way of life, of a significant portion of the world. That impact is likely to continue to expand in the early 21st century....

... I devote all this attention to McDonald's (as well as to the industry of which it is a part and that it played such a key role in spawning) because it serves here as the major example of, and the paradigm for, a wide-ranging process I call McDonaldization—that is,

> the process by which the principles of the fast-food restaurant are coming to dominate more and more sectors of American society as well as of the rest of the world....

McDonaldization has shown every sign of being an inexorable process, sweeping through seemingly impervious institutions (e.g., religion) and regions (e.g., European nations such as France) of the world....

McDONALD'S AS AN AMERICAN AND A GLOBAL ICON

McDonald's has come to occupy a central place not just in the business world but also in American and global popular culture.... The opening of a new McDonald's in a small town can be an important social event. Said one

Maryland high school student at such an opening, "Nothing this exciting ever happens in Dale City."... Even big-city and national newspapers avidly cover developments in the fast-food business.

Fast-food restaurants also play symbolic roles on television programs and in the movies. A skit on the legendary television show *Saturday Night Live* satirized specialty chains by detailing the hardships of a franchise that sold nothing but Scotch tape. In the movie *Coming to America* (1988), Eddie Murphy plays an African prince whose introduction to America includes a job at "McDowell's," a thinly disguised McDonald's. In *Falling Down* (1993), Michael Douglas vents his rage against the modern world in a fast-food restaurant dominated by mindless rules designed to frustrate customers....

McDonald's has achieved its exalted position because virtually all Americans, and many others, have passed through its golden arches (or by its drive-through windows) on innumerable occasions. Furthermore, most of us have been bombarded by commercials extolling the virtues of McDonald's, commercials tailored to a variety of audiences and that change as the chain introduces new foods, new contests, and new product tie-ins. These ever-present commercials, combined with the fact that people cannot drive or walk very far without having a McDonald's pop into view, have embedded McDonald's deeply in popular consciousness. A poll of school-age children showed that 96% of them could identify Ronald McDonald, second only to Santa Claus in name recognition....

Over the years, McDonald's has appealed to people in many ways. The restaurants themselves are depicted as spick-and-span, the food is said to be fresh and nutritious, the employees are shown to be young and eager, the managers appear gentle and caring, and the dining experience itself seems fun-filled. Through their purchases, people contribute, at least indirectly, to charities such as the Ronald McDonald Houses for sick children.

THE LONG ARM OF McDONALDIZATION

McDonald's strives continually to extend its reach within American society and beyond. As the company's chairman said, "Our goal: to totally dominate the quick service restaurant industry worldwide.... I want McDonald's to dominate."...

McDonald's began as a phenomenon of suburbs and medium-sized towns, but later it moved into smaller towns that supposedly could not support such a restaurant and into many big cities that were supposedly too sophisticated.... Today, you can find fast-food outlets in New York's Times Square. McDonald's can even be found on the Guantanamo Bay U.S. Naval Base in Cuba and in the Pentagon. Small, satellite, express, or remote outlets, opened in areas that could not support full-scale fast-food restaurants, are also expanding rapidly. They are found in small storefronts in large cities and in nontraditional settings such as museums, department stores, service stations, ... and even schools. These satellites typically offer only limited menus and may rely on larger outlets for food storage and preparation....

No longer content to dominate the strips that surround many college campuses, fast-food restaurants have moved right onto many of those campuses. The first campus fast-food restaurant opened at the University of Cincinnati in 1973. Today, college cafeterias often look like shopping mall food courts (and it's no wonder, given that campus food service is a multi-billion-dollar-a-year business).... In conjunction with a variety of "branded partners" (for example, Pizza Hut and Subway), Marriott now supplies food to many colleges and universities. The apparent approval of college administrations puts fast-food restaurants in a position to further influence the younger generation....

In other sectors of society, the influence of fast-food restaurants has been subtler but no less profound. Food produced by McDonald's and other fast-food restaurants has begun to appear in high schools and trade schools; more than 50% of school cafeterias offer popular brand-name fast foods such as McDonald's, Pizza Hut, or Taco Bell at least once a week....

The military has also been pressed to offer fast food on both bases and ships. Despite criticisms by physicians and nutritionists, fast-food outlets have turned up inside U.S. general hospitals and in children's hospitals.... While no private homes yet have a McDonald's of their own, meals at home often resemble those available in fast-food restaurants. Frozen, microwavable, and prepared foods, which bear a striking resemblance to meals available at fast-food restaurants, often find their way to the dinner table....

McDonald's is such a powerful model that many businesses have acquired nicknames beginning with "Mc." Examples include "McDentists" and "McDoctors," meaning drive-in clinics designed to deal quickly and efficiently with minor dental and medical problems; ... "McChild" care centers, meaning childcare centers such as KinderCare; "McStables," designating the nationwide racehorse-training operation of D. Wayne Lucas; and "McPaper," describing the newspaper *USA TODAY*....

THE DIMENSIONS OF McDONALDIZATION

Why has the McDonald's model proven so irresistible? Eating last food at McDonald's has certainly become a "sign"... that, among other things, one is in tune with the contemporary lifestyle. There is also a kind of magic or enchantment associated with such food and its settings. The focus here, however, is on the four dimensions that lie at the heart of the success of this model and, more generally, of McDonaldization. In short, McDonald's has succeeded because it offers consumers, workers, and managers efficiency, calculability, predictability, and control....

Efficiency

One important element of the success of McDonald's is *efficiency*, or the optimum method for getting from one point to another. For consumers, McDonald's (its drive-through is a good example) offers the best available way to get from being hungry to being full. The fast-food model offers, or at least appear to offer, an efficient method for satisfying many other needs, as well. Woody Allen's orgasmatron offered an efficient method for getting from quiescence to sexual gratification. Other institutions fashioned on the McDonald's model offer similar efficiency in exercising, losing weight, lubricating cars, getting new glasses or contacts, or completing income tax forms. Like their customers, workers in McDonaldized systems function efficiently by following the steps in a predesigned process.

Calculability

Calculability emphasizes the quantitative aspects of products sold (portion size, cost) and services offered (the time it takes to get the product). In McDonaldized systems, quantity has become equivalent to quality; a lot of something, or the quick delivery of it, means it must be good. "As a culture, we tend to believe deeply that in general 'bigger is better.'" ... People can quantify things and feel that they are getting a lot of food for what appears to be a nominal sum of money (best exemplified by the McDonald's "Dollar Menu").... In a Denny's ad, a man says, "I'm going to eat too much, but I'm never going to pay too much." ... This calculation does not take into account an important point, however: The high profit margin of fast-food chains indicates that the owners, not the consumers, get the best deal.

People also calculate how much time it will take to drive to McDonald's, be served the food, eat it, and return home; they then compare that interval to the time required to prepare food at home. They often conclude, rightly or wrongly, that a trip to the fast-food restaurant will take less time than eating at home. This sort of calculation particularly supports home delivery franchises such as Domino's, as well as other chains that emphasize saving time. A notable example of time savings in another sort of chain is LensCrafters, which promises people "Glasses fast, glasses in one hour." H&M is known for its "fast fashion."...

Workers in McDonaldized systems also emphasize the quantitative rather than the qualitative aspects of their work. Since the quality of the work is allowed to vary little, workers focus on how quickly tasks can be accomplished. In a situation analogous to that of the customer, workers are expected to do a lot of work, very quickly, for low pay.

Predictability

McDonald's also offers *predictability*, the assurance that products and services will be the same over time and in all locales. Egg McMuffins in New York will be virtually identical to those in Chicago and Los Angeles. Also, those eaten next week or next year will be about the same as those eaten today. Customers take great comfort in knowing that McDonald's offers no surprises. They know that the next Egg McMuffin they eat will not be awful, but it will not be exceptionally delicious, either. The success of the McDonald's model suggests that many people have come to prefer a world in which there are few surprises. "This is strange," notes a British observer, "considering [McDonald's is] the product of a culture which honours individualism above all."...

The workers in McDonaldized systems also behave in predictable ways. They follow corporate rules as well as the dictates of their managers. In many cases, what they do, and even what they say, is highly predictable.

Control

The fourth element in the success of McDonald's, *control*,... is exerted over the people who enter McDonald's. Lines, limited menus, few options, and uncomfortable seats all lead diners to do what management wishes them to do—eat quickly and leave. Furthermore, the drive-through window invites diners to leave before they eat. In the Domino's model, customers never enter in the first place.

The people who work in McDonaldized organizations are also controlled to a high degree, usually more blatantly and directly than customers. They are trained to do a limited number of tasks in precisely the way they are told to do them. This control is reinforced by the technologies used and the way the organization is set up to bolster this control. Managers and inspectors make sure that workers toe the line.

A CRITIQUE OF McDONALDIZATION: THE IRRATIONALITY OF RATIONALITY

McDonaldization offers powerful advantages. In fact, efficiency, predictability, calculability, and control through nonhuman technology (that is, technology that controls people rather than being controlled by them) can be thought of not only as the basic components of a rational system but also as the powerful advantages of such a system. However, rational systems inevitably spawn irrationalities....

Criticism, in fact, can be applied to all facets of the McDonaldizing world....

McDonald's and other purveyors of the fast-food model spend billions of dollars each year detailing the benefits of their system. Critics of the system, however, have few outlets for their ideas. For example, no one sponsors commercials between Saturday morning cartoons warning children of the dangers associated with fast-food restaurants.

Nonetheless, a legitimate question may be raised about this critique of McDonaldization: Is it animated by a romanticization of the past, an impossible desire to return to a world that no longer exists? Some critics do base their critiques on nostalgia for a time when life was slower and offered more surprises, when at least some people (those who were better off economically) were freer, and when one was more likely to deal with a human being than a robot or a computer. Although they have a point, these critics have undoubtedly exaggerated the positive aspects of a world without McDonald's, and they have certainly tended to forget the liabilities associated with earlier eras....

We must look at McDonaldization as both "enabling" and "constraining." ... McDonaldized systems enable us to do many things we were not able to do in the past; however, these systems also keep us from doing things we otherwise would do. McDonaldization is a "double-edged" phenomenon....

KEY CONCEPTS

bureaucracy   calculability   McDonaldization   predictability

DISCUSSION QUESTIONS

1. Look around your community. What evidence of McDonaldization do you see? How does this affect people's habits?
2. Identify what you see as both the positive and negative consequences of McDonaldization.
3. What social changes do you think have encouraged McDonaldization, and how, in turn, has McDonaldization brought other social change?

16

Racism in Toyland

CHRISTINE L. WILLIAMS

Based on a study of two different toy stores, Williams shows how the division of labor in organizations is shaped by race and ethnicity, as well as by gender.

Not long ago I had to buy a present for a six-year-old. I had at least three choices for where to shop: The Toy Warehouse, a big-box superstore with a vast array of low-cost popular toys; Diamond Toys, a high-end chain store with a more limited range of reputedly high-quality toys; or Tomatoes, a

SOURCE: Christine L. Williams, Contexts, vol. 4, issue 4, pp. 28–32, Fall 2005 by SAGE Publications. Reprinted by permission of SAGE Publications.

locally owned, neighborhood shop that sells a relatively small, offbeat assortment of traditional and politically correct toys. Can sociology offer any advice to consumers like me?

Unfortunately, many sociologists turn into utilitarian economists when it comes to analyzing shopping, assuming that customer behavior is determined only by price, convenience, and selection. But a number of social factors influence where we choose to shop, including the racial makeup of the store's workers and customers. In my book, *Inside Toyland: Working, Shopping and Social Inequality,* I argue that racial inequality (and gender and class inequality as well) influence where we choose to shop, how we shop, and what we buy. The retail industry sustains such inequality through hiring policies that favor certain kinds of workers and advertising aimed at customers from specific racial or ethnic groups.

I noticed the connection between shopping and social inequality while working as a clerk at the Toy Warehouse and Diamond Toys for three months in 2001. These stores belonged to national chains, and both employed about 70 hourly employees. At the warehouse store, I was one of only three white women on the staff; most were African American, Hispanic, or second-generation Asian American. The "guests" (as we were required to call customers) were an amazing mix from every racial and ethnic group and social class. In contrast, only three African Americans worked at the upscale toy store; most clerks were white. Most of the customers were also white and middle to upper class. My experiences taught me to notice racial diversity (or its absence) wherever I shop.

"DON'T SHOP WHERE YOU CAN'T WORK!"

This slogan, popular during the Great Depression, rallied black protesters to demand equal access to jobs in stores, and many chains responded by hiring African Americans in predominately black neighborhoods: "We employ colored salesmen" signs appeared in Sears, Walgreens, and other stores eager for black customers.

Retail work is one of the most integrated occupations in the United States today. The proportions of whites, African Americans, Asian Americans, and Hispanics employed in retail jobs more or less match their representation in the labor force. But these statistics hide segregation at the store level. More than 15 percent of employees in shoe stores and variety stores, for instance, are black, but less than 5 percent of employees are black in stores that sell liquor, gardening equipment, or needlework supplies. Inside stores, there is further segregation by task: Whites usually have the top director and manager positions, and nonwhites have the lowest-paid, often invisible backroom jobs.

In the toy stores where I worked, the two most segregated jobs were the director positions (all white men), and the janitor jobs (all Latinas subcontracted from outside firms). The Toy Warehouse employed mostly African Americans in all the other positions, including cashiers. But over time I noticed that the managers preferred to assign younger and lighter-skinned women to this position. Older African-American women who wanted to work as cashiers had to struggle to get the assignment. Lazelle, for example, who was about 35, had been asking to work as a cashier for the two months she had been working there. She worked as a merchandiser, getting items from the storeroom, pricing items, and checking prices when bar codes were missing. Lazelle finally got her chance at the registers the same day that I started. We set up next to each other, and I noticed with a bit of envy how competent and confident on the register she was. (Later she told me she had worked registers at other stores, including fast-food restaurants.) I told her I had been hoping to get assigned as a merchandiser. I liked the idea of being free to walk around the store, talk with customers, and learn more about the toys. I had mentioned to the manager that I wanted that job, but she made it clear I was destined for cashiering and service desk (and later, to my horror, computer accounting). Lazelle looked at me like I was crazy. Most workers thought merchandising was the worst job in the store because it was so physically taxing. From her point of view, I had gotten the better job, no doubt because of my race, and it seemed to her that I wanted to throw that advantage away. (The manager may also have considered my background and educational credentials in assigning me to particular jobs.)

The preference for lighter-skinned women as cashiers reflects the importance of this job in the store's general operations. In discount stores, customers seldom talk with sales clerks. The cashier is the only person most customers deal with, giving her enormous symbolic—and economic—importance for the corporation. Transactions can break down if clerks do not treat customers as they expect. The preference for white and light-skinned women as cashiers should be understood in this light: In a racist and sexist society, managers generally believe that such women are the most friendly and solicitous, and thus most able to inspire trust and confidence in a commercial transaction.

At the upscale Diamond Toys, virtually all cashiers were white. Unlike the warehouse store, where cash registers were lined up at the front of the store, the upscale store had cash registers scattered throughout the different departments. The preference for white workers in these jobs (and throughout the store) seemed consistent with the marketing of the store's workers as "the ultimate toy experts." In retail work, professional expertise is typically associated with whiteness, much as it is in domestic service.

The purported expertise of salesclerks is one of the great deceptions of the retail industry. Here, where jobs pay little and turnover rates are high (estimated at more than 100 percent per year by the National Retail Federation), many clerks know almost nothing about the products they sell. I knew nothing about toys when I was put behind the cash register, and I received no training on the merchandise at either store. Any advice I gave I literally made up. But at the upscale store I was expected to help customers with their shopping decisions. They frequently asked questions like, "What are going to be the hot toys for one-year-olds this Christmas?" or "What one item would you recommend for two sisters of different ages?" One mother asked me to help her pick out a $58 quartz watch for her seven-year-old son. A personal shopper phoned in and asked me to describe the three Britney Spears dolls we carried, help her pick out the "nicest" one, and then arrange to ship it to her employer's niece. Customers asked detailed questions about how the toys were meant to work, and they were especially curious about comparing the merits of the educational toys we offered (I was asked to compare the relative merits of the "Baby Mozart" and the "Baby Bach"). On my first day, I answered a phone call from a customer who asked me to pick out toys for a one-year-old girl and a boy who was two and a half, spend up to $100, and arrange to have the toys gift-wrapped and mailed to their recipients.

At Diamond Toys, most customers didn't mind waiting to talk with me. When the lines were long, they didn't make rude huffing noises or try to make eye contact with their fellow sufferers, as they often did at the Toy Warehouse. I couldn't help but think that the customers—mostly white—were more civil and polite at the upscale store because most of us workers were white. We were presumed to be professional, caring, and knowledgeable, even when we weren't. Like the employers of domestic workers studied by Julia Wrigley, white customers seemed less respectful of minority service workers than white workers; they were willing to pay more and wait longer for the services of whites because they apparently assumed that whites were more refined and intelligent.

On several occasions at the warehouse store, I saw customers reveal racist attitudes toward my African-American coworkers. One night, after the store had closed, I saw Doris and Selma (fiftyish African Americans) escorting several white customers out. Getting straggling customers to leave the store after closing was always a big chore. Soon after, as I was being audited in the manager's office, Selma came in very upset because one of the women she and Doris escorted out had spit out her chewing gum at her. Doris and Selma were appalled. Doris said to them, "That is really disgusting, how could you do that?" And the woman said to Doris, "What's your name?" like she was going to report her. This got Selma so angry she said, "If you take her name, take mine too," and showed her name-tag. She told the woman that she was never welcome to come back to this store. Selma was very distressed. Talking back to customers was taboo, and she knew she could be fired for what she had said. The manager told her that some people are going to be gross and disgusting and what can you do? Clearly Selma and Doris would not get in trouble over this. But I sensed it was doubly humiliating to have to fear that she might be punished for talking back to a white woman who had spit at her.

Although I suffered from plenty of customer condescension at this store, at least putting up with racism was not part of my job. Once when Tanesha, a 23-year-old African American, was training me at the service desk, two white women elbowed up to the counter to complain to me about the long wait for service (they had waited about five minutes while we were serving another customer). I said something about being in training, and they thought I meant that I was training Tanesha. So they said, "Well, call up someone else to the register!" I said I'd have to ask Tanesha to do this. They demanded that I stop training her for a moment to call another person to the exchange desk. "No you don't understand," I said, "I'm the one who is in training; she knows what she is doing, and she is the only one who can call for backup, and she is in the middle of trying to accommodate this other customer." They seemed embarrassed at having assumed that the white woman was in charge. When they realized their mistake, they looked mortified and stepped back from the counter.

SHOPPING WHILE BLACK

During the civil rights era, equal access to stores was high on the list of demands for racial justice. Before Jim Crow laws were repealed, many stores restricted their facilities to whites only. Black customers often were not allowed to try on clothes, eat at lunch counters, or use public restroom facilities in stores.

Today the worst forms of racism have been eliminated. Gone are the "whites only" signs on restrooms and drinking fountains. But some stores build to exclude. The history of suburban malls is a history of intentional racial segregation. Even today, so-called desirable retail locations are characterized by limited access. In my city of Austin, Texas, local malls have *opposed* public bus service on the grounds that it would encourage undesirable (nonwhite) patrons.

But open access is not enough to ensure racial diversity. Diamond Toys was located in a racially diverse urban shopping district, next to subway and bus lines, yet nearly all its patrons were white. I didn't fully realize this until one day when Chandrika, an 18-year-old African-American gift wrapper, told me that she thought she saw one of her friends in the store. We weren't allowed to leave our section, so she asked the plainclothes security guard to look around and see if there was a black guy in the store. I asked her about this. Was it so unusual for an African-American teenager to be in the store that one black guy would be so apparent? After all, lots of young men came in to check out the new electronic toys and collectibles. Chandrika assured me that a young black man would definitely stand out.

One way that many stores show hostility to racial/ethnic minorities is through consumer racial profiling. Like racial profiling in police work, this involves detaining, searching, and harassing such people more often than is done for whites, usually because they are suspected of stealing. Some scholars have labeled this potential violation of people's rights "shopping while black." At the stores where I worked, clerks weren't allowed to pursue anyone suspected of stealing; that was the job of the plainclothes security workers. However, relations between customers and clerks sometimes broke down, and I saw double standards in the treatment of whites and minorities.

At the warehouse store, I was told to treat shoppers as if they were my mother (most shoppers at both stores were women). At the service desk, I was told to appease them by honoring all requests for returns, even if the merchandise had been used and worn out. The goal, my manager told me, was to make these shoppers so grateful that they would return to the store and spend $20,000 per child, the amount their marketers claimed was spent on an average child's toys.

In my experience, only middle-class white women could depend on this treatment. Nevertheless, I watched many white women throw fits, loudly complaining of shoddy service and merchandise with comments like "I will never shop in this store again!" Such arrogance no doubt came from being accustomed to having their demands met. On one occasion, when a white woman threw a tantrum because the bike she had ordered was not ready as scheduled, the manager offered her a $25 gift certificate for her troubles. She refused, demanded a refund, and left shouting that she would "never come to this store again." The manager then gathered the entire staff at the front of the store and chewed us out for being disorganized and incompetent.

Members of minority groups who wanted to return used merchandise or needed special consideration were rarely accommodated. The week before the bike incident, I was on a register that broke down in the middle of a credit card transaction. A middle-class black woman in her forties was buying in-line skates for her ten-year-old daughter. The receipt came out of the register but not the slip for her to sign, so I had to call a manager, who came over and explained that she needed to go to another register and repeat the transaction. She refused, since it seemed to go through all right and she didn't want to be charged twice. She had to wait more than an hour to get this problem solved, and she wasn't offered any compensation. She didn't yell or make a scene; she waited stoically. I felt sorry for her and went to the service desk to tell a couple of my fellow workers what was happening while the managers tried to resolve it, and I said they should just give her the skates and let her go. My fellow workers thought that was the funniest thing they had ever heard. I said, "What about our policy of letting things go to make sure we keep loyal customers?" But they just laughed at me. Celeste said, "I want Christine to be the manager, she just lets the customers have whatever they want!"

Research has shown that African Americans suffer discrimination in public places, including stores; middle-class whites, on the other hand, are privileged. We do not recognize this precisely because it is so customary. Whites expect first-rate service; when they don't receive it, some feel victimized, even discriminated against, and some throw tantrums when they don't get what they want.

When African-American customers did shout or make a scene, the managers called or threatened to call the police. Each of these instances involved an African-American man complaining and demanding a refund. Once a young black man, denied a cash refund, threw a toy on the service desk and accidentally hit the telephone, which flew off the desk and hit me on the side of the head, knocking me to the floor. Within minutes, three police officers arrived and asked me if I wanted to press assault charges. I did not. After all, angry people often threw merchandise on that desk, and what happened had been an accident. But at least I was appeased. At the end of my shift, the manager gave me a "toy buck" for "taking a hit" in the line of duty, which entitled me to a free Coke.

There wasn't much shouting or throwing at the upscale store. It protected itself from conflict by catering to an upper-class clientele, much as a gated community does. This is not to say that diversity always leads to conflict, but it did at the warehouse store because race, class, and gender differences existed under a layer of power differences within the store. Clerks and customers interacted in a context where these differences had been used to shape marketing agendas, hiring practices, and labor policies—all of which benefited specific groups (especially middle-class white men and women).

WHAT ABOUT TOMATOES?

There are alternatives to shopping at large chain stores, although their numbers are dwindling. The store I'm calling Tomatoes is a small, family-owned business in an upper-middle-class neighborhood on a busy street with lots of pedestrian traffic. It's been in the neighborhood for 25 years, owned by the same family. It sells an offbeat assortment of toys, including many traditional items like kites and wooden blocks, and a variety of toys I would call "politically correct." It didn't carry Barbie, for example, but it did have "Get Real Girls," female action figures that look like G.I. Joe's sisters. I laughed when I saw a pack of plastic "multicultural" food, including spaghetti, sushi, a taco, and a bagel (all marked "made in China").

Working conditions at the store seemed very relaxed compared to what I had experienced. The owner wore shorts and a Hawaiian shirt, and the workers dressed like punk college students, including weirdly dyed hair, piercings, and tattoos. They didn't wear uniforms (as we did at the other two stores). One clerk wore her T-shirt hiked up in a knot in the front and stuffed under her bra in the back. Clerks seemed to be on a friendly, first-name basis with several of the customers, who were mostly middle-class white women.

Although I didn't get hired at Tomatoes, after several visits I noticed social patterns in the store's organization. The owner and the manager were both white men, and all the clerks were young white women. The owner was the only one who was near my age (mid-40s). You can never be sure why you aren't hired, but my impression is that I wasn't young enough or hip enough to work there. Although Tomatoes allowed more autonomy and self-expression than the stores where I worked, race, class, and gender inequality were as much a part of the social organization there as in other retail stores.

CONCLUSION

I ended up buying my gift at Tomatoes—a children's book written and autographed by Marge Piercy. My decision reflects my identity and my social relationships. But what are the implications of my choice for social inequality? My purchase supported a store that was organized around racial exclusion, gender segregation, and class distinctions.

Everyone has to shop in our consumer society, yet the way shopping is organized bolsters social divisions. The racism of shopping is reflected in labor practices, store organization, and the guidelines, explicit or unspoken, for relations between clerks and customers. When deciding where and how to shop, consumers should be aware of what their choices imply with regard to racial justice and equality. Although an individual shopper can do little to change the overall social organization of shopping, raising awareness of the inequalities that our choices support must be a first step in imagining and then creating a better alternative.

KEY CONCEPTS

discrimination
division of labor
occupational segregation
organization
segregation

DISCUSSION QUESTIONS

1. Identify an organization known to you (a store, a business, a school). How would you describe the division of labor as based on gender? On race? Are there other divisions?
2. Why does the desegregation that Williams observes occur? Is this a general pattern in other organizations?

17

Sexual Assault on Campus

A Multilevel, Integrative Approach to Party Rape

ELIZABETH A. ARMSTRONG, LAURA HAMILTON, AND BRIAN SWEENEY

In this article the authors discuss their research of a "party dorm" at a large university. The research uncovers patterns of gendered behavior that contribute to the risk of sexual assault during parties, specifically at fraternities. College women are struggling to find a balance between having fun and being in danger. The authors argue that parties are structured in such a way that puts men in control. Strategies that teach women to simply be careful fall short in their efforts to prevent sexual assault.

A 1991 National Institute of Justice study estimated that between one-fifth and one-quarter of women are the victims of completed or attempted rape while in college (Fisher, Cullen, and Turner 2000). College women "are at greater risk for rape and other forms of sexual assault than women in the general population or in a comparable age group" (Fisher et al. 2000:iii). At least half and perhaps as many as three-quarters of the sexual assaults that occur on college campuses involve alcohol consumption on the part of the victim, the perpetrator, or both (Abbey et al. 1996; Sampson 2002). The tight link between alcohol and sexual assault suggests that many sexual assaults that occur on college campuses are "party rapes." A recent report by the U.S. Department of justice defines party rape as a distinct form of rape, one that "occurs at an off-campus house or on- or off-campus fraternity and involves ... plying a woman with alcohol or targeting an intoxicated woman" (Sampson 2002:6). While party rape is classified as a form of acquaintance rape, it is not uncommon for the woman to have had no prior interaction with the assailant, that is, for the assailant to be an in-network stranger (Abbey et al. 1996).

Colleges and universities have been aware of the problem of sexual assault for at least 20 years, directing resources toward prevention and providing services to students who have been sexually assaulted. Programming has included education of various kinds, support for *Take Back the Night* events, distribution of rape whistles, development and staffing of hotlines, training of police and administrators, and other efforts. Rates of sexual assault, however, have not declined over the last five decades (Adams-Curtis and Forbes 2004:95; Bachar and Koss 2001; Marine 2004; Sampson 2002:1).

Why do colleges and universities remain dangerous places for women in spite of active efforts to prevent sexual assault? While some argue that "we know what the problems are and we know how to change them" (Adams-Curtis and Forbes 2004:115), it is our contention that we do not have a complete explanation of the problem. To address this issue we use data from a study of college life at a large midwestern university and draw on theoretical developments in the sociology of gender (Connell 1987, 1995; Lorber 1994; Martin 2004; Risman 1998, 2004). Continued high rates of sexual assault can be viewed as a case of the reproduction of gender inequality—a phenomenon of central concern in gender theory.

We demonstrate that sexual assault is a predictable outcome of a synergistic intersection of both gendered and seemingly gender neutral processes operating at individual, organizational, and interactional levels. The concentration of homogenous students with expectations of partying fosters the development of sexualized peer cultures organized around status. Residential arrangements intensify students' desires to party in male-controlled fraternities. Cultural expectations that partygoers drink heavily and trust party-mates become problematic when combined with expectations that women be nice and defer to men. Fulfilling the role of the partier produces vulnerability on the part of women, which some men exploit to extract nonconsensual sex. The party scene also produces fun, generating student investment in it. Rather than criticizing the party scene or men's behavior, students blame victims. By revealing mechanisms that lead to the persistence of sexual assault and outlining implications for policy, we hope to encourage colleges and universities to develop fresh approaches to sexual assault prevention.

APPROACHES TO COLLEGE SEXUAL ASSAULT

Explanations of high rates of sexual assault on college campuses fall into three broad categories. The first tradition, a psychological approach that we label the "individual determinants" approach, views college sexual assault as primarily a consequence of perpetrator or victim characteristics such as gender role attitudes, personality, family background, or sexual history (Flezzani and Benshoff 2003; Forbes and Adams-Curtis 2001; Rapaport and Burkhart 1984). While "situational variables" are considered, the focus is on individual characteristics (Adams-Curtis and Forbes 2004; Malamuth, Heavey, and Linz 1993). For example, Antonia Abbey and associates (2001) find that hostility toward women, acceptance of verbal pressure as a way to obtain sex, and having many consensual sexual partners distinguish men who sexually assault from men who do not. Research suggests that victims appear quite similar to other college women (Kalof 2000), except that white women, prior victims, first-year college students, and more sexually active women are more vulnerable to sexual assault (Adams-Curtis and Forbes 2004; Humphrey and White 2000).

The second perspective, the "rape culture" approach, grew out of second wave feminism (Brownmiller 1975; Buchwald, Fletcher, and Roth 1993; Lottes 1997; Russell 1975; Schwartz and DeKeseredy 1997). In this perspective, sexual assault is seen as a consequence of widespread belief in "rape myths," or ideas about the nature of men, women, sexuality, and consent that create an environment conducive to rape. For example, men's disrespectful treatment of women is normalized by the idea that men are naturally sexually aggressive. Similarly, the belief that women "ask for it" shifts responsibility from predators to victims (Herman 1989; O'Sullivan 1993). This perspective initiated an important shift away from individual beliefs toward the broader context. However, rape supportive beliefs alone cannot explain the prevalence of sexual assault, which requires not only an inclination on the part of assailants but also physical proximity to victims (Adams-Curtis and Forbes 2004:103).

A third approach moves beyond rape culture by identifying particular contexts—fraternities and bars—as sexually dangerous (Humphrey and Kahn 2000; Martin and Hummer 1989; Sanday 1990, 1996; Stombler 1994). Ayres Boswell and Joan Spade (1996) suggest that sexual assault is supported not only by "a generic culture surrounding and promoting rape," but also by characteristics of the "specific settings" in which men and women interact (p. 133). Mindy Stombler and Patricia Yancey Martin (1994) illustrate that gender inequality is institutionalized on campus by "formal structure" that supports and intensifies an already "high-pressure heterosexual peer group" (p. 180). This perspective grounds sexual assault in organizations that provide opportunities and resources.

We extend this third approach by linking it to recent theoretical scholarship in the sociology of gender. Martin (2004), Barbara Risman (1998, 2004), Judith Lorber (1994) and others argue that gender is not only embedded in individual selves, but also in cultural rules, social interaction, and organizational arrangements. This integrative perspective identifies mechanisms at each level that contribute to the reproduction of gender inequality (Risman 2004). Socialization processes influence gendered selves, while cultural expectations reproduce gender inequality in interaction. At the institutional level, organizational practices, rules, resource distributions, and ideologies reproduce gender inequality. Applying this integrative perspective enabled us to identify gendered processes at individual, interactional, and organizational levels that contribute to college sexual assault.

Risman (1998) also argues that gender inequality is reproduced when the various levels are "all consistent and interdependent" (p. 35). Processes at each level depend upon processes at other levels. Below we demonstrate how interactional processes generating sexual danger depend upon organizational resources and particular kinds of selves. We show that sexual assault results from the intersection of processes at all levels.

We also find that not all of the processes contributing to sexual assault are explicitly gendered. For example, characteristics of individuals such as age, class, and concern with status play a role. Organizational practices such as residence hall assignments and alcohol regulation, both intended to be gender neutral, also contribute to sexual danger. Our findings suggest that apparently gender neutral social processes may contribute to gender inequality in other situations.

METHOD

Data are from group and individual interviews, ethnographic observation, and publicly available information collected at a large midwestern research university. Located in a small city, the school has strong academic and sports programs, a large Greek system, and is sought after by students seeking a quintessential college experience. Like other schools, this school has had legal problems as a result of deaths associated with drinking. In the last few years, students have attended a sexual assault workshop during first-year orientation. Health and sexuality educators conduct frequent workshops, student volunteers conduct rape awareness programs, and *Take Back the Night* marches occur annually.

The bulk of the data presented in this paper were collected as part of ethnographic observation during the 2004–05 academic year in a residence hall identified by students and residence hall staff as a "party dorm." While little partying actually occurs in the hall, many students view this residence hall as one of several places to live in order to participate in the party scene on campus. This made it a good place to study the social worlds of students at high risk of sexual assault—women attending fraternity parties in their first year of college. The authors and a research team were assigned to a room on a floor occupied by 55 women students (51 first-year, 2 second-year, 1 senior, and 1 resident assistant [RA]). We observed on evenings and weekends throughout the entire academic school year. We collected in-depth background information via a detailed nine-page survey that 23 women completed and conducted interviews with 42 of the women (ranging from 1 1/4 to 2 1/2 hours). All but seven of the women on the floor completed either a survey or an interview.

With at least one-third of first-year students on campus residing in "party dorms" and one-quarter of all undergraduates belonging to fraternities or sororities, this social world is the most visible on campus. As the most visible scene on campus, it also attracts students living in other residence halls and those not in the Greek system. Dense pre-college ties among the many in-state students, class and race homogeneity, and a small city location also contribute to the dominance of this scene. Of course, not all students on this floor or at this university participate in the party scene. To participate, one must typically be heterosexual, at least middle class, white, American-born, unmarried, childless, traditional college age, politically and socially mainstream, and interested in drinking. Over three-quarters of the women on the floor we observed fit this description.

There were no non-white students among the first- and second-year students on the floor we studied. This is a result of the homogeneity of this campus and racial segregation in social and residential life. African Americans (who make up 3 to 5% of undergraduates) generally live in living-learning communities in other residence halls and typically do not participate in the white Greek party scene. We argue that the party scene's homogeneity contributes to sexual risk for white women. We lack the space and the data to compare white and African American party scenes on this campus, but in the discussion we offer ideas about what such a comparison might reveal.

We also conducted 16 group interviews (involving 24 men and 63 women) in spring 2004. These individuals had varying relationships to the white Greek party scene on campus. Groups included residents of an alternative residence hall, lesbian, gay, and bisexual students, feminists, re-entry students, academically-focused students, fundamentalist Christians, and sorority women. Eight group interviews were exclusively women, five were mixed in gender composition, and three were exclusively men. The group interviews covered a variety of topics, including discussions of social life, the transition to college, sexual assault, relationships, and the relationship between academic and social life. Participants completed a shorter version of the survey administered to the women on the residence hall floor. From these students we developed an understanding of the dominance of this party scene.

We also incorporated publicly available information about the university from informal interviews with student affairs professionals and from teaching (by all authors) courses on gender, sexuality, and introductory sociology. Classroom data were collected through discussion, student writings, e-mail correspondence, and a survey that included questions about experiences of sexual assault....

EXPLAINING PARTY RAPE

We show how gendered selves, organizational arrangements, and interactional expectations contribute to sexual assault. We also detail the contributions of processes at each level that are not explicitly gendered. We focus on each level in turn, while attending to the ways in which processes at all levels depend upon and reinforce others. We show that fun is produced along with sexual assault, leading students to resist criticism of the party scene.

Selves and Peer Culture in the Transition from High School to College

Student characteristics shape not only individual participation in dangerous party scenes and sexual risk within them but the development of these party scenes. We identify individual characteristics (other than gender) that generate interest in college partying and discuss the ways in which gendered sexual agendas generate a peer culture characterized by high-takes competition over erotic status.

Non-Gendered Characteristics Motivate Participation in Party Scenes

Without individuals available for partying, the party scene would not exist. All the women on our floor were single and childless, as are the vast majority of undergraduates at this university; many, being upper-middle class, had few responsibilities other than their schoolwork. Abundant leisure time, however, is not enough to fuel the party scene. Media, siblings, peers, and parents all serve as sources of anticipatory socialization (Merton 1957). Both partiers and nonpartiers agreed that one was "supposed" to party in college. This orientation was reflected in the popularity of a poster titled "What I Really Learned in School" that pictured mixed drinks with names associated with academic disciplines. As one focus group participant explained,

> You see these images of college that you're supposed to go out and have fun and drink, drink lots, party and meet guys. [You are] supposed to hook up with guys, and both men and women try to live up to that. I think a lot of it is girls want to be accepted into their groups and guys want to be accepted into their groups.

Partying is seen as a way to feel a part of college life. Many of the women we observed participated in middle and high school peer cultures organized around status, belonging, and popularity (Eder 1985; Eder, Evans, and Parker 1995; Milner 2004). Assuming that college would be similar, they told us that they wanted to fit in, be popular, and have friends....

Peer Culture as Gendered and Sexualized Partying was also the primary way to meet men on campus. The floor was locked to non-residents, and even men living in the same residence hall had to be escorted on the floor. The women found it difficult to get to know men in their classes, which were mostly mass lectures. They explained to us that people "don't talk" in class. Some complained they lacked casual friendly contact with men, particularly compared to the mixed-gender friendship groups they reported experiencing in high school.

Meeting men at parties was important to most of the women on our floor. The women found men's sexual interest at parties to be a source of self-esteem and status. They enjoyed dancing and kissing at parties, explaining to us that it proved men "liked" them. This attention was not automatic, but required the skillful deployment of physical and cultural assets (Stombler and Padavic 1997;

Swidler 2001). Most of the party-oriented women on the floor arrived with appropriate gender presentations and the money and know-how to preserve and refine them. While some more closely resembled the "ideal" college party girl (white, even features, thin but busty, tan, long straight hair, skillfully made-up, and well-dressed in the latest youth styles), most worked hard to attain this presentation. They regularly straightened their hair, tanned, exercised, dieted, and purchased new clothes....

The psychological benefits of admiration from men in the party scene were such that women in relationships sometimes felt deprived. One woman with a serious boyfriend noted that she dressed more conservatively at parties because of him, but this meant she was not "going to get any of the attention." She lamented that no one was "going to waste their time with me" and that, "this is taking away from my confidence." Like most women who came to college with boyfriends, she soon broke up with him.

Men also sought proof of their erotic appeal. As a woman complained, "Every man I have met here has wanted to have sex with me!" ... The women found that men were more interested than they were in having sex. These clashes in sexual expectations are not surprising: men derived status from securing sex (from high-status women), while women derived status from getting attention (from high-status men). These agendas are both complementary and adversarial: men give attention to women en route to getting sex, and women are unlikely to become interested in sex without getting attention first.

University and Greek Rules, Resources, and Procedures

Simply by congregating similar individuals, universities make possible heterosexual peer cultures. The university, the Greek system, and other related organizations structure student life through rules, distribution of resources, and procedures (Risman 2004).

Sexual danger is an unintended consequence of many university practices intended to be gender neutral. The clustering of homogeneous students intensifies the dynamics of student peer cultures and heightens motivations to party. Characteristics of residence halls and how they are regulated push student partying into bars, off-campus residences, and fraternities. While factors that increase the risk of party rape are present in varying degrees in all party venues (Boswell and Spade 1996), we focus on fraternity parties because they were the typical party venue for the women we observed and have been identified as particularly unsafe (see also Martin and Hummer 1989; Sanday 1990). Fraternities offer the most reliable and private source of alcohol for first-year students excluded from bars and house parties because of age and social networks.

University Practices as Push Factors The university has latitude in how it enforces state drinking laws. Enforcement is particularly rigorous in residence halls. We observed RAs and police officers (including gun-carrying peer police) patrolling the halls for alcohol violations. Women on our floor were "documented" within the first week of school for infractions they felt were minor. Sanctions are severe—a $300 fine, an 8-hour alcohol class, and probation for a year. As a consequence, students engaged in only minimal, clandestine alcohol consumption in their rooms. In comparison, alcohol flows freely at fraternities.

The lack of comfortable public space for informal socializing in the residence hall also serves as a push factor. A large central bathroom divided our floor. A sterile lounge was rarely used for socializing. There was no cafeteria, only a convenience store and a snack bar in a cavernous room furnished with big-screen televisions. Residence life sponsored alternatives to the party scene such as "movie night" and special dinners, but these typically occurred early in the evening. Students defined the few activities sponsored during party hours (e.g., a midnight trip to Wal-Mart) as uncool.

Intensifying Peer Dynamics The residence halls near athletic facilities and Greek houses are known by students to house affluent, party-oriented students. White, upper-middle class, first-year students who plan to rush request these residence halls, while others avoid them. One of our residents explained that "everyone knows what [the residence hall] is like and people are dying to get in here. People just think it's a total party or something." Students of color tend to live elsewhere on campus. As a consequence, our floor was homogenous in terms of age, race, sexual orientation, class, and appearance....

The homogeneity of the floor intensified social anxiety, heightening the importance of partying for making friends. Early in the year, the anxiety was palpable on weekend nights as women assessed their social options by asking where people were going, when, and with whom. One exhausted floor resident told us she felt that she "needed to" go out to protect her position in a friendship group. At the beginning of the semester, "going out" on weekends was virtually compulsory. By 11 p.m. the floor was nearly deserted.

Male Control of Fraternity Parties The campus Greek system cannot operate without university consent. The university lists Greek organizations as student clubs, devotes professional staff to Greek-oriented programming, and disbands fraternities that violate university policy. Nonetheless, the university lacks full authority over fraternities; Greek houses are privately owned and chapters answer to national organizations and the Interfraternity Council (IFC) (i.e., a body governing the more than 20 predominantly white fraternities).

Fraternities control every aspect of parties at their houses: themes, music, transportation, admission, access to alcohol, and movement of guests. Party themes usually require women to wear scant, sexy clothing and place women in subordinate positions to men. During our observation period, women attended parties such as "Pimps and Hos," "Victoria's Secret," and "Playboy Mansion"—the last of which required fraternity members to escort two scantily-clad dates. Other recent themes included: "CEO/Secretary Ho," "School Teacher/Sexy Student," and "Golf Pro/Tennis Ho."

Some fraternities require pledges to transport first-year students, primarily women, from the residence halls to the fraternity houses. From about 9 to 11 p.m. on weekend nights early in the year, the drive in front of the residence hall resembled a rowdy taxi-stand, as dressed-to-impress women waited to be carpooled to parties in expensive late-model vehicles. By allowing party-oriented first-year women to cluster in particular residence halls, the university made them easy to find. One fraternity member told us this practice was referred to as "dorm-storming."

Transportation home was an uncertainty. Women sometimes called cabs, caught the "drunk bus," or trudged home in stilettos. Two women indignantly described a situation where fraternity men "wouldn't give us a ride home." The women said, "Well, let us call a cab." The men discouraged them from calling the cab and eventually found a designated driver. The women described the men as "just dicks" and as "rude."

Fraternities police the door of their parties, allowing in desirable guests (first-year women) and turning away others (unaffiliated men). Women told us of abandoning parties when male friends were not admitted. They explained that fraternity men also controlled the quality and quantity of alcohol. Brothers served themselves first, then personal guests, and then other women. Nonaffiliated and unfamiliar men were served last, and generally had access to only the least desirable beverages. The promise of more or better alcohol was often used to lure women into private spaces of the fraternities.

Fraternities are constrained, though, by the necessity of attracting women to their parties. Fraternities with reputations for sexual disrespect have more success recruiting women to parties early in the year. One visit was enough for some of the women. A roommate duo told of a house they "liked at first" until they discovered that the men there were "really not nice."

The Production of Fun and Sexual Assault in Interaction

Peer culture and organizational arrangements set up risky partying conditions, but do not explain *how* student interactions at parties generate sexual assault. At the interactional level we see the mechanisms through which sexual assault is produced. As interactions necessarily involve individuals with particular characteristics and occur in specific organizational settings, all three levels meet when interactions take place. Here, gendered and gender neutral expectations and routines are intricately woven together to create party rape. Party rape is the result of fun situations that shift—either gradually or quite suddenly—into coercive situations. Demonstrating how the production of fun is connected with sexual assault requires describing the interactional routines and expectations that enable men to employ coercive sexual strategies with little risk of consequence....

Cultural expectations of partying are gendered. Women are supposed to wear revealing outfits, while men typically are not. As guests, women cede control of turf, transportation, and liquor. Women are also expected to be grateful for men's hospitality, and as others have noted, to generally be "nice" in ways that men are not (Gilligan 1982; Martin 2003; Phillips 2000; Stombler and Martin 1994; Tolman 2002). The pressure to be deferential and gracious may be intensified by men's older age and fraternity membership. The quandary for women, however, is that fulfilling the gendered role of partier makes them vulnerable to sexual assault.

Women's vulnerability produces sexual assault only if men exploit it. Too many men are willing to do so. Many college men attend parties looking for casual sex. A student in one of our classes explained that "guys are willing to do damn near anything to get a piece of ass." A male student wrote the following description of parties at his (non-fraternity) house:

> Girls are continually fed drinks of alcohol. It's mainly to party but my roomies are also aware of the inhibition-lowering effects. I've seen an old roomie block doors when girls want to leave his room; and other times I've driven women home who can't remember much of an evening yet sex did occur. Rarely if ever has a night of drinking for my roommate ended without sex. I know it isn't necessarily and assuredly sexual assault, but with the amount of liquor in the house I question the amount of consent a lot.

Another student—after deactivating—wrote about a fraternity brother "telling us all at the chapter meeting about how he took this girl home and she was obviously too drunk to function and he took her inside and had sex with her." Getting women drunk, blocking doors, and controlling transportation are common ways men try to prevent women from leaving sexual situations. Rape culture beliefs, such as the belief that men are "naturally" sexually aggressive, normalize these coercive strategies. Assigning women the role of sexual "gatekeeper" relieves men from responsibility for obtaining authentic consent, and enables them to view sex obtained by undermining women's ability to resist it as "consensual" (e.g., by getting women so drunk that they pass out)....

We heard many stories of negative experiences in the party scene including at least one account of a sexual assault in every focus group that included heterosexual women. Most women who partied complained about men's efforts to control their movements or pressure them to drink. Two of the women on our floor were sexually assaulted at a fraternity party in the first week of school—one was raped. Later in the semester, another woman on the floor was raped by a friend. A fourth woman on the floor suspects she was drugged; she became disoriented at a fraternity party and was very ill for the next week.

Party rape is accomplished without the use of guns, knives, or fists. It is carried out through the combination of low level forms of coercion—a lot of liquor and persuasion, manipulation of situations so that women cannot leave, and sometimes force (e.g., by blocking a door, or using body weight to make it difficult for a woman to get up). These forms of coercion are made more effective by organizational arrangements that provide men with control over how partying happens and by expectations that women let loose and trust their partymates. This systematic and effective method of extracting non-consensual sex is largely invisible, which makes it difficult for victims to convince anyone—even themselves—that a crime occurred. Men engage in this behavior with little risk of consequences.

Student Responses and the Resiliency of the Party Scene

The frequency of women's negative experiences in the party scene poses a problem for those students most invested in it. Finding fault with the party scene potentially threatens meaningful identities and lifestyles. The vast majority of heterosexual encounters at parties are fun and consensual. Partying provides a chance to meet new people, experience and display belonging, and to enhance social position. Women on our floor told us that they loved to flirt and be admired, and they displayed pictures on walls, doors, and websites commemorating their fun nights out.

The most common way that students—both women and men—account for the harm that befalls women in the party scene is by blaming victims. By attributing bad experiences to women's "mistakes," students avoid criticizing the party scene or men's behavior within it. Such victim-blaming also allows women to feel that they can control what happens to them. The logic of victim-blaming suggests that sophisticated, smart, careful women are safe from sexual assault. Only "immature," "naive," or "stupid" women get in trouble. When discussing the sexual assault of a friend, a floor resident explained that:

> She somehow got like sexually assaulted ... by one of our friends' old roommates. All I know is that kid was like bad news to start off with. So, I feel sorry for her but it wasn't much of a surprise for us. He's a shady character.

Another floor resident relayed a sympathetic account of a woman raped at knife point by a stranger in the bushes, but later dismissed party rape as nothing to worry about "'cause I'm not stupid when I'm drunk." Even a feminist focus group participant explained that her friend who was raped "made every single mistake and almost all of them had to with alcohol.... She got ridiculed when she came out and said she was raped." These women contrast "true victims" who are deserving of support with "stupid" women who forfeit sympathy (Phillips 2000). Not only is this response devoid of empathy for other women, but it also leads women to blame themselves when they are victimized (Phillips 2000).

Sexual assault prevention strategies can perpetuate victim-blaming. Instructing women to watch their drinks, stay with friends, and limit alcohol consumption implies that it is women's responsibility to avoid "mistakes" and their fault if they fail. Emphasis on the precautions women should take—particularly if not accompanied by education about how men should change their behavior—may also suggest that it is natural for men to drug women and take advantage of them. Additionally, suggesting that women should watch what they drink, trust party-mates, or spend time alone with men asks them to forgo full engagement in the pleasures of the college party scene....

Opting Out While many students find the party scene fun, others are more ambivalent. Some attend a few fraternity parties to feel like they have participated in this college tradition. Others opt out of it altogether. On our floor, 44 out of the 51 first-year students (almost 90%) participated in the party scene. Those on the floor who opted out worried about sexual safety and the consequences of engaging in illegal behavior. For example, an interviewee who did not drink was appalled by the fraternity party transport system. She explained that:

> All those girls would stand out there and just like, no joke, get into these big black Suburbans driven by frat guys, wearing like seriously no clothes, piled on top of each other. This could be some kidnapper taking you all away to the woods and chopping you up and leaving you there. How dumb can you be?

In her view, drinking around fraternity men was "scary" rather than "fun." Her position was unpopular. She, like others who did not party, was an outsider on the floor. Partiers came home loudly in the middle of the night, threw up in the bathrooms, and rollerbladed around the floor. Socially, the others simply did not exist. A few of our "misfits" successfully created social lives outside the floor. The most assertive of the "misfits" figured out the dynamics of the floor in the first weeks and transferred to other residence halls.

However, most students on our floor lacked the identities or network connections necessary for entry into alternative worlds. Life on a large university campus can be overwhelming for first-year students. Those who most needed an alternative to the social world of the party dorm were often ill-equipped to actively seek it out. They either integrated themselves into partying or found themselves alone in their rooms, microwaving frozen dinners and watching television. A Christian focus group participant described life in this residence hall: "When everyone is going out on a Thursday and you are in the room by yourself and there are only two or three other people on the floor, that's not fun, it's not the college life that you want."

DISCUSSION AND IMPLICATIONS

We have demonstrated that processes at individual, organizational, and interactional levels contribute to high rates of sexual assault. Some individual level characteristics that shape the likelihood of a sexually dangerous party scene developing are not explicitly gendered. Party rape occurs at high rates in places that cluster young, single, party-oriented people concerned about social status. Traditional beliefs about sexuality also make it more likely that one will participate in the party scene and increase danger within the scene. This university contributes to sexual danger by allowing these individuals to cluster.

However, congregating people is not enough, as parties cannot be produced without resources (e.g., alcohol and a viable venue) that are difficult for underage students to obtain. University policies that are explicitly gender-neutral—such as the policing of alcohol use in residence halls—have gendered consequences. This policy encourages first-year students to turn to fraternities to party. Only fraternities, not sororities, are allowed to have parties, and men structure parties in ways that control the appearance, movement, and behavior of female guests.

Men also control the distribution of alcohol and use its scarcity to engineer social interactions. The enforcement of alcohol policy by both university and Greek organizations transforms alcohol from a mere beverage into an unequally distributed social resource.

Individual characteristics and institutional practices provide the actors and contexts in which interactional processes occur. We have to turn to the interactional level, however, to understand *how* sexual assault is generated. Gender neutral expectations to "have fun," lose control, and trust one's party-mates become problematic when combined with gendered interactional expectations. Women are expected to be "nice" and to defer to men in interaction. This expectation is intensified by men's position as hosts and women's as grateful guests. The heterosexual script, which directs men to pursue sex and women to play the role of gatekeeper, further disadvantages women, particularly when virtually *all* men's methods of extracting sex are defined as legitimate....

Our analysis also provides a framework for analyzing the sources of sexual risk in non-university partying situations. Situations where men have a home turf advantage, know each other better than the women present know each other, see the women as anonymous, and control desired resources (such as alcohol or drugs) are likely to be particularly dangerous. Social pressures to "have fun," prove one's social competency, or adhere to traditional gender expectations are also predicted to increase rates of sexual assault within a social scene.

This research has implications for policy. The interdependence of levels means that it is difficult to enact change at one level when the other levels remain unchanged.... Without change in institutional arrangements, efforts to change cultural beliefs are undermined by the cultural commonsense generated by encounters with institutions. Efforts to educate about sexual assault will not succeed if the university continues to support organizational arrangements that facilitate and even legitimate men's coercive sexual strategies. Thus, our research implies that efforts to combat sexual assault on campus should target all levels, constituencies, and processes simultaneously. Efforts to educate both men and women should indeed be intensified, but they should be reinforced by changes in the social organization of student life.

Researchers focused on problem drinking on campus have found that reduction efforts focused on the social environment are successful (Berkowitz 2003:21). Student body diversity has been found to decrease binge drinking on campus (Wechsler and Kuo 2003); it might also reduce rates of sexual assault. Existing student heterogeneity can be exploited by eliminating self-selection into age-segregated, white, upper-middle class, heterosexual enclaves and by working to make residence halls more appealing to upper-division students. Building more aesthetically appealing housing might allow students to interact outside of alcohol-fueled party scenes. Less expensive plans might involve creating more living-learning communities, coffee shops, and other student-run community spaces.

While heavy alcohol use is associated with sexual assault, not all efforts to regulate student alcohol use contribute to sexual safety. Punitive approaches sometimes heighten the symbolic significance of drinking, lead students to drink more hard liquor, and push alcohol consumption to more private and thus more dangerous spaces. Regulation inconsistently applied—e.g., heavy policing of residence halls and light policing of fraternities—increases the power of those who can secure alcohol and host parties. More consistent regulation could decrease the value of alcohol as a commodity by equalizing access to it.

Sexual assault education should shift in emphasis from educating women on preventative measures to educating both men and women about the coercive behavior of men and the sources of victim-blaming. Mohler-Kuo and associates (2004) suggest, and we endorse, a focus on the role of alcohol in sexual assault. Education should begin before students arrive on campus and continue throughout college. It may also be most effective if high-status peers are involved in disseminating knowledge and experience to younger college students.

Change requires resources and cooperation among many people. Efforts to combat sexual assault are constrained by other organizational imperatives. Student investment in the party scene makes it difficult to enlist the support of even those most harmed by the state of affairs. Student and alumni loyalty to partying (and the Greek system) mean that challenges to the party scene could potentially cost universities tuition dollars and alumni donations. Universities must contend with Greek organizations and bars, as well as the challenges of internal coordination. Fighting sexual assault on all levels is critical, though, because it is unacceptable for higher education institutions to be sites where women are predictably sexually victimized.

REFERENCES

Abbey, Antonia, Pam McAuslan, Tina Zawacki, A. Monique Clinton, and Philip Buck. 2001. "Attitudinal, Experiential, and Situational Predictors of Sexual Assault Perpetration." *Journal of Interpersonal Violence* 16:784–807.

Abbey, Antonia, Lisa Thomson Ross, Donna McDuffie, and Pam McAuslan. 1996. "Alcohol and Dating Risk Factors for Sexual Assault among College Women." *Psychology of Women Quarterly* 20:147–69.

Adams-Curtis, Leah and Gordon Forbes. 2004. "College Women's Experiences of Sexual Coercion: A Review of Cultural, Perpetrator, Victim, and Situational Variables." *Trauma, Violence, and Abuse: A Review Journal* 5:91–122.

Bachar, Karen and Mary Koss. 2001. "From Prevalence to Prevention: Closing the Gap between What We Know about Rape and What We Do." pp. 117–42 in *Sourcebook on Violence against Women*, edited by C. Renzetti, J. Edleson, and R. K. Bergen. Thousand Oaks, CA: Sage.

Berkowitz, Alan. 2003. "How Should We Talk about Student Drinking—And What Should We Do about It?" *About Campus* May/June:16–22.

Boswell, A. Ayres and Joan Z. Spade. 1996. "Fraternities and Collegiate Rape Culture: Why Are Some Fraternities More Dangerous Places for Women?" *Gender & Society* 10:133–47.

Brownmiller, Susan. 1975. *Against Our Will: Men, Women, and Rape*. New York: Bantam Books.

Buchwald, Emilie, Pamela Fletcher, and Martha Roth, eds. 1993. *Transforming a Rape Culture*. Minneapolis, MN: Milkweed Editions.

Connell, R. W. 1987. *Gender and Power*. Palo Alto, CA: Stanford University Press.

———. 1995. *Masculinities*. Berkeley, CA: University of California Press.

Eder, Donna. 1985. "The Cycle of Popularity: Interpersonal Relations among Female Adolescents." *Sociology of Education* 58:154–65.

Eder, Donna, Catherine Evans, and Stephen Parker. 1995. *School Talk: Gender and Adolescent Culture*. New Brunswick, NJ: Rutgers University Press.

Fisher, Bonnie, Francis Cullen, and Michael Turner. 2000. "The Sexual Victimization of College Women." Washington, DC: National Institute of Justice and the Bureau of Justice Statistics.

Flezzani, James and James Benshoff. 2003. "Understanding Sexual Aggression in Male College Students: The Role of Self-Monitoring and Pluralistic Ignorance." *Journal of College Counseling* 6:69–79.

Forbes, Gordon and Leah Adams-Curtis. 2001. "Experiences with Sexual Coercion in College Males and Females: Role of Family Conflict, Sexist Attitudes, Acceptance of Rape Myths, Self-Esteem, and the Big-Five Personality Factors." *Journal of Interpersonal Violence* 16:865–89.

Gilligan, Carol. 1982. *In a Different Voice: Psychological Theory and Women's Development*. Cambridge, MA:Harvard University Press.

Herman, Diane. 1989. "The Rape Culture." pp. 20–44 in *Women: A Feminist Perspective*, edited by J. Freeman. Mountain View, CA: Mayfield.

Humphrey, John and Jacquelyn White. 2000. "Women's Vulnerability to Sexual Assault from Adolescence to Young Adulthood." *Journal of Adolescent Health* 27:419–24.

Humphrey, Stephen and Arnold Kahn. 2000. "Fraternities, Athletic Teams, and Rape: Importance of Identification with a Risky Group." *Journal of Interpersonal Violence* 15:1313–22.

Kalof, Linda. 2000. "Vulnerability to Sexual Coercion among College Women: A Longitudinal Study." *Gender Issues* 18:47–58.

Lorber, Judith. 1994. *Paradoxes of Gender*. New Haven, CT: Yale University Press.

Lottes, Ilsa L. 1997. "Sexual Coercion among University Students: A Comparison of the United States and Sweden." *Journal of Sex Research* 34:67–76.

Malamuth, Neil, Christopher Heavey, and Daniel Linz. 1993. "Predicting Men's Antisocial Behavior against Women: The Interaction Model of Sexual Aggression." pp. 63–98 in *Sexual Aggression: Issues in Etiology, Assessment, and Treatment*, edited by G. N. Hall, R. Hirschman, J. Graham, and M. Zaragoza. Washington, D.C.: Taylor and Francis.

Marine, Susan. 2004. "Waking Up from the Nightmare of Rape." *The Chronicle of Higher Education*, November 26, p. B5.

Martin, Karin. 2003. "Giving Birth Like a Girl." *Gender & Society* 17:54–72.

Martin, Patricia Yancey. 2004. "Gender as a Social Institution." *Social Forces* 82:1249–73.

Martin, Patricia Yancey and Robert A. Hummer. 1989. "Fraternities and Rape on Campus." *Gender & Society* 3:457–73.

Merton, Robert. 1957. *Social Theory and Social Structure*. New York: Free Press.

Milner, Murray. 2004. *Freaks, Geeks, and Cool Kids: American Teenagers, Schools, and the Culture of Consumption*. New York: Routledge.

Mohler-Kuo, Meichun, George W. Dowdall, Mary P. Koss, and Henry Weschler. 2004. "Correlates of Rape While Intoxicated in a National Sample of College Women." *Journal of Studies on Alcohol* 65:37–45.

O'Sullivan, Chris. 1993. "Fraternities and the Rape Culture." pp. 23–30 in *Transforming a Rape Culture*, edited by E. Buchwald, P. Fletcher, and M. Roth. Minneapolis, MN: Milkweed Editions.

Phillips, Lynn. 2000. *Flirting with Danger: Young Women's Reflections on Sexuality and Domination*. New York: New York University.

Rapaport, Karen and Barry Burkhart. 1984. "Personality and Attitudinal Characteristics of Sexually Coercive College Males." *Journal of Abnormal Psychology* 93:216–21.

Risman, Barbara. 1998. *Gender Vertigo: American Families in Transition*. New Haven, CT: Yale University Press.

———. 2004. "Gender as a Social Structure: Theory Wrestling with Activism." *Gender & Society* 18:429–50.

Russell, Diana. 1975. *The Politics of Rape*. New York: Stein and Day.

Sampson, Rana. 2002. "Acquaintance Rape of College Students." Problem-Oriented Guides for Police Series, No. 17. Washington, DC: U.S. Department of Justice, Office of Community Oriented Policing Services.

Sanday, Peggy. 1990. *Fraternity Gang Rape: Sex, Brotherhood, and Privilege on Campus*. New York: New York University Press.

———. 1996. "Rape-Prone versus Rape-Free Campus Cultures." *Violence against Women* 2:191–208.

Schwartz, Martin and Walter DeKeseredy. 1997. *Sexual Assault on the College Campus: The Role of Male Peer Support*. Thousand Oaks, CA: Sage Publications.

Stombler, Mindy. 1994. "'Buddies' or 'Slutties': The Collective Reputation of Fraternity Little Sisters." *Gender & Society* 8:297–323.

Stombler, Mindy and Patricia Yancey Martin. 1994. "Bringing Women In, Keeping Women Down: Fraternity 'Little Sister' Organizations." *Journal of Contemporary Ethnography* 23:150–84.

Stombler, Mindy and Irene Padavic. 1997. "Sister Acts: Resisting Men's Domination in Black and White Fraternity Little Sister Programs." *Social Problems* 44:257–75.

Swidler, Ann. 2001. *Talk of Love: How Culture Matters*. Chicago: University of Chicago Press.

Tolman, Deborah. 2002. *Dilemmas of Desire: Teenage Girls Talk about Sexuality*. Cambridge, MA: Harvard University Press.

Wechsler, Henry and Meichun Kuo. 2003. "Watering Down the Drinks: The Moderating Effect of College Demographics on Alcohol Use of High-Risk Groups." *American Journal of Public Health* 93:1929–33.

KEY CONCEPTS

acquaintance rape   doing gender   victimization

DISCUSSION QUESTIONS

1. Think about the most recent party you attended? Can you determine who was "in control" of the food and drink? Who controlled the guest list? What role, if any, did you have in the way the party was structured?
2. Consider the different perspectives on sexual assault presented in the article (individual determinants, rape culture, and particular contexts for sexual assault). Which do you believe offers the best explanation for sexual assault on college campuses?

Applying Sociological Knowledge: An Exercise for Students

Think of an organization with which you are familiar (a college, a work organization, or a religious organization). What are the different groups that make up this organization? Do the different groups that make up the organization have different statuses within the organization? If so, describe each group's status. Is there a hierarchy among these different groups and, if so, how does that affect how they interact with each other?

PART VII

Deviance and Crime

18

The Functions of Crime

EMILE DURKHEIM

This classic essay, written in 1895 and translated many times since, points to crime as an inevitable part of society. Durkheim's main functionalist thesis that criminal behavior exists in all social settings is still the theoretical basis for many sociological inquiries into crime and deviance.

... If there is a fact whose pathological nature appears indisputable, it is crime. All criminologists agree on this score. Although they explain this pathology differently, they nonetheless unanimously acknowledge it. However, the problem needs to be treated less summarily.

... Crime is not only observed in most societies of a particular species, but in all societies of all types. There is not one in which criminality does not exist, although it changes in form and the actions which are termed criminal are not everywhere the same. Yet everywhere and always there have been men who have conducted themselves in such a way as to bring down punishment upon their heads. If at least, as societies pass from lower to higher types, the crime rate (the relationship between the annual crime figures and population figures) tended to fall, we might believe that, although still remaining a normal phenomenon, crime tended to lose that character of normality. Yet there is no single ground for believing such a regression to be real. Many facts would rather seem to point to the existence of a movement in the opposite direction. From the beginning of the century statistics provide us with a means of following the progression of criminality. It has everywhere increased, and in France the increase is of the order of 300 percent. Thus there is no phenomenon which represents more incontrovertibly all the symptoms of normality, since it appears to be closely bound up with the conditions of all collective life. To make crime a social illness would be to concede that sickness is not something accidental, but on the contrary derives in certain cases from the fundamental constitution of the living creature. This would be to erase any distinction between the physiological and the pathological. It can certainly happen that crime itself has normal forms; this is what happens, for instance,

SOURCE: Reprinted with the permission of Simon & Schuster, Inc. from the Free Press edition of RULES OF SOCIOLOGICAL METHOD by Emile Durkheim. Edited with an introduction by Steven Lukes. Translated by W.D. Halls. Introduction and Selection,

when it reaches an excessively high level. There is no doubt that this excessiveness is pathological in nature. What is normal is simply that criminality exists, provided that for each social type it does not reach or go beyond a certain level which it is perhaps not impossible to fix in conformity with the previous rules.

We are faced with a conclusion which is apparently somewhat paradoxical. Let us make no mistake: to classify crime among the phenomena of normal sociology is not merely to declare that it is an inevitable though regrettable phenomenon arising from the incorrigible wickedness of men; it is to assert that it is a factor in public health, an integrative element in any healthy society. At first sight this result is so surprising that it disconcerted even ourselves for a long time. However, once that first impression of surprise has been overcome it is not difficult to discover reasons to explain this normality and at the same time to confirm it.

In the first place, crime is normal because it is completely impossible for any society entirely free of it to exist.

Crime … consists of an action which offends certain collective feelings which are especially strong and clear-cut. In any society, for actions regarded as criminal to cease, the feelings that they offend would need to be found in each individual consciousness without exception and in the degree of strength requisite to counteract the opposing feelings. Even supposing that this condition could effectively be fulfilled, crime would not thereby disappear; it would merely change in form, for the very cause which made the well-springs of criminality to dry up would immediately open up new ones.

Indeed, for the collective feelings, which the penal law of a people at a particular moment in its history protects, to penetrate individual consciousnesses that had hitherto remained closed to them, or to assume greater authority—whereas previously they had not possessed enough—they would have to acquire an intensity greater than they had had up to then. The community as a whole must feel them more keenly, for they cannot draw from any other source the additional force which enables them to bear down upon individuals who formerly were the most refractory….

In order to exhaust all the logically possible hypotheses, it will perhaps be asked why this unanimity should not cover all collective sentiments without exception, and why even the weakest sentiments should not evoke sufficient power to forestall any dissentient voice. The moral conscience of society would be found in its entirety in every individual, endowed with sufficient force to prevent the commission of any act offending against it, whether purely conventional failings or crimes. But such universal and absolute uniformity is utterly impossible, for the immediate physical environment in which each one of us is placed, our hereditary antecedents, the social influences upon which we depend, vary from one individual to another and consequently cause a diversity of consciences. It is impossible for everyone to be alike in this matter, by virtue of the fact that we each have our own organic constitution and occupy different areas in space. This is why, even among lower peoples where individual originality is very little developed, such originality does however exist. Thus, since there cannot be a society in which individuals do not diverge to some extent from the collective type, it is also inevitable that among these deviations some assume a criminal character. What confers upon them this character is not the intrinsic importance of the acts but the importance which the common consciousness ascribes to them. Thus if the latter is stronger and possesses sufficient authority to make these divergences very weak in absolute terms, it will also be more sensitive and exacting. By reacting against the slightest deviations with an energy which it elsewhere employs against those that are more weighty, it endues them with the same gravity and will brand them as criminal.

Thus crime is necessary. It is linked to the basic conditions of social life, but on this very account is useful, for the conditions to which it is bound are themselves indispensable to the normal evolution of morality and law.

Indeed today we can no longer dispute the fact that not only do law and morality vary from one social type to another, but they even change within the same type if the conditions of collective existence are modified. Yet for these transformations to be made possible, the collective sentiments at the basis of morality should not prove unyielding to change, and consequently should be only moderately intense. If they were too strong, they would no longer be malleable. Any arrangement is indeed an obstacle to a new arrangement; this is even more the case the more deep-seated the original arrangement. The more strongly a structure is articulated, the more it resists modification; this is as true for functional as for anatomical patterns. If there were no crimes, this condition would not be fulfilled, for such a hypothesis presumes that collective sentiments would have attained a degree of intensity unparalleled in history. Nothing is good indefinitely and without limits. The authority which the moral consciousness enjoys must not be excessive, for otherwise no one would dare to attack it and it would petrify too easily into an immutable form. For it to evolve, individual originality must be allowed to manifest itself. But so that the originality of the idealist who dreams of transcending his era may display itself, that of the criminal, which falls short of the age, must also be possible. One does not go without the other.

Nor is this all. Beyond this indirect utility, crime itself may play a useful part in this evolution. Not only does it imply that the way to necessary changes remains open, but in certain cases it also directly prepares for these changes. Where crime exists, collective sentiments are not only in the state of plasticity necessary to assume a new form, but sometimes it even contributes to determining beforehand the shape they will take on. Indeed, how often is it only an anticipation of the morality to come, a progression towards what will be! ... The freedom of thought that we at present enjoy could never have been asserted if the rules that forbade it had not been violated before they were solemnly abrogated. However, at the time the violation was a crime, since it was an offence against sentiments still keenly felt in the average consciousness. Yet this crime was useful since it was the prelude to changes which were daily becoming more necessary....

From this viewpoint the fundamental facts of criminology appear to us in an entirely new light. Contrary to current ideas, the criminal no longer appears as an utterly unsociable creature, a sort of parasitic element, a foreign, unassimilable body introduced into the bosom of society. He plays a normal role in social life. For its part, crime must no longer be conceived of as an evil which cannot be circumscribed closely enough. Far from there being cause for congratulation when it drops too noticeably below the normal level, this apparent progress assuredly coincides with and is linked to some social disturbance. Thus the number of crimes of assault never falls so low as it does in times of scarcity. Consequently, at the same time, and as a reaction, the theory of punishment is revised, or rather should be revised. If in fact crime is a sickness, punishment is the cure for it and cannot be conceived of otherwise; thus all the discussion aroused revolves round knowing what punishment should be to fulfill its role as a remedy. But if crime is in no way pathological, the object of punishment cannot be to cure it and its true function must be sought elsewhere....

KEY CONCEPTS

collective consciousness | deviance | functionalism | social facts

DISCUSSION QUESTIONS

1. According to Durkheim's theory, criminal behavior exists in all societies. Consider the possibility of a society without the ability to punish criminal behavior (no prisons, no courts, and so forth). How would individuals respond to crime? What informal social control mechanisms would help to maintain order?
2. How could you use Durkheim's theory as the basis for a research project on deviant behavior? What hypotheses could you test that would challenge or support the functionalist view of crime?

19

The Medicalization of Deviance

PETER CONRAD AND JOSEPH W. SCHNEIDER

This essay outlines the social construction of social deviance. The authors specifically refer to the medical profession as redefining certain deviant behaviors as "illness," rather than as "badness." They argue that the "medicalization of deviance changes the social response to such behavior to one of treatment rather than punishment."

SOURCE: Excerpt from Chapter 2: From Badness to Sickness from Deviance and Medicalization: From Badness to Sickness by Peter Conrad and Joseph Schneider. Used by permission of Temple University Press.

Consider the following situations. A woman rides a horse naked through the streets of Denver claiming to be Lady Godiva and after being apprehended by authorities, is taken to a psychiatric hospital and declared to be suffering from a mental illness. A well-known surgeon in a Southwestern city performs a psychosurgical operation on a young man who is prone to violent outbursts. An Atlanta attorney, inclined to drinking sprees, is treated at a hospital clinic for his disease, alcoholism. A child in California brought to a pediatric clinic because of his disruptive behavior in school is labeled hyperactive and is prescribed methylphenidate (Ritalin) for his disorder. A chronically overweight Chicago housewife receives a surgical intestinal bypass operation for her problem of obesity. Scientists at a New England medical center work on a million-dollar federal research grant to discover a heroin-blocking agent as a "cure" for heroin addiction. What do these situations have in common? In all instances medical solutions are being sought for a variety of deviant behaviors or conditions. We call this "the medicalization of deviance" and suggest that these examples illustrate how medical definitions of deviant behavior are becoming more prevalent in modern industrial societies like our own. The historical sources of this medicalization, and the development of medical conceptions and controls for deviant behavior, are the central concerns of our analysis.

Medical practitioners and medical treatment in our society are usually viewed as dedicated to healing the sick and giving comfort to the afflicted. No doubt these are important aspects of medicine. In recent years the jurisdiction of the medical profession has expanded and encompasses many problems that formerly were not defined as medical entities.... There is much evidence for this general viewpoint—for example, the medicalization of pregnancy and childbirth, contraception, diet, exercise, child development norms—but our concern here is more limited and specific. Our interests focus on the medicalization of deviant behavior: the defining and labeling of deviant behavior as a medical problem, usually an illness and mandating the medical profession to provide some type of treatment for it. Concomitant with such medicalization is the growing use of medicine as an agent of social control, typically as medical intervention. Medical intervention as social control seeks to limit, modify, regulate, isolate, or eliminate deviant behavior with medical means and in the name of health....

Conceptions of deviant behavior change, and agencies mandated to control deviance change also. Historically there have been great transformations in the definition of deviance—from religious to state-legal to medical-scientific. Emile Durkheim (1893/1933) noted in *The Division of Labor in Society* that as societies develop from simple to complex, sanctions for deviance change from repressive to restitutive or, put another way, from punishment to treatment or rehabilitation. Along with the change in sanctions and social control agent there is a corresponding change in definition or conceptualization of deviant behavior. For example, certain "extreme" forms of deviant drinking (what is now called alcoholism) have been defined as sin, moral weakness, crime, and most recently illness.... In modern industrial society there has been a substantial growth in the prestige, dominance, and jurisdiction of the medical profession (Freidson, 1970). It is only within the last century that physicians have become highly organized, consistently trained, highly paid, and sophisticated in their therapeutic techniques and abilities....

The medical profession dominates the organization of health care and has a virtual monopoly on anything that is defined as medical treatment, especially in terms of what constitutes "illness" and what is appropriate medical intervention.... Although Durkheim did not predict this medicalization, perhaps in part because medicine of his time was not the scientific, prestigious, and dominant profession of today, it is clear that medicine is the central restitutive agent in our society.

EXPANSION OF MEDICAL JURISDICTION OVER DEVIANCE

When treatment rather than punishment becomes the preferred sanction for deviance, an increasing amount of behavior is conceptualized in a medical framework as illness. As noted earlier, this is not unexpected, since medicine has always functioned as an agent of social control, especially in attempting to "normalize" illness and return people to their functioning capacity in society. Public health and psychiatry have long been concerned with social behavior and have functioned traditionally as agents of social control (Foucault, 1965; Rosen, 1972). What is significant, however, is the expansion of this sphere where medicine functions in a social control capacity. In the wake of a general humanitarian trend, the success and prestige of modern biomedicine, the technological growth of the 20th century, and the diminution of religion as a viable agent of control, more and more deviant behavior has come into the province of medicine. In short, the particular, dominant designation of deviance has changed; much of what was badness (i.e., sinful or criminal) is now sickness. Although some forms of deviant behavior are more completely medicalized than others (e.g., mental illness), recent research has pointed to a considerable variety of deviance that has been treated within medical jurisdiction: alcoholism, drug addiction, hyperactive children, suicide, obesity, mental retardation, crime, violence, child abuse, and learning problems, as well as several other categories of social deviance. Concomitant with medicalization there has been a change in imputed responsibility for deviance: with badness the deviants were considered responsible for their behavior, with sickness they are not, or at least responsibility is diminished (see Stoll, 1968). The social response to deviance is "therapeutic" rather than punitive. Many have viewed this as "humanitarian and scientific" progress; indeed, it often leads to "humanitarian and scientific" treatment rather than punishment as a response to deviant behavior....

A number of broad social factors underlie the medicalization of deviance. As psychiatric critic Thomas Szasz (1974) observes, there has been a major historical shift in the manner in which we view human conduct:

> With the transformation of the religious perspective of man into the scientific, and in particular the psychiatric, which became fully articulated during the nineteenth century, there occurred a radical shift in emphasis away from viewing man as a *responsible agent acting in and on the world* and toward viewing him *as a responsive organism being acted upon* by biological and social "forces." (p. 149)

This is exemplified by the diffusion of Freudian thought, which since the 1920s has had a significant impact on the treatment of deviance, the distribution of stigma, and the incidence of penal sanctions.

Nicholas Kittrie (1971), focusing on decriminalization, contends that the foundation of the therapeutic state can be found in determinist criminology, that it stems from the *parens patnae* power of the state (the state's right to help those who are unable to help themselves), and that it dates its origin with the development of juvenile justice at the turn of the century. He further suggests that criminal law has failed to deal effectively (e.g., in deterrence) with criminals and deviants, encouraging a use of alternative methods of control. Others have pointed out that the strength of formal sanctions is declining because of the increase in geographical mobility and the decrease in strength of traditional status groups (e.g., the family) and that medicalization offers a substitute method for controlling deviance (Pitts, 1968). The success of medicine in areas like infectious disease has led to rising expectations of what medicine can accomplish. In modern technological societies, medicine has followed a technological imperative—that the physician is responsible for doing everything possible for the patient—while neglecting such significant issues as the patient's rights and wishes and the impact of biomedical advances on society (Mechanic, 1973). Increasingly sophisticated medical technology has extended the potential of medicine as social control, especially in terms of psychotechnology (Chorover, 1973). Psychotechnology includes a variety of medical and quasimedical treatments or procedures: psychosurgery, psychoactive medications, genetic engineering, disulfiram (Antabuse), and methadone. Medicine is frequently a pragmatic way of dealing with a problem (Gusfield, 1975). Undoubtedly the increasing acceptance and dominance of a scientific world view and the increase in status and power of the medical profession have contributed significantly to the adoption and public acceptance of medical approaches to handling deviant behavior.

THE MEDICAL MODEL AND "MORAL NEUTRALITY"

The first "victories" over disease by an emerging biomedicine were in the infectious diseases in which specific causal agents—germs—could be identified. An image was created of disease as caused by physiological difficulties located *within* the human body. This was the medical model. It emphasized the internal and biophysiological environment and deemphasized the external and social psychological environment.

There are numerous definitions of "the medical model."... We adopt a broad and pragmatic definition: the medical model of deviance locates the source of deviant behavior within the individual, postulating a physiological, constitutional, organic, or, occasionally, psychogenic agent or condition that is assumed to cause the behavioral deviance. The medical model of deviance usually, although not always, mandates intervention by medical personnel with medical means as treatment for the "illness." Alcoholics Anonymous, for example, adopts a rather idiosyncratic version of the medical model—that alcoholism is a chronic disease caused by an "allergy" to alcohol—but actively discourages professional medical intervention. But by and large, adoption of the medical model legitimates and even mandates medical intervention.

The medical model and the associated medical designations are assumed to have a scientific basis and thus are treated as if they were morally neutral (Zola, 1975). They are not considered moral judgments but rational, scientifically verifiable conditions.... Medical designations *are* social judgments, and the adoption of a medical model of behavior, a political decision. When such medical designations are applied to deviant behavior, they are related directly and intimately to the moral order of society. In 1851 Samuel Cartwright, a well-known Southern physician, published an article in a prestigious medical journal describing the disease "drapetomania," which only affected slaves and whose major symptom was running away from the plantations of their white masters (Cartwright, 1851). Medical texts during the Victorian era routinely described masturbation as a disease or addiction and prescribed mechanical and surgical treatments for its cure (Comfort, 1967; Englehardt, 1974). Recently many political dissidents in the Soviet Union have been designated mentally ill, with diagnoses such as "paranoia with counterrevolutionary delusions" and "manic reformism," and hospitalized for their opposition to the political order (Conrad, 1977). Although these illustrations may appear to be extreme examples, they highlight the fact that all medical designations of deviance are influenced significantly by the moral order of society and thus cannot be considered morally neutral....

Even after a social definition of deviance becomes accepted or legitimated, it is not evident what particular type of problem it is. Frequently there are intellectual disputes over the causes of the deviant behavior and the appropriate methods of control. These battles about deviance designation (is it sin, crime, or sickness?) and control are battles over turf: Who is the appropriate definer and treater of the deviance? Decisions concerning what is the proper deviance designation and hence the appropriate agent of social control are settled by some type of political conflict.

How one designation rather than another becomes dominant is a central sociological question. In answering this question, sociologists must focus on claims-making activities of the various interest groups involved and examine how one or another attains ownership of a given type of deviance or social problem and thus generates legitimacy for a deviance designation. Seen from this perspective, public facts, even those which wear a "scientific" mantle are treated as products of the groups or organizations that produce or promote them rather than as accurate reflections of "reality." The adoption of one deviance designation or another has consequences beyond settling a dispute about social control turf....

When a particular type of deviance designation is accepted and taken for granted, something akin to a paradigm exists. There have been three major deviance paradigms: deviance as sin, deviance as crime, and deviance as sickness. When one paradigm and its adherents become the ultimate arbiter of "reality" in society, we say a hegemony of definitions exists. In Western societies, and American society in particular, anything proposed in the name of science gains great authority. In modern industrial societies, deviance designations have become increasingly medicalized. We call the change in designations from badness to sickness the medicalization of deviance....

REFERENCES

Cartwright, S. W. Report on the diseases and physical peculiarities of the negro race. *N. O. Med. Surg. J.*, 1851, 7, 691–715.

Chorover, S. Big Brother and psychotechnology. *Psychol. Today*, 1973, 7, 43–54 (Oct.).

Comfort, A. *The anxiety makers.* London: Thomas Nelson & Sons, 1967.

Conrad, P. Soviet dissidents, ideological deviance, and mental hospitalization. Presented at Midwest Sociological Society Meetings, Minneapolis, 1977.

Durkheim, E. *The division of labor in society.* New York: The Free Press, 1933. (Originally published 1893.)

Englehardt, H. T. Jr. The disease of masturbation: Values and the concept of disease. *Bull. Hist. Med.*, 1974, 48, 234–48 (Summer).

Foucault, M. *Madness and civilization.* New York: Random House, Inc. 1965.

Freidson, E. *Profession of medicine.* New York: Harper & Row Publishers Inc. 1970.

Gusfield, J. R. Categories of ownership and responsibility in social issues: Alcohol abuse and automobile use. *J. Drug Issues*, 1975, 5, 285–303 (Fall).

Kittrie, N. *The right to be different: Deviance and enforced therapy.* Baltimore: Johns Hopkins University Press, 1971.

Mechanic, D. Health and illness in technological societies. *Hastings Center Stud.*, 1973, 1(3), 7–18.

Pitts, J. Social control: The concept. In D. Sills (Ed.) *International Encyclopedia of Social Sciences.* (Vol. 14). New York: Macmillan Publishing Co., Inc. 1968.

Rosen, G. The evolution of social medicine. In H. E. Freeman, S. Levine, and L. Reeder (Eds.) *Handbook of medical sociology* (2nd ed.). Englewood Cliffs, NJ: Prentice-Hall, Inc. 1972.

Stoll, C. S. Images of man and social control. *Soc. Forces*, 1968, 47, 119–127 (Dec.).

Szasz, T. *Ceremonial chemistry.* New York: Anchor Books, 1974.

Zola, I. K. In the name of health and illness: On some socio-political consequences of medical influence. *Soc. Sci. Med.*, 1975, 9, 83–87.

KEY CONCEPTS

medicalization of deviance

social control

DISCUSSION QUESTIONS

1. Alcoholism is an example of a deviant behavior being medicalized. How has this altered the understanding and treatment of alcoholism? How does the involvement of health professionals in the treatment of alcoholism influence societal reaction to excessive drinking?
2. Some argue that rapists should be castrated. How does this illustrate the transformation of understanding rape as a move "from badness to sickness"? What assumptions guide the suggestion that rapists should be castrated as a way of stopping rape?

20

The Rich Get Richer and the Poor Get Prison

JEFFREY H. REIMAN

This essay challenges readers to view the criminal justice system from a radically different angle. Specifically, Jeffrey Reiman argues that the corrections system and broader criminal justice policy in the United States simply provide the illusion of fighting crime. In reality, he argues, criminal justice policies reinforce public fears of crimes committed by the poor. These policies, in turn, help to maintain a "criminal class" of disadvantaged people.

A criminal justice system is a mirror in which a whole society can see the darker outlines of its face. Our ideas of justice and evil take on visible form in it, and thus we see ourselves in deep relief. Step through this looking glass to view the American criminal justice system—and ultimately the whole society it reflects—from a radically different angle of vision.

In particular, entertain the idea that the goal of our criminal justice system is not to eliminate crime or to achieve justice, *but to project to the American public a visible image of the threat of crime as a threat from the poor.* To do this, the justice system must present us with a sizable population of poor criminals. To do that, it must fail

SOURCE: REIMAN, JEFFREY, THE RICH GET RICHER AND THE POOR GET PRISON, 8th, © 2007. Printed by permission of Pearson Education, Inc., Upper Saddle River, New Jersey.

in the struggle to eliminate the crimes that poor people commit, or even to reduce their number dramatically. Crime may, of course, occasionally decline, as it has recently—*but largely because of factors other than criminal justice policies....*

In recent years, we have quadrupled our prison population and, in cities such as New York, allowed the police new freedom to stop and search people they suspect. No one can deny that if you lock up enough people, and allow the police greater and greater power to interfere with the liberty and privacy of citizens, you will eventually prevent some crime that might otherwise have taken place.... I shall point out just how costly and inefficient this means of reducing crime is, in money for new prisons, in its destructive effect on inner-city life, in reduced civil liberties, and in increased complaints of police brutality. I don't deny, however, that these costly means do contribute *in some small measure* to reducing crime. Thus, when I say ... that criminal justice policy is failing, I mean that it is failing to eliminate our high crime rates. We continue to see a large population of poor criminals in our prisons and our courts, while our crime-reduction strategies do not touch on the social causes of crime. Moreover, our citizens remain fearful about criminal victimization, even after the recent declines....

Nearly 30 years ago, I taught a seminar for graduate students titled "The Philosophy of Punishment and Rehabilitation." Many of the students were already working in the field of corrections as probation officers, prison guards, or halfway-house counselors. Together we examined the various philosophical justifications for legal punishment, and then we directed our attention to the actual functioning of our correctional system. For much of the semester, we talked about the myriad inconsistencies and cruelties and the overall irrationality of the system. We discussed the arbitrariness with which offenders are sentenced to prison and the arbitrariness with which they are treated once there. We discussed the lack of privacy and the deprivation of sources of personal identity and dignity, the ever-present physical violence, as well as the lack of meaningful counseling or job training within prison walls. We discussed the harassment of parolees, the inescapability of the "ex-con" stigma, the refusal of society to let a person finish paying his or her "debt to society," and the absence of meaningful noncriminal opportunities for the ex-prisoner. We confronted time and again the bald irrationality of a society that builds prisons to prevent crime knowing full well that they do not, and one that does not seriously try to rid its prisons and postrelease practices of those features that guarantee a high rate of *recidivism,* the return to crime by prison alumni. How could we fail so miserably? We are neither an evil nor a stupid nor an impoverished people. How could we continue to bend our energies and spend our hard-earned tax dollars on cures we know are not working?

Toward the end of the semester, I asked the students to imagine that, instead of designing a criminal justice system to reduce and prevent crime, we designed one that would maintain a stable and visible "class" of criminals. What would it look like? The response was electrifying. Here is a sample of the proposals that emerged in our discussion.

First It would be helpful to have laws on the books against drug use, prostitution, and gambling—laws that prohibit acts that have no unwilling victim.

This would make many people "criminals" for what they regard as normal behavior and would increase their need to engage in *secondary crime* (the drug addict's need to steal to pay for drugs, the prostitute's need for a pimp because police protection is unavailable, and so on).

Second It would be good to give police, prosecutors, and/or judges broad discretion to decide who got arrested, who got charged, and who got sentenced to prison. This would mean that almost anyone who got as far as prison would know of others who committed the same crime but were not arrested, were not charged, or were not sentenced to prison. This would assure us that a good portion of the prison population would experience their confinement as arbitrary and unjust and thus respond with rage, which would make them more antisocial, rather than respond with remorse, which would make them feel more bound by social norms.

Third The prison experience should be not only painful but also demeaning. The pain of loss of liberty might deter future crime. But demeaning and emasculating prisoners by placing them in an enforced childhood characterized by no privacy and no control over their time and actions, as well as by the constant threat of rape or assault, is sure to overcome any deterrent effect by weakening whatever capacities a prisoner had for self-control. Indeed, by humiliating and brutalizing prisoners, we can be sure to increase their potential for aggressive violence.

Fourth Prisoners should neither be trained in a marketable skill nor provided with a job after release. Their prison records should stand as a perpetual stigma to discourage employers from hiring them. Otherwise, they might be tempted *not* to return to crime after release.

Fifth Ex-offenders' sense that they will always be different from "decent citizens," that they can never finally settle their debt to society, should be reinforced by the following means. They should be deprived for the rest of their lives of rights, such as the right to vote. They should be harassed by police as "likely suspects" and be subject to the whims of parole officers who can at any time threaten to send them back to prison for things no ordinary citizens could be arrested for, such as going out of town, or drinking, or fraternizing with the "wrong people."

And so on.

In short, when asked to design a system that would maintain and encourage the existence of a stable and visible "class of criminals," we "constructed" the American criminal justice system....

... [T]he practices of the criminal justice system keep before the public the *real* threat of crime and the *distorted* image that crime is primarily the work of the poor. The value of this *to those in positions of power* is that it deflects the discontent and potential hostility of Middle America away from the classes above them and toward the classes below them. If this explanation is hard to swallow, it should be noted in its favor that it not only explains the dismal failure of criminal justice policy to protect us against crime but also explains why the criminal justice system functions in a way that is biased against the poor at every stage from arrest to conviction. Indeed, even at an earlier stage, when crimes are defined in law, the system concentrates primarily on the predatory acts of the poor and tends to exclude or deemphasize the equally or more dangerous predatory acts of those who are well off.

In sum, I will argue that *the criminal justice system fails in the fight against crime while making it look as if crime is the work of the poor.* This conveys the image that the real danger to decent, law-abiding Americans comes from below them, rather than from above them, on the economic ladder. This image sanctifies the status quo with its disparities of wealth, privilege, and opportunity, and thus serves the interests of the rich and powerful in America—the very ones who could change criminal justice policy if they were really unhappy with it.

Therefore, it seems appropriate to ask you to look at criminal justice "through the looking glass." On the one hand, this suggests a reversal of common expectations. Reverse your expectations about criminal justice and entertain the notion that the system's real goal is the very reverse of its announced goal. On the other hand, the figure of the looking glass suggests the prevalence of image over reality. My argument is that the system functions the way it does *because it maintains a particular image of crime: the image that it is a threat from the poor.* Of course, for this image to be believable, there must be a reality to back it up. The system must actually fight crime—or at least some crime—but only enough to keep it from getting out of hand and to keep the struggle against crime vividly and dramatically in the public's view, never enough to substantially reduce or eliminate crime.

I call this outrageous way of looking at criminal justice policy the *Pyrrhic defeat* theory. A "Pyrrhic victory" is a military victory purchased at such a cost in troops and treasure that it amounts to a defeat. The Pyrrhic defeat theory argues that the failure of the criminal justice system yields such benefits to those in positions of power that it amounts to success....

The Pyrrhic defeat theory has several components. Above all, it must provide an explanation of *how* the failure to reduce crime substantially could benefit anyone—anyone other than criminals, that is.... I argue there that the failure to reduce crime substantially broadcasts a potent *ideological* message to the American people, a message that benefits and protects the powerful and privileged in our society by legitimating the present social order with its disparities of wealth and privilege, and by diverting public discontent and opposition away from the rich and powerful and onto the poor and powerless.

To provide this benefit, however, not just any failure will do. It is necessary that the failure of the criminal justice system take a particular shape. *It must fail in the fight against crime while making it look as if serious crime and thus the real danger to society are the work of the poor.* The system accomplishes this both by what it does and by what it refuses to do.... I argue that the criminal justice system refuses to label and treat as crime a large number of acts of the rich that produce as much or more damage to life and limb as the crimes of the poor.... [E]ven among the acts treated as crimes, the criminal justice system is biased from start to finish in a way that guarantees that, *for the same crimes,* members of the lower classes are much more likely than members of the middle and upper classes to be arrested, convicted, and imprisoned—thus providing living "proof" that crime is a threat from the poor….

Our criminal justice system is characterized by beliefs about what is criminal, and beliefs about how to deal with crime, that predate industrial society. Rather than being anyone's conscious plan, the system reflects attitudes so deeply embedded in tradition as to appear natural. To understand why it persists even though it fails to protect us, all that is necessary is to recognize that, on the one hand, those who are the most victimized by crime are not those in positions to make and implement policy. Crime falls more frequently and more harshly on the poor than on the better off. On the other hand, there are enough benefits to the wealthy from the identification of crime with the poor and the system's failure to reduce crime that those with the power to make profound changes in the system feel no compulsion nor see any incentive to make them. In short, the criminal justice system came into existence in an earlier epoch and persists in the present because, even though it is failing—indeed, because of the way it fails—it generates no effective demand for change. When I speak of the criminal justice system as "designed to fail," I mean no more than this. I call this explanation of the existence and persistence of our failing criminal justice system the *historical inertia* explanation….

KEY CONCEPTS

labeling theory   social class   social institution

DISCUSSION QUESTIONS

1. What does Reiman mean in arguing that the current criminal justice system works to maintain a class of criminals? Do you agree or disagree that our corrections system fails to rehabilitate and fails to deter crime?
2. If you had the power to change the corrections system in the United States, what changes would you make to help reduce and prevent crime?

Applying Sociological Knowledge: An Exercise for Students

Become a norm breaker! Think of a norm we have in society that you can go out in public and violate. Make sure it is legal! How do people treat you when you stop doing something that is implicitly expected of you? How does it feel to go against what you feel you should be doing? Was this norm something you thought about doing before or is it something that you did without even thinking? Notice how hard it is to deviate from expected norms.

PART VIII

Social Class and Social Stratification

21

The Communist Manifesto

KARL MARX AND FRIEDRICH ENGELS

The analysis of the class system under capitalism, as developed by Marx and Engels, continues to influence sociological understanding of the development of capitalism and the structure of the class system. In this classic essay, first published in 1848, Marx and Engels define the class system in terms of the relationships between capitalism, the bourgeoisie, and the proletariat. Their analysis of the growth of capitalism and its influence on other institutions continues to provide a compelling portrait of an economic system based on the pursuit of profit.

BOURGEOIS AND PROLETARIANS

The history of all hitherto existing society is the history of class struggles.... Modern industry has established the world market, for which the discovery of America paved the way. This market has given an immense development to commerce, to navigation, to communication by land. This development has, in its turn, reacted on the extension of industry; and in proportion as industry, commerce, navigation, railways extended, in the same proportion the bourgeoisie developed, increased its capital, and pushed into the background every class handed down from the Middle Ages.

We see, therefore, how the modern bourgeoisie is itself the product of a long course of development, of a series of revolutions in the modes of production and of exchange....

... [T]he bourgeoisie has at last, since the establishment of modern industry and of the world market, conquered for itself, in the modern representative state, exclusive political sway. The executive of the modern state is but a committee for managing the common affairs of the whole bourgeoisie.

The bourgeoisie, historically, has played a most revolutionary part.

The bourgeoisie, wherever it has got the upper hand, has put an end to all feudal, patriarchal, idyllic relations. It has pitilessly torn asunder the motley feudal

SOURCE: Karl Marx and Friedrich Engels, *Manifesto of the Communist Party,* with introduction by Eric Hobsbawm (New York: Verso, 1998), pp. 33–51.

ties that bound man to his "natural superiors" and has left remaining no other nexus between man and man than naked self-interest, than callous "cash payment." It has drowned the most heavenly ecstasies of religious fervour, of chivalrous enthusiasm, of philistine sentimentalism, in the icy water of egotistical calculation. It has resolved personal worth into exchange value, and in place of the numberless indefeasible chartered freedoms, has set up that single, unconscionable freedom—free trade. In one word, for exploitation, veiled by religious and political illusions, it has substituted naked, shameless, direct, brutal exploitation.

The bourgeoisie has stripped of its halo every occupation hitherto honoured and looked up to with reverent awe. It has converted the physician, the lawyer, the priest, the poet, the man of science, into its paid wage labourers.

The bourgeoisie has torn away from the family its sentimental veil, and has reduced the family relation to a mere money relation....

The need of a constantly expanding market for its products chases the bourgeoisie over the whole surface of the globe. It must nestle everywhere, settle everywhere, establish connections everywhere.

The bourgeoisie has through its exploitation of the world market given a cosmopolitan character to production and consumption in every country. To the great chagrin of reactionists, it has drawn from under the feet of industry the national ground on which it stood. All old, established national industries have been destroyed or are daily being destroyed. They are dislodged by new industries, whose introduction becomes a life and death question for all civilized nations, by industries that no longer work up indigenous raw material, but raw material drawn from the remotest zones; industries whose products are consumed, not only at home, but in every quarter of the globe. In place of the old wants, satisfied by the productions of the country, we find new wants, requiring for their satisfaction the products of distant lands and climes. In place of the old local and national seclusion and self-sufficiency, we have intercourse in every direction, universal interdependence of nations. And as in material, so also in intellectual production. The intellectual creations of individual nations become common property. National one-sidedness and narrow-mindedness become more and more impossible, and from the numerous national and local literatures, there arises a world literature.

The bourgeoisie, by the rapid improvement of all instruments of production, by the immensely facilitated means of communication, draws all, even the most barbarian, nations into civilization. The cheap prices of its commodities are the heavy artillery with which it batters down all Chinese walls, with which it forces the barbarians' intensely obstinate hatred of foreigners to capitulate. It compels all nations, on pain of extinction, to adopt the bourgeois mode of production; it compels them to introduce what it calls civilization into their midst, i.e., to become bourgeois themselves. In one word, it creates a world after its own image.

The bourgeoisie has subjected the country to the rule of the towns. It has created enormous cities, has greatly increased the urban population as compared with the rural, and has thus rescued a considerable part of the population from the idiocy of rural life. Just as it has made the country dependent on the towns, so it has made barbarian and semi-barbarian countries dependent on the civilized ones, nations of peasants on nations of bourgeois, the East on the West.

The bourgeoisie keeps more and more doing away with the scattered state of the population, of the means of production, and of property. It has agglomerated population, centralized means of production, and has concentrated property in a few hands. The necessary consequence of this was political centralization. Independent, or but loosely connected provinces, with separate interests, laws, governments and systems of taxation became lumped together into one nation, with one government, one code of laws, one national class interest, one frontier and one customs tariff....

The weapons with which the bourgeoisie felled feudalism to the ground are now turned against the bourgeoisie itself.

But not only has the bourgeoisie forged the weapons that bring death to itself; it has also called into existence the men who are to wield those weapons—the modern working class—the proletarians.

In proportion as the bourgeoisie, i.e., capital, is developed, in the same proportion is the proletariat, the modern working class, developed—a class of labourers, who live only so long as they find work, and who find work only so long as their labour increases capital. These labourers, who must sell themselves piecemeal, are a commodity, like every other article of commerce, and are consequently exposed to all the vicissitudes of competition, to all the fluctuations of the market.

Owing to the extensive use of machinery and to division of labour, the work of the proletarians has lost all individual character, and, consequently, all charm for the workman. He becomes an appendage of the machine, and it is only the most simple, most monotonous, and most easily acquired knack that is required of him. Hence, the cost of production of a workman is restricted, almost entirely, to the means of subsistence that he requires for his maintenance and for the propagation of his race. But the price of a commodity, and therefore also of labour, is equal to its cost of production. In proportion, therefore, as the repulsiveness of the work increases, the wage decreases. Nay more, in proportion as the use of machinery and division of labour increases, in the same proportion the burden of toil also increases, whether by prolongation of the working hours, by increase of the work exacted in a given time or by increased speed of the machinery, etc.

Modern industry has converted the little workshop of the patriarchal master into the great factory of the industrial capitalist. Masses of labourers, crowded into the factory, are organized like soldiers. As privates of the industrial army they are placed under the command of a perfect hierarchy of officers and sergeants. Not only are they slaves of the bourgeois class and of the bourgeois state; they are daily and hourly enslaved by the machine, by the overseer, and, above all, by the individual bourgeois manufacturer himself. The more openly this despotism proclaims gain to be its end and aim, the more petty, the more hateful and the more embittering it is.

The less the skill and exertion of strength implied in manual labour, in other words, the more modern industry becomes developed, the more is the labour of men superseded by that of women. Differences of age and sex have no longer any distinctive social validity for the working class. All are instruments of labour, more or less expensive to use, according to their age and sex....

But with the development of industry the proletariat not only increases in number; it becomes concentrated in greater masses, its strength grows, and it feels that strength more. The various interests and conditions of life within the ranks of the proletariat are more and more equalized, in proportion as machinery obliterates all distinctions of labour, and nearly everywhere reduces wages to the same low level. The growing competition among the bourgeois, and the resulting commercial crises, make the wages of the workers ever more fluctuating. The unceasing improvement of machinery, ever more rapidly developing, makes their livelihood more and more precarious; the collisions between individual workmen and individual bourgeois take more and more the character of collisions between two classes. Thereupon the workers begin to form combinations (trade unions) against the bourgeois....

This organization of the proletarians into a class, and consequently into a political party, is continually being upset again by the competition between the workers themselves. But it ever rises up again, stronger, firmer, mightier. It compels legislative recognition of particular interests of the workers, by taking advantage of the divisions among the bourgeoisie itself....

Altogether, collisions between the classes of the old society further, in many ways, the course of development of the proletariat. The bourgeoisie finds itself involved in a constant battle: at first with the aristocracy; later on, with those portions of the bourgeoisie itself, whose interests have become antagonistic to the progress of industry; at all times, with the bourgeoisie of foreign countries. In all these battles it sees itself compelled to appeal to the proletariat, to ask for its help, and thus to drag it into the political arena. The bourgeoisie itself, therefore, supplies the proletariat with its own elements of political and general education; in other words, it furnishes the proletariat with weapons for fighting the bourgeoisie.

Further, as we have already seen, entire sections of the ruling classes are, by the advance of industry, precipitated into the proletariat, or are at least threatened in their conditions of existence. These also supply the proletariat with fresh elements of enlightenment and progress.

Finally, in times when the class struggle nears the decisive hour, the process of dissolution going on within the ruling class, in fact within the whole range of old society, assumes such a violent, glaring character that a small section of the ruling class cuts itself adrift and joins the revolutionary class, the class that holds the future in its hands. Just as, therefore, at an earlier period, a section of the nobility went over to the bourgeoisie, so now a portion of the bourgeoisie goes over to the proletariat, and in particular, a portion of the bourgeois ideologists, who have raised themselves to the level of comprehending theoretically the historical movement as a whole.

Of all the classes that stand face to face with the bourgeoisie today, the proletariat alone is a really revolutionary class. The other classes decay and finally disappear in the face of modern industry; the proletariat is its special and essential product....

KEY CONCEPTS

bourgeoisie
capitalist class
communism
means of production
proletariat
working class

DISCUSSION QUESTIONS

1. What evidence do you see in contemporary society of Marx and Engels's claim that the need for a constantly expanding market means that capitalism "nestles everywhere"?
2. How do Marx and Engels depict the working class, and what evidence do you see of their argument in looking at the contemporary labor market?

22

Aspects of Class in the United States: An Introduction

JOHN BELLAMY FOSTER

John Bellamy Foster examines the increasing inequality that is characteristic of the contemporary United States. He shows that wealth is even more unevenly divided than is income and also discusses the decreased likelihood of social mobility for current generations.

If class war is continual in capitalist society, there is no doubt that in recent decades in the United States it has taken a much more virulent form. In a speech delivered at New York University in 2004 Bill Moyers pointed out that

> Class war was declared a generation ago in a powerful paperback polemic by William Simon, who was soon to be Secretary of the Treasury. He called on the financial and business class, in effect, to take back the power and privileges they had lost in the depression and

SOURCE: Aspects of Class in the United States by John Bellamy Foster from *Monthly Review* 58 no. 3 (July-August 2006). Reprinted by permission.

> the new deal. They got the message, and soon they began a stealthy class war against the rest of the society and the principles of our democracy. They set out to trash the social contract, to cut their workforces and wages, to scour the globe in search of cheap labor, and to shred the social safety net that was supposed to protect people from hardships beyond their control. *Business Week* put it bluntly at the time [in its October 12, 1974 issue]: "Some people will obviously have to do with less … it will be a bitter pill for many Americans to swallow the idea of doing with less so that big business can have more."[1]

The effects of this relentless offensive by the vested interests against the rest of the society are increasingly evident. In 2005 the *New York Times* and the *Wall Street Journal* each published a series of articles focusing on class in the United States. This rare open acknowledgement of the importance of class by the elite media can be attributed in part to rapid increases in income and wealth inequality in U.S. society over the last couple of decades—coupled with the dramatic effects of the Bush tax cuts that have primarily benefited the wealthy. But it also grew out of a host of new statistical studies that have demonstrated that intergenerational class mobility in the United States is far below what was previously supposed, and that the United States is a more class-bound society than its major Western European counterparts, with the exception of Britain. In the words of *The Wall Street Journal* (May 13, 2005):

> Although Americans still think of their land as a place of exceptional opportunity—in contrast to class-bound Europe—the evidence suggests otherwise. And scholars have, over the past decade, come to see America as a less mobile society than they once believed. As recently as the later 1980s, economists argued that not much advantage passed from parent to child, perhaps as little as 20 percent. By that measure, a rich man's grandchild would have barely any edge over a poor man's grandchild.… But over the last 10 years, better data and more number-crunching have led economists and sociologists to a new consensus: The escalators of mobility move much more slowly. A substantial body of research finds that at least 45 percent of parents' advantage in income is passed along to their children, and perhaps as much as 60 percent. With the higher estimate, it's not only how much money your parents have that matters—even your great-great grandfather's wealth might give you a noticeable edge today.

As Paul Sweezy once observed, "self-reproduction is an *essential* characteristic of a class as distinct from a mere stratum."[2] What is clear from recent data is that the upper classes in the United States are extremely effective in reproducing themselves—to a degree that invites no obvious historical comparison in modern capitalist history. According to the *New York Times* (November 14, 2002), "Bhashkar Mazumber of the Federal Reserve Bank of Chicago … found that around 65 percent of the earnings advantage of fathers was transmitted to sons." Tom Hertz, an economist at American University, states that "while few would deny that it is *possible* to start poor and end rich, the evidence suggests that this feat is more difficult to accomplish in the United States than in other high-income nations."[3]

The fact that the rich are getting both relatively and absolutely richer, and the poor are getting relatively (if not absolutely) poorer, in the United States today is abundantly clear to all—although the true extent of this trend defies the imagination. Over the years 1950 to 1970, for each additional dollar made by those in the bottom 90 percent of income earners, those in the top 0.01 percent received an additional $162. In contrast, from 1990 to 2002, for every added dollar made by those in the bottom 90 percent, those in the uppermost 0.01 percent (today around 14,000 households) made an additional $18,000.[4]

Wealth is always far more unevenly divided than income. In 2001 the top 1 percent of wealth holders accounted for 33 percent of all net worth in the United States, twice the total net worth of the bottom 80 percent of the population. Measured in terms of financial wealth (which excludes equity in owner-occupied houses), the top 1 percent in 2001 owned more than four times as much as the bottom 80 percent of the population. Between 1983 and 2001, this same top 1 percent grabbed 28 percent of the rise in national income, 33 percent of the total gain in net worth, and 52 percent of the overall growth in financial worth.[5]

Nevertheless, a considerable portion of the population still seems willing to accept substantial differentials in economic rewards on the assumption that these represent returns to merit and that all children have a fighting chance to rise to the top. The United States, the received wisdom tells us, is still the "land of opportunity." The new data on class mobility, however, indicate that this is far from the case and that the barriers separating classes are hardening.

How class advantages are passed on from one generation to the next is of course enormously difficult to determine—if only because class privileges are so various. Class inequality manifests itself in wealth, income, and occupation, but also in education, consumption, and health—and each of these are among the means by which class advantages/disadvantages are transmitted. Class inequalities, Sweezy explained,

> are not only or perhaps even primarily a matter of income: [in certain social settings] a considerable range of income differentials would be compatible with all children having substantially equal life chances. More important are a number of other factors which are less well defined, less visible, and impossible to quantify: the advantages of coming from a more "cultured" home environment, differential access to educational opportunities, the possession of "connections" in the circles of those holding positions of power and prestige, and self-confidence which children absorb from their parents—the list could be expanded and elaborated.[6]

Such intangibles are difficult to measure, but in a capitalist society they tend to interact with large differentials in income and property ownership and hence leave their quantitative trace there. It is this whole constellation of class advantages roughly correlated with income and wealth, but not simply reducible to these elements, that allows the privileged to maintain their positions of economic status and power intergenerationally even in the context of a society that on the surface appears to have many of the characteristics of a meritocracy. The well-to-do get better education, enjoy better health, have more opportunities to travel, benefit from a wide array of personal services (derived from purchase of the labor services of others), etc.—all of which translates into class advantages passed on to their children.

The fact that strong barriers restricting upward class mobility exist is of course the first point to be considered in class analysis—since without this classes would be nonexistent. However, the real historical significance of class goes far beyond this. Class is not simply about the life chances of a given individual or a family; it is the prime mover in the constitution of modern society, governing both the distribution of power and the potential for social change. It therefore permeates all aspects of social existence.

At present there is no well-developed theory of class in all of its aspects, which remains perhaps the single biggest challenge facing the social sciences. Indeed, failure to advance in this area can be seen as symptomatic of the general stagnation of the social sciences over much of the twentieth century. Nevertheless, most Marxist analyses of class take their starting point from Lenin's famous definition of class:

> Classes are large groups of people differing from each other by the place they occupy in a historically determined system of social production, by their relation (in most cases fixed and formulated in law) to the means of production, by their role in the social organization of labour, and, consequently, by the dimensions of the share of social wealth of which they dispose and the mode of acquiring it.[7]

Like all brief definitions of class, this one has its weaknesses, since it is not able to take in the dynamic nature of class relations. As Sweezy argued, a systematic treatment of class and class struggle "needs also to encompass at least the following: the formation of classes in conflict with other classes, the character and degree of their self-consciousness, their internal organizational structures, the ways in which they generate and utilize ideologies to further their interests, and their modes of reproduction and self-perpetuation."[8] If we are speaking of a "ruling class" then the ways in which this class dominates the economy and the state need to be understood. Further, it is crucial to ascertain how class articulates itself in relation to other social relations and forms of oppression, such as race and gender.

An investigation of class thus leads to the analysis of society as a whole, its relationships of power, conflict, and change.

NOTES

1. Bill Moyers, "This is the Fight of Our Lives," keynote speech, Inequality Matters Forum, New York University, June 3, 2004.
2. Paul M. Sweezy, "Paul Sweezy Replies to Ernest Mandel," *Monthly Review* 31, no. 3 (July–August 1979), 82.
3. Tom Hertz, *Understanding Mobility in America*, Center for American Progress (April 26, 2006), i, 8, http://www.americanprogress.org/.
4. Correspondents of *The New York Times, Class Matters* (New York: Times Books, 2005), 186.
5. Edward N. Wolff, "Changes in Household Wealth in the 1980s and 1990s in the U.S.," (April 27, 2004, draft), forthcoming in Edward N. Wolff, *International Perspectives on Household Wealth* (Brookfield, Vermont: Edward Elgar, 2006), http://www.econ.nyu.edu/user/wolffe/.
6. Sweezy's comments here were directed mainly at postrevolutionary societies, but he made it clear that the same issues related to the reproduction of class applied to capitalist societies. I have inserted a brief qualification in square brackets to avoid any misunderstanding related to the specific context in which he was writing. See Paul M. Sweezy, *Post-Revolutionary Society* (New York: Monthly Review Press, 1980), 79–80.
7. V. I. Lenin, *Selected Works* (Moscow: Progress Publishers, 1971), 486.
8. Sweezy, "Paul Sweezy Replies to Ernest Mandel," 79.

KEY CONCEPTS

capitalism — elites — social class (or class) — wealth
contingent worker — income — social mobility

DISCUSSION QUESTIONS

1. What is the difference between *income* and *wealth,* and why is this important for the study of class inequality?
2. In what ways does Foster's analysis challenge the idea that the United States is a nation where anyone who works hard enough can get ahead?

23

America Without a Middle Class

ELIZABETH WARREN

Elizabeth Warren argues that the American tradition of having a strong middle class is now at risk because of the economic crisis that has beset America. She shows the increasingly fragile status of many middle-class families, who are now working harder than ever just to keep up with basic expenses.

Can you imagine an America without a strong middle class? If you can, would it still be America as we know it?

Today, one in five Americans is unemployed, underemployed, or just plain out of work. One in nine families can't make the minimum payment on their credit cards. One in eight mortgages is in default or foreclosure. One in eight Americans is on food stamps. More than 120,000 families are filing for bankruptcy every month. The economic crisis has wiped more than $5 trillion from pensions and savings, has left family balance sheets upside down, and threatens to put ten million homeowners out on the street.

Families have survived the ups and downs of economic booms and busts for a long time, but the fall-behind during the busts has gotten worse while the surge-ahead during the booms has stalled out. In the boom of the 1960s, for example, median family income jumped by 33% (adjusted for inflation). But the boom of the 2000s resulted in an almost-imperceptible 1.6% increase for the typical family. While Wall Street executives and others who owned lots of stock celebrated how good the recovery was for them, middle class families were left empty-handed.

The crisis facing the middle class started more than a generation ago. Even as productivity rose, the wages of the average fully-employed male have been flat since the 1970s.

But core expenses kept going up. By the early 2000s, families were spending twice as much (adjusted for inflation) on mortgages than they did a generation ago—for a house that was, on average, only ten percent bigger and 25 years older. They also had to pay twice as much to hang on to their health insurance.

To cope, millions of families put a second parent into the workforce. But higher housing and medical costs combined with new expenses for child care,

SOURCE: Elizabeth Warren, "America Without a Middle Class." *Huffington Post*, December 3, 2009.

the costs of a second car to get to work and higher taxes combined to squeeze families even harder. Even with two incomes, they tightened their belts. Families today spend less than they did a generation ago on food, clothing, furniture, appliances, and other flexible purchases—but it hasn't been enough to save them. Today's families have spent all their income, have spent all their savings, and have gone into debt to pay for college, to cover serious medical problems, and just to stay afloat a little while longer.

Through it all, families never asked for a handout from anyone, especially Washington. They were left to go on their own, working harder, squeezing nickels, and taking care of themselves. But their economic boats have been taking on water for years, and now the crisis has swamped millions of middle class families.

The contrast with the big banks could not be sharper. While the middle class has been caught in an economic vise, the financial industry that was supposed to serve them has prospered at their expense. Consumer banking—selling debt to middle class families—has been a gold mine. Boring banking has given way to creative banking, and the industry has generated tens of billions of dollars annually in fees made possible by deceptive and dangerous terms buried in the fine print of opaque, incomprehensible, and largely unregulated contracts.

And when various forms of this creative banking triggered economic crisis, the banks went to Washington for a handout. All the while, top executives kept their jobs and retained their bonuses. Even though the tax dollars that supported the bailout came largely from middle class families—from people already working hard to make ends meet—the beneficiaries of those tax dollars are now lobbying Congress to preserve the rules that had let those huge banks feast off the middle class.

Pundits talk about "populist rage" as a way to trivialize the anger and fear coursing through the middle class. But they have it wrong. Families understand with crystalline clarity that the rules they have played by are not the same rules that govern Wall Street. They understand that no American family is "too big to fail." They recognize that business models have shifted and that big banks are pulling out all the stops to squeeze families and boost revenues. They understand that their economic security is under assault and that leaving consumer debt effectively unregulated does not work.

Families are ready for change. According to polls, large majorities of Americans have welcomed the Obama Administration's proposal for a new Consumer Financial Protection Agency (CFPA). The CFPA would be answerable to consumers—not to banks and not to Wall Street. The agency would have the power to end tricks-and-traps pricing and to start leveling the playing field so that consumers have the tools they need to compare prices and manage their money. The response of the big banks has been to swing into action against the agency, fighting with all their lobbying might to keep business-as-usual. They are pulling out all the stops to kill the agency before it is born. And if those practices crush millions more families, who cares—so long as the profits stay high and the bonuses keep coming.

America today has plenty of rich and super-rich. But it has far more families who did all the right things, but who still have no real security. Going to college and finding a good job no longer guarantee economic safety. Paying for a child's education and setting aside enough for a decent retirement have become distant dreams. Tens of millions of once-secure middle class families now live paycheck to paycheck, watching as their debts pile up and worrying about whether a pink slip or a bad diagnosis will send them hurtling over an economic cliff.

America without a strong middle class? Unthinkable, but the once-solid foundation is shaking.

KEY CONCEPTS

American dream	social stratification	middle class
class consciousness	economic crisis	social mobility

DISCUSSION QUESTIONS

1. How does Warren's argument illustrate the problems of an increasing class divide?
2. What are the social forces that are threatening the status of middle-class families in the United States?

24

The State of Poverty in America

PETER EDELMAN

The persistent belief in American society is that poverty is the result of poor work habits. Debunking this idea, Peter Edelman analyzes the underlying causes of persistent poverty in an otherwise relatively affluent society.

SOURCE: Reprinted with permission from The American Prospect (Prospect.org).

We have two basic poverty problems in the United States. One is the prevalence of low-wage work. The other concerns those who have almost no work.

The two overlap.

Most people who are poor work as much as they can and go in and out of poverty. Fewer people have little or no work on a continuing basis, but they are in much worse straits and tend to stay poor from one generation to the next.

The numbers in both categories are stunning.

Low-wage work encompasses people with incomes below twice the poverty line—not poor but struggling all the time to make ends meet. They now total 103 million, which means that fully one-third of the population has an income below what would be $36,000 for a family of three.

In the bottom tier are 20.5 million people—6.7 percent of the population—who are in deep poverty, with an income less than half the poverty line (below $9,000 for a family of three). Some 6 million people out of those 20.5 million have no income at all other than food stamps.

These dire facts tempt one to believe that there may be some truth to President Ronald Reagan's often-quoted declaration that "we fought a war against poverty and poverty won." But that is not the case. Our public policies have been remarkably successful. Starting with the Social Security Act of 1935, continuing with the burst of activity in the 1960s, and on from there, we have made great progress.

We enacted Medicaid and the Children's Health Insurance Program, and many health indicators for low-income people improved. We enacted food stamps, and the near-starvation conditions we saw in some parts of the country were ameliorated. We enacted the Earned Income Tax Credit and the Child Tax Credit, and the incomes of low-wage workers with children were lifted. We enacted Pell grants, and millions of people could afford college who otherwise couldn't possibly attend. We enacted Supplemental Security Income and thereby raised the income floor for elderly and disabled people whose earnings from work didn't provide enough Social Security. There is much more—housing vouchers, Head Start, child-care assistance, and legal services for the poor, to name a few. The Obama administration and Congress added 16 million people to Medicaid in the Affordable Care Act, appropriated billions to improve the education of low-income children, and spent an impressive amount on the least well-off in the Recovery Act.

All in all, our various public policies kept a remarkable 40 million people from falling into poverty in 2010—about half because of Social Security and half due to the other programs just mentioned. To assert that we fought a war against poverty and poverty won because there is still poverty is like saying that the Clean Air and Clean Water acts failed because there is still pollution.

Nonetheless, the level of poverty in the nation changed little between 1970 and 2000 and is much worse now. It was at 11.1 percent in 1973—the lowest level achieved since we began measuring—and after going up sharply during the Reagan and George H. W. Bush years, went back down during the 1990s to 11.3 percent in 2000, as President Bill Clinton left office.

Why didn't it fall further? The economics have been working against us for four decades, exacerbated by trends in family composition. Well-paying industrial jobs disappeared to other countries and to automation. The economy grew, but the fruits of the growth went exclusively to those at the top. Other jobs replaced the ones lost, but most of the new jobs paid much less. The wage of the median-paying job barely grew—by one measure going up only about 7 percent over the 38 years from 1973 to 2011. Half the jobs in the country now pay less than $33,000 a year, and a quarter pay less than the poverty line of $22,000 for a family of four. We have become a low-wage economy to a far greater extent than we realize.

Households with only one wage-earner—typically those headed by single mothers—have found it extremely difficult to support a family. The share of families with children headed by single mothers rose from 12.8 percent in 1970 to 26.2 percent in 2010 (and from 37.1 percent in 1971 to 52.8 percent in 2010 among African Americans). In 2010, 46.9 percent of children under 18 living in households headed by a single mother were poor.

The percentage of people in deep poverty has doubled since 1976. A major reason for this rise is the near death of cash assistance for families with children. Welfare has shrunk from 14 million recipients (too many, in my view) before the Temporary Assistance for Needy Families law (TANF) was enacted in 1996 to 4.2 million today, just 1.5 percent of the population. At last count, Wyoming had 607 people on TANF, or just 2.7 percent of its poor children. Twenty-six states have less than 20 percent of their poor children on TANF. The proportion of poor families with children receiving welfare has shrunk from 68 percent before TANF was enacted to 27 percent today.

What's the agenda going forward? The heart of it is creating jobs that yield a living income. Restoring prosperity, ensuring that the economy functions at or near full employment, is our most powerful anti-poverty weapon. We need more, though—a vital union sector and a higher minimum wage, for two. We also need work supports—health care, child care, and help with the cost of housing and postsecondary education. These are all income equivalents—all policies that will contribute to bringing everyone closer to having a living income.

There's a gigantic problem here, however: We look to be headed to a future of too many low-wage jobs. Wages in China, India, and other emerging economies may be rising, but we can't foresee any substantial increase in the prevailing wage for many millions of American jobs. That means we better start talking about wage supplements that are much bigger than the Earned Income Tax Credit. We need a dose of reality about the future of the American paycheck.

The second big problem is the crisis—and it is a crisis—posed by the 20 million people at the bottom of the economy. We have a huge hole in our safety net. In many states, TANF and food stamps combined don't even get people to half of the poverty line, and a substantial majority of poor families don't receive TANF at all.

Even worse, we have destroyed the safety net for the poorest children in the country. Seven million women and children are among the 20.5 million in deep poverty. One in four children in a household headed by a single mother is in deep poverty. We have to restore the safety net for the poorest of the poor.

Getting serious about investing in our children—from prenatal care and early-childhood assistance on through education at all levels—is also essential if we are to achieve a future without such calamitous levels of poverty. In addition, we must confront the destruction being wrought by the criminal-justice system. These are poverty issues and race issues as well. The schools and the justice system present the civil-rights challenges of this century.

Combining all of the problems in vicious interaction is the question of place—the issues that arise from having too many poor people concentrated in one area, whether in the inner city, Appalachia, the Mississippi Delta, or on Indian reservations. Such places are home to a minority of the poor, but they include a hugely disproportionate share of intergenerational and persistent poverty. Our most serious policy failing over the past four-plus decades has been our neglect of this concentrated poverty. We have held our own in other respects, but we have lost ground here.

Finally, we need to be much more forthright about how much all of this has to do with race and gender. It is always important to emphasize that white people make up the largest number of the poor, to counter the stereotype that the face of poverty is one of color. At the same time, though, we must face more squarely that African Americans, Latinos, and Native Americans are all poor at almost three times the rate of whites and ask why that continues to be true. We need as a nation to be more honest about who it is that suffers most from terrible schools and the way we lock people up. Poverty most definitely cuts across racial lines, but it doesn't cut evenly.

There's a lot to do.

KEY CONCEPTS

labor market

poverty line

DISCUSSION QUESTIONS

1. What are some of the causes that Edelman identifies as the source for persistent poverty?
2. How does Edelman's analysis of the causes of poverty differ from popular explanations among the general public?

Applying Sociological Knowledge: An Exercise for Students

Brand labels in a class-based society communicate our status to others. Make a list of all of the clothing labels you can think of and then match each label to a ranking in the class system (working class, middle class, upper-middle class, etc.). What class images are projected by each? Are there class stereotypes suggested by different labels? Whom do they affect and how? In what ways do these labels reproduce our class identities? Do they do any harm?

PART IX

Global Stratification

25

Globalization: An Introduction

D. STANLEY EITZEN AND MAXINE BACA ZINN

This article provides a foundation for understanding the basic forces behind globalization, as well as looking at its different dimensions. The authors compare global processes of the past with those affecting society now.

GLOBALIZATION DEFINED

Globalization refers to the greater interconnectedness among the world's people—to "the increasing scale, extent, variety, speed, and magnitude of international cross-border, social, economic, military, political, and cultural interrelations" (Wiarda, 2007:3). Put another way, globalization is a process whereby goods, information, people, money, communication, fashion (and other forms of culture) move across national boundaries.

There are several implications of this view of globalization. First, it signals "that we are all part of a steadily shrinking and interdependent world. Modern communications, transportation, and the Internet have all served to tie more and more countries and peoples together in newer and more complex ways" (Wiarda, 2007:3). Second, globalization is not a thing or a product, but rather a process. It involves such activities as immigration, transnational travel, using e-mail and the Internet, marketing products in one nation that are made elsewhere, moving jobs to low-wage economies, transnational investments, satellite broadcasts, the pricing of oil, coffee, wheat, and other commodities, and finding a McDonald's and drinking a Coke or Pepsi in virtually every major city in the world. Third, it follows that globalization is not simply a matter of economics, but also has far reaching political, social, and cultural implications. Fourth, globalization refers to changes that are increasingly remolding the lives of people worldwide. Globalization is not just "something out there," but is intimately connected to the everyday activities of institutions, families, and individuals within societies (Hytrek, and Zentgraf, 2008). And fifth, not everyone experiences globalization in the same way. It expands opportunities and enhances prosperity for some while leading others into poverty and hopelessness. Periods of rapid social change, we know, "threaten the familiar, destabilize old boundaries, and upset established traditions. Like the mighty Hindu god Shiva, globalization is not only a great destroyer, but also a powerful creator of new ideas, values, identities, practices, and movements" (Stegner, 2002:ix).

GLOBALIZATION THEN

Globalization is not a new phenomenon. For thousands of years people have traveled, traded, and migrated across political boundaries, exchanging food, artifacts, and knowledge. Consider the world around A.D. (the following is from Sen, 2002; and *U.S. News & World Report*, 1999). The Vikings plundered and traded, establishing settlements in northern France, Britain, Iceland, Greenland, and Russia. At the crossroads between East and West, the Byzantine Empire traded with foreigners from both sides of the globe. India was linked by maritime routes to Africa, the Middle East, and Southeast Asia. China (during the Song Empire) used sea routes to trade cotton goods, spices, and horses. The Islamic world was the first civilization to trade with the other major empires in Europe, Asia, and Africa. These cross-boundary interactions involved not only trade but the transfer of inventions, knowledge, and other cultural forms. For example, the ninth century Arab mathematician Mohammad Ibn Musa-al-Khwarismi, who gave the Western world algorithms and algebra, is "one of many non-Western contributors whose works influenced the European Renaissance and, later, the Enlightenment and the Industrial Revolution" (Sen, 2002:A3).

There have been other periods in which globalization processes accelerated. In sixteenth-century Europe, for example, trade and exploration expanded to all parts of the globe with Europeans settling in different regions. The late 1800s and early 1900s were characterized by great waves of immigration and high levels of trade and finance across national borders. The period following World War II was the precursor to contemporary globalization. The disintegration of the British, French, Dutch, Belgian, and Spanish colonial empires during this time resulted in the establishment of no fewer than eighty-eight new nations. Later, the Soviet Union collapsed, creating eighteen separate countries. These "new" countries were able to sell their raw materials and products on the world market and to purchase goods. They could now also establish local industries to compete with those in other countries. In addition, new technological innovations arising from the war laid the foundation for the transportation and communication advances of the current age. The third change in the post–World War II era was the emergence of transnational political and financial institutions. In 1945, the World Bank and the International Monetary Fund were created to help rebuild Europe and Japan. These two financial institutions continue to play an important role in the developing world. The United Nations, with its organizational units such as the World Health Organization and UNESCO, seeks to reduce tensions among nation-states and to find transnational solutions to political and social problems. The World Court exists to adjudicate international disputes and to try war criminals.

THE CHARACTERISTICS OF GLOBALIZATION NOW

Transnational connections have existed for centuries, but since the 1960s the pace of these interconnections has increased exponentially. Speed of movement (in terms of both physical travel and travel via communications technology), as well as the volume of goods, messages, and symbols transported, have increased dramatically. For example, the *daily* turnover in foreign exchange markets has risen from $800 billion in the mid-1990s to almost $4 trillion in 2009 (James, 2009:22), So, too, has space seemed to shrink as travel and communication time has decreased. The indicators of this increasingly rapid change occur along a number of dimensions (the following is dependent in part on Brecher, Costello, and Smith, 2000:2–4; Scholte, 2000:20–25; and Beynon and Dunkerley, 2000:5–7).

Production

Globalization has transformed the nature of economic activity. From the 1970s onward, transnational corporations have built factories and bought manufactured products from low-wage countries on a vastly expanded scale. A "global assembly line" has emerged in which products from athletic shoes to electronics are made by low-wage workers in low-wage countries and sold in developed countries. U.S., Japanese, and European transnational corporations have invested many billions in China and elsewhere to build state-of-the-art factories. For example, Dell, which makes personal computers, has components manufactured by hundreds of suppliers and subsuppliers in Mexico, Taiwan, Malaysia, Korea, and China. Nokia, the Finnish company known primarily for its cellular phones, uses components and assembled products produced in 10 different countries. The result has been the decline of manufacturing in developed countries and the migration of production jobs to low-wage economics. The U.S., where the avenge manufacturing wage is $16 an hour, lost one-fifth of its manufacturing jobs from 2000 to 2008 (Greenhouse, 2009). These jobs migrated to places like China, where the average manufacturing wage is 61 cents an hour. The job losses occur in a transnational domino effect as jobs that once moved to Mexico because of their low wages, for instance, have moved to even lower-wage economies. This migration of jobs has been called "the race to the bottom."

Markets

In the past, corporations limited their sales to domestic or perhaps regional markets. Now goods and services are marketed to the entire world. Nokia, for example, sells its products in 130 countries. Sometimes a transnational corporation will locate a factory in a country where it markets products. This is the strategy of the Japanese automobile manufacturers Honda and Toyota, which have major plants in the United States, where their products are popular.

Technology: The Tools of Globalization

New technologies—robotics, fiber optics, container ships, computers, communications satellites, and the Internet—have transformed information storage and retrieval, communication, production, and transportation. Microelectronic-based systems of information, for example, allow for the storage, manipulation, and retrieval of data in huge quantities (the amount of unique information generated worldwide each year is measured in exabytes, one exabyte being 1 followed by 18 zeroes). Information can be sent in microseconds via communications satellites throughout the world. As *Business Week* put it, "Anyone with a computer is a citizen of the world...." (1999:71). In short, "[t]echnological advancement in transportation and communications has not merely made the world smaller, *for many purposes it has made geography irrelevant* (emphasis added)" (Peoples and Bailey, 2003:36).

Corporate Restructuring

Major corporations have always operated internationally. Beginning in the 1980s, they reorganized internally to take advantage of the global economy. They merged with other corporations and developed strategic alliances with fellow multinationals. They arranged for companies in other societies to do various tasks ("corporate outsourcing"). Consider Boeing, which in the manufacture of its next-generation Dreamliner developed partnerships with corporations in Japan (Mitsubishi, Kawasaki, and Fuji), Sweden (Saab), Italy (Alenia Aaeronautica), France (Latecoere and Messier-Dowty), Germany (Diehl Luftfahrt Electronik) and the United Kingdom (Smiths Aerospace). Thus, Boeing is responsible for about one-third of the production of the Dreamliner while combining the remaining pieces supplied by its global partners (*Business Week*, 2009). The result is a decentralization of production for Boeing, but a consolidation of resources and power among these partnering corporations across national boundaries.

The 2000 largest global corporations in 2009 represented 62 countries, employed 76 million people, and had total sales of $30 trillion, assets of $124 trillion, and profits of $1.4 trillion (*Forbes,* 2010:96).

Neoimperialism

Following World War II, the imperialist powers gave political independence back to their colonies. Globalization, however, has kept these countries dependent economically on Western Europe, Japan, and the United States. "Globalization has taken from poor countries control of their own economic policies and concentrated their assets in the hands of first world investors. While it has enriched some third world elites, it has subordinated them to foreign corporations, international institutions, and dominant states" (Brecher, Costello, and Smith, 2000:3–4).

Changing Structure of Work

With globalization, worker security everywhere has declined. "All over the world, employers have downsized, outsourced, and made permanent jobs into contingent ones. Employers have attacked job security requirements, work rules, worker representation, healthcare, pensions, and other social benefits, and anything else that defined workers as human beings and employers as partners in a social relationship, rather than simply as buyers and sellers of labor power" (Brecher, Costello, and Smith, 2000:3; Hacker, 2006). In this pro-employer environment, labor unions have lost their power. Employers, when faced with employee demands for higher wages or better benefits, can simply threaten to move the operation to a setting where wages and benefits are lower.

Global Institutions

Organizations such as the World Trade Organization (WTO), the World Bank, and the International Monetary Fund (IMF) are involved in fostering transnational trade and providing economic development in underdeveloped countries. These kinds of organizations are powerful forces accelerating the globalization process. Whether their contributions have positive results is open to debate, as noted throughout this book.

Neoliberal Ideology and Policies

Contemporary globalization is fueled by a prevailing ideology known as neoliberalism, or the Washington Consensus. This ideology dates back to John Locke and Adam Smith, who argued that market forces will bring prosperity, liberty, and democracy if unhindered by government intervention. In terms of policy, neoliberals promote privatization, deregulation, and the dismantling of the welfare state. Most significant, neoliberalism promotes free trade—that is, the idea that state borders should be open to trade without tariffs and other restrictions. This ideology of unfettered capitalism is behind agreements such as NAFTA (the North American Free Trade Act), and it informs the policies of the World Trade Organization and the International Monetary Fund. Proponents of this ideology believe that free trade "expands economic freedom and spurs competition, [thus raising] the productivity and living standards of people in countries that open themselves to the global marketplace" (Cato Institute, cited in Ervin and Smith, 2008:2).

Governance

The sovereignty of the nation-state has for the most part been diminished by globalization (an exception is the U.S., which resists efforts by international organizations to control it). There are suprastate organizations that regulate transnational trade and international law. As a consequence of accepting neoliberal ideology, national governments do not hinder corporate decisions regarding outsourcing and the movement of capital, even though these decisions go against the welfare of their citizens. In effect, economic, political, and cultural change is now beyond the control of any national government (Benyon and Dunkerley, 2000:6).

Permeable Borders: The Transnational Movement of People, Environmental Dangers, Pandemics, and Crime

Insularity is no longer possible, as environmental pollution through the air, water, or food supply anywhere affects people elsewhere. Diseases are difficult to contain, as evidenced by the AIDS and swine flu pandemics. Criminal networks easily function and flourish when borders are permeable. They engage in the distribution of illegal drugs, prostitution, traffic in slaves, and sweatshops. Terrorism also becomes transnational when borders are porous.

Worldwide, more than 200 million people are living outside their country of birth or citizenship. Wars, droughts, floods, and changing climates are pushing people out of their homelands. So, too, is the hope of jobs luring them elsewhere.

Transnational migrant labor is typically from poor countries to rich ones. Over half of the world's legal and illegal immigrants are women. This "feminization of migration" reflects a worldwide gender revolution in which millions of women migrate across the globe to serve as nannies, maids, and sex workers. (Ehrenreich and Hochschild, 2002; Hondagneu-Sotelo, 2003).

One consequence of this flow of people across borders is the reverse flow of money, as many immigrants send money back to relatives in their native land. In 2007, this flow of money from the developed world to the developing world was an estimated $300 billion annually (DeParle, 2007).

The transnational migration of people is more than immigrant labor. It also includes involuntary migration (e.g., sex trafficking) and the transitory crossing of borders as tourists or to seek medical treatment.

Global Culture

National culture, traditionally, has been tied to place and time. It is the knowledge, symbols, and stories that people share within a national consciousness, giving identity to a nation and its people. Global culture, on the other hand, is de-ethnicized and de-territorialized, existing outside the usual reference to geographical territory. It is created and sustained by the media, corporate advertising, and the entertainment industry. The result is a single world culture "centered on consumerism, mass media, Americana, and the English language. Depending on one's perspective, this homogenization entails either progressive cosmopolitanism or oppressive imperialism" (Scholte, 2000:23). The westernized consumer lifestyle is symbolized by similar products (Nikeshoes, fashion, pop music, Disney products, movies, Coca-Cola, McDonald's, ESPN, and CNN) that are found everywhere (Benyon and Dunkerly, 2000:13–21).

The global culture is not as uniform, homogenizing, and universal as it would seem, however. There are often clashes between local and global cultures, Cultural diversity abounds. Religious fundamentalists in many parts of the world, most notably the Middle East, passionately resist modernity in general and the intrusion of the West in particular. Many people embrace their national identity and culture. Moreover, global communications and markets are often adapted to fit diverse local contexts. "Through so-called 'glocalization,' global news reports, global products, global social movements and the like take different forms and make different impacts depending on local particularities" (Scholte, 2000:23).

GLOBALIZATION: RECONFIGURING THE SOCIAL

The discipline of sociology emerged in the eighteenth and nineteenth centuries as the scholarly study of society. Understandably, since the world was divided into nations during this period, the focus of sociologists was on society as the nation-state, with geographical boundaries and social institutions unique to that society. Discussions of place included the local (community), urban and rural, and society. Social problems were examined and solutions offered for problems at the local, societal and regional levels.

Globalization—the transformation of world society in terms of flows of people, goods, capital, and ideas across national boundaries, linkages, institutions, culture, and consciousness (Lechner and Boli, 2000:2)—accelerated in the last decades of the twentieth century, resulting in sociologists and other social scientists beginning to think globally. Just as they explore the sources of social problems in the ordinary, everyday, normal workings of social institutions (e.g., institutional racism and sexism) and class relations, sociologists now also explore the ways that globalization, most particularly the world capitalist system, contributes to problems of people within and across national boundaries. For example, the various manifestations of inequality across and within societies and the degradation of the environment are consequences of the actions by transnational corporations. The pace of globalization has quickened at the beginning of the twenty-first century, and many sociologists are in the process of rethinking their eighteenth- and nineteenth-century roots to confront and understand the globalized and globalizing world. This reconceptualizing of the social is a significant shift in worldview....

REFERENCES

Beynon, John, and David Dunkerley (eds.). 2000. *Globalization: The Reader* (New York: Routledge).

Brecher, Jeremy, Tim Costello, and Brendan Smith. 2000. *Globalization from Below: The Power of Solidarity* (Cambridge, MA: South End Press).

Business Week. 1999. "The Internet Age," (October 4):71.

Business Week. 2009. "Big Changes for Boeing's Dreamliner." Online: http://articles.moneycentral.msn.com/Investing/Extra/big-changes-for-boeing.

Clifford, Mark L. 2002. "How Low Can Prices Go?" *Business Week* (December 2):60–61.

DeParle, Jason. 2007. "Migrant Money Flow: A $300 Billion Current." *New York Times* (November 18). Online: http://www.nytimes.com/2007/11/18weekinreview/18deparle.html?

Ehrenreich, Barbara, and Arlie Russell Hochschild. 2002. "Introduction." In *Global Women*, Barbara Ehrenreich and Arlie Russell Hochschild (eds.). (New York: Metropolitan Books).

Ervin, Justin, and Zachary A. Smith. 2008. *Globalization: A Reference Book*. Santa Barbara, CA: ABC-CLIO.

Forbes. 2010. "The Global 2000," (May 10):92–105.

Greenhouse, Steven. 2009. *The Big Squeeze: Tough Times for the American Worker*. New York: Anchor Hooks.

Hacker. Jacob S. 2006. *The Great Risk Shift* (New York: Oxford University Press).

Hondagneu-Sotelo, Pierrette. 2003. "Gender and Immigration: A Retrospective and Introduction." In *Gender and U.S. Immigration: Contemporary Trends,* Pierrette Hondagneu-Sotelo (ed.). (Berkeley: University of California Press), 3–19.

Hytrek, Gary, and Kristine M. Zentgraf. 2008. *America Transformed: Globalization, Inequality, and Power* (New York: Oxford University Press).

James, Harold. 2009. "The Late, Great Globalization." *Current History* 108 (January), 20–25.

Lechner, Frank J., and John Boli (eds.). 2000. *The Globalization Reader* (Oxford, UK: Blackwell Publishers).

Martin, Philip, and Jonas Widgren. 2002. "International Migration: Facing the Challenge," *Population Bulletin* 57 (March):entire issue.

Peoples, James, and Garrick Bailey. 2003. *Humanity: An Introduction to Cultural Anthropology* (Belmont, CA: Wadsworth). The section on globalization was accessed at http://www.wadsworth.com/anthropology_d/resources/terrorism/booklet.html.

Sen, Amartya. 2002. "How to Judge Globalism." *The American Prospect* (Winter): A2–A6.

Seholte, Jan Aart. 2000. *Globalization: A Critical Introduction* (New York: Palgrave).

Steger, Manfred B. 2002. *Globalism: The New Market Ideology* (Lanham, MD: Rowman & Littlefield).

U.S. News & World Report. 1999. "The Year 1000: What Life Was Like in the Last Millennium," special double issue (August 16 and 23):38–94.

KEY CONCEPTS

global assembly line	neoimperialism	transnational
globalization	neoliberalism	

DISCUSSION QUESTIONS

1. Baca Zinn and Eitzen identify five implications of globalization. List them and give an example of your own for each.
2. How is globalization now different from that of the past? What developments have most affected the process of globalization?
3. What evidence of globalization do you see in your life—for the better and for the worse?

26

Global Strategies for Workers

How Class Analysis Clarifies Us and Them and What We Need To Do

KATIE QUAN

Utilizing what she calls a class analysis, Katie Quan analyzes how a "we/they" mode of thinking has characterized the connection of American workers to other workers throughout the world. She argues that workers in the global assembly line are not the cause of workers' struggles within the United States.

INTRODUCTION

... What is the relationship of American workers to workers and their allies in other countries? Historically the position of American unions has varied widely on this question; in some periods we have accused foreign workers of being the enemy for "stealing our jobs," and at other times we have viewed them as victims of sweatshop superexploitation. But rarely have we viewed them as strategic allies with whom we share common goals and targets, as part of a big-picture analysis of power in the global economy. That perspective would lead us to organize ourselves differently than we do now, to place a much stronger priority on building a multinational labor movement in response to multinational capital and ally with other groups that are moving in the same direction.

One tool that helps us understand this relationship is class analysis. During the past twenty-five years, little has been heard about the explicit conceptualization of workers as a "class." Either people actually do not perceive reality as a matter of working class versus capitalist class, or they may avoid using the term "class" to disassociate themselves from unsuccessful Marxist governments of the last century. However, not to think in terms of class is unfortunate, since no

SOURCE: Reprinted Katie Quan, "Global Strategies for Workers: How Class Analysis Clarifies Us and them and What We Need to Do," from *What's Class Got to Do with It?: American Society in the Twenty-first Century*, edited by Michael Zweig.

matter what our ideological persuasion may be, class analysis gives us a way of viewing the world that identifies power relationships. It clarifies who has power among the global corporate ruling elite and how they are using it against the worldwide working class. It helps us see that if the working class wants to defend its interests and gain more power, it has to build strength among certain allies and target those who do hold power. Without class analysis, workers and unions have no tool for navigating complex situations that challenge us as to where our interests lie, or for formulating strategies that advance our interests. Instead, we often fall victim to pragmatic approaches, doing what seems best at the moment, or, worse yet, adopting beliefs and policies that are not in our interests….

THE GLOBAL INDUSTRY: BASIC US AND THEM

In the 1970s, I was a garment worker in New York City, sewing zippers and waistbands into hundreds of pants each day on piece rate. Piece rate is that wretched system whereby each sewing operation has a price, and the more pieces you sew, the more money you earn. The piece rate might be twenty-five cents for a zipper, and if you sewed one hundred zippers you would earn $25. I call this the system of being both the slave and the slave driver. You're the slave because you're the one who is working at a furious pace hour after hour, often straight through rest breaks and lunchtimes. And you're also the slave driver—because you are your own taskmaster, spurring yourself on to work faster and faster. In those days, piece rates in union shops were regulated by our union contract, and since I was young and worked fast, it wasn't unusual for me to make $18 an hour. That didn't mean that the other workers and I were well-off though, because the industry is seasonal, and there were many days and weeks when we lived off unemployment checks. Still, we didn't consider ourselves to be sweatshop workers—we believed that our path to freedom was to sew fast enough to beat the slave system and actually earn a decent living.

But the system beat us. We worked so hard and made so much money that our employers closed down our shops. "*We can't compete.* Labor costs are too high," they claimed. "We can't pay minimum wages and union benefits." So they laid us off and moved to areas where labor costs were lower. In the 1950s many employers ran away from the heavily unionized Northeast to the American South, and the textile and apparel unions followed the work to the southern states and organized workers there. Soon the workers in the South were told the same thing that workers in the Northeast had been told—"*We can't compete*"—and the employers went even farther south to Puerto Rico, the Caribbean, and Latin America. By the 1970s, apparel production had expanded to Asia, and by the 1990s, Asia had become the world's largest apparel-exporting region….

As our jobs left, employers told us that the next group of workers were "*stealing our jobs*" from us, and many workers believed them. I'll never forget the time a white woman from Pennsylvania came up to me and pointed her finger in my face, saying, "*You.* You Chinatown immigrants are taking *our* jobs!" I was stunned. Here I was busting my behind in the "slave and slave driver" system, barely making ends meet, and someone comes along and says I took her job. I would have told her that if she felt so strongly about it she and her co-workers could have our jobs, except that we had nowhere else to go. As immigrant women workers with poor English and few marketable skills, we had few job options, and the employers took advantage of this by forcing us to accept lower piece rates than the Pennsylvania workers.

The Pennsylvania woman and I were in a workshop together, and eventually, through discussion of piece rate dispute settlement and other workplace issues, we found that we actually had much in common. But this incident shows how an employer's message can affect workers' thinking and pit workers against each other. It leads people to believe that the problem is us, Pennsylvania workers and employers, versus them, Chinatown workers and employers—rather than us, Pennsylvania and Chinatown workers who are being exploited, versus them, employers who are moving production from place to place to find even lower labor costs. And absent a strong message from unions or other political leadership that "us" is working class people who have interests in common, no matter whether we work in Pennsylvania, Chinatown, or Latin America, workers may easily believe what the employers tell us....

The garment industry is like a chain of sweating, with multiple tiers of employers (them) who are trying to get rich off the workers (us)....

If we look at the power relationships in this chain of sweating, "them" are the retailers, manufacturers, and contractors who profit from the garment workers' labor, and who also can control their profit by squeezing those on the next lower link in the chain. The workers are "us" because we have no way of squeezing someone else's prices to benefit ourselves. It doesn't matter whether we're working in New York, Pennsylvania, or Bangkok; we're in the same boat. Our class interests are the same. The only way that we can increase piece rates is to form strong unions that can bargain collectively with our employers....

INTERNATIONAL TRADE: CONFUSING US AND THEM

Capitalists believe that international trade is generally a good thing, because it opens up markets to sell products, provides new opportunities for investment, and gives employers access to cheap labor. Generally, capitalists favor "free-trade" policies that allow them the tariff-free, quota-free ability to engage in global trade, so that they can be more competitive. However, when a group of capitalists find themselves unable to gain advantage over their competitors, then true to their opportunistic self-interests, they spurn free trade and instead erect trade barriers.... Examples are textile producers who supported tariffs to protect their markets from the 1970s through to the present, and information technology businessmen who vigorously fought for regulation of intellectual property rights in the World Trade Organization.

On the other hand, for the working class, free trade is generally a bad thing. Multinational corporations claim that consumers pay cheaper prices when businesses produce more cheaply. However, if hundreds of thousands of workers lose their jobs, then not many will have the buying power to purchase those goods. And if the workers producing the goods are being paid little, it's unlikely that they will have much buying power either. Moreover, the growth of the economy doesn't necessarily mean that benefits will trickle down to workers. Indeed, in the past twenty years, during a period of increasingly free trade, the gap between the rich and poor has widened to the point at which the richest in the nation are ten times more wealthy than the poorest, a substantial increase in disparity from earlier periods.... Without regulation of the impact of capital, such as on worker rights and environmental protection, "free trade" is not a position that is in the interests of workers.

The problem is that, in advocating for trade regulation, the policy position of the American labor movement has sometimes converged with that of capitalists who erect trade barriers to gain a competitive advantage, and in that convergence unions have adopted the rhetoric, and sometimes the ideology, of the capitalist interests as well. The most well-known example of this was "Buy American," a slogan used to champion the protectionist cause. Originally spearheaded by business interests that wanted to protect their markets, its purpose was to unite the American public, both capitalist and working class, around a common policy that was based on national unity. What was originally us, workers, and them, capitalists, became us, Americans, and them, foreigners. Unions then adopted the "Buy American" slogan and rallied their memberships around it....

The tendency of the American labor movement to pit American workers against foreign workers illustrates how strongly workers identify with nationalist interests in comparison to their class interests. We are a movement without a strong working class consciousness based on an analysis that clearly links our class interests with those of workers in other countries. When faced with global issues such as international trade, we have often reacted in ways that have hurt working class solidarity. No wonder few have thought seriously about crossing borders to organize. If we really wanted to mobilize consumers to buy things in the interests of our class, our slogan would have more accurately identified the us and them—we should have said "Buy Union," rather than "Buy American."...

In the twenty-first century, the global economy is the central battleground of class struggle.... A class analysis tells us that "them" is the very powerful forces of global capital that are changing the rules of global trade and finance to suit their corporate interests. And "us" is now defined as everyone who does not benefit from the consolidation of global capital—workers, farmers, environmentalists, and most other people in the world. To be sure, there is still a level of class struggle on the shop floor over piece rates, and there is still a level of class struggle between workers and their various employers up the chain of sweating. But this new level of struggle between the institutions of global capital and the people of the world will define the rules of all other levels of class struggle—everything from whether we will have jobs, to whether a country can have a postal service, to whether we can pass laws to protect our health. It is not just an economic struggle but also a political struggle between, on the one hand, a small group of capitalists who are not accountable to anyone, and, on the other hand, us working people.

Failure to understand this means that we will continue with a myopic vision of what needs to be done, relegating global issues to low priority when they should in fact be central. Rather than appropriating resources to unite workers and grassroots activists across borders, we might easily end up blaming each other for the problems that global capitalists have caused. And, unfortunately, if global issues don't become a central priority, we will continue to lose ground, because our strategies will be designed only to fight skirmishes, not to win the war....

CONCLUSION

Globalization is upon us, and it is unavoidable. Not only is it causing more hardship for workers in many countries but its new rules threaten the core of our democratic values and rights. The challenge for American workers is whether we will be content to watch events unfolding from a distance, or whether we will organize ourselves to unite with global allies to wage an effective response.

In this effort, a class analysis that explains the relationship between us and them is critical. We can't afford to make the mistakes of the past, when workers believed that siding with the bosses to promote the Buy American campaign was in our interests, and foreign workers like the "Japs" and immigrants ... were our enemies. We can't formulate strategies that address just our own domestic workforce when in fact labor markets are global—otherwise we will have no power. Class analysis gives us a context for understanding issues that can divide us, such as race, immigration status, and narrow thinking....

KEY CONCEPTS

class analysis　　power　　sweatshop

DISCUSSION QUESTIONS

1. Why does Quan think a class analysis is essential for understanding international patterns of labor?
2. Whose interests are served by an "us/them" framework of thinking about workers?

27

The Rise of Food Democracy

BRIAN HALWEIL

Food has become a target for those who are critical of the domination of multinational corporations in the production and distribution of food. Brian Halweil discusses the new food movements that are developing to make food more "local."

The National Touring Association, one of the largest lobbying groups in Norway, representing walkers, hikers and campers, recently joined forces with the nation's one and only celebrity chef to develop a line of foods made from indigenous ingredients to stock the country's network of camping huts. For instance, someone staying in a mountain cottage in Jotunheimen National Park would dine on cured reindeer heart, sour cream porridge and small potatoes grown only in those mountain valleys. Sekem, Egypt's largest organic food producer, has developed a line of breads, dried fruits, herbs, sauces and other items made entirely from ingredients grown in the country. The brand is recognized by 70 percent of Egyptians, and sales have doubled each of the last five years.

In Zimbabwe, six women realized that their husbands, who are peanut farmers, were literally getting paid peanuts for their crop while they bought pricey imported peanut butter. These women decided to invest in a grinder and are now producing a popular line of peanut butter from local nuts that sells for 15 percent less than mainstream brands. In Nebraska, in the United States, a group of local farmers got together and opened a farmers' grocery that stocks only foods raised in that state. They found suppliers of bacon and baked beans, sour cream and sauerkraut, and virtually all major grocery items, all from Nebraska.

What ties together these disparate enterprises from around the world? At a time when our food often travels farther than ever before, they are all evidence of "food democracy" erupting from an imperialistic food landscape. At first blush, food democracy may seem a little grandiose—a strange combination of words. But if you doubt the existence of power relations in the realm of food, consider a point made by Frances and Anna Lappé in their book *Hope's Edge* (see *UN Chronicle*, Issue 3, 2001). The typical supermarket contains no fewer than 30,000 items, about half of them produced by ten multinational food and beverage companies, with 117 men and 21 women forming the boards of directors

SOURCE: "The Rise of Food Democracy" by Brian Halweil, *UN Chronicle Online Edition*, vol. XLII, January 2005. Reprinted with permission of United Nations.

of those companies. In other words, although the plethora of products you see at a typical supermarket gives the appearance of abundant choice, much of the variety is more a matter of branding than of true agricultural variety and, rather than coming from thousands of farmers producing different local varieties, they have been globally standardized and selected for maximum profit by just a few powerful executives. Food from far-flung places has become the norm in much of the United States and the rest of the world. The value of international trade in food has tripled since 1961, while the tonnage of food shipped between countries has grown fourfold during a time when populations only doubled. For example, apples in Des Moines supermarkets come from China, even though there are apple orchards in Iowa; potatoes in Lima's supermarkets come from the United States, even though Peru boasts more varieties of potato than any other country.

The long-distance food system offers unprecedented and unparalleled choice to paying consumers—any food, any time, anywhere. At the same time, this astounding choice is laden with contradictions. Ecologist and writer Gary Nabhan wonders "what culinary melodies are being drowned out by the noise of that transnational vending machine," which often runs roughshod over local cuisines, varieties and agriculture. The choice offered by the global vending machine is often illusory, defined by infinite flavouring, packaging and marketing reformulations of largely the same raw ingredients (consider the hundreds of available breakfast cereals). The taste of products that are always available but usually out of season often leaves something to be desired.

Long-distance travel requires more packaging, refrigeration and fuel, and generates huge amounts of waste and pollution. Instead of dealing directly with their neighbours, farmers sell into a remote and complex food chain of which they are a tiny part and are paid accordingly. A whole constellation of relationships within the food shed—between neighbours, between farmers and local processors, between farmers and consumers—is lost in the process. Farmers producing for export often find themselves hungry as they sacrifice the output of their land to feed foreign mouths, while poor urbanites in both the First and Third Worlds find themselves living in neighbourhoods unable to attract most supermarkets and other food shops, and thus without healthy food choices. Products enduring long-distance transport and long-term storage depend on preservatives and additives and encounter all sorts of opportunities for contamination on their journey from farm to plate. The supposed efficiencies of the long-distance chain leave many people malnourished and underserved at both ends of the chain.

The changing nature of our food in many ways signals what the changing global economic structure means for the environment, our health and the tenor of our lives. The quality, taste and vitality of foods are profoundly affected by how and where they are produced and how they arrive at our tables. Food touches us so deeply that threats to local food traditions have sometimes provoked strong, even violent, responses. José Boyé, the French shepherd who smacked his tractor into a McDonald's restaurant to fight what he called "culinary imperialism," is one of the better-known symbols in an ascent global movement to protect and invigorate local food sheds.

It is a movement to restore rural areas, enrich poor nations, return wholesome foods to cities and reconnect suburbanites with their land by reclaiming lawns, abandoned lots and golf courses to use as local farms, orchards and gardens.

Local food is pushing through the cracks in the long-distance food system: rising fuel and transportation costs; the near extinction of family farms; loss of farmland to spreading suburbs; concerns about the quality and safety of food; and the craving for some closer connection to it. Eating local allows people to reclaim the pleasures of face-to-face interactions around food and the security that comes from knowing what one is eating. It might be the best defense against hazards intentionally or unintentionally introduced in the food supply, including E-coli bacteria, genetically modified foods, pesticide residues and bio-warfare agents. In an era of climate change and water shortages, having farmers nearby might be the best hedge against other unexpected shocks. On a more sensual level, locally grown and in-season food served fresh has a definite taste advantage—one of the reasons this movement has attracted the attention of chefs, food critics and discriminating consumers around the world.

The local alternative also offers huge economic opportunities. A study by the New Economics Foundation in London found that every £10 spent at a local food business is worth £25 for that area, compared with just £I4 when the same amount is spent in a supermarket. That is, a pound (or dollar, peso or rupee) spent locally generates nearly twice as much income for the local economy. The farmer buys a drink at the local pub; the pub owner gets a car tune-up at the local mechanic; the mechanic brings a shirt to the local tailor; the tailor buys some bread at the local bakery; the baker buys wheat for bread and fruit for muffins from the local farmer. When these businesses are not locally owned, money leaves the community at every transaction.

This sort of multiplier is perhaps most important in the developing world where the vast majority of people are still employed in agriculture. In West Africa, for example, each SI of new income for a farmer yields an average increase to other workers in the local economy, ranging from $1.96 in Niger to $2.88 in Burkina Faso. No equivalent local increases occur when people spend money on imported foods. While the idea of complete food self-sufficiency may be impractical for rich and poor nations alike, greater self-sufficiency can buffer them against the whims of international markets. To the extent that food production and distribution are relocated in the community under local ownership, more money will circulate in the local community to generate more jobs and income.

But here's what makes these declarations of food independence, despite their small size, so threatening to the agricultural status quo. They are built around certain distinctions—geographic characteristics—that global trade agreements are trying so hard to eliminate. These agreements, whether the European Union Trade Zone or the North American Free Trade Agreement, depend on erasing borders and geographic distinctions.... Multinational food companies that source the cheapest ingredients they can find also depend on erasing these distinctions....

Look around and you can glimpse the change worldwide. Farmers in Hawaii are uprooting their pineapple plantations to sow vegetables in hopes of replacing the imported salads at resorts and hotels. School districts throughout Italy have launched an impressive effort to make sure cafeterias are serving a Mediterranean diet by contracting with nearby farmers. At the rarefied levels of the World Trade Organization, officials are beginning to make room for nations to feed themselves, realizing that this might be the best hope for poor nations that cannot afford to import their sustenance. Even some of the world's biggest food companies are starting to embrace these values, a reality that raises some unsettling questions and awesome opportunities for local food advocates. Recently, officials at both Sysco—the world's largest food-service provider—and Kaiser Permanente—the largest health care provider in the United States—declared their dependence on small local farmers for certain products they cannot get anywhere else. These changes will unfold in a million different ways, but the general path will look familiar. Farmers will plant a greater diversity of crops. Less will be shipped as bulk commodity and more will be packaged, canned and prepared to be sold nearby. Small food businesses will emerge to do this work, governments will encourage new businesses, and shoppers seeking pleasure and reassurance will eat deliberately and inquire about the origins of their food. Communities worldwide all possess the capacity to regain this control and this makes the simple idea of eating local so powerful. These communities have a choice, and they are choosing instead to eat here.

KEY CONCEPTS

democracy

multinational corporation

DISCUSSION QUESTIONS

1. What does Halweil mean by food democracy? Do you see any evidence of this in your community?
2. Perhaps you were surprised by the dominance of a few multinational corporations in your local supermarket. What processes have produced this, and how is globalization part of this phenomenon?

28

Why Migration Matters

KHALID KOSER

Khalid Koser identifies the contemporary forces at work in shaping the patterns of international migration that are easy to witness.

Migration has always mattered—but today it matters more than ever before. The increasing importance of migration derives from its growing scale and its widening global reach, but also from a number of new dynamics. These include the feminization of migration, the growth of so-called irregular migration, and migration's inextricable linkages with globalization in terms of economic growth, development, and security. Climate change, moreover, is certain to raise migration still higher on nations' and international institutions' policy agendas.

The history of migration begins with humanity's very origins in the Rift Valley of Africa. It was from there that Homo sapiens emerged about 120,000 years ago, subsequently migrating across Africa, through the Middle East to Europe and Central and South Asia, and finally to the New World, reaching the Bering Strait about 20,000 years ago. Then, in the ancient world, Greek colonization and Roman expansion depended in migration; significant movements of people were also associated with the Mesopotamian, Incan, Indus, and Zhou empires. Later we see major migrations such as those involving the Vikings along the shorelines of the Atlantic and the North Sea, and the Crusaders to the Holy Land.

In more recent history—in other words, in the past two or three centuries—it is possible to discern, according to migration historian Robin Cohen, a series of major migration periods or events. In the eighteenth and nineteenth centuries, one of the most prominent migration events was the forced transportation of slaves. About 12 million people were taken, mainly from West Africa, to the New World (and also, in lesser numbers, across the Indian Ocean and the Mediterranean Sea). One of the reasons this migration is considered so important, other than its scale, is that it still resonates for descendants of slaves and for African Americans in particular. After slavery's collapse, indentured laborers from China, India, and Japan moved overseas in significant numbers—1.5 million from India alone—to work the plantations of the European powers.

SOURCE: Reprinted with permission from *Current History* magazine (April 2009).

European expansion, especially during the nineteenth century, brought about large-scale voluntary migration away from Europe, particularly to the colonies of settlement, dominions, and the Americas. The great mercantile powers—Britain, the Netherlands, Spain, and France—all promoted settlement of their nationals abroad, not just workers but also peasants, dissident soldiers, convicts, and orphans. Migration associated with expansion largely came to an end with the rise of anti-colonial movements toward the end of the nineteenth century, and indeed over the next decades some significant reverse flows back to Europe occurred, for example of the so-called *pieds noirs* to France.

The next period of migration was marked by the rise of the United States as an industrial power. Between the 1850s and the Great Depression of the 1930s, millions of workers fled the stagnant economies and repressive political regimes of northern, southern, and eastern Europe and moved to the United States. (Many fled the Irish famine as well.) Some 12 million of these migrants landed at Ellis Island in New York Harbor. Opportunities for work in the United States also attracted large numbers of Chinese migrants in the first wave of the so-called Chinese diaspora, during the last 50 years of the nineteenth century.

The next major period of migration came after World War II, when the booming postwar economies in Europe, North America, and Australia needed labor. This was the era when, for example, many Turkish migrants arrived to work in Germany and many North Africans went to France and Belgium. It was also the period when, between 1945 and 1972, about one million Britons migrated to Australia as so-called "Ten Pound Poms" under an assisted passage scheme. During the same era but in other parts of the world, decolonization continued to have an impact on migration, most significantly in the movement of millions of Hindus and Muslims after the partition of India in 1947, and of Jews and Palestinians after the creation of Israel.

By the late 1970s, and in part as a consequence of the 1973 oil crisis, the international migrant labor boom had ended in Europe, though in the United States it continued into the 1990s. Now, with the global economy's momentum shifting decisively to Asia, labor migration on that continent has grown heavily, and it is still growing. How much longer this will be true, given the current global financial crisis, is a matter open to debate.

MORE AND MORE

As even this (inevitably selective) overview of international migration's history should make clear, large movements of people have always been associated with significant global events like revolutions, wars, and the rise and fall of empires; with epochal changes like economic expansion; and with enduring challenges like conflict, persecution, and dispossession. Nevertheless, one reason to argue that migration matters more today than ever before is sheer numbers. If we define an international migrant as a person who stays outside his usual country of residence for at least one year, there are about 200 million such migrants worldwide. This is roughly equivalent to the population of the fifth-most populous country on earth, Brazil. In fact, one in every 35 people in the world today is an international migrant.

Of course, a less dramatic way to express this statistic is to say that only 3 percent of the world's population is composed of international migrants. (In migration, statistics are often used to alarm rather than to inform.) And it is also worth noting that internal migration is a far more significant phenomenon than is international migration (China alone has at least 130 million internal migrants). Still, the world total of international migrants has more than doubled in just 25 years; about 25 million were added in just the first 5 years of the twenty-first century.

And international migration affects many more people than just those who migrate. According to Stephen Castles and Mark Miller, authors of the influential volume *The Age of Migration*, "There can be few people in either industrialized or less developed countries today who do not have personal experience of migration and its effects; this universal experience has become the hallmark of the age of migration." In host countries, migrants' contributions are felt keenly in social, cultural, and economic spheres. Throughout the world, people of different national origins, who speak different languages and practice different customs and religions, are coming into unprecedented contact with each other. For some this is a threat, for others opportunity.

Migration is also a far more global process than ever before, as migrants today travel both from and to all of the world's regions. In 2005 (the most recent year for which global data are available) there were about 60 million international migrants in Europe, 44 million in Asia, 41 million in North America, 16 million in Africa, and 6 million each in Latin America and Australia. A significant portion of the world's migrants—about 35 million—lived in the United States. The Russian Federation was the second-largest host country for migrants, with about 13 million living there. Following in the ranking were Germany, Ukraine, and India, each with between 6 million and 7 million migrants.

It is much harder to say which countries migrants come from, largely because origin countries tend not to keep count of how many of their nationals are living abroad. It has been estimated that at least 35 million Chinese currently live outside their country, along with 20 million Indians and 8 million Filipinos. But in fact the traditional distinctions among migrants' countries of origin, transit, and destination have become increasingly blurred. Today almost every country in the world fulfills all three roles—migrants leave them, pass through them, and head for them.

A WORLD OF REASONS

The reasons for the recent rise in international migration and its widening global reach are complex. The factors include growing global disparities in development, democracy, and demography; in some parts of the world, job shortages that will be exacerbated by the current economic downturn; the segmentation of labor markets in high-income economies, a situation that attracts migrant workers to so-called "3D" jobs (dirty, difficult, or dangerous); revolutions in communications and transportation, which result in more people than ever before knowing about life elsewhere and having the ability to travel there; migration networks that allow existing migrant and ethnic communities to act as magnets for future migration; and a robust migration industry, including migrant smugglers and human traffickers, that profits from international migration.

In addition to being bigger, international migration today is also a more complex phenomenon than it has been in the past, as people of all ages and types move for a wide variety of reasons. For example, child migration appears to be on the increase around the world. Migrants with few skills working "3D" jobs make important contributions to the global economy, but so do highly skilled migrants and students. Some people move away from their home countries permanently, but an increasing proportion moves only temporarily, or circulates between countries. And though an important legal distinction can be made between people who move for work purposes and those who flee conflict and persecution, members of the two groups sometimes move together in so-called "mixed flows."

One trend of particular note is that women's representation among migrants has increased rapidly, starting in the 1960s and accelerating in the 1990s. Very nearly half the world's authorized migrants in 2005 were women, and more female than male authorized migrants resided in Europe, North America, Latin America and the Caribbean, the states of the former Soviet Union, and Oceania. What is more, whereas women have traditionally migrated to join their partners, an increasing proportion who migrate today do so independently. Indeed, they are often primary breadwinners for families that they leave behind.

A number of reasons help explain why women comprise an increasing proportion of the world's migrants. One is that global demand for foreign labor, especially in more developed countries, is becoming increasingly gender-selective. That is, more jobs are available in the fields typically staffed by women—services, health care, and entertainment. Second, an increasing number of countries have extended the right of family reunion to migrants, allowing them to be joined by their spouses and children. Third, in some countries of origin, changes in gender relations mean that women enjoy more freedom than previously to migrate independently. Finally, in trends especially evident in Asia, there has been growth in migration of women for domestic work (this is sometimes called the "maid trade"); in organized migration for marriage (with the women sometimes referred to as "mail-order brides"); and in the trafficking of women, above all into the sex industry.

MOST IRREGULAR

Another defining characteristic of the new global migration is the growth of irregular migration and the rapid rise of this phenomenon in policy agents. Indeed, of all the categories of international migrants, none attracts as much attention or divides opinion as consistently as irregular migrants—people often described as "illegal," "undocumented," or "unauthorized."

Almost by definition, irregular migration defies enumeration (although most commentators believe that its scale is increasing). A commonly cited estimate holds that there are around 40 million irregular migrants worldwide, of whom perhaps one-third are in the United States. There are between 3.5 million and 5 million irregular migrants in the Russian Federation, and perhaps 5 million in Europe. Each year, an estimated 2.5 million to 4 million migrants are thought to cross international borders without authorization.

One reason that it is difficult to count irregular migrants is that even this single category covers people in a range of different situations. It includes migrants who enter or remain in a country without authorization; those who are smuggled or trafficked across an international border; those who seek asylum, are not granted it, and then fail to observe a deportation order; and people who circumvent immigration controls, for example by arranging bogus marriages or fake adoptions.

What is more, an individual migrant's status can change—often rapidly. A migrant can enter a country in an irregular fashion but then regularize her status, for example by applying for asylum or entering a regulation program. Conversely, a migrant can enter regularly then become irregular by working without a permit or overstaying a visa. In Australia, for example, British citizens who have stayed beyond the expiration of their visas account for by far the largest number of irregular migrants.

THE RICH GET RICHER

International migration matters more today than ever because of its new dimensions and dynamics, but even more because of its increased impact—on the global economy, on international politics, and on society. Three impacts are particularly worth noting: international migration's contribution to the global economy; the significance of migration and security.

Kodak, Atlantic Records, RCA, NBC, Google, Intel, Hotmail, Sun, Microsoft, Yahoo, eBay—all these U.S. firms were founded or cofounded by migrants. It has been estimated that international migrants make a net contribution to the U.S. economy of $60 billion, and that half of the scientists, engineers, and holders of Ph.D. degrees in the United States were born overseas. It is often suggested (though this is hard to substantiate) that migrants are worth more to the British economy than is North Sea oil. Worldwide migrant labor is thought to earn at least $20 trillion. In some of the Gulf States, migrants comprise 90 percent of the labor force.

Such a selection of facts and figures can suggest a number of conclusions about international migration's significance for the global economy. First, migrants are often among the most dynamic and entrepreneurial members of society. This has always been the case. In many ways the history of U.S. economic growth is the history of migrants: Andrew Carnegie (steel), Adolphus Busch (beer), Samuel Goldwyn (movies), and Helena Rubenstein (cosmetics) were all migrants. Second, migrants fill labor market gaps both at the top end and the bottom end—a notion commonly captured in the phrase "migrants do the work that natives are either unable or unwilling to do." Third, the significance of migrant labor varies across countries but more importantly across economic sectors. In the majority of advanced economies, migrant workers are overrepresented in agriculture, construction, heavy industry, manufacturing, and services—especially food, hospitality, and domestic services. (It is precisely these sectors that the global financial crisis is currently hitting hardest.) Finally, migrant workers contribute significantly more to national economies than they take away (through, for example, pensions and welfare benefits). That is, migrants tend to be young and they tend to work.

This last conclusion explains why migration is increasingly considered one possible response to the demographic crisis that affects increasing numbers of advanced economies (though it does not affect the United States yet). In a number of wealthy countries, a diminishing workforce supports an expanding retired population, and a mismatch results between taxes that are paid into pension and related programs and the payments that those programs must make. Importing youthful workers in the form of migrants appears at first to be a solution—but for two reasons, it turns out to be only a short-term response. First, migrants themselves age and eventually retire. Second, recent research indicates that, within a generation, migrants adapt their fertility rates to those that prevail in the countries where they settle. In other words, it would not take long for migrants to exacerbate rather than relieve a demographic crisis.

THE POOR GET RICHER

International migration does not affect only the economies of countries to which migrants travel—it also strongly affects the economies of countries from which migrants depart, especially in the realm of development in poorer countries. The World Bank estimates that each year migrants worldwide send home about $300 billion. This amounts to triple the value of official development assistance, and is the second-largest source of external funding for developing countries after foreign direct investment. The most important recipient countries for remittances are India ($27 billion), China ($26 billion), Mexico ($25 billion), and the Philippines ($17 billion). The top countries from which remittances are dispatched are the United States ($42 billion), Saudi Arabia ($16 billion), Switzerland ($14 billion), and Germany ($12 billion).

The impact of remittances on development is hotly debated, and to an extent the impact depends on who receives the money and how it is spent. It is indisputable that remittances can lift individuals and families out of poverty: annual household incomes in Somaliland are doubled by remittances. Where remittances are spent on community projects such as wells and schools, as is often the case in Mexico, they also have a wider benefit. And remittances make a significant contribution to gross domestic product (GDP) at the national level, comprising for example 37 percent of GDP in Tonga and 27 percent in both Jordan and Lesotho.

Most experts emphasize that remittances should not be viewed as a substitute for official development assistance. One reason is that remittances are private monies, and thus it is difficult to influence how they are spent or invested. Also, remittances fluctuate over time, as is now becoming apparent in the context of the global economic crisis. Finally, it has been suggested that remittances can generate a "culture of migration," encouraging further migration, and even provide a disincentive to work where families come to expect money from abroad. It has to be said, even so, that the net impact of remittances in developing countries is positive.

International migration, moreover, can contribute to development through other means than remittances. For example, it can relieve pressure on the labor markets in countries from which migrants originate, reducing competition and unemployment. Indonesia and the Philippines are examples of countries that deliberately export labor for this reason (as well as to obtain remittance income). In addition, migrants can contribute to their home countries when they return by using their savings and the new skills they have acquired—although the impact they can have really depends on the extent to which necessary infrastructure is in place for them to realize their potential.

At the same time, however, international migration can undermine development through so-called "brain drain." This term describes a situation in which skills that are already in short supply in a country depart that country through migration. Brain drain is a particular problem in sub-Saharan Africa's health sector, as significant numbers of African doctors and nurses work in the United Kingdom and elsewhere in Europe. Not only does brain drain deprive a country of skills that are in high demand—it also undermines that country's investment in the education and training of its own nationals.

SAFETY FIRST

A third impact of international migration—one that perhaps more than any other explains why it has risen toward the top of policy agendas—is the perception that migration constitutes a heightened security issue in the era after 9/11. Discussions of this issue often revolve around irregular migration—which, in public and policy discourses, is frequently associated with the risk of terrorism, the spread of infectious diseases, and criminality.

Such associations are certainly fair in some cases. A strong link, for example, has been established between irregular migrants from Morocco, Algeria, and Syria and the Madrid bombings of March 2004. For the vast majority of irregular migrants, however, the associations are not fair. Irregular migrants are often assigned bad intentions without any substantiation. Misrepresenting evidence can criminalize and demonize all irregular migrants, encourage them to remain underground—and divert attention from those irregular migrants who actually are criminals and should be prosecuted, as well as those who are suffering from disease and should receive treatment.

Irregular migration is indeed associated with risks, but not with the risks most commonly identified. One legitimate risk is irregular migration's threat to the exercise of sovereignty. States have a sovereign right to control who crosses their borders and remains on their territory, and irregular migration challenges this right. Where irregular migration involves corruption and organized crime, it can also become a threat to public security. This is particularly the case when illegal entry is facilitated by migrant smugglers and human traffickers, or when criminal gangs compete for control of migrants' labor after they have arrived.

When irregular migration results in competition for scarce jobs, this can generate xenophobic sentiments within host populations. Importantly, these sentiments are often directed not only at migrants with irregular status but also at established migrants, refugees, and ethnic minorities. When irregular migration receives a great deal of media attention, it can also undermine public confidence in the integrity and effectiveness of a state's migration and asylum policies.

In addition, irregular migration can undermine the "human security" of migrants themselves. The harm done to migrants by irregular migration is often underestimated—in fact, irregular migration can be very dangerous. A large number of people die each year trying to cross land and sea borders while avoiding detection by the authorities. It has been estimated, for example, that as many as 2,000 migrants die each year trying to cross the Mediterranean from Africa to Europe, and that about 400 Mexicans die annually trying to cross the border into the United States.

People who enter a country or remain in it without authorization are often at risk of exploitation by employers and landlords. Female migrants with irregular status, because they are confronted with gender-based discrimination, are often obliged to accept the most menial jobs in the informal sector, and they may face specific health-related risks, including exposure to HIV/AIDS. Such can be the level of human rights abuses involved in contemporary human trafficking that some commentators have compared it to the slave trade.

Migrants with irregular status are often unwilling to seek redress from authorities because they fear arrest and deportation. For the same reason, they do not always make use of public services to which they are entitled, such as emergency health care. In most countries, they are also barred from using the full range of services available both to citizens and to migrants with regular status. In such situations, already hard-pressed nongovernmental organizations, religious bodies, and other civil society institutions are obliged to provide assistance, at times compromising their own legality.

IN HARD TIMES

What might the future of international migration look like? Tentatively at least, the implications of the current global economic crisis for migration are beginning to emerge. Already a slowdown in the movement of people at a worldwide level has been reported, albeit with significant regional and national variations, and this appears to be largely a result of declining job opportunities in destination countries. The economic sectors in which migrants tend to be overrepresented have been hit first; as a consequence migrant workers around the world are being laid off in substantial numbers.

Interestingly, it appears that most workers are nevertheless not returning home, choosing instead to stay and look for new jobs. Those entitled to draw on social welfare systems can be expected to do so, thus reducing their net positive impact on national economies. (It remains to be seen whether national economic stimulus packages, such as the one recently enacted in the United States, will help migrant workers get back to work.) Scattered cases of xenophobia have been reported around the world, as anxious natives increasingly fear labor competition from migrant workers.

In the last quarter of 2008 remittances slowed down. Some project that in 2009 remittances, for the first time in decades, may shrink. Moreover, changes in exchange rates mean that even if the volume of remittances remains stable, their net value to recipients may decrease. These looming trends hold worrying implications for households, communities, and even national economies in poor countries.

Our experience of previous economic downturns and financial crises—including the Great Depression, the oil crisis of the early 1970s, and the Asian, Russian, and Latin American financial crises between 1997 and 2000—tells us that such crises' impact on international migration is relatively short-lived and that migration trends soon rebound. Few experts are predicting that the current economic crisis will fundamentally alter overall trends toward increased international migration and its growing global reach.

HOT IN HERE

In the longer term, what will affect migration patterns and processes far more than any financial crisis is climate change. One commonly cited prediction holds that 200 million people will be forced to move as a result of climate change by 2050, although other projections range from 50 million to a startling 1 billion people moving during this century.

The relationship that will develop between climate change and migration appears complex and unpredictable. One type of variable is in climate change events themselves—a distinction is usually made between slow-onset events like rising sea levels and rapid-onset events like hurricanes and tsunamis. In addition, migration is only one of a number of possible responses to most climate change events. Protective measures such as erecting sea walls may reduce the impact. Societies throughout history have adapted to climate change by altering their agricultural and settlement practices.

Global warming, moreover, will make some places better able to support larger populations, as growing seasons are extended, frost risks reduced, and new crops sustained. Where migration does take place, it is difficult to predict whether the movement will mainly be internal or cross-border, or temporary or permanent. And finally, the relationship between climate change and migration may turn out to be indirect. For example, people may flee conflicts that arise over scarce resources in arid areas, rather than flee desertification itself.

Notwithstanding the considerable uncertainty, a consensus has emerged that, within the next 10 years, climate-related international migration will become observably more frequent, and the scale of overall international migration will increase significantly. Such migration will add still further complexity to the migration situation, as the new migrants will largely defy current classifications.

One immediately contentious issue is whether people who cross borders as a result of the effects of climate change should be defined as "climate refugees" or "climate migrants." The former conveys the fact that at least some people will literally need to seek refuge from the impacts of climate change, will find themselves in situations as desperate as those of other refugees, and will deserve international assistance and protection. But the current definition of a refugee in international law does not extend to people fleeing environmental pressures, and few states are willing to amend the law. Equally, the description "climate migrant" underestimates the involuntariness of the movement, and opens up the possibility for such people to be labeled and dealt with as irregular migrants.

Another legal challenge arises with the prospect of the total submergence by rising sea levels of low-lying island states such as the Maldives—namely, how to categorize people who no longer have a state. Will their national flags be lowered outside UN headquarters in New York, and will they be granted citizenship in another country?

The complexities of responding to climate-related movements of people illustrate a more general point, that new responses are required to international migration as it grows in scale and complexity. Most of the legal frameworks and international institutions established to govern migration were established at the end of World War II, in response to a migration reality very different from that existing today, and as a result new categories of migrants are falling into gaps in protection. New actors have also emerged in international migration, including most importantly the corporate sector, and they have very little representation in migration policy decisions at the moment.

Perhaps most fundamentally, a shift in attitude is required, away from the notion that migration can be controlled, focusing instead on trying to manage migration and maximize its benefits.

KEY CONCEPTS

assimilation immigration migration

DISCUSSION QUESTIONS

1. What are the social changes that have resulted in the migration patterns that Koser identifies?
2. How is contemporary migration shaped by the process of globalization?
3. Ask yourself under what conditions you would migrate (permanently) to another country? How are the conditions you noted related to the contemporary patterns of migration that Koser identifies?

Applying Sociological Knowledge: An Exercise for Students

Take a look at the clothes in your closet. Where are they made? Do some research online into the living environment in some of these countries. Go to the CIA World Fact Book (at www.cia.gov). Using the links to these different countries, check under "Economy," "People," and "Government" to answer these questions:

1. What is the *life expectancy* of people in this nation?
2. What is the *infant mortality rate*?
3. What percent of people live below the *poverty line*?
4. What is the unemployment rate?
5. What percentage of household income is held by the highest and lowest income groups?
6. What transnational issues (see the bottom of the page) does the nation face?

Having answered these questions, what would you now say about *global stratification*?

PART X

Race and Ethnicity

29

The Souls of Black Folk

W. E. B. DUBOIS

W. E. B. DuBois, the first African American Ph.D. from Harvard University, is a classic sociological analyst. In this well-known essay, he develops the idea that African Americans have a "double consciousness"—one that they must develop as a protective strategy to understand how Whites see them. Originally writing this essay in 1903, DuBois also reflects on the long struggle for African American freedom.

Between me and the other world there is ever an unasked question: unasked by some through feelings of delicacy; by others through the difficulty of rightly framing it. All, nevertheless, flutter round it. They approach me in a half-hesitant sort of way, eye me curiously or compassionately, and then, instead of saying directly, How does it feel to be a problem? they say, I know an excellent colored man in my town; or, I fought at Mechanicsville; or, Do not these Southern outrages make your blood boil? At these I smile, or am interested, or reduce the boiling to a simmer as the occasion may require. To the real question, How does it feel to be a problem? I answer seldom a word....

After the Egyptian and Indian, the Greek and Roman, the Teuton and Mongolian, the Negro is a sort of seventh son, born with a veil, and gifted with second-sight in this American world,—a world which yields him no true self-consciousness, but only lets him see himself through the revelation of the other world. It is a peculiar sensation, this double-consciousness, this sense of always looking at one's self through the eyes of others, of measuring one's soul by the tape of a world that looks on in amused contempt and pity. One ever feels his twoness,—an American, a Negro; two souls, two thoughts, two unreconciled strivings; two warring ideals in one dark body, whose dogged strength alone keeps it from being torn asunder.

The history of the American Negro is the history of this strife—this longing to attain self-conscious manhood, to merge his double self into a better and truer self. In this merging he wishes neither of the older selves to be lost. He would not Africanize America, for America has too much to teach the world and Africa. He would not bleach his Negro soul in a flood of white Americanism, for he knows that Negro blood has a message for the world. He simply wishes to

SOURCE: W. E. B. DuBois, *The Souls of Black Folk*, edited and with an introduction by Donald B. Gibson (New York: Penguin, 1989), pp. 3–12.

make it possible for a man to be both a Negro and an American, without being cursed and spit upon by his fellows, without having the doors of Opportunity closed roughly in his face.

This, then, is the end of his striving: to be a co-worker in the kingdom of culture, to escape both death and isolation, to husband and use his best powers and his latent genius. These powers of body and mind have in the past been strangely wasted, dispersed, or forgotten. The shadow of a mighty Negro past flits through the tale of Ethiopia the Shadowy and of Egypt the Sphinx. Throughout history, the powers of single black men flash here and there like falling stars, and die sometimes before the world has rightly gauged their brightness. Here in America, in the few days since Emancipation, the black man's turning hither and thither in hesitant and doubtful striving has often made his very strength to lose effectiveness, to seem like absence of power, like weakness. And yet it is not weakness—it is the contradiction of double aims. The double-aimed struggle of the black artisan—on the one hand to escape white contempt for a nation of mere hewers of wood and drawers of water, and on the other hand to plough and nail and dig for a poverty-stricken horde—could only result in making him a poor craftsman, for he had but half a heart in either cause. By the poverty and ignorance of his people, the Negro minister or doctor was tempted toward quackery and demagogy; and by the criticism of the other world, toward ideals that made him ashamed of his lowly tasks. The would-be black *savant* was confronted by the paradox that the knowledge people needed was a twice-told tale to his white neighbors, while the knowledge which would teach the white world was Greek to his own flesh and blood. The innate love of harmony and beauty that set the ruder souls of his people a-dancing and a-singing raised but confusion and doubt in the soul of the black artist; for the beauty revealed to him was the soul-beauty of a race which his larger audience despised, and he could not articulate the message of another people. This waste of double aims, this seeking to satisfy two unreconciled ideals, has wrought sad havoc with the courage and faith and deeds of ten thousand thousand people,—has sent them often wooing false gods and invoking false means of salvation, and at times has even seemed about to make them ashamed of themselves....

The Nation has not yet found peace from its sins; the freedman has not yet found in freedom his promised land. Whatever of good may have come in these years of change, the shadow of a deep disappointment rests upon the Negro people—a disappointment all the more bitter because the unattained ideal was unbounded save by the simple ignorance of a lowly people....

Merely a concrete test of the underlying principles of the great republic is the Negro Problem, and the spiritual striving of the freedmen's sons is the travail of souls whose burden is almost beyond the measure of their strength, but who bear it in the name of an historic race, in the name of this the land of their fathers' fathers and in the name of human opportunity....

KEY CONCEPTS

caste system

double consciousness

DISCUSSION QUESTIONS

1. What does DuBois mean by "double consciousness," and how does this affect how African American people see themselves and others?
2. In the contemporary world, what examples do you see that black people are still defined as "a problem," as DuBois notes? How does this affect the black experience?

30

Toward a Framework for Understanding Forces That Contribute to or Reinforce Racial Inequality

WILLIAM JULIUS WILSON

William Julius Wilson identifies the social structural processes that shape racial inequality. His analysis shows how outcomes, such as poverty and unemployment, are shaped by forces that go beyond individual attitudes and behaviors.

UNDERSTANDING THE IMPACT OF STRUCTURAL FORCES

Two types of structural forces contribute directly to racial group outcomes such as differences in poverty and employment rate: social acts and social processes. "Social acts" refers to the behavior of individuals within society. Examples of social acts are stereotyping; stigmatization; discrimination in hiring, job promotions, housing, and admission to educational institutions—as well as exclusion from unions, employers' associations, and clubs—when any of these are the act of an individual or group exercising power over others.

SOURCE: Toward a Framework for Understanding Forces That Contribute to or Reinforce Racial Inequality, by William Julius Wilson, from *Race and Social Problems* (2009), vol. 1, pp. 3–11. Reprinted with permission of Springer Science and Business Media.

"Social processes" refers to the "machinery" of society that exists to promote ongoing relations among members of the larger group. Examples of social processes that contribute directly to racial group outcomes include laws, policies, and institutional practices that exclude people on the basis of race or ethnicity. These range from explicit arrangements such as Jim Crow segregation laws and voting restrictions to more subtle institutional processes, such as school tracking that purports to be academic but often reproduces traditional segregation, racial profiling by police that purports to be about public safety but focuses solely on minorities, and redlining by banks that purports to be about sound fiscal policy but results in the exclusion of people of color from home ownership. In all of these cases, ideologies about group differences are embedded in organizational arrangements.

However, many social observers who are sensitive to and often outraged by the direct forces of racism, such as discrimination and segregation, have paid far less attention to those political and economic forces that *indirectly* contribute to racial inequality. I have in mind political actions that have an impact on racial group outcomes, even though they are not explicitly designed or publicly discussed as matters involving race, as well as impersonal economic forces that reinforce longstanding forms of racial inequality. These structural forces are classified as indirect because they are mediated by the racial groups' position in the system of social stratification (the extent to which the members of a group occupy positions of power, influence, privilege, and prestige). In other words, economic changes and political decisions may have a greater adverse impact on some groups than on others simply because the former are more vulnerable as a consequence of their position in the social stratification system....

Take, for instance, impersonal economic forces, which sharply increased joblessness and declining real wages among many poor African Americans in the last several decades. As with all other Americans, the economic fate of African Americans is inextricably connected with the structure and functioning of a much broader, globally influenced modern economy. In recent years, the growth and spread of new technologies and the growing internationalization of economic activity have changed the relative demand for different types of workers. The wedding of emerging technologies and international competition has eroded the basic institutions of the mass production system and eradicated related jobs in manufacturing in the United States. In the last several decades, almost all of the improvements in productivity have been associated with technology and human capital, thereby drastically reducing the importance of physical capital and natural resources. The changes in technology that are producing new jobs are making many others obsolete.

Although these trends tend to benefit highly educated or highly skilled workers, they have contributed to the growing threat of job displacement and eroding wages for unskilled workers. This development is particularly problematic for African Americans who have a much higher proportion of workers in low-skilled jobs than whites....

The workplace has been revolutionized by technological changes that range from mechanical development like robotics, to advances in information technology like computers and the internet. While even educated workers are struggling to keep pace with technological changes, lower-skilled workers with less education are falling behind with the increased use of information-based technologies and computers and face the growing threat of job displacement in certain industries. To illustrate, in 1962 the employment-to-population ratio—the percentage of adults who are employed—was 52.5% for those with less than a high school diploma, but by 1990 it had plummeted to 37.0%. By 2006 it rebounded slightly to 43.2%, possibly because of the influx of low-skilled Latino immigrants in low-wage service-sector jobs.

In the new global economy, highly educated, well-trained men and women are in demand, as illustrated most dramatically in the sharp differences in employment experiences among men. Compared to men with lower levels of education, college-educated men spend more time working, not less. The shift in the demand for labor is especially devastating for those low-skilled workers whose incorporation into the mainstream economy has been marginal or recent. Even before the economic restructuring of the nation's economy, low-skilled African Americans were at the end of the employment line, often the last to be hired and the first to be let go.

The computer revolution is a major reason for the shift in the demand for skilled workers. Even "unskilled" jobs such as fast food service require employees to work with computerized systems, even though they are not considered skilled workers. Whereas only one-quarter of U.S. workers directly used a computer on their jobs in 1984, by 2003 that figure had risen to more than half (56.1%) the workforce....

The shift in the United States away from low-skilled workers can also be related to the growing internationalization of economic activity, including increased trade with countries that have large numbers of low-skilled, low-wage workers. Two developments facilitated the growth in global economic activity: (1) advances in information and communication technologies, which enabled companies to shift work to areas around the world where wages for unskilled work are much lower than in the "first world;" and (2) the expansion of free trade, which reduced the price of imports and raised the output of export industries. But increasing imports that compete with labor-intensive industries (e.g., apparel, textile, toys, footwear, and some manufacturing) hurts unskilled labor.

Since the late 1960s international trade has accounted for an increasing share of the U.S. economy, and, beginning in the early 1980s, imports of manufactured goods from developing countries have soared. According to economic theory, the expansion of trade with countries that have a large proportion of relatively unskilled labor will result in downward pressure on the wages of low-skilled Americans because of the lower prices of the goods those foreign workers produce. Because of the concentration of low-skilled black workers in vulnerable labor-intensive industries (e.g., 40% of textile workers are African American even though blacks are only about 13% of the general population; this overrepresentation is typical in many low-skill industries), developments in international trade are likely to further exacerbate their declining labor market experiences.

Note that the sharp decline in the relative demand for low-skilled labor has had a more adverse effect on blacks than on whites in the United States because a substantially larger proportion of African Americans are unskilled. Indeed, the disproportionate percentage of unskilled African Americans is one of the legacies of historic racial subjugation. Black mobility in the economy was severely impeded by job discrimination as well as failing segregated public schools, where per-capita expenditures to educate African American children were far below amounts provided for white public schools. While the more educated and highly trained African Americans, like their counterparts among other racial groups, have very likely benefited from the shifts in labor demand, those with lesser skills have suffered....

The economic situation for many African Americans has now been further weakened because they tend not only to reside in communities that have higher jobless rates and lower employment growth—for example, places like Detroit or Philadelphia—but also they lack access to areas of higher employment growth....

The growing suburbanization of jobs means that labor markets today are mainly regional, and long commutes in automobiles are common among blue-collar as well as white-collar workers. For those who cannot afford to own, operate, and insure a private automobile, the commute between inner-city neighborhoods and suburban job locations becomes a Herculean task. For example, Boston welfare recipients found that only 14% of the entry-level jobs in the fast-growth areas of the Boston metropolitan region could be accessed via public transit in less than 1 hour. And in the Atlanta metropolitan area, fewer than one-half of the entry level jobs are located within a quarter mile of a public transit systems. To make matters worse, many inner-city residents lack information about suburban job opportunities. In the segregated inner-city ghettos, the breakdown of the informal job information network magnifies the problems of *job spatial mismatch*—the notion that work and people are located in two different places.

Although racial discrimination and segregation exacerbate the labor-market problems of the low-skilled African Americans, many of these problems are currently driven by shifts in the economy. Between 1947 and the early 1970s, all income groups in America experienced economic advancement. In fact, poor families enjoyed higher growth in annual real income than did other families. In the early 1970s, however, this pattern began to change. American families in higher income groups, especially those in the top 20%, continued to enjoy steady income gains (adjusted for inflation), while those in the lowest 40% experienced declining or stagnating incomes. This growing disparity in income, which continued through the mid-1990s, was related to a slowdown in productivity growth and the resulting downward pressure on wages....

More than any other group, low-skilled workers depend upon a strong economy, particularly a sustained tight labor market—that is, one in which there are ample jobs for all applicants. In a slack labor market—a labor market with high unemployment—employers can afford to be more selective in recruiting and granting promotions. With fewer jobs to award, they can inflate job requirements, pursuing workers with college degrees, for example, in jobs that have traditionally been associated with high school-level education. In such an economic climate, discrimination rises and disadvantaged minorities, especially those with low levels of literacy, suffer disproportionately.

Conversely, in a tight labor market job vacancies are numerous, unemployment is of short duration, and wages are higher. Moreover, in a tight labor market the labor force expands because increased job opportunities not only reduce unemployment but also draw in workers who previously dropped out of the labor force altogether during a slack labor market period. Thus, in a tight labor market the status of all workers—including disadvantaged minorities—improves....

Undoubtedly, if the robust economy could have been extended for several more years, rather than coming to an abrupt halt in 2001, joblessness and concentrated poverty in inner cities would have declined even more. Nonetheless, many people concerned about poverty and rising inequality have noted that productivity and economic growth are only part of the picture.

Thanks to the Clinton-era economic boom, in the latter 1990s there were signs that America's rising economic inequality that began in the early 1970s was finally in remission. Nonetheless, worrisome questions were raised by many observers at that time: Will this new economy eventually produce the sort of progress that prevailed in the two and a half decades prior to 1970—a pattern in which a rising tide did indeed lift all boats? Or would the government's social and economic policies prevent us from duplicating this prolonged pattern of broadly equal economic gains? In other words, the future of ordinary families, especially poor working families, depends a great deal on how the government decides to react to changes in the economy, and often this reaction has a profound effect on racial outcomes....

During Bill Clinton's 8 years in office, redistribution measures were taken to increase the minimum wage. But the George W. Bush administration halted increases in the minimum wage for nearly a decade, until the Democrats regained control of Congress in 2006 and voted to again increase the minimum wage in 2007. All of these political acts contributed to the decline in real wages experienced by the working poor. Because people of color are disproportionately represented among the working poor, these political acts have reinforced their position in the bottom rungs of the racial stratification ladder. In short, in terms of structural factors that contribute to racial inequality, there are indeed nonracial political forces that definitely have to be taken into account....

KEY CONCEPTS

race social process social act

DISCUSSION QUESTIONS

1. Explain what Wilson means by social processes, and identify one such process that has contributed to persistent racial inequality.
2. What social policy changes does Wilson's argument suggest for alleviating racial inequality?

31

Racial Formation

MICHAEL OMI AND HOWARD WINANT

Omi and Winant coined the term racial formation *to signify the process by which groups come to be defined as races. They emphasize that race is a socially constructed concept, even though it has significant and lasting effects.*

In 1982–83, Susie Guillory Phipps unsuccessfully sued the Louisiana Bureau of Vital Records to change her racial classification from black to white. The descendant of an 18th-century white planter and a black slave, Phipps was designated "black" in her birth certificate in accordance with a 1970 state law which declared anyone with at least 1/32nd "Negro blood" to be black.

The Phipps case raised intriguing questions about the concept of race, its meaning in contemporary society, and its use (and abuse) in public policy. Assistant Attorney General Ron Davis defended the law by pointing out that some type of racial classification was necessary to comply with federal recordkeeping requirements and to facilitate programs for the prevention of genetic diseases. Phipps's attorney, Brian Begue, argued that the assignment of racial categories on birth certificates was unconstitutional and that the 1/32nd designation was inaccurate. He called on a retired Tulane University professor who cited research indicating that most Louisiana whites have at least 1/20th "Negro" ancestry.

In the end, Phipps lost. The court upheld the state's right to classify and quantify racial identity....

Phipps's problematic racial identity, and her effort to resolve it through state action, is in many ways a parable of America's unsolved racial dilemma. It illustrates the difficulties of defining race and assigning individuals or groups to racial categories. It shows how the racial legacies of the past—slavery and bigotry—continue to shape the present. It reveals both the deep involvement of the state in the organization and interpretation of race, and the inadequacy of state institutions to carry out these functions. It demonstrates how deeply Americans both as individuals and as a civilization are shaped, and indeed haunted, by race.

Having lived her whole life thinking that she was white, Phipps suddenly discovers that by legal definition she is not. In U.S. society, such an event is

SOURCE: Racial Formation, by Michael Omi and Howard Winant, from *Racial Formation in the United States*, From the 1960s to the 1990s, 2e, pp. 53–61, Rutledge 1994. Reprinted with permission of Taylor & Francis Group.

indeed catastrophic. But if she is not white, of what race is she? The state claims that she is black, based on its rules of classification ... and another state agency, the court, upholds this judgment. But despite these classificatory standards which have imposed an either-or logic on racial identity, Phipps will not in fact "change color." Unlike what would have happened during slavery times if one's claim to whiteness was successfully challenged, we can assume that despite the outcome of her legal challenge, Phipps will remain in most of the social relationships she had occupied before the trial. Her socialization, her familial and friendship networks, her cultural orientation, will not change. She will simply have to wrestle with her newly acquired "hybridized" condition. She will have to confront the "Other" within.

The designation of racial categories and the determination of racial identity is no simple task. For centuries, this question has precipitated intense debates and conflicts, particularly in the U.S.—disputes over natural and legal rights, over the distribution of resources, and indeed, over who shall live and who shall die.

A crucial dimension of the Phipps case is that it illustrates the inadequacy of claims that race is a mere matter of variations in human physiognomy, that it is simply a matter of skin color. But if race cannot be understood in this manner, how can it be understood? We cannot fully hope to address this topic—no less than the meaning of race, its role in society, and the forces which shape it—in one [article], nor indeed in one book. Our goal in this [article], however, is far from modest: we wish to offer at least the outlines of a theory of race and racism.

WHAT IS RACE?

There is a continuous temptation to think of race as an essence, as something fixed, concrete, and objective. And there is also an opposite temptation: to imagine race as a mere illusion, a purely ideological construct which some ideal non-racist social order would eliminate. It is necessary to challenge both these positions, to disrupt and reframe the rigid and bipolar manner in which they are posed and debated, and to transcend the presumably irreconcilable relationship between them.

The effort must be made to understand race as an unstable and "decentered" complex of social meanings constantly being transformed by political struggle. With this in mind, let us propose a definition: race is a concept which signifies and symbolizes social conflicts and interests by referring to different types of human bodies. Although the concept of race invokes biologically based human characteristics (so-called "phenotypes"), selection of these particular human features for purposes of racial signification is always and necessarily a social and historical process. In contrast to the other major distinction of this type, that of gender, there is no biological basis for distinguishing among human groups along the lines of race.... Indeed, the categories employed to differentiate among human groups along racial lines reveal themselves, upon serious examination, to be at best imprecise, and at worst completely arbitrary.

If the concept of race is so nebulous, can we not dispense with it? Can we not "do without" race, at least in the "enlightened" present? This question has been posed often, and with greater frequency in recent years.... An affirmative answer would of course present obvious practical difficulties: it is rather difficult to jettison widely held beliefs, beliefs which moreover are central to everyone's identity and understanding of the social world. So the attempt to banish the concept as an archaism is at best counterintuitive. But a deeper difficulty, we believe, is inherent in the very formulation of this schema, in its way of posing race as a *problem,* a misconception left over from the past, and suitable now only for the dustbin of history.

A more effective starting point is the recognition that, despite its uncertainties and contradictions, the concept of race continues to play a fundamental role in structuring and representing the social world. The task for theory is to explain this situation. It is to avoid both the utopian framework which sees race as an illusion we can somehow "get beyond," and also the essentialist formulation which sees race as something objective and fixed, a biological datum. Thus we should think of race as an element of social structure rather than as an irregularity within it; we should see race as a dimension of human representation rather than an illusion. These perspectives inform the theoretical approach we call racial formation.

RACIAL FORMATION

We define *racial formation* as the sociohistorical process by which racial categories are created, inhabited, transformed, and destroyed. Our attempt to elaborate a theory of racial formation will proceed in two steps. First, we argue that racial formation is a process of historically situated projects in which human bodies and social structures are represented and organized. Next we link racial formation to the evolution of hegemony, the way in which society is organized and ruled. Such an approach, we believe, can facilitate understanding of a whole range of contemporary controversies and dilemmas involving race, including the nature of racism, the relationship of race to other forms of differences, inequalities, and oppression such as sexism and nationalism, and the dilemmas of racial identity today.

From a racial formation perspective, race is a matter of both social structure and cultural representation. Too often, the attempt is made to understand race simply or primarily in terms of only one of these two analytical dimensions.... For example, efforts to explain racial inequality as a purely social structural phenomenon are unable to account for the origins, patterning, and transformation of racial difference.

Conversely, many examinations of racial difference—understood as a matter of cultural attributes à la ethnicity theory, or as a society-wide signification system, à la some poststructuralist accounts—cannot comprehend such structural phenomena as racial stratification in the labor market or patterns of residential segregation.

An alternative approach is to think of racial formation processes as occurring through a linkage between structure and representation. *Racial projects do the ideological "work" of making these links. A racial project is simultaneously an interpretation, representation, or explanation of racial dynamics, and an effort to recognize and redistribute resources along particular racial lines.* Racial projects connect what race means in a particular discursive practice and the ways in which both social structures and everyday experiences are racially *organized*, based upon that meaning. Let us consider this proposition, first in terms of large-scale or macro-level social processes, and then in terms of other dimensions of the racial formation process.

Racial Formation as a Macro-Level Social Process

To interpret the meaning of race is to frame it social structurally. Consider, for example, this statement by Charles Murray on welfare reform:

> My proposal for dealing with the racial issue in social welfare is to repeal every bit of legislation and reverse every court decision that in any way requires, recommends, or awards differential treatment according to race, and thereby put us back onto the track that we left in 1965. We may argue about the appropriate limits of government intervention in trying to enforce the ideal, but at least it should be possible to identify the ideal: Race is not a morally admissible reason for treating one person differently from another. Period....

Here there is a partial but significant analysis of the meaning of race: it is not a morally valid basis upon which to treat people "differently from one another." We may notice, someone's race, but we cannot act upon that awareness. We must act in a "color-blind" fashion. This analysis of the meaning of race is immediately linked to a specific conception of the role of race in the social structure: it can play no part in government action, save in "the enforcement of the ideal." No state policy can legitimately require, recommend, or award different status according to race. This example can be classified as a particular type of racial project in the present-day U.S.—a "neoconservative" one.

Conversely, *to recognize the racial dimension in social structure is to interpret the meaning of race.* Consider the following statement by the late Supreme Court Justice Thurgood Marshall on minority "set-aside" programs:

> A profound difference separates governmental actions that themselves are racist, and governmental actions that seek to remedy the effects of prior racism or to prevent neutral government activity from perpetuating the effects of such racism....

Here the focus is on the racial dimensions of social structure—in this case of state activity and policy. The argument is that state actions in the past and present have treated people in very different ways according to their race, and thus the government cannot retreat from its policy responsibilities in this area. It cannot suddenly declare itself "color-blind" without in fact perpetuating the same type of differential, racist treatment.... Thus, race continues to signify difference and structure inequality. Here, racialized social structure is immediately linked to an interpretation of the meaning of race. This example too can be classified as a particular type of racial project in the present-day U.S.—a "liberal" one.

To be sure, such political labels as "neoconservative" or "liberal" cannot fully capture the complexity of racial projects, tor these are always multiply determined, politically contested, and deeply shaped by their historical context. Thus encapsulated within the neoconservative example cited here are certain egalitarian commitments which derive from a previous historical context in which they played a very different role, and which are rearticulated in neoconservative racial discourse precisely to oppose a more open-ended, more capacious conception of the meaning of equality. Similarly, in the liberal example, Justice Marshall recognizes that the contemporary state, which was formerly the architect of segregation and the chief enforcer of racial difference, has a tendency to reproduce those patterns of inequality in a new guise. Thus he admonishes it (in dissent, significantly) to fulfill its responsibilities to uphold a robust conception of equality. These particular instances, then, demonstrate how racial projects are always concretely framed, and thus are always contested and unstable. The social structures they uphold or attack, and the representations of race they articulate, are never invented out of the air, but exist in a definite historical context, having descended from previous conflicts. This contestation appears to be permanent in respect to race.

These two examples of contemporary racial projects are drawn from mainstream political debate; they may be characterized as center-right and center-left expressions of contemporary racial politics.... We can, however, expand the discussion of racial formation processes far beyond these familiar examples. In fact, we can identify racial projects in at least three other analytical dimensions: first, the political spectrum can be broadened to include radical projects, on both the left and right, as well as along other political axes. Second, analysis of racial projects can take place not only at the macro-level of racial policy-making, state activity, and collective action, but also at the micro-level of everyday experience. Third, the concept of racial projects can be applied across historical time, to identity racial formation dynamics in the past....

KEY CONCEPTS

one drop rule

racial formation

DISCUSSION QUESTIONS

1. What do Omi and Winant mean by *racial formation*? Do you see evidence of this process for groups other than African Americans?
2. Under what historical conditions does a given group come to be defined as a race? Is this happening for Latinos and Latinas?

32

Color-Blind Privilege

The Social and Political Functions of Erasing the Color Line in Post Race America

CHARLES A. GALLAGHER

Charles A. Gallagher discusses the problem of a color-blind approach to race and race relations in this country. By denying race as a structural basis for inequality, we fail to recognize the privilege of whiteness. With the blurring of racial lines, white college students lack a clear understanding of how the existing social, political, and economic systems advantage or privilege whites.

INTRODUCTION

An adolescent white male at a bar mitzah wears a FUBU shirt while his white friend preens his tightly set, perfectly braided corn rows. A black model dressed in yachting attire peddles a New England yuppie boating look in Nautica advertisements. It is quite unremarkable to observe white, Asian or African-Americans with dyed purple, blond or red hair. White, black and Asian students decorate their bodies with tattoos of Chinese characters and symbols. In cities and suburbs young adults across the color line wear hip-hop clothing and listen to white rapper Eminem and black rapper Jay-Z. A north Georgia branch of the NAACP installs a white biology professor as its president. The music of Jimi Hendrix is used to sell Apple Computers. Du-Rag kits, complete with bandana headscarf and elastic headband, are on sale for $2.95 at hip-hop clothing stores and family centered theme parks like Six Flags. Salsa has replaced ketchup as the best selling condiment in the United States. Companies as diverse as Polo,

SOURCE: Charles A. Gallagher, "Color-Blind Privilege: The Social and Political Functions of Erasing the Color Line in Post Race America." *Race, Gender and Class* 10 (June 2003). Reprinted with permission of the author.

McDonalds, Tommy Hilfiger, Walt Disney World, Master Card, Skechers sneakers, IBM, Giorgio Armani and Neosporin antibiotic ointment have each crafted advertisements that show a balanced, multiracial cast of characters interacting and consuming their products in a post-race, color-blind world....

Americans are constantly bombarded by depictions of race relations in the media which suggest that discriminatory racial barriers have been dismantled. Social and cultural indicators suggest that America is on the verge, or has already become, a truly color-blind nation. National polling data indicate that a majority of whites now believe discrimination against racial minorities no longer exists. A majority of whites believe that blacks have "as good a chance as whites" in procuring housing and employment or achieving middle class status while a 1995 survey of white adults found that a majority of whites (58%) believed that African Americans were better off finding jobs than whites.[1] Much of white America now see a level playing field, while a majority of black Americans see a field which is still quite uneven.... The color-blind or race neutral perspective holds that in an environment where institutional racism and discrimination have been replaced by equal opportunity, one's qualifications, not one's color or ethnicity, should be the mechanism by which upward mobility is achieved. Whites and blacks differ significantly, however, in their support for affirmative action, the perceived fairness of the criminal justice system, the ability to acquire the "American Dream," and the extent to which whites have benefited from past discrimination.[2]

This article examines the social and political functions color-blindness serves for whites in the United States. Drawing on interviews and focus groups with whites from around the country, I argue that color-blindness maintains white privilege by negating racial inequality. Embracing post-race, color-blind perspective provides whites with a degree of psychological comfort by allowing them to imagine that being white or black or brown has no bearing on an individual's or a group's relative place in the socio-economic hierarchy. My interviews included seventeen focus group and thirty individual interviews with whites around the country. While my sample is not representative of the total white population, I used personal contacts and snowball sampling to purposively locate respondents raised in urban, suburban and rural environments. Twelve of the seventeen focus groups were conducted in a university setting, one in a liberal arts college in the Rocky Mountains and the other at a large urban university in the Northeast. Respondents in these focus groups were selected randomly from the student population. The occupational range for my individual interviews was quite eclectic and included a butcher, construction worker, hair stylist, partner in a prestigious corporate law firm, executive secretary, high school principal, bank president from a small town, retail workers, country lawyer and custodial workers. Twelve of the thirty individual interviews were with respondents who were raised in rural and/or agrarian settings. The remaining respondents lived in suburbs of large cities or in urban areas.

What linked this rather disparate group of white individuals together was their belief that race-based privilege had ended. As a majority of my respondents saw it, color-blindness was now the norm in the United States. The illusion of racial equality implicit in the myth of color-blindness was, for many whites, a form of comfort. This aspect of pleasure took the form of political empowerment ("what about whites' rights") and moral gratification from being liberated from "oppressor" charges ("we are not responsible for the past"). The rosy picture that color-blindness presumes about race relations and the satisfying sense that one is part of a period in American history that is morally superior to the racist days of the past is, quite simply, a less stressful and more pleasurable social place for whites to inhabit.

THE NORM OF COLOR-BLINDNESS

The perception among a majority of white Americans that the socio-economic playing field is now level, along with whites' belief that they have purged themselves of overt racist attitudes and behaviors, has made color-blindness the dominant lens through which whites understand contemporary race relations. Color-blindness allows whites to believe that segregation and discrimination are no longer an issue because it is now illegal for individuals to be denied access to housing, public accommodations or jobs because of their race.... Individuals from any racial background can wear hip-hop clothing, listen to rap music (both purchased at Wal-Mart) and root for their favorite, majority black, professional sports team. Within the context of racial symbols that are bought and sold in the market, color-blindness means that one's race has no bearing on who can ... live in an exclusive neighborhood, attend private schools or own a Rolex.

The passive interaction whites have with people of color through the media creates the impression that little, if any, socio-economic difference exists between the races. Research has found that whites who are exposed to images of upper-middle class African Americans ... believe that blacks have the same socio-economic opportunities as whites. Highly visible and successful racial minorities like Secretary of State Colin Powell and National Security Advisor Condoleezza Rice are further proof to white America that the state's efforts to enforce and promote racial equality has been accomplished. Reflecting on the extent to which discrimination is an obstacle to socio-economic advancement and the perception of seeing African-Americans in leadership roles, Tom explained:

> If you look at some prominent black people in society today and I don't really see [racial discrimination]. I don't understand how they can keep bringing this problem onto themselves. If they did what society would want them to I don't see that society is making problems for them. I don't see it.

... The new color-blind ideology does not, however, ignore race; it acknowledges race while ignoring racial hierarchy by taking racially coded styles and products and reducing these symbols to commodities or experiences which whites and racial minorities can purchase and share. It is through such acts of shared consumption that race becomes nothing more than an innocuous cultural signifier. Large corporations have made American culture more homogeneous through the ubiquitousness of fast food, television, and shopping malls but this trend has also created the illusion that we are all the same through consumption. Most adults eat at national fast food chains like McDonalds, shop at mall anchor stores like Sears and J. C. Penney's and watch major league sports, situation comedies or television drama. Defining race only as cultural symbols that are for sale allows whites to experience and view race as nothing more than a benign cultural marker that has been stripped of all forms of institutional, discriminatory or coercive power. The post-race, color-blind perspective allows whites to imagine that depictions of racial minorities working in high status jobs and consuming the same products, or at least appearing in commercials for products whites desire or consume, is the same as living in a society where color is no longer used to allocate resources or shape group outcomes. By constructing a picture of society where racial harmony is the norm, the color-blind perspective functions to make white privilege invisible while removing from public discussion the need to maintain any social programs that are race-based.

… Starting with the deeply held belief that America is now a meritocracy, whites are able to imagine that the socio-economic success they enjoy relative to racial minorities is a function of individual hard work, determination, thrift and investments in education. The color-blind perspective removes from personal thought and public discussion any taint or suggestion of white supremacy or white guilt while legitimating the existing social, political and economic arrangements which privilege whites. This perspective insinuates that class and culture, and not institutional racism, are responsible for social inequality. Color-blindness allows whites to define themselves as politically progressive and racially tolerant as they proclaim their adherence to a belief system that does not see or judge individuals by the "color of their skin." This perspective ignores, as Ruth Frankenberg puts it, how whiteness is a "location of structural advantage societies structured in racial dominance."[3] Frankenberg uses the term "color and power evasiveness" rather than color-blindness to convey how the ability to ignore race by members of the dominant group reflects a position of power and privilege. Color-blindness hides white privilege behind a mask of assumed meritocracy while rendering invisible the institutional arrangements that perpetuate racial inequality. The veneer of equality implied in color-blindness allows whites to present their place in the racialized social structure as one that was earned.

Given this norm of color-blindness it was not surprising that respondents in this study believed that using race to promote group interests was a form of racism.

Joe, a student in his early twenties from a working class background, was quite adamant that the opportunity structure in the United States did not favor one racial group over another.

> I mean, I think that the black person of our age has as much opportunity as me, maybe he didn't have the same guidance and that might hurt him. But I mean, he's got the same opportunities that I do to go to school, maybe even more, to get more money. I can't get any aid … I think that blacks have the same opportunities as whites nowadays and I think it's old hat.

Not only does Joe believe that young blacks and whites have similar educational experiences and opportunity but it is his contention that blacks are more likely or able to receive money for higher education. The idea that race matters in any way, according to Joe, is anachronistic; it is "old hat" in a color-blind society to blame one's shortcomings on something as irrelevant as race.

Believing and acting as if America is now color-blind allows whites to imagine a society where institutional racism no longer exists and racial barriers to upward mobility have been removed. The use of group identity to challenge the existing racial order by making demands for the amelioration of racial inequities is viewed as racist because such claims violate the belief that we are a nation that recognizes the rights of individuals not rights demanded by groups. Sam, an upper middle class respondent in his 20's, draws on a pre- and post-civil rights framework to explain racial opportunity among his peers:

> I guess I can understand in my parents' generation. My parents are older, my dad is almost 60 and my mother is in her mid 50's, ok? But the kids I'm going to school with, the minorities I'm going to school with, I don't think they should use racism as an excuse for not getting a job. Maybe their parents, sure, I mean they were discriminated against. But these kids have every opportunity that I do to do well.

In one generation, as Sam sees it, the color line has been erased. Like Sam's view that opportunity structure is open there is, according to Tara, a reason to celebrate the current state of race relations:

> I mean, like you are not the only people that have been persecuted—I mean, yea, you have been, but so has every group. I mean if there's any time to be black in America it's now.

Seeing society as race-neutral serves to decouple past historical practices and social conditions from present day racial inequality as was the case for a number of respondents who pointed out that job discrimination had ended. Michelle was quite direct in her perception that the labor market is now free of discrimination stating that "I don't think people hire and fire because someone is black and white now." Ken also believed that discrimination in hiring did not occur since racial minorities now have legal recourse if discrimination occurs:

> I think that pretty much we got past that point as far as jobs. I think people realize that you really can't discriminate that way because you will end up losing ... because you will have a lawsuit against you.

... The logic inherent in the color-blind approach is circular; since race no longer shapes life chances in a color-blind world there is no need to take race into account when discussing differences in outcomes between racial groups. This approach erases America's racial hierarchy by implying that social, economic and political power and mobility are equally shared among all racial groups. Ignoring the extent or ways in which race shapes life chances validates whites' social location in the existing racial hierarchy while legitimating the political and economic arrangements which perpetuate and reproduce racial inequality and privilege....

THE COST OF RACIALIZED PLEASURES

Being able to ignore or being oblivious to the ways in which almost all whites are privileged in a society cleaved on race has a number of implications. Whites derive pleasure in being told that the current system for allocating resources is fair and equitable. Creating and internalizing a color-blind view of race relations reflects how the dominant group is able to use the mass media, immigration stories of upward mobility, rags-to-riches narratives and achievement ideology to make white privilege invisible. Frankenberg argues that whiteness can be "displaced," as is the case with whiteness hiding behind the veil of color-blindness. It can also be made "normative" rather than specifically "racial," as is the case when being white is defined by white respondents as being no different than being black or Asian.[4] Lawrence Bobo and associates have advanced a theory of laissez-faire racism that draws on the color-blind perspective. As whites embrace the equality of opportunity narrative they suggest that

> laissez-faire racism encompasses an ideology that blames blacks themselves for their poorer relative economic standing, seeing it as a function of perceived cultural inferiority. The analysis of the bases of laissez-faire racism underscores two central components: contemporary stereotypes of blacks held by whites, and the denial of societal (structural) responsibility for the conditions in black communities.[5]

As many of my respondents make clear, if the opportunity structure is open ("It doesn't matter what color you are"), there must be something inherently wrong with racial minorities or their culture that explains group level differences....

... [T]he form color-blindness takes as the nation's hegemonic political discourse is a variant of laissez-faire racism. Historian David Roediger contends that in order for the Irish to be absorbed into the white race in the mid-nineteenth century "the imperative to define themselves as whites came from the particular 'public and psychological wages' whiteness offered" these new immigrants.[6] There is still a "wage" to whiteness, that element of ascribed status whites automatically receive because of their membership in the dominant group. But within the framework of color-blindness the imperative has switched from whites overtly defining themselves or their interests as white, to one where they claim that color is irrelevant; being white is the same as being black, yellow, brown or red....

My interviews with whites around the country suggest that in this post-race era of color-blind ideology [Ralph] Ellison's keen observations about race relations need modification. The question now is what are we to make of a young white man from the suburbs who listens to hip-hop, wears baggy hip-hop pants, a baseball cap turned sideways, unlaced sneakers and an oversized shirt emblazoned with a famous NBA player who, far from shouting racial epithets, lists a number of racial minorities as his heroes? It is now possible to define oneself as not being racist because of the clothes you wear, the celebrities you like or the music you listen to while believing that blacks or Latinos are disproportionately poor or over-represented in low pay, dead end jobs because they are part of a debased, culturally deficient group. Having a narrative that smoothes over the cognitive dissonance and oft time schizophrenic dance that whites must do when they navigate race relations is an invaluable source of pleasure.

NOTES

1. The Gallup Organization, "Black/White Relations in the U.S.," *The Gallup Poll Monthly* (June 10, 1997): 1–5; David Shipler, *A Country of Strangers: Blacks and Whites in America* (New York: Vintage, 1998).
2. David Moore, "Americans' Most Important Sources of Information: Local News," *The Gallup Poll Monthly* (September 1995): 2–5; David Moore and Lydia Saad, "No Immediate Signs that Simpson Trial Intensified Racial Animosity," *The Gallup Poll Monthly* (October 1995): 2–5; Kaiser Foundation, *The Four Americas: Government and Social Policy through the Eyes of America's Multi-Racial and Multi-Ethnic Society* (Menlo Park, CA: Kaiser Family Foundation, 1995).
3. O. Ruth Frankenberg, "The Mirage of an Unmarked Whiteness," in *The Making and Unmaking of Whiteness*, eds. Birget Brander Rasmussen, Eric Klineberg, Irene Nexica, and Matt Wray (Durham: Duke University Press, 2001).
4. Ibid., 76.
5. Lawrence Bobo and James R. Kluegel, "Status, Ideology, and Dimensions of Whites' Racial Beliefs and Attitudes: Progress and Stagnation," in *Racial Attitudes in the 1990s: Continuity and Change*, eds. Steven A. Tuch and Jack K. Martin, 95 (Westport, CT: Praeger Publishers, 1997).
6. David Roediger, *The Wages of Whiteness: Race and the Making of the American Working Class*, 137 (New York: Verso Press, 1991).

KEY CONCEPTS

color-blind racism	prejudice	stereotype
post-racial society	racism	white privilege

DISCUSSION QUESTIONS

1. Summarize Gallagher's argument for why a color-blind attitude is still a privileged attitude. What does color-blindness *not* see when viewing race relations in the United States?
2. What is the problem with a generation of individuals who do not judge others by the "color of their skin"? Can the individualistic ideology of color-blindness coexist with a society of racist practices?

33

Subordinating Myth: Latino/a Immigration, Crime, and Exclusion

JAMIE LONGAZEL

Unauthorized immigration (or undocumented workers) has been a heated national discussion, one that indicates the need for reform in our national immigration policies. In this article, the author discusses how immigrants have been framed by perceptions of them as criminals, misrepresenting their actual experience.

INTRODUCTION

Political and popular rhetoric in the contemporary United States has virtually synonymized the terms "immigrant" and "criminal" thus contributing to the passage and implementation of laws at various levels of government that punish immigrants, primarily those from Mexico and other parts of Latin America. Based in large part on assumptions of entrenched immigrant criminality, the U.S.-Mexico border is now staffed with substantially more agents, reinforced with walls and fences, and patrolled with cameras, flood-lights motion sensors, and drones (e.g. Inda 2006; Nevins 2010). Immigration law has become so punitive that it has converged with criminal law to form a hybrid doctrine legal scholars have termed "crimmigration law" (Stumpf 2006; see also Welch 2002; Newton 2008). And many states and local municipalities have gone so far as to implement laws of their own, many of which rely on an "attrition through enforcement" strategy designed to make life so difficult for supposedly criminal immigrants that they ultimately choose to "self deport."...

All the while, a "scholarly consensus" (Lee and Martinez 2009) has been emerging that runs contrary to this conventional wisdom: *Immigrants do not increase crime; they may actually help reduce it.* Scholars conducting research on various levels, amongst diverse immigrant groups, and using multiple methodologies have found negative relationships between immigration and crime in the United States (e.g. Hagan and Palloni 1999; Nielsen et al. 2005; Reid et al. 2005; Sampson et al. 2005; Rumbaut and Ewing 2007; Sampson 2008; Butcher and Morrison Piehl 2008; Lee and Martinez 2009; Ousey and Kubrin 2009). This also holds true for Latina/o immigrants and undocumented immigrants, groups who have been subjected to vilification in recent years (Sampson et al. 2005). In fact, research finds that first generation immigrants are less crime-prone than their second- and third-generation counterparts, suggesting that, if anything, it is *increased "Americanness"* rather than a failure to assimilate that would lead immigrants to crime (e.g. Sampson 2006; Rumbaut and Ewing 2007). "Cities of concentrated immigration," in short, "are some of the safest places around" (Sampson 2008: 30).

To explain what emerges as a glaring contradiction—low rates of immigrant criminality on one hand, hyper-enforcement driven primarily by the assumption that immigrants are crime-prone on the other—this article will discuss the criminalization of Latina/o immigrants as a *subordinating myth*, a social construction based on a false premise that "forcefully contributes to preserving racial [and ethnic] stratification through exclusion ... [by misallocating] political, cultural, and material goods" (Haney López 2010: 1045). The exploitation of immigrant laborers is indeed widely acknowledged as a means by which primarily white elites reap the material benefits produced by immigrants of color (e.g. Estrada et al. 1981; Calavita 1992, 1994; Massey 2007), but what is often overlooked is the extent to which criminalization serves similar purposes. More specifically, by reviewing recent research on Latino/a immigration, crime, and social control in the context of racial/ethnic stratification, this article will reveal how the false premise of entrenched immigrant criminality creates profits, engenders political benefits for elites, degrades minority groups, and controls exploitable populations.

Despite rapidly growing rates of immigrants and refugees arriving in the United States from other global regions (e.g., Semple 2012), my primary focus is on immigrants from Mexico and other Latin American countries, whom I'll refer to throughout as simply "Latino/a immigrants." This group has bore the brunt of immigrant criminalization in recent years and the American public perceives them more negatively in comparison to other groups (Timberlake and Williams 2012)....

RACIAL STRATIFICATION AND IMMIGRATION

Understanding race/ethnicity from the perspective of racial stratification ... entails seeing race/ethnicity "not simply as an emotional eruption, efficient shortcut, cognitive habit, or institutional tic, but as a central means of ordering and rationalizing the distribution of resources, broadly conceived" (Haney López 2010: 104). To that end, racial/ethnic stratification is achieved through *exploitation*—"when people in one social group expropriate a resource produced by members of another social group" (Massey 2007: 6)—and *opportunity hoarding*—where "beneficiaries do not enlist the efforts of outsiders but instead exclude them from access to relevant resources" (Tilly 1998: 91).

Racial stratification theory serves as a useful framework for helping us to understand the contemporary subordination of immigrants. It is widely acknowledged that the history of immigration into the United States, particularly Latino/a immigration, is in many respects a history of exploitation, as this group has historically "formed a reserve labor pool that could be called up as the situation dictated" (Estrada et al. 1981: 110).... And although the exploitive intentions of modern capitalists may be less explicit than those of the past, there is no denying that such practices are widespread today (e.g., Massey 2007). My focus here, however, is on the relevance of hoarding; I seek to explicate how criminalization has served to exclude Latina/o immigrants from access to material, cultural, and political resources.

A useful jumping off point then is Ian Haney López's (2010) recent trenchant analysis, *Post-Racial Racism*, which links the war on crime and attendant mass incarceration to racial hoarding. He notes that punitive crime policies contribute to racial exclusion in principally four ways: through *profit* (i.e. the reemergence of prison labor and the privatization of prisons, both of which have produced massive profits for corporate elites at the expense of communities of color), *politics* (i.e. felon disenfranchisement laws and the political tendency to play to the racial anxieties of whites), *degradation* (i.e. the forging of symbolic linkages between people of color and crime which alter the "cultural meanings tied to racial categories" [1048]), and *population control* (i.e. exerting direct control over minority populations by transforming the prison into a "surrogate ghetto" [Wacquant 2009: 195])....

CRIMINALIZING FOR PROFIT

Accounting for the increased punitiveness in immigration law is the emergence of an "immigration industrial complex" which, much like the pervasive 'prison industrial complex,' is characterized by laws and policies that serve the interests of government and corporate elites (Golash-Boza 2009). Uncovering the workings of privately operated immigration detention facilities puts the immigration industrial complex clearly on display. Currently, just under 400,000 immigrants are detained per year in the United States (Department of Homeland Security 2011), an all-time high, and almost half of these detainees—most of whom are Latina/o—are housed in private facilities (Kirkham 2012). In fact, the private prison industry was pulled out of financial peril in the late 1990s by a surge in immigrant detention (Kirkham 2012). Today, thanks to increased border enforcement and contracts with the federal government, the industry now thrives on immigrant detention....

Immigrants have thus quite literally become the "raw materials" of a now booming for-profit prison industry (see Christie 2000: 87). Under these arrangements, private prisons and their stockholders desire increased incarceration *regardless* of whether it is empirically misguided (Selman and Leighton 2010). This often means lobbying intensely for punitive laws that will assure prisons remain "well-stocked" and thus profitable....

THE POLITICS OF CRIMINALIZATION

Racial threat theorists have long acknowledged that rising minority populations coupled with economic decline tend to provoke anxiety amongst the white majority (Blumer 1958; Quillian 1995). As immigrant populations increase and white workers confront painful economic shifts such as demanufacturing, the stage is thus set for harsh anti-immigrant backlash.... Thus in addition to disenfranchisement, which obviously precludes undocumented immigrants from political participation, potential class-based political allies become formidable foes as white workers are encouraged to point their fingers at working class immigrants as the source of their declining economic and social position....

CRIMINALIZATION AS DEGRADATION

Considering how criminalization contributes to degradation and enables population control unveils how punitive immigration policies play out "on the ground." Frighteningly, scholars have empirically documented how anti-immigrant ordinances stoke resentment and lead to increases in the actual discrimination experienced by Latino/as....

In discussing group degradation as a mechanism of exclusion, Haney López (2010) invokes the concept of microaggression—"stunning, automatic acts of disregard that stem from unconscious attitudes of white superiority and constitute a verification of black inferiority" (Davis 1989: 1576). This concept brings the devastating effects of discrimination more clearly into view: degrading day-to-day encounters breed humiliation and painfully reaffirm one's subordinate status....

CONTROLLING CRIMINALIZED IMMIGRANT POPULATIONS

The criminalization of immigrants also engenders population control to the extent that immigrants find themselves, as Susan Coutin (2007: 9) so eloquently put it in her ethnography of Salvadoran migrants, "physically present but legally absent, existing in a space that is not actually "elsewhere" or beyond borders but that is rather a hidden dimension of social reality." Latina/o immigrants, that is to say, experience "entrapment," as Guillermina Núñez and Josiah Heyman (2009: 354) similarly point out in their study of how migrants (documented and undocumented) "are not just enclosed inside the country as a whole, but are also impeded from moving around locally to access vital resources and join with loved ones."

Much of this restriction stems from precarious legal status and hyperenforcement—what Nicholas De Genova (2002) terms *deportability*—which puts migrants and their family members in a constant state of fear. A recent Center for

American Progress report reveals the astounding portion of immigrants who are understandably anxious about engaging in seemingly safe, mundane tasks such as driving a car (67%) and walking in public (64%). Indeed, most respondents report avoiding public places at all costs, and, when that is not possible, going to great lengths to blend in (García and Keyes 2012)….

CONCLUSION

Beginning with the criminalization of Latino/a immigrants on one hand and scholarly research that refutes connections between immigration and crime on the other, I have discussed in this article how what appears as a glaring contradiction can be understood as a *subordinating myth*. Using racial stratification theory as a guide, I reviewed recent research on Latino/a immigration, crime, and social control as a way to illuminate how criminalization has contributed to exclusion in the areas of profit, politics, degradation, and population control. Specifically, I discussed how private detention facilities and other governmental and corporate elites have capitalized financially on the increased incarceration of predominately Latina/o immigrant populations; I described the political benefits afforded to those who scapegoat Latina/o immigrants as criminals; I drew attention to the degrading features of immigration enforcement; and I highlighted how harsh immigration laws enable population control….

Exposing the myth of immigrant criminality is thus only a first step. As scholars, we must continue to unpack the relationship between criminalization and exclusion. This entails envisioning crime not just as an act an individual chooses to commit, but rather a political tool that serves racial purposes. Such an approach has the potential of building a powerful counter-narrative around issues of immigration and crime that can make the simplistic, empirically inaccurate immigrant-as-criminal narrative appear derisory in comparison. That is to say, if we are to *subordinate the myth* of immigrant criminality, we must place the criminalization of immigrants within a broader historical, racial, and sociological context that illuminates how criminalization contributes to exclusion.

REFERENCES

Blumer, Herbert. 1958. 'Race Prejudice as a Sense of Group Position.' *Pacific Sociological Review* 1:3–7.

Butcher, Kristin F and Anne Morrison Piehl. 2008. 'Crime, Corrections, and California: What Does Immigration Have to Do with It?.' *California Counts* 9(3):1–23.

Calavita, Kitty. 1992. *Inside the State: The Bracero Program*, Immigration, and the I.N.S. New York, NY: Routledge.

Calavita, Kitty. 1994. 'U.S. Immigration and Policy Responses: The Limits of Legislation.' Pp. 55–82 in *Controlling Immigration: A Global Perspective*, edited by

Wayne A. Cornelius, Phillip L. Martin and James Frank Hollifield. Standford, CA: Stanford University Press.

Calavita, Kitty. 2005. *Immigrants at the Margins: Law Race, and Exclusion in Southern Europe*, United Kingdom: Cambridge University Press.

Christie, Nils. 2000. *Crime Control as Industry: Towards Gulags, Western Style* (3rd Edition). New York, NY: Routledge.

Coutin, Susan Bibler. 2007. *Nation of Emigrants: Shifting Boundaries of Citizenship in El Salvador and The United States*. Ithaca, NY: Cornell University Press.

Davis, Peggy C. 1989. 'Law as Microaggression.' *The Yale Law Journal* 98:1559–77.

De Genova, Nicholas P. 2002. 'Migrant "Illegality" and Deportability in Everyday Life.' *Annual Review of Anthropology* 31:419–47.

Department of Homeland Security. 2011. 'Immigration Enforcement Actions: 2010.' *Annual Report*. [Online]. Retrieved on 10 July 2012 from: http://www.dhs.gov/xlibrary/assets/statistics/publications/enforcement-ar-2010.pdf

Estrada, Leobardo F, Chris Garcia, Reynaldo Flores Macis and Lionel Maldonado. 1981. 'Chicanos in the United States.' *Daedalus* 110:103–13.

Garcia, Angela S. and David G. Keyes. 2012. 'Life as an Undocumented Immigrant: How Restrictive Local Immigration Policies Affect Daily Life.' *Center for American Progress*. [Online]. Retrieved on 10 July 2012 from: http://www.americanprogress.org/issues/2012/03/life_as_undocumented.html/

Golash-Boza, Tanya. 2009. 'A Confluence of Interests in Immigration Enforcement: How Politicians, the Media, and Corporations Profit from Immigration Policies Destined to Fail.' *Sociology Compass* 3(2):283–94.

Hagan, John and Alberto Palloni. 1999. 'Sociological Criminology and the Mythology of Hispanic Immigrant Crime.' *Social Problems* 46:617–32.

Haney López, Ian. 2010. 'Post-Racial Racism: Racial Stratification and Mass Incarceration in the Age of Obama.' *California Law Review* 98:1023–74.

Inda, Jonathan Xavier. 2006. *Targeting Immigrants: Government, Technology, and Ethics*. Malden, MA: Blackwell Publishing.

Kirkham, Chris. 2012. 'Private Prisons Profit From Immigration Crackdown, Federal and Local Law Enforcement Partnerships.' *Huffington Post*. [Online]. Retrieved on 10 July 2012 from: http://www.huffingtonpost.com/2012/06/07/private-prisons-immigration-federal-law-enforcement_n_1569219.html

Lee, Matthew and Ramiro Martinez. 2009. 'Immigration Reduces Crime: An Emerging Scholarly Consensus.' Pp. 3–16 in *Sociology of Crime, Law, and Deviation* edited by William McDonald. New York, NY: Elsevier.

Massey, Douglas S. 2007. *Categorically Unequal: The American Stratification System*. New York, NY: Sage.

Nevins, Joseph. 2010. *Operation Gatekeeper and Beyond: The War on "Illegals" and the Remaking of the U.S.-Mexico Boundary*. New York, NY: Routledge.

Newton, Lina. 2008. *Illegal, Alien, or Immigrant: The Politics of Immigration Reform*. New York, NY: New York University Press.

Nielsen, Amy L., Matthew T. Lee and Ramiro Martinez Jr. 2005. 'Integrating Race, Place, and Motive in Social Disorganization Theory: Lessons from a Comparison of Black and Latino Homicide Types in Two Immigrant Destination Cities.' *Criminology* 43:837–72.

Núñez, Guillermina Gina and Josiah McC. Heyman. 2007. 'Entrapment Processes and Immigrant Communities in a Time of Heightened Border Vigilance.' *Human Organization* 66(4):354–65.

Ousey, Graham C. and Charis E. Kubrin. 2009. 'Exploring the Connection Between Immigration and Violent Crime Rates in U.S. Cities, 1980–2000.' *Social Problems* 56(3):447–73.

Quillian, Lincoln. 1995. 'Prejudice as a Response to Perceived Group Threat: Population Composition and Anti-Immigrant and Racial Prejudice in Europe.' *American Sociological Review* 60(4):586–11.

Reid, Lesley Williams, Harold E. Weiss, Robert M. Adelman and Charles Jaret. 2005. 'The Immigration-Crime Relationship: Evidence Across US Metropolitan Areas." *Social Science Research* 34:757–80.

Rumbaut, Rubén G, and Walter A. Ewing. 2007. 'The Myth of Immigrant Criminality and the Paradox of Assimilation: Incarceration Rates among Native and Foreign-Born Men.' *Immigration Policy Center*. [Online]. Retrieved on 10 July 2012 from: http://www.ime.gob.mx/2007/immigrant_criminality.pdf

Sampson, Robert J. 2006. 'Open Doors Don't Invite Criminals: Is Increased Immigration Behind the Drop in Crime?' *New York Times*. [Online], Retrieved on 10 July 2012 from: http://www.nytimes.com/2006/03/11/opinion/11sampson.html

Sampson, Robert J. 2008. 'Rethinking Crime and Immigration.' *Contexts* 7(1):28–33.

Sampson, Robert J, Jeffrey D. Morenoff and Stephen Raudenbush. 2005. 'Social Autonomy of Racial and Ethnic Disparities in Violence.' *American Journal of Public Health* 95:224–32.

Selman, Donna and Paul Leighton. 2010. *Punishment for Sale: Private Prisons*, Big Business, and the Incarceration Binge. Lanham, MD: Rowman and Littlefield.

Semple, Kirk. 2012. 'In a Shift Biggest Wave of Migrants is now Asian.' *New York Times*. [Online]. Retrieved on 10 July 2012 from: http://www.nytimes.com/2012/06/19/us/asians-surpass-hispanics-as-biggest-immigrant-wave.html?_r=1

Stumpf, Juliet P. 2006. 'The Crimmigration Crisis: Immigrants, Crime, and Sovereign Power.' *bepress Legal Series*. [Online]. Retrieved on 10 July 2012 from: http://law.bepress.com/cgi/viewcontent.cgi?article=7625&context=expresso

Tilly, Charles. 1998. *Durable Inequality*, Berkeley, CA: University of California Press.

Timberlake, Jeffery M. and Rhys H. Williams. 2012. 'Stereotypes of U.S. Immigrants from Four Global Regions.' *Social Science Quarterly*. [Online]. Retrieved on 10 July 2012 from: http://onlinelibrary.wiley.com/doi/10.1111/j.1540-6237.2012.00860.x/pdf.

Wacquant, Loïc. 2009. *Punishing the Poor: The Neoliberal Government of Social Insecurity*. Durham, NC: Duke University Press.

Welch, Michael. 2002. *Detained: Immigration Laws and the Expanding I.N.S. Jail Complex*. Philadelphia, PA: Temple University Press.

KEY CONCEPTS

assimilation ethnic group immigration

DISCUSSION QUESTIONS

1. How do the authors see the media as having demonized unauthorized immigration? What evidence have you seen of this?
2. Why are the semantics (that is, words used) surrounding immigration so important? What different connotation is suggested by using the term "illegal alien" versus "unauthorized immigrant?"

Applying Sociological Knowledge: An Exercise for Students

Consider how racial and ethnic groups are portrayed in the media. Take one type of television show (sitcom, sports broadcast, news, or drama) and watch two such shows for a specified period of time (say, one hour each). As you watch, systematically observe how different racial and ethnic groups (including whites) are depicted. You should keep a tally of how many characters from different racial or ethnic groups are shown and how they are portrayed. Then ask yourself how images in the media shape racial stereotypes and group prejudice.

PART XI

Gender

34

The Social Construction of Gender

MARGARET L. ANDERSEN

In this essay, Margaret Andersen outlines the meaning of the phrase "social construction of gender." She discusses the difference between the terms sex *and* gender *and defines sexuality as it relates to both. After a brief discussion of the cultural basis of gender, the essay outlines the difference between conceptualizing gender as a role versus a gendered institutions approach.*

To understand what sociologists mean by the phrase the social construction of *gender*, watch people when they are with young children. "Oh, he's such a boy!" someone might say as he or she watches a 2-year-old child run around a room or shoot various kinds of play guns. "She's so sweet," someone might say while watching a little girl play with her toys. You can also see the social construction of gender by listening to children themselves or watching them play with each other. Boys are more likely to brag and insult other boys (often in joking ways) than are girls; when conflicts arise during children's play, girls are more likely than boys to take action to diffuse the conflict (McCloskey and Coleman, 1992; Miller, Danaber, and Forbes, 1986).

To see the social construction of gender, try to buy a gender-neutral present for a child—that is, one not specifically designed with either boys or girls in mind. You may be surprised how hard this is, since the aisles in toy stores are highly stereotyped by concepts of what boys and girls do and like. Even products such as diapers, kids' shampoos, and bicycles are gender stereotyped. Diapers for boys are packaged in blue boxes; girls' diapers are packaged in pink. Boys wear diapers with blue borders and little animals on them; girls wear diapers with pink borders with flowers. You can continue your observations by thinking about how we describe children's toys. Girls are said to play with dolls; boys play with action figures!

When sociologists refer to the social construction of gender, they are referring to the many different processes by which the expectations associated with being a boy (and later a man) or being a girl (later a woman) are passed on through society. This process pervades society, and it begins the minute a child is born. The exclamation "It's a boy!" or "It's a girl!" in the delivery room sets a course that from

SOURCE: ANDERSEN, MARGARET L., THINKING ABOUT WOMEN: SOCIOLOGICAL PERSPECTIVES ON SEX AND GENDER, 6th Edition, © 2003.
Reprinted by permission of Pearson Education, Inc., Upper Saddle River, NJ.

that moment on influences multiple facets of a person's life. Indeed, with the modern technologies now used during pregnancy, the social construction of gender can begin even before one is born. Parents or grandparents may buy expected children gifts that reflect different images, depending on whether the child will be a boy or a girl. They may choose names that embed gendered meanings or talk about the expected child in ways that are based on different social stereotypes about how boys and girls behave and what they will become. All of these expectations—communicated through parents, peers, the media, schools, religious organizations, and numerous other facets of society—create a concept of what it means to be a "woman" or be a "man." They deeply influence who we become, what others think of us, and the opportunities and choices available to us. The idea of the social construction of gender sees society, not biological sex differences, as the basis for gender identity. To understand this fully, we first need to understand some of the basic concepts associated with the social construction of gender.

SEX, GENDER, AND SEXUALITY

The terms *sex, gender,* and *sexuality* have related, but distinct, meanings within the scholarship on women. Sex refers to the biological identity and is meant to signify the fact that one is either male or female. One's biological sex usually establishes a pattern of gendered expectations, although, … biological sex identity is not always the same as gender identity; nor is biological identity always as clear as this definition implies.

Gender is a social, not biological, concept, referring to the entire array of social patterns that we associate with women and men in society. Being "female" and "male" are biological facts; being a woman or a man is a social and cultural process—one that is constructed through the whole array of social, political, economic, and cultural experiences in a given society. Like race and class, gender is a social construct that establishes, in large measure, one's life chances and directs social relations with others. Sociologists typically distinguish sex and gender to emphasize the social and cultural basis of gender, although this distinction is not always so clear as one might imagine, since gender can even construct our concepts of biological sex identity.

Making this picture even more complex, sexuality refers to a whole constellation of sexual behaviors, identities, meaning systems, and institutional practices that constitute sexual experience within society. This is not so simple a concept as it might appear, since sexuality is neither fixed nor unidimensional in the social experience of diverse groups. Furthermore, sexuality is deeply linked to gender relations in society. Here, it is important to understand that sexuality, sex, and gender are intricately linked social and cultural processes that overlap in establishing women's and men's experiences in society.

Fundamental to each of these concepts is understanding the significance of culture. Sociologists and anthropologists define culture as "the set of definitions of reality held in common by people who share a distinctive way of life" (Kluckhohn, 1962:52). Culture is, in essence, a pattern of expectations about what are appropriate behaviors and beliefs for the members of the society; thus, culture provides prescriptions for social behavior. Culture tells us what we ought to do, what we ought to think, who we ought to be, and what we ought to expect of others....

The cultural basis of gender is apparent especially when we look at different cultural contexts. In most Western cultures, people think of *man* and *woman* as dichotomous categories—that is, separate and opposite, with no overlap between the two. Looking at gender from different cultural viewpoints challenges this assumption, however. Many cultures consider there to be three genders, or even more. Consider the Navaho Indians. In traditional Navaho society, the *berdaches* were those who were anatomically normal men but who were defined as a third gender and were considered to be intersexed. Berdaches married other men. The men they married were not themselves considered to be berdaches; they were defined as ordinary men. Nor were the berdaches or the men they married considered to be homosexuals, as they would be judged by contemporary Western culture....

Another good example for understanding the cultural basis of gender is the *hijras* of India (Nanda, 1998). Hijras are a religious community of men in India who are born as males, but they come to think of themselves as neither men nor women. Like berdaches, they are considered a third gender. Hijras dress as women and may marry other men; typically, they live within a communal subculture. An important thing to note is that hijras are not born so; they choose this way of life. As male adolescents, they have their penises and testicles cut off in an elaborate and prolonged cultural ritual—a rite of passage marking the transition to becoming a hijra....

These examples are good illustrations of the cultural basis of gender. Even within contemporary U.S. society, so-called "gender bending" shows how the dichotomous thinking that defines men and women as "either/or" can be transformed. Cross-dressers, transvestites, and transsexuals illustrate how fluid gender can be and, if one is willing to challenge social convention, how easily gender can be altered. The cultural expectations associated with gender, however, are strong, as one may witness by people's reactions to those who deviate from presumed gender roles....

In different ways and for a variety of reasons, all cultures use gender as a primary category of social relations. The differences we observe between men and women can be attributed largely to these cultural patterns.

THE INSTITUTIONAL BASIS OF GENDER

Understanding the cultural basis for gender requires putting gender into a sociological context. From a sociological perspective, gender is systematically structured in social institutions, meaning that it is deeply embedded in the social structure of society. Gender is created, not just within family or interpersonal relationships (although these are important sources of gender relations), but also within the structure of all major social institutions, including schools, religion, the economy, and the state (i.e., government and other organized systems of authority such as the police and the military). These institutions shape and mold the experiences of us all.

Sociologists define institutions as established patterns of behavior with a particular and recognized purpose; institutions include specific participants who share expectations and act in specific roles, with rights and duties attached to them. Institutions define reality for us insofar as they exist as objective entities in our experience....

Understanding gender in an institutional context means that gender is not just an attribute of individuals; instead, institutions themselves *are gendered*. To say that an institution is gendered means that the whole institution is patterned on specific gendered relationships. That is, gender is "present in the processes, practices, images and ideologies, and distribution of power in the various sectors of social life" (Acker, 1992:567). The concept of a gendered institution was introduced by Joan Acker, a feminist sociologist. Acker uses this concept to explain not just that gender expectations are passed to men and women within institutions, but that the institutions themselves are structured along gendered lines. Gendered institutions are the total pattern of gender relations—stereotypical expectations, interpersonal relationships, and men's and women's different placements in social, economic, and political hierarchies. This is what interests sociologists, and it is what they mean by the social structure of gender relations in society.

Conceptualizing gender in this way is somewhat different from the related concept of gender roles. Sociologists use the concept of social roles to refer to culturally prescribed expectations, duties, and rights that define the relationship between a person in a particular position and the other people with whom she or he interacts. For example, to be a mother is a specific social role with a definable set of expectations, rights, and duties. Persons occupy multiple roles in society; we can think of social roles as linking individuals to social structures. It is through social roles that cultural norms are patterned and learned. Gender roles are the expectations for behavior and attitudes that the culture defines as appropriate for women and men.

The concept of gender is broader than the concept of gender roles. *Gender* refers to the complex social, political, economic, and psychological relations between women and men in society. Gender is part of the social structure—in other words, it is institutionalized in society. *Gender roles* are the patterns through which gender relations are expressed, but our understanding of gender in society cannot be reduced to roles and learned expectations.

The distinction between gender as institutionalized and gender roles is perhaps most clear in thinking about analogous cases—specifically, race and class. Race relations in society are seldom, if ever, thought of in terms of "race roles." Likewise, class inequality is not discussed in terms of "class roles." Doing so would make race and class inequality seem like matters of interpersonal interaction. Although race, class, and gender inequalities are experienced within interpersonal interactions, limiting the analysis of race, class, or gender relations to this level of social interaction individualizes more complex systems of inequality; moreover, restricting the analysis of race, class, or gender to social roles hides the power relations that are embedded in race, class, and gender inequality (Lopata and Thorne, 1978).

Understanding the institutional basis of gender also underscores the interrelationships of gender, race, and class, since all three are part of the institutional framework of society. As a social category, gender intersects with class and race; thus, gender is manifested in different ways, depending on one's location in the race and class system. For example, African American women are more likely than White women to reject gender stereotypes for women, although they are more accepting than White women of stereotypical gender roles for children. Although this seems contradictory, it can be explained by understanding that African American women may reject the dominant culture's view while also hoping their children can attain some of the privileges of the dominant group (Dugger, 1988).

Institutional analyses of gender emphasize that gender, like race and class, is a part of the social experience of us all—not just of women. Gender is just as important in the formation of men's experiences as it is in women's (Messner, 1998). From a sociological perspective, class, race, and gender relations are systemically structured in social institutions, meaning that class, race, and gender relations shape the experiences of all. Sociologists do not see gender simply as a psychological attribute, although that is one dimension of gender relations in society. In addition to the psychological significance of gender, gender relations are part of the institutionalized patterns in society. Understanding gender, as well as class and race, is central to the study of any social institution or situation. Understanding gender in terms of social structure indicates that social change is not just a matter of individual will—that if we changed our minds, gender would disappear. Transformation of gender inequality requires change both in consciousness and in social institutions....

REFERENCES

Acker, Joan. 1992. "Gendered Institutions: From Sex Roles to Gendered Institutions." *Contemporary Sociology* 21 (September): 565–569.

Dugger, Karen. 1988. "The Social Location of Black and White Women's Attitudes." *Gender & Society* 2 (December): 425–448.

Kluckhohn, C. 1962. *Culture and Behavior.* New York: Free Press.

Lopata, Helene Z., and Barne Thorne. 1978. "On the Term 'Sex Roles.'" *Signs* 3 (Spring): 718–721.

McCloskey, Laura A., and Lerita M. Coleman. 1992. "Difference Without Dominance Children's Talk in Mixed- and Same-Sex Dyads." *Sex Roles* 27 (September): 241–258.

Messner, Michael A. 1998. "The Limits of 'The Male Sex Role' An Analysis of the Men's Liberation and Men's Rights Movements' Discourse." *Gender & Society* 12 (June): 255–276.

Miller, D., D. Danaber, and D. Forbes. 1986. "Sex-related Strategies for Coping with Interpersonal Conflict in Children Five and Seven." *Developmental Psychology* 22: 543–548.

Nanda, Serena. 1998. *Neither Man Nor Woman: The Hijras of India.* Belmont, CA: Wadsworth.

KEY CONCEPTS

feminism
gender
gendered institution
gender role
gender socialization
sex
sexism

DISCUSSION QUESTIONS

1. Walk through a baby store. Can you easily identify products for girls and for boys? Could you easily purchase gender-neutral clothing?
2. Consider an occupation that is traditionally men's work or traditionally women's work. What happens when a member of the other gender works in that field? What stereotypes and derogatory assumptions do we make about a woman working in a man's occupation or a man working in a woman's occupation?

35

What It Means to Be Gendered Me: Life on the Boundaries of a Dichotomous System

BETSY LUCAL

Betsy Lucal is a female sociology professor who is often mistaken for a man. In this piece she describes her experiences interacting with others in public spaces where her gender is misinterpreted. The social construction of gender is clear in the cues she provides by way of her appearance and in how they are received. Her analysis concludes that the "two and only two" gender model is problematic.

SOURCE: Betsy Lucal, Contexts, vol. 4, issue 4, pp. 28–32, Fall 2005 by SAGE Publications. Reprinted by permission of SAGE Publications.

I understood the concept of "doing gender" (West and Zimmerman 1987) long before I became a sociologist. I have been living with the consequences of inappropriate "gender display" (Goffman 1976; West and Zimmerman 1987) for as long as I can remember.

My daily experiences are a testament to the rigidity of gender in our society, to the real implications of "two and only two" when it comes to sex and gender categories (Garfinkel 1967; Kessler and McKenna 1978). Each day, I experience the consequences that our gender system has for my identity and interactions. I am a woman who has been called "Sir" so many times that I no longer even hesitate to assume that it is being directed at me. I am a woman whose use of public rest rooms regularly causes reactions ranging from confused stares to confrontations over what a man is doing in the women's room. I regularly enact a variety of practices either to minimize the need for others to know my gender or to deal with their misattributions.

I am the embodiment of Lorber's (1994) ostensibly paradoxical assertion that the "gender bending" I engage in actually might serve to preserve and perpetuate gender categories. As a feminist who sees gender rebellion as a significant part of her contribution to the dismantling of sexism, I find this possibility disheartening.

In this article, I examine how my experiences both support and contradict Lorber's (1994) argument using my own experiences to illustrate and reflect on the social construction of gender. My analysis offers a discussion of the consequences of gender for people who do not follow the rules as well as an examination of the possible implications of the existence of people like me for the gender system itself. Ultimately, I show how life on the boundaries of gender affects me and how my life, and the lives of others who make similar decisions about their participation in the gender system, has the potential to subvert gender.

Because this article analyzes my experiences as a woman who often is mistaken for a man, my focus is on the social construction of gender for women. My assumption is that, given the gendered nature of the gendering process itself, men's experiences of this phenomenon might well be different from women's.

THE SOCIAL CONSTRUCTION OF GENDER

It is now widely accepted that gender is a social construction, that sex and gender are distinct, and that gender is something all of us "do." This conceptualization of gender can be traced to Garfinkel's (1967) ethnomethodological study of "Agnes."[1] In this analysis, Garfinkel examined the issues facing a male who wished to pass as, and eventually become, a woman. Unlike individuals who perform gender in culturally expected ways, Agnes could not take her gender for granted and always was in danger of failing to pass as a woman (Zimmerman 1992).

This approach was extended by Kessler and McKenna (1978) and codified in the classic "Doing Gender" by West and Zimmerman (1987). The social constructionist approach has been developed most notably by Lorber (1994, 1996)....

Taken as a whole, this work provides a number of insights into the social processes of gender, showing how gender(ing) is, in fact, a process.

We apply gender labels for a variety of reasons; for example, an individual's gender cues our interactions with her or him. Successful social relations require all participants to present, monitor, and interpret gender displays (Martin 1998; West and Zimmerman 1987). We have, according to Lorber, "no social place for a person who is neither woman nor man" (1994, 96); that is, we do not know how to interact with such a person. There is, for example, no way of addressing such a person that does not rely on making an assumption about the person's gender ("Sir" or "Ma'am"). In this context, gender is "omnirelevant" (West and Zimmerman 1987). Also, given the sometimes fractious nature of interactions between men and women, it might be particularly important for women to know the gender of the strangers they encounter; do the women need to be wary, or can they relax (Devor 1989)?

According to Kessler and McKenna (1978), each time we encounter a new person, we make a gender attribution. In most cases, this is not difficult. We learn how to read people's genders by learning which traits culturally signify each gender and by learning rules that enable us to classify individuals with a wide range of gender presentations into two and only two gender categories. As Weston observed, "Gendered traits are called attributes for a reason: People attribute traits to others. No one possesses them. Traits are the product of evaluation" (1996, 21). The fact that most people use the same traits and rules in presenting genders makes it easier for us to attribute genders to them.

We also assume that we can place each individual into one of two mutually exclusive categories in this binary system. As Bem (1993) notes, we have a polarized view of gender; there are two groups that are seen as polar opposites. Although there is "no rule for deciding 'male' or 'female' that will always work" and no attributes "that always and without exception are true of only one gender" (Kessler and McKenna 1978, 158, 1), we operate under the assumption that there are such rules and attributes.

Kessler and McKenna's analysis revealed that the fundamental schema for gender attribution is to "See someone as female only when you cannot see [the person] as male" (1978, 158). Individuals basically are assumed to be male/men until proven otherwise, that is, until some obvious marker of conventional femininity is noted. In other words, the default reading of a nonfeminine person is that she or he is male; people who do not deliberately mark themselves as feminine are taken to be men. Devor attributed this tendency to the operation of gender in a patriarchal context: "Women must mark themselves as 'other'," whereas on the other hand, "few cues [are required] to identify maleness" (1989, 152). As with language, masculine forms are taken as the genetically human; femininity requires that something be added. Femininity "must constantly reassure its audience by a willing demonstration of difference" (Brownmiller 1984, 15)....

Not only do we rely on our social skills in attributing genders to others, but we also use our skills to present our own genders to them. The roots of this understanding of how gender operates lie in Goffman's (1959) analysis of the "presentation of self in everyday life," elaborated later in his work on "gender display" (Goffman 1976). From this perspective, gender is a performance, "a stylized repetition of acts" (Butler 1990, 140, emphasis removed). Gender display refers to "conventionalized portrayals" of social correlates of gender (Goffman 1976). These displays are culturally established sets of behaviors, appearances, mannerisms, and other cues that we have learned to associate with members of a particular gender.

In determining the gender of each person we encounter and in presenting our genders to others, we rely extensively on these gender displays. Our bodies and their adornments provide us with "texts" for reading a person's gender (Bordo 1993). As Lorber noted, "Without the deliberate use of gendered clothing, hairstyles, jewelry, and cosmetics, women and men would look far more alike" (1994, 18–19). Myhre summarized the markers of femininity as "having longish hair; wearing makeup, skirts, jewelry, and high heels; walking with a wiggle; having little or no observable body hair; and being in general soft, rounded (but not too rounded), and sweet-smelling" (1995, 135). (Note that these descriptions comprise a Western conceptualization of gender.) Devor identified "mannerisms, language, facial expressions, dress, and a lack of feminine adornment" (1989, x) as factors that contribute to women being mistaken for men.

A person uses gender display to lead others to make attributions regarding her or his gender, regardless of whether the presented gender corresponds to the person's sex or gender self-identity. Because gender is a social construction, there may be differences among one's sex, gender self-identity (the gender the individual identifies as), presented identity (the gender the person is presenting), and perceived identity (the gender others attribute to the person).[2] For example, a person can be female without being socially identified as a woman, and a male person can appear socially as a woman. Using a feminine gender display, a man can present the identity of a woman and, if the display is successful, be perceived as a woman.

But these processes also mean that a person who fails to establish a gendered appearance that corresponds to the person's gender faces challenges to her or his identity and status. First, the gender nonconformist must find a way in which to construct an identity in a society that denies her or him any legitimacy (Bem 1993). A person is likely to want to define herself or himself as "normal" in the face of cultural evidence to the contrary. Second, the individual also must deal with other people's challenges to identity and status—deciding how to respond, what such reactions to their appearance mean, and so forth.

Because our appearances, mannerisms, and so forth constantly are being read as part of our gender display, we do gender whether we intend to or not. For example, a woman athlete, particularly one participating in a nonfeminine sport such as basketball, might deliberately keep her hair long to show that, despite actions that suggest otherwise, she is a "real" (i.e., feminine) woman. But we also do gender in less conscious ways such as when a man takes up more space when sitting than a woman does. In fact, in a society so clearly organized around gender, as ours is, there is no way in which to not do gender (Lorber 1994).

Given our cultural rules for identifying gender (i.e., that there are only two and that masculinity is assumed in the absence of evidence to the contrary), a person who does not do gender appropriately is placed not into a third category but rather into the one with which her or his gender display seems most closely to fit; that is, if a man appears to be a woman, then he will be categorized as "woman," not as something else. Even if a person does not want to do gender or would like to do a gender other than the two recognized by our society, other people will, in effect, do gender for that person by placing her or him in one and only one of the two available categories. We cannot escape doing gender or, more specifically, doing one of two genders....

People who follow the norms of gender can take their genders for granted....

People who, for whatever reasons, do not adhere to the rules, risk gender misattribution and any interactional consequences that might result from this misidentification. What are the consequences of misattribution for social interaction? When must misattribution be minimized? What will one do to minimize such mistakes?...

For me, the social processes and structures of gender mean that, in the context of our culture, my appearance will be read as masculine. Given the common conflation of sex and gender, I will be assumed to be a male. Because of the two-and-only-two genders rule, I will be classified, perhaps more often than not, as a man—not as an atypical woman, not as a genderless person. I must be one gender or the other; I cannot be neither, nor can I be both. This norm has a variety of mundane and serious consequences for my everyday existence. Like Myhre (1995), I have found that the choice not to participate in femininity is not one made frivolously.

My experiences as a woman who does not do femininity illustrate a paradox of our two-and-only-two gender system. Lorber argued that "bending gender rules and passing between genders does not erode but rather preserves gender boundaries" (1994, 21). Although people who engage in these behaviors and appearances do "demonstrate the social constructedness of sex, sexuality, and gender" (Lorber 1994, 96), they do not actually disrupt gender. Devor made a similar point: "When gender blending females refused to mark themselves by publicly displaying sufficient femininity to be recognized as women, they were in no way challenging patriarchal gender assumptions" (1989, 142).... I have found that my own experiences both support and challenge this argument....

MY SELF AS DATA

This analysis is based on my experiences as a person whose appearance and gender/sex are not, in the eyes of many people, congruent. How did my experiences become my data?... I am using my "unique biography, life experiences, and/or situational familiarity to understand and explain social life" (Riemer 1988, 121; see also Riemer 1977). It is an analysis of "unplanned personal experience," that is, experiences that were not part of a research project but instead are part of my daily encounters (Reinharz 1992)....

GENDERED ME

Each day, I negotiate the boundaries of gender. Each day, I face the possibility that someone will attribute the "wrong" gender to me based on my physical appearance.

I am six feet tall and large-boned. I have had short hair for most of my life. For the past several years, I have worn a crew cut or flat top. I do not shave or otherwise remove hair from my body (e.g., no eyebrow plucking). I do not wear dresses, skirts, high heels, or makeup. My only jewelry is a class ring, a "men's" watch (my wrists are too large for a "women's" watch), two small earrings (gold hoops, both in my left ear), and (occasionally) a necklace. I wear jeans or shorts, T-shirts, sweaters, polo/golf shirts, button-down collar shirts, and tennis shoes or boots. The jeans are "women's" (I do have hips) but do not look particularly "feminine." The rest of the outer garments are from men's departments. I prefer baggy clothes, so the fact that I have "womanly" breasts often is not obvious (I do not wear a bra). Sometimes, I wear a baseball cap or some other type of hat. I also am white and relatively young (30 years old).[3]

My gender display—what others interpret as my presented identity—regularly leads to the misattribution of my gender. An incongruity exists between my gender self-identity and the gender that others perceive. In my encounters with people I do not know, I sometimes conclude, based on our interactions, that they think I am a man. This does not mean that other people do not think I am a man, just that I have no way of knowing what they think without interacting with them.

Living with It

I have no illusions or delusions about my appearance. I know that my appearance is likely to be read as "masculine" (and male) and that how I see myself is socially irrelevant. Given our two-and-only-two gender structure, I must live with the consequences of my appearance. These consequences fall into two categories: issues of identity and issues of interaction.

My most common experience is being called "Sir" or being referred to by some other masculine linguistic marker (e.g., "he," "man"). This has happened for years, for as long as I can remember, when having encounters with people I do not know.[4] Once, in fact, the same worker at a fast-food restaurant called me "Ma'am" when she took my order and "Sir" when she gave it to me.

Using my credit cards sometimes is a challenge. Some clerks subtly indicate their disbelief, looking from the card to me and back at the card and checking my signature carefully. Others challenge my use of the card, asking whose it is or demanding identification. One cashier asked to see my driver's license and then asked me whether I was the son of the cardholder. Another clerk told me that my signature on the receipt "had better match" the one on the card. Presumably, this was her way of letting me know that she was not convinced it was my credit card.

My identity as a woman also is called into question when I try to use women-only spaces. Encounters in public rest rooms are an adventure. I have been told countless times that "This is the ladies' room." Other women say nothing to me, but their stares and conversations with others let me know what they think. I will hear them say, for example, "There was a man in there." I also get stares when I enter a locker room. However, it seems that women are less concerned about my presence there, perhaps because, given that it is a space for changing clothes, showering, and so forth, they will be able to make sure that I am really a woman. Dressing rooms in department stores also are problematic spaces. I remember shopping with my sister once and being offered a chair outside the room when I began to accompany her into the dressing room.

Women who believe that I am a man do not want me in women-only spaces. For example, one woman would not enter the rest room until I came out, and others have told me that I am in the wrong place. They also might not want to encounter me while they are alone....

I, on the other hand, am not afraid to walk alone, day or night. I do not worry that I will be subjected to the public harassment that many women endure (Gardner 1995). I am not a clear target for a potential rapist. I rely on the fact that a potential attacker would not want to attack a big man by mistake. This is not to say that men never are attacked, just that they are not viewed, and often do not view themselves, as being vulnerable to attack.

Being perceived as a man has made me privy to male-male interactional styles of which most women are not aware. I found out, quite by accident, that many men greet, or acknowledge, people (mostly other men) who make eye contact with them with a single nod. For example, I found that when I walked down the halls of my brother's all-male dormitory making eye contact, men nodded their greetings at me. Oddly enough, these same men did not greet my brother; I had to tell him about making eye contact and nodding as a greeting ritual. Apparently, in this case I was doing masculinity better than he was!...

There is, however, a negative side to being assumed to be a man by other men. Once, a friend and I were driving in her car when a man failed to stop at an intersection and nearly crashed into us. As we drove away, I mouthed "stop sign" to him. When we both stopped our cars at the next intersection, he got out of his car and came up to the passenger side of the car, where I was sitting. He yelled obscenities at us and pounded and spit on the car window. Luckily, the windows were closed. I do not think he would have done that if he thought I was a woman. This was the first time I realized that one of the implications of being seen as a man was that I might be called on to defend myself from physical aggression from other men who felt challenged by me. This was a sobering and somewhat frightening thought.

Recently, I was verbally accosted by an older man who did not like where I had parked my car. As I walked down the street to work, he shouted that I should park at the university rather than on a side street nearby. I responded that it was a public street and that I could park there if I chose. He continued to yell, but the only thing I caught was the last part of what he said: "Your tires are going to get cut!" Based on my appearance that day—I was dressed casually and carrying a backpack, and I had my hat on backward—I believe he thought that I was a young male student rather than a female professor. I do not think he would have yelled at a person he thought to be a woman—and perhaps especially not a woman professor.

Given the presumption of heterosexuality that is part of our system of gender, my interactions with women who assume that I am a man also can be viewed from that perspective. For example, once my brother and I were shopping when we were "hit on" by two young women. The encounter ended before I realized what had happened. It was only when we walked away that I told him that I was pretty certain that they had thought both of us were men. A more common experience is realizing that when I am seen in public with one of my women friends, we are likely to be read as a heterosexual dyad. It is likely that if I were to walk through a shopping mall holding hands with a woman, no one would look twice, not because of their open-mindedness toward lesbian couples but rather because of their assumption that I was the male half of a straight couple. Recently, when walking through a mall with a friend and her infant, my observations of others' responses to us led me to believe that many of them assumed that we were a family on an outing, that is, that I was her partner and the father of the child.

Dealing with It

Although I now accept that being mistaken for a man will be a part of my life so long as I choose not to participate in femininity, there have been times when I consciously have tried to appear more feminine. I did this for a while when I was an undergraduate and again recently when I was on the academic job market. The first time, I let my hair grow nearly down to my shoulders and had it permed. I also grew long fingernails and wore nail polish. Much to my chagrin, even then one of my professors, who did not know my name, insistently referred to me in his kinship examples as "the son." Perhaps my first act on the way to my current stance was to point out to this man, politely and after class, that I was a woman.

More recently, I again let my hair grow out for several months, although I did not alter other aspects of my appearance. Once my hair was about two and a half inches long (from its original quarter inch), I realized, based on my encounters with strangers, that I had more or less passed back into the category of "woman." Then, when I returned to wearing a flat top, people again responded to me as if I were a man.

Because of my appearance, much of my negotiation of interactions with strangers involves attempts to anticipate their reactions to me. I need to assess whether they will be likely to assume that I am a man and whether that actually matters in the context of our encounters. Many times, my gender really is irrelevant, and it is just annoying to be misidentified. Other times, particularly when my appearance is coupled with something that identifies me by name (e.g., a check or credit card) without a photo, I might need to do something to ensure that my identity is not questioned. As a result of my experiences, I have developed some techniques to deal with gender misattribution.

In general, in unfamiliar public places, I avoid using the rest room because I know that it is a place where there is a high likelihood of misattribution and where misattribution is socially important. If I must use a public rest room, I try to make myself look as nonthreatening as possible. I do not wear a hat, and I try to rearrange my clothing to make my breasts more obvious. Here, I am trying to use my secondary sex characteristics to make my gender more obvious rather than the usual use of gender to make sex obvious. While in the rest room, I never make eye contact, and I get in and out as quickly as possible. Going in with a woman friend also is helpful; her presence legitimizes my own....

I stopped trying on clothes before purchasing them a few years ago because my presence in the changing areas was met with stares and whispers....

My strategy with credit cards and checks is to anticipate wariness on a clerk's part. When I sense that there is some doubt or when they challenge me, I say, "It's my card." I generally respond courteously to requests for photo ID, realizing that these might be routine checks because of concerns about increasingly widespread fraud. But for the clerk who asked for ID and still did not think it was my card, I had a stronger reaction. When she said that she was sorry for embarrassing me, I told her that I was not embarrassed but that she should be. I also am particularly careful to make sure that my signature is consistent with the back of the card. Faced with such situations, I feel somewhat nervous about signing my name—which, of course, makes me worry that my signature will look different from how it should.

Another strategy I have been experimenting with is wearing nail polish in the dark bright colors currently fashionable. I try to do this when I travel by plane. Given more stringent travel regulations, one always must present a photo ID. But my experiences have shown that my driver's license is not necessarily convincing. Nail polish might be. I also flash my polished nails when I enter airport rest rooms, hoping that they will provide a clue that I am indeed in the right place.

There are other cases in which the issues are less those of identity than of all the norms of interaction that, in our society, are gendered. My most common response to misattribution actually is to appear to ignore it, that is, to go on with the interaction as if nothing out of the ordinary has happened....

These experiences with gender and misattribution provide some theoretical insights into contemporary Western understandings of gender and into the social structure of gender in contemporary society. Although there are a number of ways in which my experiences confirm the work of others, there also are some ways in which my experiences suggest other interpretations and conclusions.

WHAT DOES IT MEAN?

Gender is pervasive in our society. I cannot choose not to participate in it. Even if I try not to do gender, other people will do it for me. That is, given our two-and-only-two rule, they must attribute one of two genders to me. Still, although I cannot choose not to participate in gender, I can choose not to participate in femininity (as I have), at least with respect to physical appearance.

That is where the problems begin. Without the decorations of femininity, I do not look like a woman. That is, I do not look like what many people's commonsense understanding of gender tells them a woman looks like. How I see myself, even how I might wish others would see me, is socially irrelevant.

It is the gender that I *appear* to be (my "perceived gender") that is most relevant to my social identity and interactions with others. The major consequence of this fact is that I must be continually aware of which gender I "give off" as well as which gender I "give" (Goffman 1959)....

This reality brings me to a paradox of my experiences. First, not only do others assume that I am one gender or the other, but I also insist that I *really am* a member of one of the two gender categories. That is, I am female; I self-identify as a woman. I do not claim to be some other gender or to have no gender at all. I simply place myself in the wrong category according to stereotypes and cultural standards; the gender I present, or that some people perceive me to be presenting, is inconsistent with the gender with which I identify myself as well as with the gender I could be "proven" to be. Socially, I display the wrong gender; personally, I identify as the proper gender.

Second, although I ultimately would like to see the destruction of our current gender structure, I am not to the point of personally abandoning gender. Right now, I do not want people to see me as genderless as much as I want them to see me as a woman. That is, I would like to expand the category of "woman" to include people like me. I, too, am deeply embedded in our gender system, even though I do not play by many of its rules. For me, as for most people in our society, gender is a substantial part of my personal identity (Howard and Hollander 1997). Socially, the problem is that I do not present a gender display that is consistently read as feminine. In fact, I consciously do not participate in the trappings of femininity. However, I do identify myself as a woman, not as a man or as someone outside of the two-and-only-two categories....

The subversive potential of my gender might be strongest in my classrooms. When I teach about the sociology of gender, my students can see me as the embodiment of the social construction of gender. Not all of my students have transformative experiences as a result of taking a course with me; there is the chance that some of them see me as a "freak" or as an exception. Still, after listening to stories about my experiences with gender and reading literature on the subject, many students begin to see how and why gender is a social product. I can disentangle sex, gender, and sexuality in the contemporary United States for them. Students can begin to see the connection between biographical experiences and the structure of society. As one of my students noted, I clearly live the material I am teaching. If that helps me to get my point across, then perhaps I am subverting the binary gender system after all. Although my gendered presence and my way of doing gender might make others—and sometimes even me—uncomfortable, no one ever said that dismantling patriarchy was going to be easy.

NOTES

1. Ethnomethodology has been described as "the study of commonsense practical reasoning" (Collins 1988, 274). It examines how people make sense of their everyday experiences. Ethnomethodology is particularly useful in studying gender because it helps to uncover the assumptions on which our understandings of sex and gender are based.

2. I thank an anonymous reviewer for suggesting that I use these distinctions among the parts of a person's gender.
3. I obviously have left much out by not examining my gendered experiences in the context of race, age, class, sexuality, region, and so forth. Such a project clearly is more complex....
4. In fact, such experiences are not always limited to encounters with strangers. My grandmother, who does not see me often, twice has mistaken me for either my brother-in-law or some unknown man.

REFERENCES

Bern, S. L. 1993. *The lenses of gender*. New Haven, CT: Yale University Press.

Bordo, S. 1993. *Unbearable weight*. Berkeley: University of California Press.

Brownmilter, C. 1984. *Femininity*. New York: Fawcett.

Butler, J. 1990. *Gender trouble*. New York: Routledge.

Collins, R. 1988. *Theoretical sociology*. San Diego: Harcourt Brace Jovanovich.

Devor, H. 1989. *Gender blending: Confronting the limits of duality*. Bloomington: Indiana University Press.

Gardner, C. B. 1995. *Passing by: Gender and public harassment*. Berkeley: University of California.

Garfinkel, H. 1967. *Studies in ethnomethodology*. Englewood Cliffs, NJ: Prentice Hall.

Goffman, E. 1959. *The presentation of self in everyday life*. Garden City, NY: Doubleday.

_____. 1976. Gender display. *Studies in the Anthropology of Visual Communication* 3:69–77.

Howard, J. A., and J. Hollander. 1997. *Gendered situations, gendered selves*. Thousand Oaks, CA: Sage.

Kessler, S. J., and W. McKenna. 1978. *Gender: An ethnomethodological approach*. New York: John Wiley.

Lorber, J. 1994. *Paradoxes of gender*. New Haven, CT: Yale University Press.

_____. 1996. Beyond the binaries: Depolarizing the categories of sex, sexuality, and gender. *Sociological Inquiry* 66:143–59.

Martin, K. A. 1998. Becoming a gendered body: Practices of preschools. *American Sociological Review* 63:494–511.

Myhre, J. R. M. 1995. One bad hair day too many, or the hairstory of an androgynous young feminist. In *Listen up: Voices from the next feminist generation*, edited by B. Findlen. Seattle, WA: Seal Press.

Reinharz, S. 1992. *Feminist methods in social research*. New York: Oxford University Press.

Riemer, J. W. 1977. Varieties of opportunistic research. *Urban Life* 5:467–77.

_____. 1988. Work and self. In *Personal sociology*, edited by P. C. Higgins and J. M. Johnson. New York: Praeger.

West, C., and D. H. Zimmerman. 1987. Doing gender. *Gender & Society* 1:125–51.

Weston, K. 1996. *Render me, gender me*. New York: Columbia University Press.

Zimmerman, D. H. 1992. They were all doing gender, but they weren't all passing: Comment on Rogers. *Gender & Society* 6:192–98.

KEY CONCEPTS

doing gender ethnomethodology gender identity

DISCUSSION QUESTIONS

1. Think of a time when you encountered an individual whose gender was not readily apparent. What did you do? How did you address him/her? Did this make you feel uncomfortable?
2. Can you envision a society where we do not rely on a "two and only two" gender system? What advantages would this bring? What disadvantages?

36

Examining Media Contestation of Masculinity and Head Trauma in the National Football League

ERIC ANDERSON AND EDWARD M. KIAN

This piece is an analysis of sports news coverage of an NFL event in which one of the star players (Aaron Rodgers) took himself out of a game after a severe hit to the head. The discussion in the sports media revealed a long-standing expectation that masculinity involves toughness, the absence of femininity, and the willingness to play through pain. The way the event was discussed and the amount of attention it received revealed a possible shift in the way masculinity and sports are viewed.

... On February 6, 2011, the single most watched television event in US history occurred (Deggans 2011). Over 111 million viewers watched the Green Bay Packers defeat the Pittsburgh Steelers in Super XLV. Leading the Packers to

SOURCE: Eric Anderson and Edward M. Kian, Examining Media Contestation of Masculinity and Head Trauma in the National Football League, by SAGE Publications. Reprinted by permission of SAGE Publications.

victory was quarterback Aaron Rodgers, who was honored as the game's most valuable player after passing for three touchdowns. The event maintained significance, not only for the size of the audience it drew, but for the way athletes and sport media discuss head injuries in America's most masculinity generating sport.

Rodgers may not have been able to play at all, however, if it were it not for veteran teammate, Donald Driver. Earlier in the season, Rodger's head was slammed to the field by an opposing player. Driver—a wide receiver who at the time had the longest running continuous service among all current Green Bay players—talked to Rodgers on the pitch. This was not a "man up" style pep talk encouraging Rodgers to return to the gridiron. Instead, it was a discussion that contravened the warrior masculinity script long associated with football (Kian et al. 2011; Adams, Anderson, and McCormack 2010): Driver implored Rodgers to take himself out of the game.

Mark Viera, of *The New York Times*, quoted Driver on that conversation with his quarterback as having said:

> I was very concerned about him. I kind of whispered in his ear, walked behind him during the time he was sitting on the bench and kind of told him: 'This is just a game. Your life is more important than this game.' I told him I love him to death, and you've got to make the choice, but this game is not that important (Viera 2010, 2).

Rodgers elected not to re-enter the game, and his team ultimately lost to the Detroit Lions, 7-3. It was a rare moment of putting health before victory, in what is otherwise a game celebrated for orthodox notions of masculine sacrifice.

After the Packers' medical staff diagnosed Rodgers with a concussion (his second of the season), he also sat out the following game. With two defeats in a row, the Packers' dream of a Super Bowl appearance began to fade. Yet, when a healthy Rodgers returned two weeks later, he led his team to six consecutive victories to close out the season, including the Super Bowl, in all likelihood because he had not "man upped" and continued to play through a concussion against the Lions.

LITERATURE REVIEW

Hegemonic Masculinity in American Football

Men's competitive, organized team sports—particularly contact sports—have long maintained utility in shaping and defining acceptable forms of heterosexual masculinity in Western cultures (Hargreaves 1994; Messner 1992). American football has been deemed the most masculine and violent team sport in US culture since the early part of the twentieth century (Rader 2008). Violence in the game has escalated over recent years, as improved strength and conditioning techniques and nutritional advancements have pushed younger athletes to be bigger, stronger, and faster (Sanderson 2002).

The iconography of the American football player—young, muscle-bound, and willing to commit violence to himself and others—has placed this category of athlete at the top of a masculine hierarchy of men in the United States. This is true of boys, many of who view professional and college football players as role models (Messner 1992).

Connell (1995) described the social process that exalts this idealized form of masculinity as hegemonic. Thus, hegemonic masculinity was described as a social process in which one form of institutionalized masculinity is culturally exalted above all others (Connell 1995). Key to understanding the operation of hegemonic masculinity is that almost all men exhibit either a complicit, subordinate or marginalized form (archetype) of masculinity. Yet despite very few men achieving the requisites of hegemonic masculinity, many men desire to obtain or at least be associated with the hegemonic form. Essentially, the process of hegemony influences the oppressed to maintain the rightfulness or naturalization of their oppression (Gramsci 1971).

Men at the top of a hegemonic stratification must obey certain rules in order to maintain their privileged position. Heterosexuality, for example, is compulsory (Pronger 1990); and the frequent expression or action of things associated with femininity are also taboo (Messner 1992). But it is not enough for one to simply adhere to the rules of masculinity; one must also be heard advocating them. Adams, Anderson, and McCormack (2010) called this *masculinity-establishing discourse*. Here, athletes use familiar expressions invoking masculinity, denying weakness, and/or using femphobia or homophobia to "motivate" others. Putting this discourse into action serves to establish/reestablish football as a masculine sport. Through a process of regulating, disciplining, and policing it defines the perimeters of warrior behaviors and attitudes that constitute hegemonic masculinity.

It is these same discourses, including phrases like "man up," "no pain, no gain," and "pain is temporary, pride is forever," that encourage men to position their own bodies as an expendable weapon of athletic war. The discourse encourages athletes to conceal all fear in the pursuit of glory. Similarly, in the event of injury, football players must not show signs of pain or distress; instead, they must talk about returning to the game as soon as possible. Much of the iconic imagery of the American football player is derived from this willingness and ability to play through pain (McDonough et al. 1999)....

Sport Journalists Upholding the Masculine Warrior Narrative

The image of the emotionally and physically impenetrable football player has been reified by the dominant sporting media (Nylund 2004). Media-portrayed sporting narratives of heroic disposition, even in the face of debilitating injury or risk of death, are produced as part of orthodox notions of commitment to sport and victory (e.g., Pedersen 2002; Vincent and Crossman 2008). This is for several reasons. Foremost it is because the preponderance of individuals in sport media is men (Hardin 2005; Kian 2007).

Lapchick, Moss II, Russell, and Scearce (2011) surveyed 320 daily newspapers and popular sport Web sites in the U.S., finding that men comprised 94% of sports editors, 90% of assistant sports editors, 89% of reporters, 90% of columnists, and 84% of copy editors/designers. Similarly, Nylund (2004) documented that 80 percent of US sports-talk radio hosts are men.

These male sports journalists have been shown to uphold hegemonic masculinity by primarily covering men's sports (which are construed as more masculine) and providing negative stereotypes of female athletes (Duncan 2006; Kian and Hardin 2009). Sports journalists have also provided less coverage of male athletes and men's sports that do not meet ideal characteristics of orthodox masculinity, such as figure skating and gymnastics (Vincent et al. 2002)....

Heavily invested in masculinity-making practices, rather than countering the cultural practice of praising strong-willed male athletes who put health before victory, sport journalists have long used the stories of fallen athletes in order to promote their own sporting prowess and masculinity....

Head Trauma as a Social Problem in Sport

Whereas this study examines the narratives US print media used to cover the Rodgers–Driver concussion conversation, it is important to contextualize the study in light of rapidly emerging research into the detrimental effects of concussions, as well as the simultaneous academic and cultural awakening of viewing concussions not as temporary incidents resulting in having "your bell rung," but instead for what they are, medically: traumatic brain injuries in the form of CTE.

Many sport fans associate the life-long, debilitating, and deadly head trauma caused in sport primarily with boxing and other combat sports, with the false assumption that the helmets worn in American football protect from such trauma. Whereas improved helmet design does reduce the risk of concussion (Viano, Casson, and Pellman 2007), a phenomenon known as *risk compensation* suggests that athletes wearing headgear may in fact play more recklessly, based on the misguided belief that helmets protect them from injury (Hagel and Meuwisse 2004). Thus, this suggests that helmets could even increase injury rates, particularly in younger athletes, who are less skilled in tackling.

CTE is a progressive neurological disorder caused from blows to one's head that can occur even when wearing a helmet for protection (Viano, Casson, and Pellman 2007). Having similar symptoms as Alzheimer's, CTE begins with behavioral and personality changes. It is followed by disinhibition and irritability, before the individual moves into dementia. It takes years for the initial trauma to give rise to nerve-cell breakdown and death, but CTE is not the result of an endogenous disease like Alzheimer's (McKee et al. 2009). It is the result of traumatic brain injury—the type routinely occurring in contact sports....

Social Reaction to CTE

There has been a swift but limited response to practice and policies concerning using the head as a weapon in sport. The American Association of Pediatrics has issued new guidelines for concussions. Likely in response to these publicized studies showing significant long-term brain damage for former NFL players

(Smith 2009), the NFL has instituted new rules on tackling, and recently changed helmet design. The NFL also revised, multiple times, its concussion policies between 2009 and 2011. As it stands, a player who is diagnosed with a concussion during a game is prohibited from playing again in that same game and he may only play in a following game once cleared by a physician that is not employed by the NFL (Smith 2011). This is because a player who is concussed before a previous concussion is healed stands at risk of death....

With an understanding that head trauma is not an inevitable part of American football; that it is instead produced by a masculinization of the game which encourages tackling high and hard, this article investigates the role of media in reproducing masculine discourse in football. Whereas media coverage of sport has long assisted in formulating the notion of heroic athletes who overcome pain and adversity (Messner 2002; Trujillo 1991), it is unclear how modern sport media might frame narratives regarding concussions in relation to three social factors: (1) emerging evidence of the damage caused by using the head in American football; (2) a cultural softening of orthodox requisites required to be considered masculine, and; (3) the increasing threat of civil litigation against the NFL by players affected by CTE.

METHODOLOGY

Textual Analysis

In this research, we conducted a textual analysis of all print media articles that mentioned Rodgers, Driver, Rodgers' concussion, and the conversation between the two on Rodgers' concussion that were published in three national newspapers and seven popular sport Internet sites....

SAMPLING SELECTION

The three newspapers examined in this study were *USA Today, The New York Times,* and *Los Angeles Times* which, per the Audit Bureau of Circulations (Shea 2010), ranked second, third, and fourth, among the most circulated newspapers in the United States. Topping that list was *The Wall Street Journal,* which was excluded from this study as it is a business-specific newspaper.

The seven Internet sites were chosen because all provide a great deal of coverage on American professional football. Four of those sites, *AOL Sports, CBSSports.com, Fox Sports.com,* and *ESPN Internet* are considered mainstream, broad-based sites that provide detailed coverage for a variety of sports. In contrast, *Deadspin* is a non-traditional sport media site that has emerged as one of the most influential and popular sport blogs in the world. The final two sites examined in this study were popular blogs, *Pro Football Talk* and *Kissing Suzy Kolber.* These were selected due to their singular and extensive focus on American professional football.

Two researchers searched for and downloaded all articles published in the 10 outlets that included references toward players, Rodgers and Driver, Rodgers' concussion, and made some mention of the conversation between the two players on Rodgers' concussion. Subscriptions allowing for access to all content were attained for ESPN Internet, the only online media outlet examined that did not make all of its content freely available on its Website.

The time frame covered December 12, 2010, the day that Rodgers sustained his concussion against the Lions, through December 20, 2010, which included all articles published after the Packers' next game against the Patriots, in which Rodgers ultimately sat out due to the concussion he sustained against the Lions. However, because Green Bay advanced through the post-season to the Super Bowl, where the mainstream US media are famous for lavishing a plethora of coverage during Super Bowl "media week," we also examined articles falling under the above-mentioned criteria published from Sunday January 30, 2011, through February, 7, 2011, which was the day after the Packers defeated the Pittsburgh Steelers in Super Bowl XLV.

CODING PROCEDURES AND DATA ANALYSIS

The two researchers examined how masculinity was constructed and framed in articles on Rodgers' concussion and his conversation with Driver through media articles....

All words of each article were read by the two coders who initially worked independent of each other. Content related to the broad headings under masculinities in articles were highlighted and each coder wrote theoretical memos to better explain commonalities within text they highlighted (Turner 1981)....

References and particularly themes relating to masculinity within articles were given greater importance in the coding process....

RESULTS

All but one of the ten articles published within the examined periods focused on the significance of Driver urging Rodgers not to continue to play against Detroit. Four dominant themes emerged from our analysis: (1) media attitudes toward concussions are changing; (2) media contend players' attitudes generally lag behind others; (3) the event was not newsworthy; and (4) views of risk widely vary.

Media Attitudes toward Concussions Are Changing

The dominant theme from the data was that Driver received universal praise from the media. Journalists portrayed this as a selfless act—a commendable feat of putting the health of his quarterback over his own interest (of making the playoffs and posting his best possible statistics as a receiver). *The New York Times* writer Mark Viera framed the discussion as trendsetting for football as a whole, writing in his *The Fifth Down* notebook that, "It can be seen as another sign that N.F.L. players' attitudes toward concussions are changing" (Viera 2010, 2). Tim Keown of ESPN Internet went further, writing:

> with one act of humanity and one public display of perspective, Packers receiver Donald Driver did more than a thousand studies or a million speeches. He accomplished what no one had yet managed to do in this era of heightened awareness of head injuries in the NFL: He made it OK for a teammate to leave the game because of a concussion (2010, 1).

Keown argued that, for exhibiting such selflessness, Driver should be considered for the NFL's Walter Payton Man of the Year, an annual award honoring a player for his volunteer and charity work, as well as excellence on the field....

The praise Driver and Rodgers received for pulling out of play is more notable because it is likely that Rodgers' removal cost the team a victory over the Lions. Indeed, most of the articles were clear to highlight this.... This makes the transgressive act of voluntary withdrawal from a game a more serious violation of orthodox masculinity than had the team been points ahead.

Media Contend Players' Attitudes Generally Lag Behind Others

Whereas writers lavished much praise upon Driver and Rodgers for the courage they exhibited in recognizing the seriousness of Rodgers' concussion, some also noted that change will likely occur slowly in a sport that values intrepid men playing through pain.

> Players are competitive. They're also stubborn and independent. They believe they should have a say over whether they can play with an injury, working under the premise that no one can look inside their heads and determine how they feel. That libertarian streak does not lend itself well to looking down the huddle and telling a teammate to leave the field because his eyes look a little glossy.
>
> It's the sport, it's the culture, and it's the people who reside within it. The biggest obstacle to the NFL's concussion policy is persuading players to leave the field or seek treatment when they know they—like Rodgers—will be finished for that game and quite possibly the next. The blackout concussions are an easy call because there's no hiding them. Brett Favre flat-out on the frozen plastic of TCF Stadium, for example, can be diagnosed from the living room. But those that are less severe, when a player gets his bell rung but manages to hide it well enough to stay in the game, present a loophole that can lead to multiple concussions and more serious problems (Keown 2010, 15–16).

This theme was best exemplified not directly from a journalist, but through a quote one author used from Packers' defensive captain and standout cornerback,

Charles Woodson, who said he expected Rodgers to play against the Patriots the following week despite his concussion.

> "He's a football player, and that's how we are," Woodson said. "We're going to play. I know if the doctors say, 'Hey, we don't think there's a problem,' he's going to play. [If not], I guarantee he'll lobby to play. That's how we are. That's how we're cut" (WR Says He Urged QB to Sit 2010, 19).

However, based on their actions, players' desire to play is evidently becoming less important than their long-term health....

One reason players' attitudes may be slow to change would be due to media discounting the heroism of athletes who freely elect not to play when concussed. One way of doing this is by ignoring athletes who voluntarily sit themselves out.

This Event Was Not Newsworthy

Sport journalists not only frame the narratives that assist media consumers in defining the news, but they also shape attitudes by determining what events are worthy of news coverage or not (Kuypers 2002). Sport media "symbolically annihilate" news and people they deem less important merely by not covering them (Tuchman 1978, 17). Thus, sport media tend to ignore stories that challenge dominant and long-standing social constructions (Vincent 2004).

Whereas authors who wrote about this event unanimously praised Driver's discussion with his quarterback that led to Rodgers taking himself out of the game, most media gatekeepers evidently did not believe this event newsworthy. Even though most online sites publish hundreds of wire stories (such as one Associated Press story on the conversation between Driver and Rodgers included in our study), the three newspapers and seven Internet sites in this research combined to only produce nine articles during the week after the Green Bay–Detroit game that mentioned the conversation between Driver and his quarterback....

Views of Risk Widely Vary

Concussions in professional American football were a major news story throughout the 2010 season, even before that December game between the Packers and Lions. However, there was considerable backlash from many players, and some media members, against the NFL's new "get-tough" policy on illegal hits against "defenseless" players; hits which league officials claimed result in a greater chance of major injuries, including concussions (Battista 2010). At the forefront of this debate were the Steelers, the only team to win six Super Bowls, as well as a franchise whose players and national fan base have long relished their blue-collar, physical, and tough reputations (O'Brien 2004).

No member of the Steelers received more media attention in stories on concussions during the 2010 season than standout linebacker James Harrison, the 2008 Associated Press NFL Defensive Most Valuable Player. In helping lead Pittsburgh to Super Bowl XLV, Harrison was fined four different times for a total of $125,000 in 2010 by the NFL's league office for "illegal on-field hits." This was highlighted by a $75,000 fine levied against Harrison for a helmet-to-helmet hit he delivered on Mohammad Massaquoi of the Cleveland Browns. Both Massaquoi and fellow Cleveland receiver Josh Cribbs left that game due to concussions derived from Harrison's hits, although Harrison's hit on Cribbs was deemed legal by league officials (Bouchette 2010).

After the Steelers' win over the Browns, but before he was fined, Harrison was quoted by ESPN as having said, "I don't want to injure anybody. There's a big difference between being hurt and being injured. You get hurt, you shake it off and come back the next series or the next game. I try to hurt people" (James Harrison OK with dishing pain 2010). Harrison also publicly threatened to retire from the NFL after being fined by the league office for the Massaquoi hit....

The league's focus on stricter enforcement of rules to better protect player safety was a major news story during Super Bowl media week, with much debate centering on whether the Steelers were being signaled out and unfairly punished for rough plays, which Pittsburgh players deemed legal hits. Whereas Harrison's fines were frequently mentioned in articles published during Super Bowl week, only one article appeared that week in any of the ten examined media outlets that specifically mentioned the concussion Rodgers sustained against the Lions and the resulting conversation between Driver and Rogers.

In the *New York Times* article published two days before the Super Bowl, the author Alan Schwarz opened by praising Rodgers and Driver for their conversation, "A professional player telling another to put his long-term health ahead of the team—a once and, to some, still-heretical idea—thrilled those who are trying to temper the sport's win-now, regret-later ideology. Neurologists nodded. Parents cheered" (Schwarz 2011c, 3).

Schwarz then contrasted the Packers with the Steelers, a team he framed as having an archaic attitude toward concussions:

> As for the rebuttal in football's continuing debate, that was gladly delivered this week by none other than the Packers' opponent in Sunday's Super Bowl—the Pittsburgh Steelers, whose stars stumped as football's defiant traditionalists. The hard-hitting linebacker James Harrison mocked the NFL's crackdown on head-to-head tackles, suggesting that the league "lay a pillow down where I'm going to tackle them, so they don't hit the ground too hard." Receiver Hines Ward questioned all the fuss about brain injuries, and said that advising his own oft-concussed quarterback, Ben Roethlisberger, about health was all but preposterous (Schwarz 2011c, 4–5).

After the first five paragraphs, Schwarz framed the remainder of the article decidedly in favor of the Packers' actions and attitudes, quoting a sport psychologist who warned of the negative impact on youth players who model their games after Steelers' players and Dustin Fink, a high school athletics trainer who said he was rooting for the Packers due to messages on concussions delivered by each team.

DISCUSSION

American football has long been the most masculinized team sport in the United States. This is a status supported by sport media who exist as a group of heterosexual men desiring to be associated with a hegemonic form of masculinity (Nylund 2004). Part of the production of this hegemonic form of masculinity has traditionally included self-sacrifice to keep players in line with the ethos of sacrificing tomorrow's health for today's glory, even in the face of debilitating injury or risk of death (e.g., Pedersen 2002; Vincent and Crossman 2008). There is, however, growing cultural awareness about the use of contact sports in promoting chronic brain injury; and in this article, we have shown that some of this comes from sport media. We suggest that this is a reflection of a larger cultural shift in the type of masculinity that young men value; and this includes team sport athletes.

Anderson (2009) described four recent trends among heterosexual university athletes: (1) increased physical tactility among same-sex peers; (2) increased emotional bonding between same-sex peers; (3) less peer-aggression; and (4) the adoption of feminized clothing styles and body posture. Anderson (2009) described these attitudes and behaviors as "inclusive masculinity" and theorized there emergence through his notion of decreasing cultural homohysteria (p. 7). Principally Anderson's theory maintains that whereas there are a number of social factors that shape/reshape cultural constructions of idealized heterosexual masculinity, and homohysteria is the most significant. Homohysteria—described as heterosexual men's fear of being publicly homosexualized by violating rigid gendered boundaries—situates levels of homophobia temporally and spatially, recognizing that cultural homophobia has different effects dependent on the social context. Accordingly, homohysteria is a useful theoretical tool for understanding the significance that homophobia maintains within particular cultures.

Anderson (2009) contended that in temporal–cultural moments with high levels of homohysteria, masculinity and homosexuality are viewed as incompatible, meaning that heterosexual men go to great lengths to avoid being perceived as gay. Here, homophobia is used as a weapon to stratify men in deference to a dominant hegemonic force (Connell 1995). This is particularly effective because anyone can be suspected of being gay (Anderson 2008)....

In light of these findings, we suggest that the practice of accepting traumatic injury for the sake of team victory may be under assault for three reasons. First, there is growing cultural awareness as to the significant, debilitating, and oftentimes life-ending impact that concussions (and even repeated hits to the head that do not result in immediate concussions) have on players (Colvin et al. 2009). Highlighted by the suicides of several ex-NFL players, and the appearance of CTE in the brain of a university player who had never suffered a concussion, media, players, coaches, and especially NFL officials are seemingly awakening to the very real dangers that using the head for sport causes.

Second, young athletes have been socialized into a rapidly changing culture: one that fosters emotional intimacy between men and the expression of feelings, including fear and pain (Anderson 2008). This means that NFL players might increasingly be risk adverse, in part because inclusive masculinity does not require men to sacrifice health for the sake of sport (Adams, Anderson, and McCormack 2010; Anderson and McGuire 2010), the way previous researchers (Messner 1992; Trujillo, 1991) found decades ago. Accordingly, as the once orthodox image of the team sport athlete loses its cultural hegemony, multiple types of masculinity are permitted to flourish without the hierarchy necessary in a hegemonic system (McCormack 2010). This means that those who once used to flirt with hegemonic masculinity, those Connell (1995, p. 79) described as "complicit," have less reason to build their masculine capital by upholding the "heroism" of playing through concussion as they once did. Accordingly, men who write about sport maintain more freedom in decrying this type of self-violence, too.

Third, and largely out of scope of this article, it is likely that governing bodies of sport are increasingly concerned about being held accountable for the health and safety of the players they govern. Injury in the NFL can, for example, be viewed as an occupational hazard. If the NFL does not protect players from this, they could be opening the league up to considerable punitive damage, should the courts agree with players who file suit....

Collectively, this means that the warrior scripts of masculine sacrifice, once so valorized in the NFL, may be contested through a triangulated causal model—concern for safety, a weakening hegemonic model, and liability issues. Whatever the causes, change is already stemming from official NFL policies, which during the 2010 season made progressive changes in an attempt to lessen the likelihood of concussions occurring. The promotion of softer attitudes toward violence we document in this research is also significant. This is because they are promulgated by many of the foremost sports media venues, including ESPN.

Still, our research found an absence of discussion about this issue from the majority of sport media sources. Four of the ten examined outlets never wrote about this discussion and only four of the ten had more than one article that mentioned the conversation between Rodgers and Drivers. Whereas we cannot ascribe other journalists as shunning the Driver/Rodgers incident, their silence on the matter is significant, as sport media tend to ignore stories that challenge dominant and long-standing social constructions (Duncan 2006; Kian, Mondello and Vincent 2009)....

If a division does exist—or is beginning to emerge—between players, the duo of Driver (a celebrated veteran of the Green Bay Packers) and Rodgers (a Super Bowl MVP) is important in that it provides the voices of two individuals, both of whom maintain high masculine capital because they are champions. This gives them more social capital for the promotion of health over complicit violence.

The same is true of sport media. The importance of the position of the sport journalists, who did report (positively) on the Rodgers–Driver discussion, should not be underestimated. As one ESPN reporter commented, "He made it OK for a teammate to leave the game because of a concussion," and that likely made it okay for journalists to find Driver's act so refreshing (Keown 2010, 1).

It is too early to determine what broader implications the Rodger's incident might have for other levels of play. However, in light of this emerging evidence, we should highlight that some positive change to youth sport has already occurred. For example, the national youth sport league, i9 Sports, already plays flag football (where there is no tackling). Whereas the NFL will most likely always be a contact-based sport, it does appear that players' safety is beginning to be taken more seriously....

REFERENCES

Adams, A., E. Anderson, and M. McCormack. 2010. "Establishing and Challenging Masculinity: He Influence of Gendered Discourses in Organized Sport." *Journal of Language and Social Psychology* 29:278–300.

Anderson, E. 2008. "'Being Masculine is not About who you Sleep with...': Heterosexual Athletes Contesting Masculinity and the One-Time Rule of Homosexuality." *Sex Roles* 58:104–15.

Anderson, E. 2009. *Inclusive Masculinity: The Changing Nature of Masculinities*. New York, NY: Routledge.

Anderson, E., and R. McGuire, 2010. "Inclusive Masculinity and the Gendered Politics of Men's Rugby." *The Journal of Gender Studies* 19:249–61.

Battista, J. 2010, October 20. "Defenders Criticize NFL for Helmet-to-Helmet Fines." *The New York Times*, accessed May 3, 2011, http://www.nytimes.com/2010/10/21/sports/football/21hits.html.

Bouchette, E. 2010, October 19. "NFL Fines Steelers' Harrison $75,000; Harrison: 'It was a legal hit'." *Pittsburgh Post-Gazette*, accessed May 6, 2011, http://www.post-gazette.com/pg/10292/1096447-100.stm.

Colvin, A. C., J. Mullen, M. R. Lovell, R. V. West, M. W. Collins, and M. Groh. 2009. "The Role of Concussion History and Gender in Recovery from Soccer-Related Concussion." *American Journal of Sports Medicine* 37:1699–704.

Connell, R. W. 1995. *Masculinities*. Sydney: Allen and Unwin.

Deggans. 2011, February 7. "It's Official: Super Bowl Most-Watched Event in TV History with Average 111-Million Viewers." *St. Petersburg Times*. accessed February 26, 2011. http://www.tampabay.com/blogs/media/content/its-official-super-bowl-most-watched-event-tv-history-average-111-million-viewers.

Duncan, M. C. 2006. "Gender Warriors in Sport: Women and the Media." In *Handbook of Sports and Media*, eds. A. A. Raney and J. Bryant, 231–52. Mahwah, NJ: Lawrence Erlbaum Associates.

Gramsci, A. 1971. *Selections from the Prison Notebooks*. New York, NY: International Publishers.

Hagel, B., and W. Meuwisse. 2004. "A 'Side Effect' of Sport Injury Prevention?" *Clinical Journal of Sports Medicine* 14:193–6.

Hardin, M. 2005. "Stopped at the Gate: Women's Sports, 'Reader Interest,' and Decision-making by Editors." *Journalism & Mass Communication Quarterly* 82:62–77.

Hargreaves, J. 1994. *Sporting Females: Critical Issues in the History and Sociology and Women's Sports*. New York, NY: Routledge.

James Harrison OK with Dishing Pain. 2010, October 19. *ESPN Internet*. accessed May 7, 2011, http://sports.espn.go.com/nfl/news/story?id=5699976.

Keown, T. 2010, February 4. "Donald Driver to the NFL's Rescue." ESPN Internet. accessed June 5, 2011, http://sports.espn.go.com/espn/commentary/news/story?page¼keown/101221.

Kian, E. M. 2007. "Gender in Sports Writing by the Print Media: An Exploratory Examination of Writers' Experiences and Attitudes." *The SMART Journal* 4:5–26.

Kian, E. M., and M. Hardin. 2009. "Framing of Sport Coverage based on the Sex of Sports Writers: Female Journalists Counter the Traditional Gendering of Media Content." *International Journal of Sport Communication* 2:185–204.

Kian, E. M., G. Clavio, J. Vincent, and S. D. Shaw. 2011. "Homophobic and Sexist yet Uncontested: Examining Football Fan Postings on Internet Message Boards." *Journal of Homosexuality* 58:680–99.

Kian, E. M., M. Mondello, and J. Vincent. 2009. "ESPN–The Women's Sports Network? A Content Analysis of Internet Coverage of March Madness." *Journal of Broadcasting & Electronic Media* 53:477–95.

Kuypers, J. A. 2002. *Press Bias and Politics: How the Media Frame Controversial Issues.* Westport, CT: Praeger.

Lapchick, R., A. Moss II, C. Russell, and R. Scearce. 2011. The 2010-11 Associated Press Sports Editors racial and gender report card, accessed August 8, 2011, University of Central Florida, Institute for Diversity and Ethics in Sport Web site: http://www.tidesport.org/RGRC/2011/2011_APSE_RGRC_FINAL.pdf?page=lapchick/110517.

McCormack, M. 2010. "Hierarchy Without Hegemony: Locating Boys in an Inclusive School Setting." *Sociological Perspectives* 54:83–102.

McDonough, W., P. King, P. Zimmerman, V. Carucci, G. Garber, K. Lamb, et al. 1999. *The NFL Century*. Los Angeles, CA: National Football League Properties, Inc.

McKee, A. C., B. E. Gavett, R. A. Stern, C. J. Nowinski, R. C. Cantu, N. W. Kowall, D. Perl, A. S. Hedley-McIntosh, P. McCory, C. F. Finchs, J. Best, D. J. Chalmers, and R. Wolfe. 2009. "Does Padded Headgear Prevent Head Injury in Rugby Union Football?" *Medicine and Science in Sports and Exercise* 41:306–13.

Messner, M. 1992. *Power at Play: Sports and the Problem of Masculinity*. Boston, MA: Beacon Press.

Messner, M. A. 2002. *Taking the Field: Women, Men, and Sports*. Minneapolis, MN: University of Minnesota Press.

Nylund. D. 2004. "When in Rome: Heterosexism, Homophobia, and Sports Talk Radio." *Journal of Sport & Social Issues* 28:136–68.

O'Brien, J. 2004. *Lambert: The Man in the Middle … and Other Outstanding Linebackers.* Pittsburgh, PA: James P. O'Brien Publishing.

Pedersen, P. M. 2002. "Examining Equity in Newspaper Photographs: A Content Analysis of the Print Media Photographic Coverage of Interscholastic Athletics." *International Review for the Sociology of Sport* 37:303–18.

Pronger, B. 1990. *The Arena of Masculinity: Sports, Homosexuality, and the Meaning of Sex.* New York, NY: St. Martin's Press.

Rader, B. G. 2008. *American Sports: From the Age of Folk Games to the Age of Televised sports* (6th ed.). Upper Saddle River, NJ: Prentice Hall.

Sanderson, A. R. 2002. "The Many Dimensions of Competitive Balance." *Journal of Sports Economics* 3:204–28.

Schwarz, A. 2011c, February 4. "The two teams show divide in debate on safety." *The New York Times*. accessed May 2, 2011, http://www.nytimes.com/2011/02/04/sports/football/04rodgers.html?_r=1.

Shea, D. 2010, April 26. "Top 25 newspapers by circulation: *Wall Street Journal* trounces *USA Today.*" *HuffintonPost.com*. accessed May 4, 2011, http://www.huffingtonpost.com/2010/04/26/top-25-newspapers-by-circ_n_552051.html#s84769&title=2_USA_Today.

Smith, M. D. 2011, April 19. "NFL points to Stewart Bradley as reason for changes on concussions." *Pro Football Talk*. accessed May 2, 2011, http://profootballtalk.nbcsports.com/2011/04/19/nfl-points-to-stewart-bradley-as-reason-for-changes-on-concussions/.

Smith, S. 2009, January 27. "Dead athletes' brains show damage from concussions." *CNNHealth.com*. accessed May 2, 2011, http://www.cnn.com/2009/HEALTH/01/26/athle-te.brains/index.html.

Trujillo, N. 1991. "Hegemonic Masculinity on the Mound: Media Representations of Nolan Ryan and American sports culture." *Critical Studies in Mass Communication* 8:290–303.

Tuchman, G. 1978. "Introduction: The Symbolic Annihilation of Women by the Mass Media." In *Hearth and Home: Images of Women in the Mass Media*, eds. G. Tuchman, A. K. Daniels, & J. Benét, 3–29. New York, NY: Oxford University Press.

Turner, B. A. 1981. "Some Practical Aspects of Qualitative Data Analysis: One Way of Organizing the Cognitive Processes Associated with the Generation of Grounded Theory." *Quantity and Quality* 15:225–47.

Viano, D. C., I. R. Casson, and E. J. Pellman. 2007. "Concussion in Professional Football: Biomechanics of the Struck Player–part 14." *Neurosurgery* 61:313.

Viera, M. 2010, December 17. "After injury, receiver told Rodgers to stay out." *The New York Times*. accessed February 15, 2011, http://www.nytimes.com/2010/12/18/sports/football/18nfl.html.

Vincent, J. 2004. "Game, Sex, and Match: The Construction of Gender in British Newspaper Coverage of the 2000 Wimbledon Championships." *Sociology of Sport Journal* 21:435–56.

Vincent, J., and Grossman, J. 2008. "Champions, a Celebrity Crossover, and a Capitulator: The Construction of Gender in Broadsheet Newspapers' Narratives about Selected Competitors During "The Championships.'" *International Journal of Sport Communication* 1:78–102.

Vincent, J., C. Imwold, V. Masemann, and J. T. Johnson. 2002. "A Comparison of Selected 'Serious' and 'Popular' British, Canadian, and United States Newspaper Coverage of Female and Male Athletes Competing in the Centennial Olympic Games." *International Review for the Sociology of Sport* 37:319–35.

WR Says he Urged QB to sit After Hits. 2010, December 16. *ESPN Internet*. accessed April 21, 2011, http://m.espn.go.com/nfl/story?storyId=5927308.

KEY CONCEPTS

gender hierarchy   hegemonic masculinity   masculinity

DISCUSSION QUESTIONS

1. Anderson and Kian mention that sports coverage is primarily about men's sports. Consider coverage of women's sports. What differences do you see in the way media portray female athletes and male athletes? What explanations can you offer to why these differences exist?
2. Given the research on concussions and head injuries, do you see a change in all sports regarding playing while injured? While clear positive consequences exist for players' health, what negative consequences might come from a change in the way athletes are expected to behave when injured?

37

The Many Faces of Gender Inequality

AMARTYA SEN

In this piece, Sen examines his home country of India for patterns of gender inequality. In the beginning he outlines several different kinds of gender inequality. He goes on to explain in detail the inequality in mortality rates in South Asia and how this leads to natality inequality with fewer women than men in Asian countries.

It was more than a century ago, in 1870, that Queen Victoria wrote to Sir Theodore Martin complaining about "this mad, wicked folly of 'Woman's Rights.'" The formidable empress certainly did not herself need any protection that the acknowledgment of women's rights might offer. Even at the age of eighty, in 1899, she could write to Arthur James Balfour that "we are not interested in the possibilities of defeat; they do not exist." Yet that is not the way most people's lives go, reduced and defeated as they frequently are by adversities. And within every community, nationality, and class, the burden of hardship often falls disproportionately on women.

The afflicted world in which we live is characterized by a deeply unequal sharing of the burden of adversities between women and men. Gender inequality exists in most parts of the world, from Japan to Morocco, from Uzbekistan

SOURCE: Amartya Sen, "The Many Faces of Gender Inequality," *The New Republic* 17 (2001). Reprinted with permission of the author.

to the United States. Yet inequality between women and men is not everywhere the same. It can take many different forms. Gender inequality is not one homogeneous phenomenon, but a collection of disparate and inter-linked problems. I will discuss just a few of the varieties of the disparity between the genders.

MORTALITY INEQUALITY

In some regions in the world, inequality between women and men directly involves matters of life and death, and takes the brutal form of unusually high mortality rates for women and a consequent preponderance of men in the total population, as opposed to the preponderance of women found in societies with little or no gender bias in health care and nutrition. Mortality inequality has been observed and documented extensively in North Africa and in Asia, including China and South Asian nations.

NATALITY INEQUALITY

Given the preference for boys over girls that characterizes many male-dominated societies, gender inequality can manifest itself in the form of parents' wanting a baby to be a boy rather than a girl. There was a time when this could be no more than a wish—a daydream or a nightmare, depending on one's perspective. But with the availability of modern techniques to determine the gender of a fetus sex-selective abortion has become common in many countries. It is especially, prevalent in East Asia, in China and South Korea in particular; but it is found also in Singapore and Taiwan, and it is beginning to emerge as a statistically significant phenomenon in India and in other parts of South Asia as well. This is high-tech sexism.

BASIC-FACILITY INEQUALITY

Even when demographic characteristics do not show much anti-female bias or any at all, there are other ways in which women can get less than a square deal. Afghanistan may be the only country in the world where the government is keen on actively excluding girls from schooling (the Taliban regime combines this with other features of massive gender inequality); but there are many countries in Asia and Africa, and also in Latin America, where girls have far less opportunity for schooling than do boys. And there are other deficiencies in basic facilities available to women, varying from encouragement to cultivate one's natural talents to fair participation in social functions of the community.

SPECIAL-OPPORTUNITY INEQUALITY

Even when there is relatively little difference in basic facilities including schooling, the opportunities for higher education may be far fewer for young women than for young men. Indeed, gender bias in higher education and professional training can be observed even in some of the richest countries in the world, in Europe and North America. Sometimes this type of asymmetry has been based on the superficially innocuous idea that the respective "provinces" of men and women are just different. This thesis has been championed in different forms over the centuries, and it has always enjoyed a great implicit, as well as explicit, following. It was presented with particular directness more than one hundred years before Queen Victoria's complaint about "woman's rights" by the Reverend James Fordyce in his *Sermons to Young Women* (1766), a book that, as Mary Wollstonecraft noted in *A Vindication of the Rights of Woman* (1792), had been "long made a part of woman's library." Fordyce warned the young women to whom his sermons were addressed against "those masculine women that would plead for your sharing any part of their province with us," identifying the province of men as including not only "war," but also "commerce, politics, exercises of strength and dexterity, abstract philosophy and all the abstruser sciences." Such clear-cut beliefs about the provinces of men and women are now rather rare, but the presence of extensive gender asymmetry can be seen in many areas of education, training, and professional work even in Europe and North America.

PROFESSIONAL INEQUALITY

In employment as well as promotion in work and occupation, women often face greater handicaps than men. A country such as Japan may be quite egalitarian in matters of demography or basic facilities, and even to a great extent in higher education, and yet progress to elevated levels of employment and occupation seems to be much more problematic for women than for men....

OWNERSHIP INEQUALITY

In many societies, the ownership of property can also be very unequal. Even basic assets such as homes and land may be very asymmetrically shared. The absence of claims to property can not only reduce the voice of women, it can also make it harder for women to enter and to flourish in commercial, economic, and even some social activities. Inequality in property ownership is quite widespread across the world, but its severity can vary with local rules. In India, for example, traditional inheritance laws were heavily weighed in favor of male children (until the legal reforms after independence), but the community of Nairs (a large caste in Kerala) has had matrilineal inheritance for a very long time.

HOUSEHOLD INEQUALITY

Often there are fundamental inequalities in gender relations within the family or the household. This can take many different forms. Even in cases in which there are no overt signs of anti-female bias in, say, mortality rates, or male preference in births, or in education, or even in promotion to higher executive positions, family arrangements can be quite unequal in terms of sharing the burden of housework and child care. It is quite common in many societies to take for granted that men will naturally work outside the home, whereas women could do so if and only if they could combine such work with various inescapable and unequally shared household duties. This is sometimes called a "division of labor," though women could be forgiven for seeing it as an "accumulation of labor." The reach of this inequality includes not only unequal relations within the family, but also derivative inequalities in employment and recognition in the outside world. Also, the established persistence of this type of "division" or "accumulation" of labor can also have far-reaching effects on the knowledge and the understanding of different types of work in professional circles. In the 1970s when I first started working on gender inequality, I remember being struck by the fact that the *Handbook of Human Nutrition Requirements of the World Health Organization*, in presenting "calorie requirements" for different categories of people, chose to classify household work as "sedentary activity," requiring very little deployment of energy. I was not able to determine precisely how this remarkable bit of information had been collected....

It is important to take note of the implications of the varieties of gender inequality. The variations entail that inequality between women and men cannot be confronted and overcome by one all-purpose remedy. Over time, moreover, the same country can move from one type of gender inequality to another. I shall presently argue that there is new evidence that India, my own country, is undergoing just such a transformation at this time. The different forms of gender inequality may also impose adversities on the lives of men and boys, in addition to those of women and girls. In understanding the different aspects of the evil of gender inequality, we have to look beyond the predicament of women and examine the problems created for men as well by the asymmetrical treatment of women. These causal connections can be very significant, and they can vary with the form of gender inequality. Finally, inequalities of different kinds can frequently nourish one another, and we have to be aware of their linkages.

In what follows, a substantial part of my empirical focus will be on two of the most elementary kinds of gender inequality: mortality inequality and natality inequality. I shall be concerned particularly with gender inequality in South Asia, the so-called Indian subcontinent. While I shall separate out the subcontinent for special attention, I must warn against the smugness of thinking that the United States and Western Europe are free from gender bias simply because some of the empirical generalizations that can be made about other regions of the world would not hold in the West. Given the many faces of gender inequality, much depends on which face we look at.

Consider the fact that India, along with Bangladesh, Pakistan, and Sri Lanka, has had female heads of government, which the United States and Japan have not yet had.... Indeed, in the case of Bangladesh, where both the prime minister and the leader of the opposition are women, one might begin to wonder whether any man could soon rise to a leadership position there. To take another bit of anecdotal evidence against Western complacence in this matter, I had a vastly larger proportion of tenured women colleagues when I was a professor at Delhi University—as long ago as the 1960s—than I had in the 1990s at Harvard University or presently have at Trinity College, Cambridge.... In the scale of mortality inequality, India is close to the bottom of the league in gender disparity, along with Pakistan and Bangladesh; and natality inequality is also beginning to rear its ugly head very firmly and very fast in the subcontinent in our own day.

In the bulk of the subcontinent, with only a few exceptions (such as Sri Lanka and the state of Kerala in India), female mortality rates are very significantly higher than what could be expected given the mortality patterns of men (in the respective age groups). This type of gender inequality need not entail any conscious homicide, and it would be a mistake to try to explain this large phenomenon by invoking the cases of female infanticide that are reported from China or India: those are truly dreadful events, but they are relatively rare. The mortality disadvantage of women works, rather, mainly through the widespread neglect of health, nutrition, and other interests of women that influence their survival....

It has been widely observed that given similar health care and nutrition, women tend typically to have lower age-specific mortality rates than men. Indeed, even female fetuses tend to have a lower probability of miscarriage than male fetuses. Everywhere in the world, more male babies are born than female babies (and an even higher proportion of male fetuses are conceived compared with female fetuses); but throughout their respective lives the proportion of males goes on falling as we move to higher and higher age groups, due to typically greater male mortality rates. The excess of females over males in the populations of Europe and North America comes about as a result of this greater survival chance of females in different age groups.

In many parts of the world, however, women receive less attention and health care than do men, and girls in particular often receive very much less support than boys. As a result of this gender bias, the mortality rates of females often exceed those of males in these countries. The concept of the "missing women" was devised to give some idea of the enormity of the phenomenon of women's adversity in mortality by focusing on the women who are simply not there, owing to mortality rates that are unusually high compared with male mortality rates. The basic idea is to find some rough and ready way to understand the quantitative difference between the actual number of women in these countries and the number of women that we could expect to see if the gender pattern of mortality were similar there to the patterns in other regions of the world that do not demonstrate a significant bias against women in health care and other attentions relevant for survival....

The problem of gender bias in life and death has been much discussed, but there are other issues of gender inequality that are sorely in need of greater investigation. I will note four substantial phenomena that happen to be quite widely observed in South Asia.

There is, first, the problem of the undernourishment of girls as compared with boys. At the time of birth, girls are obviously no more nutritionally deprived than boys, but this situation changes as society's unequal treatment takes over from the non-discrimination of nature. There has been plenty of aggregative evidence on this for quite some time now; but it has been accompanied by some anthropological skepticism about the appropriateness of using aggregate statistics with pooled data from different regions to interpret the behavior of individual families. Still there have also been more detailed and concretely local studies on this subject, and they confirm the picture that emerges on the basis of aggregate statistics. One case study from India, which I myself undertook in 1983 along with Sunil Sengupta, involved weighing every child in two large villages. The time pattern that emerged from this study, which concentrated particularly on weight-for-age as the chosen indicator of nutritional level for children under five, showed clearly how an initial neonatal condition of broad nutritional symmetry turns gradually into a situation of significant female disadvantage. The local investigations tend to confirm rather than contradict the picture that emerges from aggregate statistics.

In interpreting the causal process that leads to this female disadvantage, it is important to emphasize that the lower level of nourishment of girls may not relate directly to their being underfed as compared with boys. Often enough, the differences may arise more from the neglect of heath care of girls compared with what boys receive. Indeed, there is some direct information about comparative medical neglect of girls vis-à-vis boys in South Asia. When I studied, with Jocelyn Kynch, admissions data from two large public hospitals in Bombay, it was very striking to find clear evidence that the admitted girls were typically more ill than the boys, suggesting that a girl has to be more stricken and more ill before she is taken to the hospital. Undernourishment may well result from a greater incidence of illness, which can adversely affect both the absorption of nutrients and the performance of bodily functions.

There is, secondly, a high incidence of maternal undernourishment in South Asia. Indeed, in this part of the world, maternal undernutrition is much more common than in most other regions. Comparisons of body mass index (BMI), which is essentially a measure of weight for height, bring this out clearly enough, as do statistics of such consequential characteristics as the incidence of anemia.

Thirdly, there is the problem of the prevalence of low birth weight. In South Asia, as many as 21 percent of children are born clinically underweight (by accepted medical standards), more than in any other substantial region in the world. The predicament of being low in weight in childhood seems often enough to begin at birth in the case of South Asian children. In terms of weight for age, around 40 to 60 percent of the children in South Asia are undernourished, compared with 20 to 40 percent undernourishment even in sub-Saharan Africa. The children start deprived and stay deprived. Finally, there is also a high incidence of cardiovascular diseases. Generally, South Asia stands out as having more cardiovascular diseases than any other part of the Third World. Even when other countries, such as China, show a greater prevalence of the standard predisposing conditions to such illness, the subcontinental population seems to have more heart problems than these other countries.

It is not difficult to see that the first three of these problems are very likely connected causally. The neglect of the care of girls and women, and the underlying gender bias that their experience reflects, would tend to yield more maternal undernourishment; and this in turn would tend to yield more fetal deprivation and distress, and underweight babies, and child undernourishment....

These biological connections illustrate a more general point: gender inequality can hurt the interests of men as well as women. Indeed, men suffer far more from cardiovascular diseases than do women. Given the uniquely critical role of women in the reproductive process, it would be hard to imagine that the deprivation to which women are subjected would not have some adverse impact on the lives of all people—men as well as women, adults as well as children—who are "born of a woman," as the Book of Job says. It would appear that the extensive penalties of neglecting the welfare of women rebound on men with a vengeance.

But there are also other connections between the disadvantage of women and the general condition of society—non-biological connections—that operate through women's conscious agency. The expansion of women's capabilities not only enhances women's own freedom and well-being, it also has many other effects on the lives of all. An enhancement of women's active agency can contribute substantially to the lives of men as well as women, children as well as adults; many studies have demonstrated that the greater empowerment of women tends to reduce child neglect and mortality, to decrease fertility and overcrowding, and more generally to broaden social concern and care....

There is something to cheer in the developments that I have been discussing, and there is considerable evidence of a weakened hold of gender disparity in several fields in the subcontinent; but the news is not, alas, all good. There is also evidence of a movement in the contrary direction, at least with regard to natality inequality.... For India as a whole, the female-male ratio of the population under age six has fallen from 94.5 girls per 100 boys in 1991 to 92.7 girls per 100 boys in 2001. While there has been no such decline in some parts of the country (most notably Kerala), it has fallen very sharply in Punjab, Haryana, Gujarat, and Maharashtra, which are among the richer Indian states.

Taking together all the evidence that exists, it is clear that this change reflects not a rise in female child mortality, but a fall in female births vis-à-vis male births; and it is almost certainly connected with the increased availability and the greater use of gender determination of fetuses. Fearing that sex-selective abortion might occur in India, the Indian parliament some years ago banned the use of sex determination techniques for fetuses, except as a by-product of other necessary medical investigation. But it appears that the enforcement of this law has been comprehensively neglected....

I do not believe that this need be an insurmountable difficulty (other types of evidence can in fact be used for prosecution), but the reluctance of the mothers to give evidence brings out perhaps the most disturbing aspect of this natality inequality. I refer to the "son preference" that many Indian mothers themselves seem to harbor. This form of gender inequality cannot be removed, at least in the short run, by the enhancement of women's empowerment and agency, since that agency is itself an integral part of the cause of natality inequality.

Policy initiatives have to take adequate note of the fact that the pattern of gender inequality seems to be shifting in India, right at this time, from mortality inequality (the female life expectancy at birth has now become significantly higher than male life expectancy) to natality inequality. And, worse, there is clear evidence that the traditional routes of combating gender inequality, such as the use of public policy to influence female education and female economic participation, may not, on their own, serve as a path to the eradication of natality inequality. A sharp pointer in that direction comes from the countries in East Asia that have high levels of female education and economic participation....

Still, there are reasons for concern. For a start, these may be early days, and it has to be asked whether with the spread of sex-selective abortion India may catch up with—and perhaps even go beyond—Korea and China. Moreover, even now there are substantial variations within India, and the all-India average hides the fact that there are states in India where the female-male ratio for children is very much lower than the Indian average....

Gender inequality, then, has many distinct and dissimilar faces. In overcoming some of its worst manifestations, especially in mortality rates, the cultivation of women's empowerment and agency, through such means as women's education and gainful employment, has proved very effective. But in dealing with the new form of gender inequality, the injustice relating to natality, there is a need to go beyond the question of the agency of women and to look for a more critical assessment of received values. When anti-female bias in behavior (such as sex-specific abortion) reflects the hold of traditional masculinist values from which mothers themselves may not be immune, what is needed is not just freedom of action but also freedom of thought—the freedom to question and to scrutinize inherited beliefs and traditional priorities. Informed critical agency is important in combating inequality of every kind, and gender inequality is no exception.

KEY CONCEPTS

mortality rate | mortality inequality | natality inequality

DISCUSSION QUESTIONS

1. Look at the most recent census data for the United States and consider the birth rates and mortality rates as compared to those in India presented in this reading. Are the rates drastically different? What explanations can you offer for the mortality rate in the United States?
2. Of the different kinds of gender inequality outlined in the beginning of this reading, which ones do you see as most prevalent? Have you personally experienced any of these types of inequality?

Applying Sociological Knowledge: An Exercise for Students

Imagine that you wake up tomorrow morning and are a member of the other gender. Make a list of all the things about yourself that you think would have changed. After making your list, make a note of which characteristics are biological or physical, which are attitudinal, which involve behavior, and which are institutional. Then ask yourself how individual and institutional forms of gender are related.

PART XII

Sexuality and Intimate Relationships

38

"Dude, You're a Fag": Adolescent Masculinity and the Fag Discourse

C. J. PASCOE

In this piece, Pascoe summarizes the results from her field research in a working-class, suburban high school in California. She examines the use of the word fag *as an insult among and between adolescent males. Using gender and queer theory, the author presents evidence that there is a discourse that uses the word* fag *and negative homosexual stereotypes in interactions among high school boys. The term* fag *is not often used to refer to actual homosexual boys. Instead, heterosexual high school boys use the term to mock or tease other heterosexual boys. The "fag discourse," then, is a tool for establishing and highlighting the masculinity of the person using the language.*

"There's a faggot over there! There's a faggot over there! Come look!" yelled Brian, a senior at River High School, to a group of 10-year-old boys. Following Brian, the 10-year-olds dashed down a hallway. At the end of the hallway Brian's friend, Dan, pursed his lips and began sashaying towards the 10-year-olds. He minced towards them, swinging his hips exaggeratedly and wildly waving his arms. To the boys Brian yelled, "Look at the faggot! Watch out! He'll get you!" In response the 10-year-olds raced back down the hallway screaming in terror. (From author's field notes)

The relationship between adolescent masculinity and sexuality is embedded in the specter of the faggot. Faggots represent a penetrated masculinity in which "to be penetrated is to abdicate power" (Bersani, 1987: 212). Penetrated men symbolize a masculinity devoid of power, which, in its contradiction, threatens both psychic and social chaos. It is precisely this specter of penetrated masculinity that functions as a regulatory mechanism of gender for contemporary American adolescent boys.

Feminist scholars of masculinity have documented the centrality of homophobic insults to masculinity (Lehne, 1998; Kimmel, 2001) especially in school settings (Wood, 1984; Smith, 1998; Burn, 2000; Plummer, 2001; Kimmel, 2003). They argue that homophobic teasing often characterizes masculinity in adolescence and early adulthood, and that anti-gay slurs tend to primarily be directed at other gay boys.

SOURCE: Pascoe, C. J. 2005. "'Dude, You're a Fag': Adolescent Masculinity and the Fag Discourse." *Sexualities* 8: 329–346 by SAGE Publications. Reprinted by permission of SAGE Publications.

This article both expands on and challenges these accounts of relationships between homophobia and masculinity. Homophobia is indeed a central mechanism in the making of contemporary American adolescent masculinity. This article both critiques and builds on this finding by (1) pointing to the limits of an argument that focuses centrally on homophobia, (2) demonstrating that the fag is not only an identity linked to homosexual boys but an identity that can temporarily adhere to heterosexual boys as well and (3) highlighting the racialized nature of the fag as a disciplinary mechanism.

"Homophobia" is too facile a term with which to describe the deployment of "fag" as an epithet. By calling the use of the word "fag" homophobia—and letting the argument stop with that point—previous research obscures the gendered nature of sexualized insults (Plummer, 2001). Invoking homophobia to describe the ways in which boys aggressively tease each other overlooks the powerful relationship between masculinity and this sort of insult. Instead, it seems incidental in this conventional line of argument that girls do not harass each other and are not harassed in this same manner. This framing naturalizes the relationship between masculinity and homophobia, thus obscuring the centrality of such harassment in the formation of a gendered identity for boys in a way that it is not for girls.

"Fag" is not necessarily a static identity attached to a particular (homosexual) boy. Fag talk and fag imitations serve as a discourse with which boys discipline themselves and each other through joking relationships. Any boy can temporarily become a fag in a given social space or interaction. This does not mean that those boys who identify as or are perceived to be homosexual are not subject to intense harassment. But becoming a fag has as much to do with failing at the masculine tasks of competence, heterosexual prowess and strength or in anyway revealing weakness or femininity, as it does with a sexual identity. This fluidity of the fag identity is what makes the specter of the fag such a powerful disciplinary mechanism. It is fluid enough that boys police most of their behaviors out of fear of having the fag identity permanently adhere and definitive enough so that boys recognize a fag behavior and strive to avoid it.

The fag discourse is racialized. It is invoked differently by and in relation to white boys' bodies than it is by and in relation to African-American boys' bodies. While certain behaviors put all boys at risk for becoming temporarily a fag, some behaviors can be enacted by African-American boys without putting them at risk of receiving the label. The racialized meanings of the fag discourse suggest that something more than simple homophobia is involved in these sorts of interactions. An analysis of boys' deployments of the specter of the fag should also extend to the ways in which gendered power works through racialized selves.

THEORETICAL FRAMING

The sociology of masculinity entails a "critical study of men, their behaviors, practices, values and perspectives" (Whitehead and Barrett, 2001: 14). Recent studies of men emphasize the multiplicity of masculinity (Connell, 1995) detailing the ways in which different configurations of gender practice are promoted, challenged or reinforced in given social situations....

Heeding Timothy Carrigan's admonition that an "analysis of masculinity needs to be related as well to other currents in feminism" (Carrigan et al., 1987: 64), in this article I integrate queer theory's insights about the relationships between gender, sexuality, identities and power with the attention to men found in the literature on masculinities. Like the sociology of gender, queer theory destabilizes the assumed naturalness of the social order (Lemert, 1996). Queer theory is a "conceptualization which sees sexual power as embedded in different levels of social life" and interrogates areas of the social world not usually seen as sexuality (Stein and Plummer, 1994). In this sense queer theory calls for sexuality to be looked at not only as a discrete arena of sexual practices and identities, but also as a constitutive element of social life (Warner, 1993; Epstein, 1996)....

... This article does not seek to establish that there are homosexual boys and heterosexual boys and the homosexual ones are marginalized. Rather this article explores what happens to theories of gender if we look at a *discourse* of sexualized identities in addition to focusing on seemingly static identity categories inhabited by men. This is not to say that gender is reduced only to sexuality, indeed feminist scholars have demonstrated that gender is embedded in and constitutive of a multitude of social structures—the economy, places of work, families and schools. In the tradition of post-structural feminist theorists of race and gender who look at "border cases" that explode taken-for-granted binaries of race and gender (Smith, 1994), queer theory is another tool which enables an integrated analysis of sexuality, gender and race.

As scholars of gender have demonstrated, gender is accomplished through day-to-day interactions (Fine, 1987; Hochschild, 1989; West and Zimmerman, 1991; Thorne, 1993). In this sense gender is the "activity of managing situated conduct in light of normative conceptions of attitudes and activities appropriate for one's sex category" (West and Zimmerman, 1991: 127). Similarly, queer theorist Judith Butler argues that gender is accomplished interactionally through "a set of repeated acts within a highly rigid regulatory frame that congeal over time to produce the appearance of substance, of a natural sort of being" (Butler, 1999: 43). Specifically she argues that gendered beings are created through processes of citation and repudiation of a "constrictive outside" (Butler, 1993: 3) in which is contained all that is cast out of a socially recognizable gender category. The "constitutive outside" is inhabited by abject identities, unrecognizably and unacceptably gendered selves. The interactional accomplishment of gender in a Butlerian model consists, in part, of the continual iteration and repudiation of this abject identity. Gender, in this sense, is "constituted through the force of exclusion and abjection, on which produces a constitutive outside to the subject, an abjected outside, which is, after all, 'inside' the subject as its own founding repudiation" (Butler, 1993: 3). This repudiation creates and reaffirms a "threatening specter" (Butler, 1993: 3) of failed, unrecognizable gender, the existence of which must be continually repudiated through interactional processes.

I argue that the "fag" position is an "abject" position and, as such, is a "threatening specter" constituting contemporary American adolescent masculinity. The fag discourse is the interactional process through which boys name and repudiate this abjected identity. Rather than analyzing the fag as an identity for homosexual boys, I examine uses of the discourse that imply that any boy can become a fag, regardless of his actual desire or self-perceived sexual orientation. The threat of the abject position infuses the faggot with regulatory power. This article provides empirical data to illustrate Butler's approach to gender and indicates that it might be a useful addition to the sociological literature on masculinities through highlighting one of the ways in which a masculine gender identity is accomplished through interaction.

METHOD

Research Site

I conducted fieldwork at a suburban high school in north-central California which I call River High. River High is a working class, suburban 50-year-old high school located in a town called Riverton. With the exception of the median household income and racial diversity (both of which are elevated due to Riverton's location in California), the town mirrors national averages in the percentages of white collar workers, rates of college attendance, and marriages, and age composition (according to the 2000 census). It is a politically moderate to conservative, religious community. Most of the students' parents commute to surrounding cities for work.

On average Riverton is a middle-class community. However, students at River are likely to refer to the town as two communities: "Old Riverton" and "New Riverton." A busy highway and railroad tracks bisect the town into these two sections. River High is literally on the "wrong side of the tracks," in Old Riverton. Exiting the freeway, heading north to Old Riverton, one sees a mix of 1950s-era ranch-style homes, some with neatly trimmed lawns and tidy gardens, others with yards strewn with various car parts, lawn chairs and appliances. Old Riverton is visually bounded by smoke-puffing factories. On the other side of the freeway New Riverton is characterized by wide sidewalk-lined streets and new walled-in home developments. Instead of smokestacks, a forested mountain, home to a state park, rises majestically in the background. The teens from these homes attend Hillside High, River's rival.

River High is attended by 2000 students. River High's racial/ethnic breakdown roughly represents California at large: 50 percent white, 9 percent African-American, 28 percent Latino and 6 percent Asian (as compared to California's 46, 6, 32, and 11 percent respectively, according to census data and school records). The students at River High are primarily working class.

Research

I gathered data using the qualitative method of ethnographic research. I spent a year and a half conducting observations, formally interviewing 49 students at

River High (36 boys and 13 girls), one male student from Hillside High, and conducting countless informal interviews with students, faculty and administrators. I concentrated on one school because I explore the richness rather than the breadth of data.

I recruited students for interviews by conducting presentations in a range of classes and hanging around at lunch, before school, after school at various events talking to different groups of students about my research, which I presented as "writing a book about guys." The interviews usually took place at school, unless the student had a car, in which case he or she met me at one of the local fast food restaurants where I treated them to a meal. Interviews lasted anywhere from half an hour to two hours.

The initial interviews I conducted helped me to map a gendered and sexualized geography of the school, from which I chose my observation sites. I observed a "neutral" site—a senior government classroom, where sexualized meanings were subdued. I observed three sites that students marked as "fag" sites—two drama classes and the Gay/Straight Alliance. I also observed two normatively "masculine" sites—auto-shop and weightlifting. I took daily field notes focusing on how students, faculty and administrators negotiated, regulated and resisted particular meanings of gender and sexuality. I attended major school rituals such as Winter Ball, school rallies, plays, dances and lunches. I would also occasionally "ride along" with Mr Johnson (Mr J.), the school's security guard, on his battery-powered golf cart to watch which, how and when students were disciplined. Observational data provided me with more insight to the interactional processes of masculinity than simple interviews yielded. If I had relied only on interview data I would have missed the interactional processes of masculinity which are central to the fag discourse.

Given the importance of appearance in high school, I gave some thought as to how I would present myself, deciding to both blend in and set myself apart from the students. In order to blend in I wore my standard graduate student gear—comfortable, baggy cargo pants, a black t-shirt or sweater and tennis shoes. To set myself apart I carried a messenger bag instead of a backpack, didn't wear makeup, and spoke slightly differently than the students by using some slang, but refraining from uttering the ubiquitous "hecka" and "hella."

The boys were fascinated by the fact that a 30-something white "girl" (their words) was interested in studying them. While at first many would make sexualized comments asking me about my dating life or saying that they were going to "hit on" me, it seemed eventually they began to forget about me as a potential sexual/romantic partner. Part of this, I think, was related to my knowledge about "guy" things. For instance, I lift weights on a regular basis and as a result the weightlifting coach introduced me as a "weight-lifter from U.C. Berkeley," telling the students they should ask me for weight-lifting advice. Additionally, my taste in movies and television shows often coincided with theirs. I am an avid fan of the movies "Jackass" and "Fight Club," both of which contain high levels of violence and "bathroom" humor. Finally, I garnered a lot of points among boys because I live off a dangerous street in a nearby city famous for drug deals, gang fights and frequent gun shots.

WHAT IS A FAG?

"Since you were little boys you've been told, 'hey, don't be a little faggot,'" explained Darnell, an African-American football player, as we sat on a bench next to the athletic field. Indeed, both the boys and girls I interviewed told me that "fag" was the worst epithet one guy could direct at another. Jeff, a slight white sophomore, explained to me that boys call each other fag because "gay people aren't really liked over here and stuff." Jeremy, a Latino Junior told me that this insult literally reduced a boy to nothing, "To call someone gay or fag is like the lowest thing you can call someone. Because that's like saying that you're nothing."

Most guys explained their or others' dislike of fags by claiming that homophobia is just part of what it means to be a guy. For instance Keith, a white soccer-playing senior, explained, "I think guys are just homophobic." However, it is not just homophobia, it is a *gendered* homophobia. Several students told me that these homophobic insults only applied to boys and not girls. For example, while Jake, a handsome white senior, told me that he didn't like gay people, he quickly added, "Lesbians, okay that's *good*." Similarly Cathy, a popular white cheerleader, told me "Being a lesbian is accepted because guys think 'oh that's cool.'" Darnell, after telling me that boys were told not to be faggots, said of lesbians, "They're [guys are] fine with girls. I think it's the guy part that they're like ewwww!" In this sense it is not strictly homophobia, but a gendered homophobia that constitutes adolescent masculinity in the culture of this school. However, it is clear, according to these comments, that lesbians are "good" because of their place in heterosexual male fantasy, not necessarily because of some enlightened approach to same-sex relationships. It does, however, indicate that using only the term homophobia to describe boys' repeated use of the word "fag" might be a bit simplistic and misleading.

Additionally, girls at River High rarely deployed the word "fag" and were never called "fags." I recorded girls uttering "fag" only three times during my research. In one instance, Angela, a Latina cheerleader, teased Jeremy, a well-liked white senior involved in student government, for not ditching school with her, "You wouldn't 'cause you're a faggot." However, girls did not use this word as part of their regular lexicon. The sort of gendered homophobia that constitutes adolescent masculinity does not constitute adolescent femininity. Girls were not called dykes or lesbians in any sort of regular or systematic way. Students did tell me that "slut" was the worst thing a girl could be called. However, my field notes indicate that the word "slut" (or its synonym "ho") appears one time for every eight times the word "fag" appears. Even when it does occur, "slut" is rarely deployed as a direct insult against another girl.

Highlighting the difference between the deployment of "gay" and "fag" as insults brings the gendered nature of this homophobia into focus. For boys and girls at River High "gay" is a fairly common synonym for "stupid." While this word shares the sexual origins of "fag," it does not *consistently* have the skew of gender-loaded meaning. Girls and boys often used "gay" as an adjective referring to inanimate objects and male or female people, whereas they used "fag" as a noun that denotes only un-masculine males. Students used "gay" to describe anything from someone's clothes to a new school rule that the students did not like, as in the following encounter:

> In auto-shop Arnie pulled out a large older version black laptop computer and placed it on his desk. Behind him Nick said "That's a gay laptop! It's five inches thick!"

A laptop can be gay, a movie can be gay or a group of people can be gay. Boys used "gay" and "fag" interchangeably when they refer to other boys, but "fag" does not have the non-gendered attributes that "gay" sometimes invokes.

While its meanings are not the same as "gay," "fag" does have multiple meanings which do not necessarily replace its connotations as a homophobic slur, but rather exist alongside. Some boys took pains to say that "fag" is not about sexuality. Darnell told me "It doesn't even have anything to do with being gay." J.L., a white sophomore at Hillside High (River High's cross-town rival) asserted "Fag, seriously, it has nothing to do with sexual preference at all. You could just be calling somebody an idiot you know?" I asked Ben, a quiet, white sophomore who wore heavy metal t-shirts to auto-shop each day, "What kind of things do guys get called a fag for?" Ben answered "Anything … literally, anything. Like you were trying to turn a wrench the wrong way, 'dude, you're a fag.' Even if a piece of meat drops out of your sandwich, 'you fag!'" Each time Ben said "you fag" his voice deepened as if he were imitating a more masculine boy. While Ben might rightly *feel* like a guy could be called a fag for "anything … literally, anything," there are actually specific behaviors which, when enacted by most boys, can render him more vulnerable to a fag epithet. In this instance Ben's comment highlights the use of "fag" as a generic insult for incompetence, which in the world of River High, is central to a masculine identity. A boy could get called a fag for exhibiting any sort of behavior defined as non-masculine (although not necessarily behaviors aligned with femininity) in the world of River High: being stupid, incompetent, dancing, caring too much about clothing, being too emotional or expressing interest (sexual or platonic) in other guys. However, given the extent of its deployment and the laundry list of behaviors that could get a boy in trouble, it is no wonder that Ben felt like a boy could be called "fag" for "anything."

One-third (13) of the boys I interviewed told me that, while they may liberally insult each other with the term, they would not actually direct it at a homosexual peer. Jabes, a Filipino senior, told me

> I actually say it [fag] quite a lot, except for when I'm in the company of an actual homosexual person. Then I try not to say it at all. But when I'm just hanging out with my friends I'll be like, "shut up, I don't want to you hear you any more, you stupid fag."

Similarly J.L. compared homosexuality to a disability, saying there is "no way" he'd call an actually gay guy a fag because

> There's people who are the retarded people who nobody wants to associate with. I'll be so nice to those guys and I hate it when people make fun of them. It's like, "bro do you realize that they can't help that?" And then there are gay people. They were born that way.

According to this group of boys, gay is a legitimate, if marginalized, social identity. If a man is gay, there may be a chance he could be considered masculine by other men (Connell, 1995). David, a handsome white senior dressed smartly in khaki pants and a white button-down shirt said, "Being gay is just a lifestyle. It's someone you choose to sleep with. You can still throw around a football and be gay." In other words there is a possibility, however slight, that a boy can be gay and masculine. To be a fag is, by definition, the opposite of masculine, whether or not the word is deployed with sexualized or non-sexualized meanings. In explaining this to me, Jamaal, an African-American junior, cited the explanation of popular rap artist, Eminem,

> Although I don't like Eminem, he had a good definition of it. It's like taking away your title. In an interview they were like, "you're always capping on gays, but then you sing with Elton John." He was like "I don't mean gay as in gay."

This is what Riki Wilchins calls the "Eminem Exception. Eminem explains that he doesn't call people 'faggot' because of their sexual orientation but because they're weak and unmanly" (Wilchins, 2003). This is precisely the way in which this group of boys at River High uses the term "faggot." While it is not necessarily acceptable to be gay, at least a man who is gay can do other things that render him acceptably masculine. A fag, by the very definition of the word, indicated by students' usages at River High, cannot be masculine. This distinction between "fag" as an unmasculine and problematic identity and "gay" as a possibly masculine, although marginalized, sexual identity is not limited to a teenage lexicon, but is reflected in both psychological discourses (Sedgwick, 1995) and gay and lesbian activism.

BECOMING A FAG

"The ubiquity of the word faggot speaks to the reach of its discrediting capacity" (Corbett, 2001: 4). It is almost as if boys cannot help but shout it out on a regular basis—in the hallway, in class, across campus as a greeting, or as a joke. In my fieldwork I was amazed by the way in which the word seemed to pop uncontrollably out of boys' mouths in all kinds of situations. To quote just one of many instances from my field notes:

> Two boys walked out of the P.E. locker room and one yelled "fucking faggot!" at no one in particular.

This spontaneous yelling out of a variation of fag seemingly apropos of nothing happened repeatedly among boys throughout the school.

The fag discourse is central to boys' joking relationships. Joking cements relationships between boys (Kehily and Nayak, 1997; Lyman, 1998) and helps to manage anxiety and discomfort (Freud, 1905). Boys invoked the specter of the fag in two ways: through humorous imitation and through lobbing the epithet at one another. Boys at River High imitated the fag by acting out an exaggerated "femininity," and/or by pretending to sexually desire other boys. As indicated by the introductory vignette in which a predatory "fag" threatens the little boys, boys at River High link these performative scenarios with a fag identity. They lobbed the fag epithet at each other in a verbal game of hot potato, each careful to deflect the insult quickly by hurling it toward someone else. These games and imitations make up a fag discourse which highlights the fag not as a static but rather as a fluid identity which boys constantly struggle to avoid.

In imitative performances the fag discourse functions as a constant reiteration of the fag's existence, affirming that the fag is out there; at any moment a boy can become a fag. At the same time these performances demonstrate that the boy who is invoking the fag is *not* a fag. By invoking it so often, boys remind themselves and each other that at any point they can become fags if they are not sufficiently masculine.

> Mr. McNally, disturbed by the noise outside of the classroom, turned to the open door saying "We'll shut this unless anyone really wants to watch sweaty boys playing basketball." Emir, a tall skinny boy, lisped "I wanna watch the boys play!" The rest of the class cracked up at his imitation.

Through imitating a fag, boys assure others that they are not a fag by immediately becoming masculine again after the performance. They mock their own performed femininity and/or same-sex desire, assuring themselves and others that such an identity is one deserving of derisive laughter. The fag identity in this instance is fluid, detached from Emir's body. He can move in and out of this "abject domain" while simultaneously affirming his position as a subject.

Boys also consistently tried to put another in the fag position by lobbing the fag epithet at one another.

> Going through the junk-filled car in the auto-shop parking lot, Jay poked his head out and asked "Where are Craig and Brian?" Neil, responded with "I think they're over there," pointing, then thrusting his hips and pulling his arms back and forth to indicate that Craig and Brian might be having sex. The boys in auto-shop laughed.

This sort of joke temporarily labels both Craig and Brian as faggots. Because the fag discourse is so familiar, the other boys immediately understand that Neil is indicating that Craig and Brian are having sex. However these are not necessarily identities that stick. Nobody actually thinks Craig and Brian are homosexuals. Rather the fag identity is a fluid one, certainly an identity that no boy wants, but one that a boy can escape, usually by engaging in some sort of discursive contest to turn another boy into a fag. However, fag becomes a hot potato that no boy wants to be left holding. In the following example, which occurred soon after the "sex" joke, Brian lobs the fag epithet at someone else, deflecting it from himself:

> Brian initiated a round of a favorite game in auto-shop, the "cock game." Brian quietly, looking at Josh, said, "Josh loves the cock," then slightly louder, "Josh loves the cock." He continued saying this until he was yelling "JOSH LOVES THE COCK!" The rest of the boys

> laughed hysterically as Josh slinked away saying "I have a bigger dick than all you mother fuckers!"

These two instances show how the fag can be mapped, momentarily, on to one boy's body and how he, in turn, can attach it to another boy, thus deflecting it from himself.

These examples demonstrate boys invoking the trope of the fag in a discursive struggle in which the boys indicate that they know what a fag is—and that they are not fags. This joking cements bonds between boys as they assure themselves and each other of their masculinity through repeated repudiations of a non-masculine position of the abject.

RACING THE FAG

The fag trope is not deployed consistently or identically across social groups at River High. Differences between white boys' and African-American boys' meaning making around clothes and dancing reveal ways in which the fag as the abject position is racialized.

Clean, oversized, carefully put together clothing is central to a hip-hop identity for African-American boys who identify with hip-hop culture. Richard Majors calls this presentation of self a "cool pose" consisting of "unique, expressive and conspicuous styles of demeanor, speech, gesture, clothing, hairstyle, walk, stance and handshake," developed by African-American men as a symbolic response to institutionalized racism (Majors, 2001: 211). Pants are usually several sizes too big, hanging low on a boy's waist, usually revealing a pair of boxers beneath. Shirts and sweaters are similarly oversized, often hanging down to a boy's knees. Tags are frequently left on baseball hats worn slightly askew and sit perched high on the head. Meticulously clean, unlaced athletic shoes with rolled up socks under the tongue complete a typical hip-hop outfit.

This amount of attention and care given to clothing for white boys not identified with hip-hop culture (that is, most of the white boys at River High) would certainly cast them into an abject, fag position. White boys are not supposed to appear to care about their clothes or appearance, because only fags care about how they look. Ben illustrates this:

> Ben walked in to the auto-shop classroom from the parking lot where he had been working on a particularly oily engine. Grease stains covered his jeans. He looked down at them, made a face and walked toward me with limp wrists, laughing and lisping in a high pitch sing-song voice "I got my good panths all dirty!"

Ben draws on indicators of a fag identity, such as limp wrists, as do the boys in the introductory vignette to illustrate that a masculine person certainly would not care about having dirty clothes. In this sense, masculinity, for white boys, becomes the carefully crafted appearance of not caring about appearance, especially in terms of cleanliness.

However, African-American boys involved in hip-hop culture talk frequently about whether or not their clothes, specifically their shoes, are dirty:

> In drama class both Darnell and Marc compared their white Adidas basketball shoes. Darnell mocked Marc because black scuff marks covered his shoes, asking incredulously "Yours are a week old and they're dirty—I've had mine for a month and they're not dirty!" Both laughed.

Monte, River High's star football player echoed this concern about dirty shoes when looking at the fancy red shoes he had lent to his cousin the week before, told me he was frustrated because after his cousin used them, the "shoes are hella scuffed up." Clothing, for these boys, does not indicate a fag position, but rather defines membership in a certain cultural and racial group (Perry, 2002).

Dancing is another arena that carries distinctly fag associated meanings for white boys and masculine meanings for African-American boys who participate in hip-hop culture. White boys often associate dancing with "fags." J.L. told me that guys think "'NSync's gay" because they can dance. 'NSync is an all white male singing group known for their dance moves. At dances white boys frequently held their female dates tightly, locking their hips together. The boys never danced with one another, unless engaged in a round of "hot potato." White boys often jokingly danced together in order to embarrass each other by making someone else into a fag:

> Lindy danced behind her date, Chris. Chris's friend, Matt, walked up and nudged Lindy aside, imitating her dance moves behind Chris. As Matt rubbed his hands up and down Chris's back, Chris turned around and jumped back startled to see Matt there instead of Lindy. Matt cracked up as Chris turned red.

However dancing does not carry this sort of sexualized gender meaning for all boys at River High. For African-American boys, dancing demonstrates membership in a cultural community (Best, 2000). African-American boys frequently danced together in single sex groups, teaching each other the latest dance moves, showing off a particularly difficult move or making each other laugh with humorous dance moves. Students recognized K.J. as the most talented dancer at the school. K.J. is a sophomore of African-American and Filipino descent who participated in the hip-hop culture of River High. He continually wore the latest hip-hop fashions. K.J. was extremely popular. Girls hollered his name as they walked down the hall and thrust urgently written love notes folded in complicated designs into his hands as he sauntered to class. For the past two years K.J. won first place in the talent show for dancing. When he danced at assemblies the room reverberated with screamed chants of "Go K.J.! Go K.J.! Go K.J.!" Because dancing for African-American boys places them within a tradition of masculinity, they are not at risk of becoming a fag for this particular gendered practice. Nobody called K.J. a fag. In fact in several of my interviews, boys of multiple racial/ethnic backgrounds spoke admiringly of K.J.'s dancing abilities.

IMPLICATIONS

These findings confirm previous studies of masculinity and sexuality that position homophobia as central to contemporary definitions of adolescent masculinity. These data extend previous research by unpacking multi-layered meanings that boys deploy through their uses of homophobic language and joking rituals. By attending to these meanings I reframe the discussion as one of a fag discourse, rather than simply labeling this sort of behavior as homophobia. The fag is an "abject" position, a position outside of masculinity that actually constitutes masculinity. Thus, masculinity, in part, becomes the daily interactional work of repudiating the "threatening specter" of the fag.

The fag extends beyond a static sexual identity attached to a gay boy. Few boys are permanently identified as fags; most move in and out of fag positions. Looking at "fag" as a discourse rather than a static identity reveals that the term can be invested with different meanings in different social spaces. "Fag" may be used as a weapon with which to temporarily assert one's masculinity by denying it to others. Thus "fag" becomes a symbol around which contests of masculinity take place.

The fag epithet, when hurled at other boys, may or may not have explicit sexual meanings, but it always has gendered meanings. When a boy calls another boy a fag, it means he is not a man, not necessarily that he is a homosexual. The boys in this study know that they are not supposed to call homosexual boys "fags" because that is mean. This, then, has been the limited success of the mainstream gay rights movement. The message absorbed by some of these teenage boys is that "gay men can be masculine, just like you." Instead of challenging gender inequality, this particular discourse of gay rights has reinscribed it. Thus we need to begin to think about how gay men may be in a unique position to challenge gendered as well as sexual norms.

This study indicates that researchers who look at the intersection of sexuality and masculinity need to attend to the ways in which racialized identities may affect how "fag" is deployed and what it means in various social situations.... It is important to look at when, where and with what meaning "the fag" is deployed in order to get at how masculinity is defined, contested, and invested in among adolescent boys.

REFERENCES

Bersani, Leo. (1987). "Is the Rectum a Grave?" *October* 43: 197–222.

Best, Amy. (2000). *From Night: Youth, Schools and Popular Culture*. New York: Routledge.

Burn, Shawn M. (2000). "Heterosexuals' Use of 'Fag' and 'Queer' to Deride One Another: A Contributor to Heterosexism and Stigma." *Journal of Homosexuality* 40: 1–11.

Butler, Judith. (1993). *Bodies that Matter*. New York: Routledge.

Butler, Judith. (1999). *Gender Trouble*. New York: Routledge.

Carrigan, Tim, Connell, Bob and Lee, John. (1987). "Toward a New Sociology of Masculinity," in Harry Brod (ed.), *The Making of Masculinities: The New Men's Studies*, pp. 188–202. Boston, MA: Allen & Unwin.

Connell, R.W. (1995). *Masculinities*. Berkeley: University of California Press.

Corbett, Ken. (2001). "Faggot—Loser." *Studies in Gender and Sexuality* 2: 3–28.

Epstein, Steven. (1996). 'A Queer Encounter', in Steven Seidman (ed.). *Queer Theory/Sociology*, pp. 188–202. Cambridge, MA: Blackwell.

Fine, Gary. (1987). *With the Boys: Little League Baseball and Preadolescent Culture*. Chicago, IL: University of Chicago Press.

Freud, Sigmund. (1905). *The Basic Writings of Sigmund Freud* (translated and edited by A.A. Brill). New York: The Modern Library.

Hochschild, Arlie. (1989). *The Second Shift*. New York: Avon.

Kehily, Mary Jane and Nayak, Anoop. (1997). "Lads and Laughter: Humour and the Production of Heterosexual Masculinities," *Gender and Education* 9: 69–87.

Kimmel, Michael. (2001). "Masculinity as Homophobia: Fear, Shame, and Silence in the. Construction of Gender Identity," in Stephen Whitehead and Frank Barrett (eds). *The Masculinities Reader*, pp. 266–187. Cambridge: Polity.

Kimmel, Michael. (2003). "Adolescent Masculinity, Homophobia, and Violence: Random School Shootings, 1982–2001," *American Behavioral Scientist* 46: 1439–58.

Lemert, Charles. (1996). "Series Editor's Preface," in Steven Seidman (ed.), *Queer Theory/Sociology*. Cambridge, MA: Blackwell.

Lyman, Peter. (1998). "The Fraternal Bond as a Joking Relationship: A Case Study of the Role of Sexist Jokes in Male Group Bonding," in Michael Kimmel and Michael Messner (eds.), *Men's lives*, pp. 171–93. Boston, MA: Allyn and Bacon.

Majors, Richard. (2001). "Cool Pose: Black Masculinity and Sports," in Stephen Whitehead and Frank Barrett (eds.), *The Masculinities Reader*, pp. 208–17. Cambridge: Polity.

Perry, Pamela. (2002). *Shades of White: White Kids and Racial Identities in High School*. Durham, NC: Duke University Press.

Plummer, David C. (2001). "The Quest for Modern Manhood: Masculine Stereotypes, Peer Culture and the Social Significance of Homophobia," *Journal of Adolescence* 24: 15–23.

Sedgwick, Eve K. (1995). "Gosh, Boy George, You Must be Awfully Secure in Your Masculinity!" in Maurice Berger, Brian Wallis and Simon Watson (eds.), *Constructing Masculinity*, pp. 11–20. New York: Routledge.

Smith, George W. (1998). "The Ideology of 'Fag': The School Experience of Gay Students," *The Sociological Quarterly* 39: 309–35.

Smith, Valerie. (1994). "Split Affinities: The Case of Interracial Rape," in Anne Herrmann and Abigail Stewart (eds.). *Theorizing Feminism*, pp. 155–70. Boulder, CO: Westview Press.

Stein, Arlene and Plummer, Ken. (1994). "'I Can't Even Think Straight': 'Queer' Theory and the Missing Sexual Revolution in Sociology," *Sociological Theory* 12: 178 ff.

Thorne, Barrie. (1993). *Gender Play: Boys and Girls in School.* New Brunswick, NJ: Rutgers University Press.

Warner, Michael. (1993). "Introduction," in Michael Warner (ed.), *Fear of a Queer Planet: Queer Politics and Social theory*, pp. vii–xxxi. Minneapolis: University of Minnesota Press.

West, Candace and Zimmerman, Don. (1991). "Doing Gender," in Judith Lorber (ed.), *The Social Construction of Gender*, pp. 102–21. Newbury Park: Sage.

Whitehead, Stephen and Barrett, Frank. (2001). "The Sociology of Masculinity," in Stephen Whitehead and Frank Barrett (eds.), *The Masculinities Reader*, pp. 472–6. Cambridge: Polity.

Wilchins, Riki. (2003). "Do You Believe in Fairies?" *The Advocate*, 4 February.

KEY CONCEPTS

adolescence

ethnography

homophobia

participant observation

sexual orientation

DISCUSSION QUESTIONS

1. Pascoe explains that the term *fag* is rarely, if ever, directed at someone who is actually a homosexual. Why do you think this is? How does the term *fag* get used among heterosexuals, but not as a label for homosexuals?
2. Pascoe states that very rarely did she overhear girls using the *fag* language? Can you explain why? Is this consistent with your own high school experiences?

39

Strategic Ambiguity: Protecting Emphasized Femininity and Hegemonic Masculinity in the Hookup Culture

DANIELLE M. CURRIER

In this reading, Currier summarizes her findings from interviews with college students about the experiences of "hooking up." She specifically looks at how hooking up has no clear definition and can involve various different sexual activities. The term "strategic ambiguity" is used to explain that by not having a consistent and clear definition of hooking up behavior, men and women are able to be vague about their sexual behaviors. Additionally, the author uncovers clear gender distinctions in what are accepted sexual behaviors for men and for women.

"Hooking up" is a term commonly used in contemporary American society to refer to sexual activity between two people who are not in a committed romantic relationship (historically known as "casual sex"). Despite the apparent ubiquity of hookups, there is an underlying ambiguity to the term. When someone says a hookup has occurred, there is no clarity about precisely what happened unless details are offered.

In this article, I use interview data to analyze how and why heterosexual college students use ambiguous language to describe their and others' casual sexual activities. I use the concept of "strategic ambiguity" to explore the intentionality and usefulness of the vagueness of this term. Historically, the term "strategic ambiguity" was used by policy analysts to refer to a country's policy of intentionally misleading or hiding information from other countries about its foreign policy intentions or strategies.[1] In a sociological context, I use this term to refer to the impression management strategy used to protect one's social identity and/or self-image by using ambiguous language to describe one's activities in a given situation. Specific to hookups, strategic ambiguity is employed when individuals use the term "hookup" to describe their sexual activities rather than give details about their sexual activities. This behavior helps maintain the existing

SOURCE: Danielle M. Currier, Strategic Ambiguity: Protecting Emphasized Femininity and Hegemonic Masculinity in the Hookup Culture, by SAGE Publications. Reprinted by permission of SAGE Publications.

(and seemingly appropriate) gender order. I argue that it allows women to protect their social identities as feminine and men to protect their social identities as masculine.

My analysis focuses on the following questions: How do gender norms and heterosexism affect how college students describe and interpret hookups, and how do social prescriptions of emphasized femininity and hegemonic masculinity influence the gendered meaning-making regarding this topic? The components of emphasized femininity and hegemonic masculinity on which I focus are the ones most frequently found in the interview data: how women walk a fine line between hooking up "enough" but not "too much" and exhibit a level of sexual compliance to men by downplaying their own sexual desires, and how men have a hyper-focus on heterosexual sexual activity and bonding with other men....

... [T]his analysis combines previously disparate topics, adds the lens of strategic ambiguity, and proposes that maintaining the current gender order is motivation for this intentional vagueness.

LITERATURE REVIEW

Hegemonic Masculinity and Emphasized Femininity

The most often cited theorizing on hegemonic masculinity and emphasized femininity comes from R. W. Connell (1987, 1990, 1995, 2001), who introduced a complex analysis of these concepts. Hegemonic masculinity is the form of masculinity that is most highly valued in a society and is rooted in the social dominance of men over women and nonhegemonic men (particularly homosexual men). Emphasized femininity is "the pattern of femininity which is given most cultural and ideological support ... patterns such as sociability ... compliance ... [and] sexual receptivity [to men]" (Connell 1987, 24). Both are multilayered processes of acting out gender publicly in culturally and socially accepted, defined, and appropriate ways (see West and Zimmerman 1987). Although there are variations of hegemonic masculinity and emphasized femininity, there are commonly accepted and expected displays of both, grounded in time and culturally specific contexts....

Casual Sex and Hookup Literature

Extramarital and casual sex (sex between people not in a committed or monogamous relationship) have been studied extensively since the 1970s. After the sexual and feminist revolutions of the 1960s–1970s, many researchers assumed that premarital sex in the context of a committed and/or monogamous relationship was socially acceptable but that sexual activity in casual or uncommitted contexts was less straightforward from both an interactional and a research perspective. The research highlighted micro aspects and focused on two topics—long-term, uncommitted partnerships and single events (often termed "one-night stands"). Some research focused on general attitudes about, personal experiences with, and interpretations of casual sex, other researchers examined individual, parental, or family characteristics affecting sexual decision-making, and still others looked at the frequency of casual sex, numbers of sexual partners, or choices surrounding birth control (Cates 1991; Gerrard and Gibbons 1982; Maticka-Tyndale 1991; Regan and Dreyer 1999; Sherwin and Corbett 1985). Research results varied depending on research questions, methodologies, and participant pools, but most researchers concluded that as of the early 1990s, casual sex was limited in scope among young people and there were usually negative reasons for or repercussions to this type of activity.

It is only in recent years that research has shown that casual sex is, in fact, common, and is not always socially constructed as negative (Bogle 2007, 2008; Daniel and Fogarty 2007; England, Fitzgibbons Shafer, and Fogarty 2007; England and Thomas 2006; Manning, Longmore, and Giordano 2005; Martinez, Copen, and Abma 2011; McGinn 2004; Paul, McManis, and Hayes 2000). The term "hooking tip" has emerged in academic and popular discourse to refer to a wide range of sexual interactions outside committed romantic relationships. There are two clear perspectives in the examination of hookups—one negative and one more neutral. Much popular writing/reporting and some academic literature are disapproving and moralistic, focusing on the negative aspects of hooking up, particularly for women.... In contrast, there is a growing body of academic research that is more neutral and focuses on the social meanings and gendered implications of hookups. This literature offers a nuanced and complex picture of social-sexual interactions and shows how hookups are simultaneously affecting and being affected by traditional/historic patterns of intimate partnering.

Without exception, these researchers have found that hookups are ubiquitous and normative among college students....

Researchers have also found that women and men report engaging in hookups at approximately similar rates, although men do report slightly higher numbers than do women (Bogle 2008; England and Thomas 2006; Fielder and Carey 2009; Kahn et al. 2000; Kimmel 2008; Lambert, Kahn, and Apple 2003; Glenn and Marquardt 2001; Paul, McManis, and Hayes 2000). Differences in reported numbers can be attributed, at least in part, to the fact that heterosexual men tend to overreport and heterosexual women tend to underreport their sexual activities and that men may be hooking up with more different partners than do women (Brown and Sinclair 1999; Jonason and Fisher 2009; Kimmel 2008).

Research has also consistently shown that hookups are "normal" among contemporary college students, whether or not they engage in or approve of such activities, that hookups have become the dominant sexual and intimacy script on many college campuses, and that young people are not engaging in traditional "dating and mating" patterns as much as in the past....

These findings do not, however, indicate a wholesale rejection of traditional moral sexual standards. Recent research shows that many young people still have their primary sexual relations within romantic relationships, many will engage in hookups for a certain period of time during high school and/or college before settling into more traditional dating routines later in college or postcollege, and many still seek emotional connectedness in their sexual relationships, even when participating in casual hookups (Armstrong, England, and Fogarty 2012; Bogle 2004, 2007, 2008; Daniel and Fogarty 2007; M. E. Eisenberg et al. 2009; England and Thomas 2006; Epstein et al. 2009; Eshbaugh and Gute 2008; Garcia and Reiber 2008; Hamilton and Armstrong 2009; Kimmel 2008; Manning, Giordano, and Longmore 2006; Martinez, Copen, and Abma 2011; Reid, Elliott, and Webber 2011). In addition, recent national research has shown that the percentage of teenagers reporting having intercourse has declined in recent years. In 1998, 51 percent of females and 60 percent of males ages 15–19 reported having engaged in sexual intercourse. In the period 2006–2010, those rates dropped to 43 percent of females and 42 percent of males (Martinez, Copen, and Abma 2011). These findings could indicate less sexual activity in general among teenagers in recent years, or more oral sex or "non-intercourse" activities as a substitute for intercourse, but additional research and analysis must be done to ascertain the causes and consequences.

For the most part, researchers find that hookups are perpetuating a sexual double standard in which men receive more sexual and social benefits from hooking up than do women (Bogle 2004, 2007, 2008; Eshbaugh and Gute 2008; Fielder and Carey 2009; Glenn and Marquardt 2001; Kimmel 2008; McGinn 2004; Ramage 2007; Sessions Stepp 2008)....

I add the concept of strategic ambiguity to the existing hookup literature and offer a gendered analysis without focusing on the negative repercussions for women. I analyze how ambiguously defining both "hookup" and "sex" is an interactional strategy (within a larger cultural social-sexual context) that reinforces a gender order still based in a sexual double standard. Specifically, I examine the gendered interactional dynamics in hookups through the lens of emphasized femininity and hegemonic masculinity, highlighting how men and women protect their sexual and social identities by relying on some historic/traditional definitions of gender while simultaneously challenging the concept that women should not be sexual beings.

METHODS

The data used for this analysis come from in-depth interviews with 78 full-time, heterosexual students at a co-ed, public university in the South (SU) between March 2006 and December 2007. These data are part of a larger research study that included information from non-heterosexual students, which will be analyzed in a future article. At the time of the study, there were approximately 8,300 undergraduates. The student body comprised 60 percent women and 40 percent men; 90 percent white, 5 percent black, and 5 percent "other"; about 10 percent of the students belonged to fraternities and sororities; and 5 percent were intercollegiate athletes....

The one-on-one interviews were conducted in private, ranged from 30 minutes to four hours, were taped, and transcribed by me or a professional transcriptionist who signed a confidentiality agreement. All participants chose a unique pseudonym (used in all paperwork and reporting of the data) to ensure confidentiality. A combination of convenience, purposive, and snowball sampling was used to gain participants. No class credit or monetary incentive was offered for participation.... Each interview included the open-ended questions "How do you define a hookup?" "Have you hooked up?" "If so, how much?" "If not; why not?" and "What is your opinion of hookups?" All participants were asked about their sexual histories, social and sexual identities, birth control use, violence in dating relationships, and attitudes and beliefs about gender, sex, hookups, and the "hookup culture."

Interview data were coded by me and three trained undergraduate research assistants, using extensive cross-verification to create intercoder reliability. For this particular analysis, I coded for terms and concepts that related to definitions of hookup and sex, ambiguity or vagueness, normativity of hookups, and conceptions of femininity and masculinity. All members of the research team searched the interview transcripts independently; those terms and concepts found by at least three of the team were included for this analysis.

FINDINGS

The following examination is grounded in two findings that mirror previous research: hooking up as common among college students, and the sexual double standard in hookups. While these topics have been considered elsewhere, my analysis is unique because I analyze them through the lens of strategic ambiguity and incorporate a discussion of emphasized femininity or hegemonic masculinity and the underlying gender order.

I discuss two components each of emphasized femininity and hegemonic masculinity that were the most often repeated by participants. Emphasized femininity was evident in how the women walked a fine line between "enough" and "too much" hooking up, and displayed a level of sexual compliance by downplaying their own sexual pleasure and desires in hookups. Hegemonic masculinity was evident in the active hyper-heteronormativity expected of men and in how men's descriptions and interpretations of hookups often revolved around their relationships with other men rather than their relationships with women.

"Everyone Is Out There to Hook Up"

All of my participants reported that hookups are perceived as both ever-present and normative in college and are a central component of social life at SU, whether or not they participated in hookups. The large majority of my participants (84.6 percent) reported having hooked up at least once in their lives. (See Table 1 for specifics of reported activities.)

> If you're out on the weekends, if you're not with a girlfriend or boyfriend, you're looking for it. Maybe the person you're trying to hook

TABLE 1 Reported Sexual Activities of Interview Participants

	% of Sample (Number of Cases)	% of Sample (N)	% of Sample (N)
Characteristic	All	Women	Men
Those reporting hooking up at least once in their lives	84.6% (66)	80% (40)	93% (26)
Those reporting hooking up at least once at SU	71.7% (56)	64% (32)	86% (24)
Those reporting being virgins by choice	15.4% (12)	20% (10)	7% (2)

> up with isn't going to hook up with you, but it's likely she's trying to hook up with someone else. Everyone is out there to hook up. And sometimes it happens, sometimes it doesn't. But, I would say it's on just about everyone's mind, the desire. (Bruce, white man, 22)

As Bruce indicates, the social-sexual scene is based on both actual hookups ("sometimes it happens") and the perceived potential and desire ("everyone is out there to hook up"), and this potential is a constant undercurrent in the social interactions on college campuses.

"Anything from Kissing to Having Sex ..."

The first question in every interview was "How do you define a hookup?" Despite the consensus that "everyone is doing it," there was no consensus on what exactly "it" is. I found two interesting patterns in the definitions—first, there were no significant differences in how women and men defined (or failed to define) "hookup." Second, both sexes struggled equally and most participants used similar language. I heard a plethora of definitions, from both women and men, and all were aware of the relative nature of how they were addressing the topic:

> It depends—I guess it is in the eye of the beholder.... If I met a girl at the bar and even if we didn't have sex, we just went to her place and made out for an hour and then left, then I would consider that ... yeah, we hooked up ... and then it can go the other way, like, if I said we had intercourse, because that's their definition. So I guess that's kind of the double-edged sword, the defining it. (Bodie, white man, 22)

Clearly, a hookup is "anything from kissing to having sex" (Casper, white woman, 22) and is "in the eye of the beholder" revealing the subjectivity and underlying ambiguity of the definition. Many descriptive differences were framed in the context of levels of alcohol consumption (usually much), how well they knew the people with whom they had hooked up (often little), and the context in which hookups occurred (often a party)....

Ambiguity and "Sex"

I found two components to the relationship between the definition of "hookup" and the definition of "sex": whether or not a hookup included sex, and what sex actually is. When asked to define a hookup, all participants referred to either the presence or absence of sex. However, a few were adamant that a hookup always included sex:

> That's the goal ... to have sex. Like, have sex first and then ... maybe you'll talk again. (Angela, white woman, 21)

> Basically just randomly meeting somebody and going back to their room and having sex. (Kayleigh, white woman, 19)

Most participants, however, said that hookups and sex are not synonymous:

> Hooking up and having sex can be two different things. Because I don't really consider having sex hooking up ... that's a different thing. Like, having sex is separate from hooking up. (Brianna, white woman, 22)

> I would not define it as having sex necessarily. I think it is sexual contact that goes beyond kissing. (Carl, white man, 20)

These comments reveal the ambiguous relationship between hookups and sex. They also relate to the underlying ambiguity about what constitutes "sex," an ambiguity rooted in heteronormative conceptions of sexual activity. When asked to define "sex" both women and men participants usually gave heterosexist definitions, primarily vaginal/penile penetration:

> Sex? Penetration. That is sex. Intercourse is sex. (Styx, white man, 20)

> In my definition it's intercourse. (Angela, white woman, 21)

> Having sex ... intercourse. I don't consider foreplay sex ... anything that doesn't include the actual penis going in the vagina is foreplay. (Trinity, white woman, 20)

This unambiguously heterosexist definition was in stark contrast to the fuzziness of the plethora of definitions provided for "hookup." This heterosexist bias revealed the complex relationship between "sex" and "hookup." Most respondents contended that oral sex is normative behavior in many hookups, and for most of them was not "sex":

> Oral sex is, like, third base ... you're not having sex ... you're just hooking up. (Casper, white woman, 19)

> Oral sex is oral sex. It's foreplay, that's not sex. (Styx, white man, 20)

> Sex is vaginal. Oral sex is just with the hookups. (Lee, white woman, 20)

When I asked for details about why oral sex was not "sex," most could only respond "because it isn't" or "it just isn't," if they could come up with a response at all....

A clearly gendered aspect of oral sex was the giver–receiver dynamic. Although overtly stated in only a few interviews, the implication was that women performed it on men, and rarely the other way around. A couple of women indicated that if they were going to perform oral sex on a man, he had to be willing to reciprocate, but most of the women did not expect it:

> I think it's really common for girls to do it to guys, but I don't think it's common at all for guys to do it to girls. (Hannah, white woman, 22)

> I think it's more expected for girls to do it to guys. Because it still—I don't know, it still has that demeaning look towards it—guys doing it, you know. (Donald, Hispanic man, 19)

Both Hannah and Donald were clear about how oral sex was often a oneway proposition. And Donald's comments indicate that part of that proposition related to conceptions of masculinity and what was appropriate behavior for men....

I have concluded that defining oral sex as "not sex" helps maintain the gender order privileging men and perpetuates the dominant constructions of masculinity and femininity whereby men are sexually dominant and women are (at least to some degree) sexually submissive. The fact that women perform oral sex on men more than vice versa helps illuminate the patterns of emphasized femininity evident in the interviews. Women expect to satisfy men's sexual needs and desires by helping them achieve orgasm, but do not expect the same in return, thus privileging the men's desires over their own.

Protecting and Reinforcing Emphasized Femininity

I have identified three ways in which women participants conformed to prescriptions of emphasized femininity: disregarding their own sexual desires by performing oral sex on men without reciprocation, ignoring their right to sexual pleasure in hookups, and being intentionally ambiguous about their sexual activities during a hookup by not talking about or downplaying their sexual activity.

When I asked women directly about why they did not expect to receive oral sex, all but two indicated it was because they did not expect (or ask for) sexual pleasure in a hookup:

> I think [girls don't do it because they] don't think they have to get pleasure themselves. (Hannah, white woman, 22)

This led to discussions about pleasure in general, unrelated to oral sex. Both women and men expressed the belief that pleasure in hookups was primarily focused on men:

> DC: So is there a lot of pleasure involved in it?
>
> Angela: In a hookup? Not as much. Because sex is defined as over when the guy climaxes. So ... guys get off more during hookups than women, (white woman, 21)

DC: How much do you think sexual pleasure is involved on both sides?

Bodie: Oh, God, way more for guys. Based on what I've heard and what I've seen and experienced firsthand, I think for girls it takes more of an emotional attachment to orgasm, whereas for guys it's a lot easier. The guys are guaranteed every time, and if the girl gets hers, [it's a] "lucky her" kind of thing, (white man, 22)

An interesting dynamic seen in Angela's and Bodie's comments is that pleasure can be interpreted as synonymous with orgasm, rather than as a spectrum of physical and emotional experiences to be enjoyed by both parties. This highlights a central aspect of the gender asymmetry I found in both subtle and overt ways throughout the interviews—hookups were expected to be more sexually satisfying for men than for women, and pleasure was often seen as a unidimensional achievement of orgasm for the man.

However, my discussions of pleasure with a few participants were more complex and multilayered than the norm. While many indicated that pleasure or satisfaction meant orgasm, some suggested that there were aspects of pleasure that were nonphysical and resulted from attention, affection, and intimacy. This confirms recent research showing that sexual pleasure is a complicated gendered dynamic that is just beginning to be negotiated openly in uncommitted heterosexual relationships (Armstrong, England, and Fogarty 2012).

Ultimately, the sex asymmetry evident in oral sex participation and physical pleasure continues the historic pattern of women being submissive to men's sexual desires, and segues into discussion of the second pattern I identified among women participants—intentional and strategic ambiguity, ultimately as a way of avoiding the label "slut." The word "slut" was used repeatedly in interviews with both women and men when referring to women who had "too much" sexual activity, and being labeled as such was a deep fear for most of my women participants:

> Obviously, because girls get labeled if you hook up with too many people or you just hook up with a random person like the first night you meet 'em. They get labeled a slut.... And obviously you don't want that reputation of guys thinking they can just get in your pants whenever, (Lee, white woman, 20)

Lee's use of the word "obviously" reveals underlying cultural assumptions that are made about what is defined as appropriate femininity, and acting in ways that result in being labeled "slut" is clearly not appropriate. However, a challenge for women is the lack of consensus on how much sexual activity is necessary for a woman to be labeled as a slut. Clearly, hooking up with "too many people" is bad, but when pressed to define "too many," few were able to give a definitive number. The answers ranged from one to "never too much."

To navigate this catch-22, the women engaged in two patterns of strategic ambiguity: not talking about or being vague about the details of their hookups:

> I don't think girls admit to things as much as guys do. Or at least they don't go around telling everyone. I know my old roommate, she was the hookup queen! I mean, she would hook up with a different guy

> literally every night. But she would never admit to it. She just didn't want to say it because she didn't want to be a slut or whatever. (Brianna, white woman, 22)

> I definitely think girls hook up a lot more than they let on. Guys are constantly talking about hooking up. "I hooked tip with this girl, I hooked up with this girl...." Well, where are all the girls? Where are all the girls that all these guys are hooking up with? (Trinity, white woman, 20)

Both Brianna and Trinity reveal how many women internalize the idea that there are boundaries around how much and what kind of "sex" they can/should be having. If women admit to transgressing the unspoken, often unclear, boundaries ("hook up a lot"), they will be negatively sanctioned, so they sidestep this potential by actively steering clear of concrete discussions of what they are doing. Note that the label "slut" was applied only to women—evidence of the underlying double standard vis-à-vis labeling people regarding their amount of sexual activity. This pattern of differential gendered labeling has been recently noted by Amy Sehalet (2010).

Other women used strategic ambiguity by remaining vague about their sexual activity:

> If you don't want to say you had sex, just say you hooked up, but then people can think oh, they just fooled around a little bit. (Angela, white woman, 21)

> When people say, "We hooked up," you don't really know what they mean by that. They could be having sex every night. And you're assuming that they probably just made out or something like that. (Brianna, while woman, 22)

Angela and Brianna are examples of the many women who strategically and intentionally used the ambiguity of the term "hookup" while they were simultaneously frustrated with what they perceived to be social restrictions on their activities. Ironically (from an analytic standpoint), one of the common complaints I heard from women was their perception that having sex may not be a positive thing for them ("don't want to say you had sex"), or at least not too much sex ("having sex every night" needs to be hidden by women), despite the perceived prevalence of sexual activity among their peers. And although they were able to protect their individual social identities and status by downplaying sexual activity, they were perpetuating the very cultural double standard against which they were chaffing, the dynamic that makes hooking up so complex a negotiation for women.

These patterns of impression management among many women reveal an underlying dynamic of the sexual double standard—that women *don't* and men *do* talk about their sexual activities. In general, while many women used strategic ambiguity to imply that they *are not* having sex, many men used the ambiguity to imply that they *are* having sex, and always with women, not men.

Protecting and Reinforcing Hegemonic Masculinity

The clearest patterns I found in the data were those apparent in the relationships among hooking up, certain conceptions of masculinity, and strategic ambiguity. I identified two patterns of perceived pressure on men to achieve appropriate "masculinity," both of which were supported by the ambiguity of "hookup" and "sex": a hyper-focus on heterosexuality and sexual activity, and the importance of bonding with or impressing other men, much more than bonding with or impressing women. In trying to achieve accountability and social status with other men, men used strategic ambiguity in two ways: overt exaggerations or more subtle avoidance or duplicity (note that the latter is similar to the strategy employed by women, but with different aims).

In many interviews with the men, there was a subtle and indirect yet constant reference to the connection between masculinity and an active pursuit and expression of heterosexual sexual activity. This relates to a core aspect of Connell's conceptualization of hegemonic masculinity: "The most important feature of contemporary masculinity is that it is heterosexual" (1987, 24). Many men wanted to avoid the label "gay" and maintain a heterosexual social image:

> For guys, I guess sexuality is kind of questioned if you're not seen talking to a girl or making out with a girl. (Henry, black man, 23)

> If you're not having sex—with my definition of sex—people might think there's something wrong ... You might be thought to be gay. (Lance, white man, 22)

An interesting aspect of the men's concern with their public image with "people" was that it centered on what other men thought of them, not what women thought of them:

> People talk to impress their guy friends, to make conversation ... not only to impress but to maybe make a name for themselves, a reputation that they would want or one that would make them more popular.... It's definitely [with other] guys. Because I think the standard for acceptance among men on a college campus is ... a great deal higher than men to women. (Lance, white man, 22)

No men in my study expressed worry that women would think they were gay, and they were not trying to impress women by hooking up a lot. The men were doing masculinity in ways that made them "accountable" to other men (West and Zimmerman 1987). Not only did they want to emphasize their heterosexuality and impress other men, but they also assumed that hooking up with or having sex with various women would "make a name" for themselves. This is a stark contrast to the women's active strategies to avoid getting a "name" (usually "slut").

Most interviewees recognized the cultural assumption that college men are, or should be, sexually active....

... This again relates to the pressure on men to prove that they are "real men" in the eyes of other men by hooking up with women and conforming to proscriptions of hegemonic masculinity. The ambiguity of defining hookups and sex is part of this dynamic and allows men to appear to be performing "appropriate" masculinity, even if in reality they are not.

Unfortunately, this desire to be "a man" in the eyes of other men clearly comes at the expense of both the women they are treating as sexual objects and of their potential to have emotionally satisfying relationships with women (both are addressed at length in Kimmel 2008). It also reveals how the continuing sexual double standard informs the gender order and how an actively heterosexual definition of masculinity affects the social-sexual interactions of men, with both women and men.

CONCLUSION

In this analysis, I have applied the concept of "strategic ambiguity" to the sexual experiences and interpretations of hookups among heterosexual college women and men. In this context, strategic ambiguity is an interactional strategy to maintain social status by using vague language or rhetoric. By keeping the definition of "hookup" ambiguous, women are able to protect their status as "good girls" (sexual but not promiscuous) and men are able to protect their social status as "real men" (heterosexual, highly sexually active).

This study demonstrates how the current sexual double standard is deeply nuanced and complex. It is not the historic model, which unambiguously restricts women's and encourages men's sexual activity. The young women in my study reported walking a fine line between hooking up "enough" but not "too much," always trying to avoid the label "slut." In this sexual context, they have the "out" of oral sex, which allows them to be sexually active but not engage in "sex" because it is not socially constructed as "real sex." In addition, most men in my study reported experiencing both subtle and overt pressure from other men to achieve and maintain a socially acceptable level of "masculine" behavior and displays. This pressure comes with the expectation of "real sex," which can be problematic for men who do not want to engage in that activity. Interestingly, the differential peer pressure combined with heterosexism begs the question of with whom the men are expected to hook up if women are not "allowed" to do so as much.

However, my participants reported that, in their experiences, hookups revolve more around men and their desires than around women and their desires, and doing femininity still often means reacting to men and cultural definitions of masculinity. Ultimately, the patterns discussed in this article are tangible evidence of how emphasized femininity is often a reaction to or an offshoot of hegemonic masculinity, and in the context of hookups, doing femininity still often means reacting to men and cultural definitions of masculinity....

NOTE

1. For more information on the history of this term, see E. Eisenberg (2007).

REFERENCES

Armstrong, Elizabeth A., Paula England, and Alison Fogarty. 2012. Accounting for women's orgasm and sexual enjoyment in college hookups and relationships. *American Sociological Review* 77:435–62.

Bogle, Kathleen A. 2004. From dating to hooking up: The emergence of a new sexual script. *Dissertation Abstracts International. A, The Humanities and Social Sciences* 64.

Bogle, Kathleen A. 2007. The shift from dating to hooking up in college: What scholars have missed. *Sociology Compass*, http://onlinelibrary.wiley.com/journal/10.1111/%28ISSN%291751-9020.

Bogle, Kathleen A. 2008. *Hooking up: Sex, dating, and relationships on campus*. New York: New York University Press.

Brown, N. R., and R. C. Sinclair. 1999. Estimating lifetime sexual partners: Men and women do it differently. *Journal of Sex Research* 36:292–97.

Cates, Willard. 1991. Teenagers and sexual risk taking: The best of times and the worst of times. *Journal of Adolescent Health* 12:84–94.

Connell, Racwyn. 1987. *Gender and power: Society, the person and sexual politics*. Stanford, CA: Stanford University Press.

Connell, Racwyn. 1990. An iron man: The body and some contradictions of hegemonic masculinity. In *Sport, men, and the gender order: Critical feminist perspectives*, edited by Michael A. Messner and Donald Sabo. Champaign, IL: Human Kinetics Books.

Connell, Racwyn. 1995. *Masculinities*. Berkeley: University of California Press.

Connell, Racwyn. 2001. *The men and the boys*. Berkeley: University of California Press.

Daniel, Christy, and Kate Fogarty. 2007. "Hooking up" and hanging out: Casual sexual behavior among adolescent young adults today. Document #FCS2279 for the Institute of Food and Agricultural Sciences, University of Florida.

Eisenberg, Eric. 2007. *Strategic ambiguities: Essays on communication, organization, and identity*. Thousand Oaks, CA: Sage.

Eisenberg, Marla E., Dianne M. Ackard, Michael D. Resnick, and Dianne Neumark-Sztainer. 2009. Casual sex and psychological health among young adults: Is having "friends-with-benefits" emotionally damaging? *Perspectives on Sexual and Reproductive Health* 41:231–37.

England, Paula, Emily Fitzgibbons Shafer, and Alison C. K. Fogarty. 2007. Hooking up and forming romantic relationships on today's college campuses. In *The gendered society reader*, edited by M. Kimmel. New York: Oxford University Press.

England, Paula, and Reuben J. Thomas. 2006. The decline of the date and the rise of the college hook up. In *The family in transition*, edited by Arlene S. Skolnick and Jerome H. Skolnick. Boston: Allyn & Bacon.

Epstein, Marina, Jerel P. Calzo, Andrew P. Smiler, and L. Monique Ward. 2009. "Anything from making out to having sex": Men's negotiations of hooking up and friends-with-benefits scripts. *Journal of Sex Research* 46:414–24.

Eshbaugh, Elaine M., and Gary Gute. 2008. Hookups and sexual regret among college women. *Journal of Social Psychology* 148:77–89.

Fielder, Robyn L., and Michael P. Carey. 2009. Predictors and consequences of sexual "hookups" among college students: A short-term prospective study. *Archives of Sexual Behavior*, http://link.springer.com/joumal/10508?utm_campaign=FTA2012-10508&utm_medium=banner&utm_source=springer&wl_me=springer.banner.FTA2012-10508.

Garcia, Justin R., and Chris Reiber. 2008. Hook-up behavior: A biopsychosocial perspective. *Journal of Social, Evolutionary, and Cultural Psychology* 2:192–208.

Gerrard, Meg, and F. X. Gibbons. 1982. Sexual experience, sex guilt, and sexual moral reasoning. *Journal of Personality* 50:345–59.

Glenn, Norval, and Elizabeth Marquardt. 2001. *Hooking up, hanging out, and hoping for Mr. Right: College women on dating and mating today*. New York: Institute for American Values.

Hamilton, Laura, and Elizabeth A. Armstrong. 2009. Gendered sexuality in young adulthood: Double binds and flawed options. *Gender & Society* 23:589–616.

Jonason, Peter K., and Terri D. Fisher. 2009. The power of prestige: Why young men report having more sex partners than young women. *Sex Roles* 60:151–59.

Kahn, A. S., K. Fricker, J. Hoffman, T. Lambert, M. Tripp, K. Childress, et al. 2000. Hooking up: Dangerous new dating methods? In *Sex, unwanted sex, and sexual assault on college campuses,* Symposium conducted at the annual meeting of the American Psychological Association, A. S. Kahn, Chair, Washington, DC, August.

Kimmel, Michael. 2008. *Guyland: The perilous world where boys become men: Understanding the critical years between 16 and 26.* New York: HarperCollins.

Lambert, Tracy A., Arnold S. Kahn, and Kevin J. Apple. 2003. Pluralistic ignorance and hooking up. *Journal of Sex Research* 40:129–34.

Manning, W. D., Monica A. Longmore, and Peggy C. Giordano. 2005. Adolescents' involvement in non-romantic sexual activity. *Social Science Research* 34:384–407.

Manning, Wendy D., Peggy C. Giordano, and Monica A. Longmore. 2006. Hooking up: The relationship contexts of nonrelationship sex. *Journal of Adolescent Research* 21:459–83.

Martinez, G., C. E. Copen, and J. C. Abma. 2011. Teenagers in the United States: Sexual activity, contraceptive use, and childbearing, 2006–2010 National Survey of Family Growth. National Center for Health Statistics. *Vital Health Statistics* 23(31).

Maticka-Tyndale, Eleanor. 1991. Modification of sexual activities in the era of AIDS: A trend analysis of adolescent sexual activities. *Youth Society* 23:31–49.

McGinn, Daniel. 2004. Mating behavior 101: Social scientists have recently begun to study sex on campus, in search of the truth about "hooking up." *Newsweek*, 4 October.

Paul, Elizabeth L. Brian McManis, and Alison Hayes. 2000. "Hookups": Characteristics and correlates of college students' spontaneous and anonymous sexual experiences. *Journal of Sex Research* 37:76–88.

Ramage, Stephanie. 2007. Hooked: As hooking up replaces casual dating, young women are caught in a trap. *The Sunday Paper*, December 16, http://www.sundaypaper.com.

Regan, Pamela C. and Carla S. Dreyer. 1999. Lust? Love? Status? Young adults' motives for engaging in casual sex. *Journal of Psychology and Human Sexuality* 11: 1–24.

Reid, Julie A., Sinikka Elliott, and Gretchen R. Webber. 2011. Casual hookups lo formal dates: Refining the boundaries of the sexual double standard. *Gender & Society* 25:545–68.

Schalet, Amy. 2010. Sex, love, and autonomy in the teenage sleepover. *Contexts* 9(3): 16–21.

Sessions Stepp, Laura. 2008. *Unhooked: How young women pursue sex, delay love and lose at both*. New York: Riverhead Books.

Sherwin, Robert, and Sherry Corbett. 1985. Campus sexual norms and dating relationships: A trend analysis. *Journal of Sex Research* 21:258–74.

West, Candace, and Don H. Zimmerman. 1987. Doing gender. *Gender & Society* 1:125–51.

KEY CONCEPTS

emphasized femininity

hegemonic masculinity

hooking up

strategic ambiguity

DISCUSSION QUESTIONS

1. This research finds that traditional dating behaviors are not common on college campuses. Is this still true? What do you observe among college students with regard to dating versus "hooking up"? Are there advantages and disadvantages to dating? Are there advantages and disadvantages to hooking up? How are these advantages and disadvantages different for men and for women?
2. Consider the debate about how women are hurt by hooking up versus how women are empowered by hooking up. What are the arguments on both sides? What evidence can you provide on each side of the debate?

40

Sex, Love, and Autonomy in the Teenage Sleepover

AMY SCHALET

In this reading, Amy Schalet compares attitudes about sex among teenagers in the Netherlands and in the United States. While the Dutch have seemingly more permissive attitudes about their adolescent children's sexual activity, both American and Dutch families navigate gender expectations about sexuality. In the United States, the more restrictive views about sex lead to less education about sexual independence, contraception, and overall sexual health among American teens.

Karel Doorman, a soft-spoken civil servant in the Netherlands, keeps tabs on his teenage children's computer use and their jobs to make sure neither interferes with school performance or family time. But Karel wouldn't object if his daughter Heidi were to have a sexual relationship.

"No," he explains. "She is sixteen, almost seventeen. I think she knows very well what matters, what can happen. If she is ready [for sex], I would let her be ready." Karel would also let his daughter spend the night with a steady boyfriend in her room, if the boyfriend had come over to the house regularly before-hand and did not show up "out of the blue." That said, Karel suspects his daughter might prefer a partner of her own sex. If so, Karel would accept her orientation, he says, though "the adjustment process" might take a little longer.

Karel's approach stands in sharp contrast to that of Rhonda Fursman, a northern California homemaker and former social worker. Rhonda tells her kids that pre-marital sex "at this point is really dumb." It's on the list with shoplifting, she explains, "sort of like the Ten Commandments: don't do any of those because if you do, you know, you're going to be in a world of hurt." Rhonda responds viscerally when asked whether she would let her fifteen-year-old son spend the night with a girlfriend. "No way, Jose!" She elaborates: "That kind of recreation … is just not something I would feel comfortable with him doing here." She might change her mind "if they are engaged or about to be married."

SOURCE: Amy Schalet, Sex, Love, and Autonomy in the Teenage Sleepover, by SAGE Publications. Reprinted by permission of SAGE Publications.

Karel and Rhonda illustrate a puzzle: the vast majority of American parents oppose a sleepover for high-school-aged teenagers, while Dutch teenagers who have steady boyfriends or girlfriends are typically allowed to spend the night with them in their rooms. This contrast is all the more striking when we consider the trends toward a liberalization of sexual behavior and attitudes that have taken place throughout Europe and the United States since the 1960s. In similar environments, both parents and kids are experiencing adolescent sex, gender, and relationships very differently. A sociological exploration of these contrasts reveals as much about the cultural differences between these two countries as it does about views on adolescent sexuality and child rearing.

ADOLESCENT SEXUALITY IN CONTEMPORARY AMERICA

Today, most adolescents in the U.S., like their peers across the industrialized world, engage in intercourse—either opposite or same-sex—before leaving their teens (usually around seventeen). Initiating sex and exploring romantic relationships, often with several successive partners before settling into long-term cohabitation or marriage, are now normative parts of adolescence and young adulthood in the developed world. But in the U.S., teenage sex has been fraught with cultural ambivalences, heated political struggles, and poor health outcomes, generating concern among the public, policy makers, scholars, and parents. American adolescent sexuality has been dramatized rather than normalized.

In some respects, the problems associated with adolescent sexuality in America are surprising. Certainly, age at first intercourse has dropped in the U.S. since the sexual revolution, but not as steeply as often assumed. In a recent survey of the adult American population, sociologist Edward Laumann and colleagues found that even in the 1950s and '60s, only a quarter of men and less than half of women were virgins at age nineteen. The majority of young men had multiple sexual partners by age 20. And while women especially were supposed to enter marriage as virgins, demographer Lawrence Finer has shown that women who came of age in the late 1950s and early '60s almost never held to that norm. Still, a 1969 Gallup poll found that two thirds of Americans said it was wrong for "a man and women to have sex relations before marriage."

But by 1985, Gallup found that a slim majority of Americans no longer believed such relations were wrong. Analyzing shifts in public opinion following the sexual revolution, sociologists Larry Petersen and Gregory Donnenwerth showed that among Americans with a religious affiliation, only conservative Protestants who attended church frequently remained unchanged. Among all other religious groups, acceptance of pre-marital sex actually grew, although Laumann and colleagues reported a majority of the Americans continued to believe sex among *teenagers* was always wrong. Even youth agreed: six in ten fifteen to nineteen-year-olds surveyed in the 2002 National Survey for Family Growth said sixteen-year-olds with strong feelings for one another shouldn't have sex.

Part of the opposition to adolescent sexuality is its association with unintended consequences such as pregnancy and sexually transmitted diseases. In the U.S., the rate of unintended pregnancies among teenagers rose during the 1970s and '80s, dropping only in the early '90s. However, despite almost a decade and a half of impressive decreases in pregnancy and birth rates, the teen birth rate remains many times higher in the U.S. than it is in most European countries. In 2007, births to American teens (aged fifteen to nineteen) were eight times as high as in the Netherlands.

One would imagine the predominant public policy approach would be to improve education about, and access to, contraception. But "abstinence-only-until-marriage" programs, initiated in the early 1980s, have received generous federal funding over the past fifteen years, and were even written into the recent U.S. health reform law (which also supports comprehensive sex education). For years, schools funded under the federal "abstinence-only" policy were prohibited from educating teens about condoms and contraception and required to teach that sex outside of heterosexual marriage was damaging. A 2004 survey by NPR, the Kaiser Family Foundation, and Harvard University found that most parents actually thought that contraception and condom education should be included, but two thirds still agreed sex education should teach that abstinence outside of marriage is "the accepted standard for school-aged children." And for most parents, abstinence means no oral sex or intimate touching.

While American parents of the post-Sexual Revolution era have wanted minors to abstain, few teens have complied. Many American teenagers have had positive and enriching sexual experiences; however, researchers have also documented intense struggles. Comparing teenage boys and girls, for example, University of Michigan sociologist Karin Martin found that puberty and first sex empowered boys but decreased self-esteem among girls. Psychologist Deborah Tolman found the girls she interviewed confronted dilemmas of desire because of a double standard that denies or stigmatizes their sexual desires, making girls fear being labeled "sluts." Analyzing the National Longitudinal Survey of Adolescent Health, researchers Kara Joyner and Richard Udry found that even without sex, first romance brings girls "down" because their relationship with their parents deteriorates.

Nor are American girls of the post-Sexual Revolution era the only ones who must navigate gender dilemmas. Sociologist Laura Carpenter found that many of the young men she interviewed in the 1990s viewed their virginity as a stigma which they sought to cast off as rapidly as possible. And in her ethnography, *Dude, You're a Fag,* C.J. Pascoe found boys are pressured by other boys to treat girls as sex objects and sometimes derided for showing affection for their girlfriends. But despite public pressures, privately boys are as emotionally invested in relationships as girls, found Peggy Giordano and her associates in a recent national study out of Toledo, Ohio. Within those relationships, however, boys are less confident.

In the 1990s, the National Longitudinal Study for Adolescent Health found that steady romantic relationships are common among American teenagers.

Girls and boys typically have their first intercourse with people they are dating. But the Toledo group found that once they are sexually experienced, the majority of boys and girls also have sex in non-dating relationships, often with a friend or acquaintance. And even when they have sex in dating relationships, a quarter of American girls and almost half of boys say they are "seeing other people" (which may or may not include sexual intercourse).

TEEN SEXUALITY IN THE NETHERLANDS

In a late 1980s qualitative study with 120 parents and older teenagers, Dutch sociologist Janita Ravesloot concluded that in most families, parents accepted that sexuality "from the first kiss to the first coitus" was part of the youth phase. In middle class families, teenagers reported that parents accepted their sexual autonomy, but didn't engage in elaborate conversations with them because of lingering feelings of shame. Working-class parents were more likely to use their authority to impose norms, including that sex belonged only in steady relationships. In a few strongly religious families—Christian or Islamic—parents categorically opposed sex before marriage: here there were "no overnights with steady boy- or girlfriends at home."* But such families remain a minority. A 2003 survey by *Statistics Netherlands* found that two thirds of Dutch fifteen to seventeen-year-olds with steady boy- or girlfriends are allowed to spend the night with them in their bedrooms, and that boys and girls are equally likely to get permission for a sleepover.

This could hardly have been predicted in the 1950s. Then, women *and* men typically initiated intercourse in their early twenties, usually in a serious relationship (if not engagement or marriage). In the late '60s, a national survey conducted by sociologist G.A. Kooy found most respondents still rejected premarital sex when a couple was not married or planning to do so very shortly. But by the early 1980s, the same survey found that six out of ten respondents no longer objected to a girl having intercourse with a boy as long as she was in love with him. Noting the shift in attitudes since the 1950s, Kooy spoke of a "moral landslide." His colleague, sociologist Evert Ketting, even went as far as to speak of a "moral revolution."

What changed was not just a greater acceptance of sex outside of the context of heterosexual marriage. There was also serious new deliberation among the general public, health professionals, and the media about the need to adjust the moral rules governing sexual life to real behavior. As researchers for the Guttmacher Institute later noted, "One might say the entire society has experienced a course in sex education." The new moral rules cast sexuality as a part of life that should be governed by self-determination, mutual respect, frank conversation, and the prevention of unintended consequences. Notably, these new rules were applied to minors and institutionalized in Dutch health care policies that removed financial and emotional barriers to accessing contraceptives—including the requirements for a pelvic examination and parental consent.

Indeed, even as the age of first sexual intercourse was decreasing, the rate of births among Dutch teenagers dropped steeply between 1970 and 1996 to one of the lowest in the world. What distinguished the very low Dutch teenage birth rate from, for instance, that of their Swedish counterparts, was that it was accompanied by a very low teen abortion rate....

Sex education has played a key role. Sociologists Jane Lewis and Trudie Knijn find that Dutch sex education curricula are more likely than programs elsewhere to openly discuss female sexual pleasure, masturbation, and homosexuality. The Dutch curricula also emphasize the importance of self-reliance and mutual respect in negotiating enjoyable and healthy sexual relationships during adolescence.

A 2005 survey of Dutch youth, ages twelve to twenty-five, found the majority described their first sexual experiences—broadly defined—as well-timed, within their control, and fun. About first intercourse, 86 percent of women and 93 percent of men said, "We both were equally eager to have it." This doesn't mean that gender doesn't matter. Researcher Janita Ravelsoot found that more girls than boys reported that their parents expected them to only have intercourse in relationships. Girls were also aware that they might be called sluts for having sex too soon or with too many successive partners. And although most of the 2005 respondents said they were (very) satisfied with the pleasure and contact they felt with their partner during sex, men were much more likely to usually or always orgasm during sex and less likely to report having experienced pain.

It also appears that having sex outside of the context of monogamous romantic relationships isn't as common among Dutch adolescents, especially older ones, as among their American counterparts. Again in the 2005 survey, two thirds of male youth and 81 percent of Dutch females had their last sex in a monogamous steady relationship, usually with a partner with whom they were "very much in love." ...

EXPLAINING THE DIFFERENCES

So why do parents in two countries with similar levels of development and reproductive technologies have such different attitudes toward the sexual experiences of teenagers? Two factors immediately spring to mind. The first is religion. As the Laumann team found, Americans who do not view religion as a central force in their decision-making are much less likely to categorically condemn teenage sex. And devout Christians and Muslims in the Netherlands are more likely to exhibit attitudes towards sexuality and marriage that are similar to those of their American counterparts. That Americans are far more likely to be religiously devout than the Dutch, many of whom left their houses of worship in the 1960s and '70s, explains part of the difference between the two countries.

A second factor is economic security. Like most European countries, the Dutch government provides a range of what sociologists call "social" and what reproductive health advocates call "human" rights: the right to housing, healthcare, and a minimum income. Not only do such rights ensure access, if need be, to free contraceptive and abortion services, government supports make coming of age less perilous for both teenagers and parents. This might make the prospect of sex derailing a child's life less haunting. Ironically, the very lack of such rights and high rates of childhood poverty in the U.S. contributes to high rates of births among teenagers. Without adequate support systems or educational and job opportunities, young people are simply more likely to start parenthood early in life.

While they no doubt contribute, neither religion nor economics can solve the whole puzzle. Even Dutch and American families matched on these dimensions still have radically divergent views of teenage sexuality and the sleepover. After interviewing 130 white middle-class Dutch and American teenagers (mostly 10th graders) and parents, I became convinced that a fuller solution is to look at the different cultures of independence and control that characterize these two middle classes.

In responding to adolescent sexuality, American parents emphasize its dangerous and conflicted elements, describing it in terms of "raging hormones" that are difficult for young people to control and in terms of antagonistic relationships between the sexes (girls and boys pursue love and sex respectively, and girls are often the losers of the battle). Moreover, American parents see it as their obligation to encourage adolescents' separation from home before accepting their sexual activity....

Dutch parents, by contrast, downplay the dangerous and difficult sides of teenage sexuality, tending to normalize it. They speak of readiness (*er aan toe zijn*), a process of becoming physically and emotionally ready for sex that they believe young people can self-regulate, provided they've been encouraged to pace themselves and prepare adequately. Rather than emphasizing gender battles, Dutch parents talk about sexuality as emerging from relationships and are strikingly silent about gender conflicts. And unlike Americans who are often skeptical about teenagers' capacities to fall in love, they assume that even those in their early teens fall in love. They permit sleepovers, even if that requires an "adjustment" period to overcome their feelings of discomfort, because they feel obliged to stay connected and accepting as sex becomes part of their children's lives.

These different approaches to adolescent sexuality are part of the different cultures of independence and control. American middle-class culture conceptualizes the self and (adult) society as inherently oppositional during adolescence. Breaking away from the family is necessary for autonomy, as is the occasional use of parental control (for instance, in the arena of sexuality), until teenagers are full adults. Dutch middle-class culture, in contrast, conceptualizes the self and society as interdependent. Based upon the assumption that young people develop autonomy in the context of ongoing relationships of interdependence, Dutch parents don't see teenage sexuality in the household as a threat to their children's autonomy or to their own authority. To the contrary, allowing teenage sexuality in the home—"domesticating" it, as it were—allows Dutch parents to exert more informal social control.

WHAT IT MEANS FOR KIDS

The acceptance of adolescent sexuality in the family creates the opportunity for Dutch girls to integrate their sexual selves with their roles as family members, even if they may be subject to a greater level of surveillance. Karel's daughter, Heidi, for example, told me she knows that her parents would permit a boyfriend to spend the night, but they wouldn't be happy unless they knew the boy and felt comfortable with him. By contrast, many American girls must physically and psychically bifurcate their sexual selves and their roles as daughters. Caroline's mother loves her boyfriend. Still, Caroline, who is seventeen, says her parents would "kill" her if she asked for a sleepover. They know she has sex, but "it's really overwhelming for them to know that their little girl is in their house having sex with a guy. That is just scary to them."

American boys receive messages ranging from blanket prohibition to open encouragement. One key message is that sex is a symbol and a threat—in the event of pregnancy—to their adult autonomy. Jesse has a mother who is against premarital sex and a father who believes boys just want to get laid. But like Caroline, Jesse knows there will be no sleepovers: "They have to wait for me to break off from them, to be doing my own thing, before they can just handle the fact that I would be staying with my girlfriend like that," he says. By contrast, Dutch boys are, or anticipate being, allowed a sleepover. And like their female counterparts, they say permission comes with a social control that encourages a relational sexuality and girlfriends their parents like. Before Frank's parents would permit a sleepover, they would first have "to know someone well." Gert-Jan says his parents are lenient, but "my father is always judging, 'That's not a type for you'."

These different templates for adolescent sex, gender, and autonomy also affect boys' and girls' own navigation of the dilemmas of gender. The category "slut" appears much more salient in the interviews with American girls than Dutch girls. One reason may be that the cultural assumption that teenagers can and do fall in love lends credence to Dutch girls' claims to being in love, while the cultural skepticism about whether they can sustain the feelings and form the attachments that legitimate sexual activity put American girls on the defensive. Kimberley, an American, had her first sex with a boy she loves, but she knows that people around her might discount such claims, saying "You're young, you can't fall in love." By contrast, in the Netherlands, Natalie found her emotions and relationship validated: her mother was happy to hear about her first intercourse because "she knows how serious we are."

In both countries, boys confront the belief and sometimes the reality that they are interested in sex but not relationships. But there is evidence in both countries that boys are often emotionally invested. The American boys I have interviewed tend to view themselves as unique for their romantic aspirations and describe themselves, as Jesse does, as "romantic rebels." "The most important thing to me is maintaining love between me and my girlfriend," while "most guys are pretty much in it for the sex," he says. The Dutch boys I interviewed did not perceive themselves as unusual for falling in love (or for wanting to) before having sex. Sam, for instance, believes that "everyone wants [a relationship]." He explains why: "Someone you can talk to about your feelings and such, a feeling of safety, I think that everyone, the largest percentage of people wants a relationship."

CULTURE'S COST

How sexuality, love, and autonomy are perceived and negotiated in parent-child relationships and among teenagers depends on the cultural templates people have available. Normalization and dramatization each have "costs" and "benefits." On balance, however, the dramatization of adolescent sexuality makes it more difficult for parents to communicate with teenagers about sex and relationships, and more challenging for girls and boys to integrate their sexual and relational selves. The normalization of adolescent sexuality does not eradicate the tensions between parents and teenagers or the gender constructs that confine both girls and boys. But it does provide a more favorable cultural climate in which to address them....

NOTE

*Note, this quote and subsequent quotes from Dutch sources are the author's translations. Names have been changed to protect anonymity.

KEY CONCEPTS

adolescence | cross-cultural | sexual revolution

DISCUSSION QUESTIONS

1. Some argue that American attitudes about sex have become more permissive in the past half century. Schalet argues, however, that Americans are still more restrictive regarding sex among teens. Consider your own attitudes about sex and sexuality as they compare to those of your parents and grandparents. What changes do you see? How can you explain generational differences regarding sexuality?
2. Why is age a factor when discussing sexuality? What makes sex among adolescents unique? What complications exist for teenagers that are different for adults?

Applying Sociological Knowledge: An Exercise for Students

Sexuality is generally thought to be a private matter, and yet public social norms very much shape and regulate sexuality. For a period of one week, observe and keep track of every comment you hear that seems to enforce certain sexual scripts. What have you heard? What assumptions does everyday talk make about heterosexuality? About gays and lesbians? How does everyday public talk shape social norms about sexuality?

PART XIII

Family

41

Beyond the Nuclear Family: The Increasing Importance of Multigenerational Bonds

VERN L. BENGSTON

Family structure has changed over the years. In this piece, Bengston outlines the research of Ernest W. Burgess, a sociologist who studied family in the first half of the 20th century. The changing form of American families has continued since the work of Burgess and, the argument is made in this reading, the mutigenerational family is now much more prevalent than in years past. Sociological research on the institution of family must consider how grandparents and parents work together in the care of children.

During the past decade, sociologists have been engaged in an often heated debate about family change and family influences in contemporary society. This debate in many ways reflects the legacy of Ernest W. Burgess (1886–1965), the pioneer of American family sociology. It can be framed in terms of four general hypotheses, each of which calls attention to significant transitions in the structure and functions of families over the 20th century.

The first and earliest hypothesis concerns the *emergence of the "modern" nuclear family form* following the Industrial Revolution. This transition (suggested by Burgess in 1916 and elaborated by Ogburn, 1932, and Parsons, 1944) proposed that the modal structure of families had changed from extended to nuclear, and its primary functions had changed from social-institutional to emotional-supportive. The second hypothesis concerns the *decline of the modern nuclear family* as a social institution, a decline said to be attributable to the fact that its structure has been truncated (because of high divorce rates) and its functions further reduced (Popenoe, 1993). A third hypothesis can be termed the *increasing heterogeneity of family forms*, relations that extend beyond biological or conjugal relationship boundaries. Growing from the work of feminist scholars (Coontz, 1991; Skolnick, 1991; Stacey, 1990), and research on racial and ethnic minority families

(Burton, 1995; Collins, 1990; Stack, 1974), this perspective suggests that family structures and relationships should be redefined to include both "assigned" and "created" kinship systems (Cherlin, 1999). I suggest a fourth hypothesis for consideration: The *increasing importance of multigenerational bonds.* I propose that relations across more than two generations are becoming increasingly important to individuals and families in American society; that they are increasingly diverse in structure and functions; and that in the early 21st century, these multigenerational bonds will not only enhance but in some cases replace nuclear family functions, which have been so much the focus of sociologists during the 20th century....

The Emergence of the "Modern" Nuclear Family Form

... Burgess' hypothesis was that families had changed. He broke from late-19th-century views of the extended family structure as the bedrock of social organization and progress to say, "The family in historical times has been, and at present is, in transition from an institution to a companionship" (Burgess, 1926, p. 104). He focused on the nuclear family and its changing functions as the consequence of industrialization and modernization, arguments echoed later by Ogburn (1932), Davis (1941), and Parsons (1944). His thesis was that urbanization, increased individualism and secularism, and the emancipation of women had transformed the family from a social institution based on law and custom to one based on companionship and love.

Burgess advanced his position very quietly in a number of scholarly journal publications. These appeared to have escaped notice by the popular press at that time, quite unlike today's debates about the family. He argued that the family had become more specialized in its functions and that structural and objective aspects of family life had been supplanted by more emotional and subjective functions. This he termed the "companionship" basis of marriage, which he suggested had become the underlying basis of the "modern" family form.

But Burgess went further. He proposed that the most appropriate way to conceptualize and study the family was as "a unity of interacting personalities" (Burgess, 1926). By this he meant three things: First, "the family" is essentially a *process*, an interactional system influenced by each of its members; it not merely a *structure*, or a household. Second, the behaviors of one family member—a troubled child, a detached father—could not be understood except in *relationship* to other family members, their ongoing patterns of interactions, and personalities developing and changing through such interactions. This conceptualization provided the intellectual basis for the first marriage and family counseling programs in the United States. Third, the central *functions* of families had changed from being primarily structural units of social organization to being relationships supporting individuals' needs. Marriage was transformed from a primarily economic union to one based on sentiment and companionship.

Thus, Burgess represented a bridge between 19th-century conceptions of the family as a unit in social evolution to 20th-century ideas of families as supporting individuals' needs. His work also provided a bridge in sociological theory, from structural-functionalism to symbolic interactionism and phenomenology. But in all this, Burgess' focus was on the nuclear family, a White, middle-class, two-generation family; and the family forms emerging in the 21st century will, as I argue below, look much different than the family that Burgess observed.

The Decline of the Modern Nuclear Family Form

The "decline of the family" in American society is a theme that has become the focus of increasingly heated debates by politicians, pundits, and family sociologists during the last decade. David Popenoe (1993), the most articulate proponent of this position, has argued that there has been a striking decline in the family's structure and functions in American society, particularly since 1960. Moreover, his hypothesis is that recent family decline is "more serious" than any decline in the past, because "what is breaking up is the nuclear family, the fundamental unit stripped of relations and left with two essential functions that cannot be performed better elsewhere: Childrearing and the provision to its members of affection and companionship" (Popenoe, p. 527). Supporters of the family decline hypothesis have focused on the negative consequences of changing family structure, resulting from divorce and single parenting, on the psychological, social, and economic wellbeing of children. Furthermore, they suggest that social norms legitimating the pursuit of individual over collective goals and the availability of alternate social groups for the satisfaction of basic human needs have substantially weakened the social institution of the family as an agent of socialization and as a source of nurturance for family members (Popenoe).

There is much to support Popenoe's hypothesis. There has been a significant change in nuclear family structure over the past 50 years, starting with the growing divorce rate in the 1960s, which escalated to over half of first marriages in the 1980s (Amato & Booth, 1997; Bumpass, Sweet, & Martin, 1990). There also has been an increase in single-parent families, accompanied by an increase in poverty for the children living in mother-headed families (McLanahan, 1994). The absence of fathers in many families today has created problems for the economic and emotional wellbeing of children (Popenoe, 1996).

At the same time, the "family decline" hypothesis is limited, and to some critics flawed, by its preoccupation with the family as a coresident household and the nuclear family as its primary representation. Popenoe defined the family as "a relatively small domestic group of kin (or people in a kinlike relationship) consisting of at least one adult and one dependent person" (Popenoe, 1993, p. 529). Although this might be sufficient as a demographic definition of a "family household," it does not include important aspects of family functions that extend beyond boundaries of coresidence. There is nothing in Popenoe's hypothesis to reflect the function of multigenerational influences on children—the role of grandparents in socializing or supporting grandchildren, particularly after the divorce of middle-generation parents (Johnson & Barer, 1987; Minkler & Rowe, 1993). Nor is there any mention of what Riley and Riley (1993) have called the "latent matrix of kin connections," a web of "continually shifting linkages that provide the potential for activating and intensifying close kin relationships" in times of need by family members (Riley & Riley, p. 169). And there is no consideration of the longer years of shared lives between generations, now extending into many decades, and their consequences for the emotional and economic support for family members across several generations (Bengtson & Allen, 1993; Silverstein & Litwak, 1993).

The Increasing Heterogeneity of Family Forms

A third hypothesis has been generated by feminist scholars (Coontz, 1991; Osmond & Thorne, 1993; Skolnick, 1991; Stacey, 1993, 1996; Thorne & Yalom, 1992) and researchers studying minority families (Burton, 1995; Collins, 1990; Stack, 1974). This hypothesis can be summarized as follows: Families are changing in both forms and meanings, expanding beyond the nuclear family structure to involve a variety of kin and nonkin relationships. Diverse family forms are emerging, or at least being recognized for the first time, including the matriarchal structure of many African American families. Stacey (1996) argued that the traditional nuclear family is increasingly ill-suited for a postindustrial, postmodern society. Women's economic and social emancipation over the past century has become incongruent with the nuclear "male breadwinner" family form and its traditional allocation of power, resources, and labor. We have also seen a normalization of divorce and of stepparenting in recent years. Many American families today are what Ahrons (1994) has described as "binuclear." Following divorce and remarriage of the original marital partners and parents, a stable, child-supportive family context may emerge. Finally, because some four million children in the United States are being raised by lesbian or gay parents (Stacey & Biblarz, in press), these and other alternative family forms "are here ... and let's get used to it!" (Stacey, 1996, p. 105).

In responding to Popenoe, Stacey (1996) argued that the family is indeed in decline—if what we mean by "family" is the nuclear form of dad, mom, and their biological or adopted kids. This form of the family rose and fell with modern industrial society. In the last few decades, with the shift to a postindustrial domestic economy within a globalized capitalist system and with the advent of new reproductive technologies, the modern family system has been replaced by what Stacey has called "the postmodern family condition," a pluralistic, fluid, and contested domain in which diverse family patterns, values, and practices contend for legitimacy and resources. Stacey suggested that family diversity and fluidity are now "normal," and the postmodern family condition opens the possibility of egalitarian, democratic forms of intimacy, as well as potentially threatening levels of insecurity.

The Increasing Importance of Multigenerational Bonds

I want to suggest a fourth hypothesis about family transitions during the 20th century that builds on those of Burgess, Popenoe, Stack, and Stacey but reflects the recent demographics development of much greater longevity. It is this: *Relations across more than two generations are becoming increasingly important to individuals and families in American society*. Considering the dramatic increase in life expectancy over the past half century, this is not a particularly radical departure from conventional wisdom. But I suggest a corollary to this hypothesis, which I hope will lead to spirited debate: *For many Americans, multigenerational bonds are becoming more important than nuclear family ties for well-being and support over the course of their lives....*

THE MACROSOCIOLOGY OF INTERGENERATIONAL RELATIONSHIPS

The demographic structure of American families has changed significantly in recent years. We hear most about two trends: The increase in divorce rates since the 1960s, with one out of two first marriages ending in divorce (Cherlin, 1992); and the increasing number of children living in single-parent households, often accompanied by poverty (McLanahan & Sandefur 1994; Walker & McGraw, 2000). But there is a third trend that has received much less attention: The increased longevity of family members and the potential resource this represents for the well-being of younger generations in the family.

Multigenerational Family Demography: From Pyramids to Beanpoles

First consider how much the age structure of the U.S. population has changed over the past 100 years. Treas (1995b) provided a valuable overview of these changes and their consequences for families. In 1900, the shape of the American population structure by age was that of a pyramid, with a large base (represented by children under age 5) progressively tapering into a narrow group of those aged 65 and older. This pyramid characterized the shape of the population structure by age in most human societies on record, from the dawn of civilization through the early Industrial Revolution and into the early 20th century (Laslett, 1976; Myers, 1990). But by 1990, the age pyramid for American society had come to look more like an irregular triangle. By 2030, it will look more like a rectangle, with strikingly similar numbers in each age category starting from children and adolescents through those above the age of 60. The story here is that because of increases in longevity and decreases in fertility, the population age structure of the United States, like most industrialized societies, has changed from a pyramid to a rectangle in just over a century of human historical experience.

Second, consider the implications of these macrosocietal changes in age distribution for the generational structure of families in American society. At the same time, there have been increases in life expectancy over the 20th century, decreases in fertility have occurred, and the population birthrate has decreased from 4.1 in 1900 to 1.9 in 1990 (Cherlin, 1999). This means that the age structure of most American families has changed from a pyramid to what might be described as a "bean-pole" (Bengtson, Rosenthal, & Burton, 1990), a family structure in which the shape is long and thin, with more family generations alive but with fewer members in each generation. Whether the "beanpole" structure adequately describes a majority of families today has been debated (Farkas & Hogan, 1995; Treas, 1995a). Nevertheless, the changes in demographic distribution by age since 1900 are remarkable, and the progression "from pyramids to beanpoles" has important implications for family functions and relationships into the 21st century.

... What might be lost in a review of macrosocial demographic trends are the consequences for individual family members and their chances of receiving family support.... The increasing availability of extended intergenerational kin (grandparents, great-grandparents, uncles, and aunts) has become a resource for children as they grow up and move into young adulthood....

... Other implications of these demographic changes over the 20th century should be noted. First, we now have more years of "cosurvivorship between generations" than ever before in human history (Bengtson, 1996; Goldscheider, 1990). This means that more and more aging parents and grandparents are available to provide for family continuity and stability across time (Silverstein, Giarrusso, & Bengtson, 1998)....

When Parenting Goes Across Several Generations

A function not addressed by Burgess was the importance of grandparents to family members' well-being, an understandable oversight given the historical period when he was writing, when the expected life span of individuals was almost 3 decades shorter than today. Popenoe (1993) also did not discuss the importance of grandparents in the potential support they represent for younger generation members.

Grandparents provide many unacknowledged functions in contemporary families (Szinovacz, 1998). They are important role models in the socialization of grandchildren (Elder, Rudkin, & Conger, 1994; King & Elder, 1997). They provide economic resources to younger generation family members (Bengtson & Harootyan, 1994). They contribute to cross-generational solidarity and family continuity over time (King, 1994; Silverstein et al., 1998). They also represent a bedrock of stability for teenage moms raising infants (Burton & Bengtson, 1985).

Perhaps most dramatic is the case in which grandparents (or great-grandparents) are raising grandchildren (or great-grandchildren). Over four million children under age 18 are living in a grandparent's household. Frequently this is because these childrens' parents are incapacitated (by imprisonment, drug addiction, violence, or psychiatric disorders) or unable to care for their offspring without assistance (Minkler & Rowe, 1993)....

When Parents Divorce and Remarry, Divorce and Remarry

The rising divorce rate over the last half of the 20th century has generated much concern about the fate of children (McLanahan & Sandefur, 1994). The probability that a marriage would end in divorce doubled between the 1960s and the 1970s, and half of all marriages since the late 1970s ended in divorce (Cherlin, 1992). About 40% of American children growing up in the 1980s and 1990s experienced the breakup of their parents' marriages (Bengtson, Rosenthal, & Burton, 1995; Furstenberg & Cherlin, 1991), and a majority of these also experienced their parents' remarriage and the challenges of a "blended family."

In the context of marital instability, the breakup of nuclear families, and the remarriage of parents, it is clear that grandparents and step-grandparents are becoming increasingly important family connections (Johnson & Barer, 1987)....

When Help Flows Across Generations, It Flows Mostly Downward

An unfortunate stereotype of the older generation today is of "greedy geezers" who are spending their children's inheritance on their own retirement pleasures (Bengtson, 1993). This myth is not in accord with the facts. Intergenerational patterns of help and assistance flow mostly from the older generations to younger generations in the family. For example, McGarry and Schoeni (1995) have shown that almost one third of U.S. parents gave a gift of $500 or more to at least one of their adult children during the past year; however, only 9% of adult children report providing $500 to their aging parents. Similar results are reported by Bengtson and Harootyan (1994) and Soldo and Hill (1993)....

THE MICROSOCIOLOGY OF INTERGENERATIONAL RELATIONSHIPS

Although there have been important changes in the demography of intergenerational relationships since the 19th century, population statistics about family and household structure tell only one part of the story. At the behavioral level, these changes have more immediate consequences in the ways family members organize their lives and pursue their goals in the context of increasing years of intergenerational "shared lives." How to conceptualize and measure these intergenerational interactions has become increasingly important since Burgess (1926) put forth his definition of the family as "a unity of interacting personalities." ...

The Strength of Intergenerational Relationships Over Time

Using longitudinal data from the Longitudinal Study of Generations we have been able to chart the course of intergenerational solidarity dimensions over time. Our design allows consideration of the development and aging of each of the three and now four generations in our sample, as well as the sociohistorical context of family life as it has changed over the years of the study (Bengtson et al., in press).

One consistent result concerns the high levels of affectual solidarity (reflecting the emotional bonds between generations) that have been found over six times of measurement, from 1971 to 1997 (Bengtson et al., 2000). Three things should be noted. We find that the average solidarity scores between grandparents and parents, parents and youth, grandparents and grandchildren are high, considerably above the expected midpoint of the scale. Second, these scores are remarkably stable over the 26 years of measurement.... Third, there is a "generational bias" in these reports: Parents consistently report higher affect than their children do over time, as do grandparents compared with grandchildren.... The older generation has a greater psychosocial investment, or "stake," in their joint relationship than does their younger generation, and this influences their perceptions and evaluations of their common intergenerational relationships.

These results indicate the high level of emotional bonding across generations and the considerable stability of parent-child affectual relationships over long periods of time....

The Diversity of Intergenerational Relationships

... We found five types or classes of intergenerational family relationships.... One type we labeled "Tight-Knit," characterized by high emotional closeness, living fairly close to each other, interacting frequently, and having high levels of mutual help and support. This seems similar to what Parsons (1944) described as the ideal "modern family" type of relationship. At the other extreme is the "Detached" type, with low levels of connectedness in all of the observed measures of solidarity. This appears similar to what would be predicted by the "decline of the family" hypothesis. Between the Tight-Knit and the Detached are three classes which we called variegated types (Silverstein & Bengtson, 1997, p. 442). The "Sociable" and "Intimate-but-Distant" types seem similar to what Litwak (1960a) described as the "modified extended family" in which functional exchange is low or absent, but there are high levels of affinity that suggest the potential for future support and exchange. The "Obligatory" type suggests a high level of structural connectedness (proximity and interaction) with an average level of functional exchange, but a low level of emotional attachment....

These results suggest the folly of using a "one size fits all" model of intergenerational relationships. There is considerable diversity among the types; there is no one model type. Our findings reinforce the message of Burton (1995) in her enumeration of 16 structural types of relationships between teenage mothers and older generation family members: Diversity and complexity are inherent features of family networks across generations.

The Effects of Changing Family Forms on Intergenerational Influence

Situating multigenerational families in socio-historical context allows us to broaden our inquiry about their importance and functionality. How have intergenerational influences changed over recent historical time? Are families still important in shaping the developmental outcomes of its youth? What have been the effects of changing family structures and roles, the consequences of divorce and maternal employment, on intergenerational influences? ...

Our analysis suggests that today's Generation Xers are surprisingly similar to what their baby boomer parents were on these measures at the same age, almost 30 years ago. This suggests that despite changes in family structure and socioeconomic context, intergenerational influences on youths' achievement orientations remain strong. Generation Xers whose parents divorced were slightly less advantaged in terms of achievement orientations than Generation Xers who came from nondivorced families but were nevertheless higher on these outcome measures than were their baby boomer parents at the same age, regardless of family structure. We also found that maternal employment has not negatively affected the aspirations, values, and self-esteem of youth across these two generations, despite the dramatic increase in women's labor force participation over the past 3 decades. Finally, we found that Generation Xer women have considerably higher educational and occupational aspirations in 1997 than did their baby-boomer mothers almost 30 years before. In fact, Generation X young women's aspirations were higher than Generation X young men's.

These findings challenge the hypothesis that families are declining in function and influence and that "alternative" family structures spell the downfall of American youth. Multigenerational families continue to perform their functions in the face of recent social change and varied family forms....

MULTIGENERATIONAL FAMILY BONDS: MORE IMPORTANT THAN EVER BEFORE?

My hypothesis is that multigenerational family bonds are important, more so than family research has acknowledged to date. I have argued that demographic changes over the 20th century ("from pyramids to beanpoles" and "longer years of shared lives") have important implications for families in the 21st century, particularly with regard to the "latent network" of family support across generations. I have suggested that multigenerational relationships are increasingly diverse in structure and functions within American society. I propose that because the increase in marital instability and divorce have weakened so many nuclear families, these multigenerational bonds will not only enhance but in some cases replace some of the nuclear family functions that have been the focus of so much recent debate....

CONCLUSION: BEYOND THE NUCLEAR FAMILY

Are families declining in importance within American society? Seven decades ago, Ernest W. Burgess addressed this question from the standpoint of family transformations across the 19th and 20th centuries. His hypothesis was that families and their functions had changed from a social institution based on law and custom to a set of relationships based on emotional affect and companionship (Burgess, 1926). But this did not mean a loss of social importance. He suggested that the modern family should be considered as "a unity of interacting personalities" (Burgess) and that future research should focus on the interactional dynamics within families. In all this, Burgess' focus, and that of those who followed him (Ogburn, 1932; Parsons, 1944), was on the nuclear family form.

Seven decades later this question—are families declining in importance?—has resurfaced. Some family experts have hypothesized that families have lost most of their social functions along with their diminished structures because of high divorce rates and the growing absence of fathers in the lives of many children (Popenoe, 1993). A contrasting hypothesis is that families are becoming more diverse in structure and forms (Skolnick, 1991; Stacey, 1996).

In this article, I have suggested another hypothesis, one that goes beyond our previous pre-occupation with the nuclear or two-generation family structure. This concerns the increasing importance of multigenerational bonds and the multigenerational extension of family functions....

Burgess was right, many decades ago: The American family is in transition. But it is not only in transition "from institution to companionship," as he argued. Over the century, there have been significant changes in the family's structure and functions. Prominent among them has been the extension of family bonds, of affection and affirmation, of help and support, across several generations, whether these be biological ties or the creation of kinlike relationship. But as families have changed, they have not necessarily declined in importance. The increasing prevalence and importance of multigenerational bonds represents a valuable new resource for families in the 21st century....

REFERENCES

Ahrons, C. R. (1994). *The good divorce.* New York: Harper Collins.

Amato. P., & Booth, A. (1997). *A generation at risk: Growing up in an era of family upheaval.* Cambridge, MA: Harvard.

Bengtson, V. L. (1993). Is the "contract across generations" changing? Effects of population aging on obligations and expectations across age groups. In. V. L. Bengtson & W. A. Achenbaum (Eds.), *The changing contract across generations* (pp. 3–24). New York: Aldine de Gruyter.

Bengtson, V. L. (1996). Continuities and discontinuities in intergenerational relationships over time. In V. L. Bengtson (Ed.). *Adulthood and aging: Research on continuities and discontinuities.* New York: Springer.

Bengtson, V. L., & Allen. K. R. (1993). The life course perspective applied to families over time. In P. Boss, W. Doherty, R. LaRossa, W. Schumm, & S. Steinmetz (Eds.), *Sourcebook of family theories and methods: A contextual approach* (pp. 469–498). New York: Plenum Press.

Bengtson, V. L., Biblarz, T., Clarke, E., Giarrusso, R., Roberts. R. F. L., Richlin-Klonsky, J., & Silverstein, M. (2000). Intergenerational relationships and aging: Families, cohorts, and social change. In J. M. Claire & R. M. Allman (Eds.), *The gerontological prism: Developing interdisciplinary bridges*. Amityville. NY: Baywood.

Bengtson, V. L., Biblarz, T. L., & Roberts. R. E. L. (in press). *Generation X and their elders*. New York: Cambridge University Press.

Bengtson, V. L., & Harootyan, R. (Eds.) (1994). *Intergenerational linkages: Hidden connections in American society*. New York: Springer.

Bengtson, V. L., Rosenthal, C. J., & Burton, L. M. (1990). Families and aging: Diversity and heterogeneity. In R. Binstock & L. George (Eds.), *Handbook of aging and the social sciences* (3rd ed., pp. 263–287). New York: Academic Press.

Bengtson, V. L. Rosenthal, C. J., & Burton, L. M. (1995). Paradoxes of families and aging. In R. H. Binstock & L. K. George (Eds.), *Handbook of aging and the social sciences* (4th ed., pp. 253–282). San Diego, CA: Academic Press.

Bumpass, L. L., Sweet, J., & Martin, C. (1990). Changing patterns of remarriage. *Journal of Marriage and the Family*, 52, 747–756.

Burgess, E. W. (1916). *The function of socialization in social evolution*. Chicago: University of Chicago Press.

Burgess, E. W. (1926). The family as a unity of interacting personalities. *The Family*, 7, 3–9.

Burton, L. (1995). Intergenerational patterns of providing care in African-American families with teenage childbearers: Emergent patterns in an ethnographic study. In V. L. Bengtson, K. W. Schaie. & L. M. Burton (Eds.). *Adult intergenerational relations* (pp. 79–97). New York: Springer.

Burton, L. M., & Bengtson, V. L. (1985). Black grandmothers: Issues of timing and continuity in roles. In V. Bengtson & J. Robertson (Eds.), *Grandparenthood* (pp. 304–338). Beverly Hills, CA: Sage.

Cherlin, A. J. (1992). *Marriage, divorce, remarriage* (rev, and enlarged ed). Cambridge, MA: Harvard University Press.

Cherlin, A. J. (1999). *Public and private families*. Boston: McGraw-Hill.

Collins, P. H. (1990). *Black feminist thought*. New York: Routledge.

Coontz, S. (1991). *The way we never were*. New York: Basic Books.

Davis, K. (1941). Family structure and functions. *American Sociological Review*, 8, 311–320.

Elder, G. H., Rudkin, L., & Conger, R. D. (1994). Intergenerational continuity and change in rural America. In K. W. Schaie, V. Bengtson, & L. Burton (Eds.). *Societal impact on aging: Intergenerational perspectives*. New York: Springer.

Farkas, J., & Hogan, D. (1995). The demography of changing intergenerational relationships. In V. L. Bengtson, K. Warner Schaie, & L. M. Burton (Eds.), *Adult intergenerational relations: Effects of societal change* (pp. 1–19). New York: Springer.

Furstenberg, F., & Cherlin, A. (1991). *Divided families: What happens to children when parents part*. Cambridge, MA: Harvard University Press.

Goldscheider, F. K. (1990). The aging of the gender revolution: What do we know and what do we need to know? *Research on Aging*, 12, 531–545.

Johnson, C. L., & Barer, B. M. (1987). Marital instability and the changing kinship networks of grandparents. *Gerontologist*, 27, 330–335.

King, V. (1994). Variation in the consequences of nonresident father involvement for children's well-being. *Journal of Marriage and the Family*, 56, 963–972.

King, V., & Elder, G. H. Jr. (1997). The legacy of grand-parenting: Childhood experiences with grandparents and current involvement with grandchildren. *Journal of Marriage and the Family*, 59, 848–859.

Laslett, P. (1976). Societal development and aging. In R. Binstock & E. Shanas (Eds.), *Handbook of aging and the social sciences* (pp. 87–116). New York: Van Nostrand Reinhold.

Litwak, E. (1960a). Geographic mobility and extended family cohesion. *American Sociological Review*, 25, 385–394.

McGarry, K., & Schoeni, R. F. (1995). Transfer behavior in the health and retirement study. *Journal of Human Resources*, 30 (Suppl.), S184–S226.

McLanahan, S. S. (1994). The consequences of single motherhood. *American Prospect*, 18, 94–58.

McLanahan, S. S., & Sandefur, G. (1994). *Growing up with a single parent: What helps, what hurts*. Cambridge, MA: Harvard University Press.

Minkler, M., & Roe, J. (1993). *Grandparents as caregivers*. Newbury Park, CA: Sage.

Myers, G. (1990). Demography of aging. In R. Binstock & L. George (Eds.), *Handbook of aging and the social sciences* (3rd ed., pp. 19–44). New York: Academic Press.

Ogburn, W. F. (1932). The family and its functions. In W. F. Ogburn's, *Recent social trends*. New York: McGraw-Hill.

Osmond, M. W., & Thorne, B. (1993). Feminist theories: The social construction of gender in families and society (pp. 591–623). In P. G. Boss, W. J. Doherty. R. La Rossa, W. R. Schumm, & S. K. Steinmetz, *Sourcebook of family methods and theories*. New York: Plenum Press.

Parsons, T. (1944). The social structure of the family. In R. N. Anshen (Ed.), *The family: Its function and destiny* (pp. 173–201). New York: Harper.

Popenoe, D. (1993). American family decline, 1960–1990: A review and appraisal. *Journal of Marriage and the Family*, 55, 527–555.

Riley, M. W., & Riley, J. W. (1993). Connections: Kin and cohort. In V. L. Bengtson & W. A. Achenbaum (Eds.), *The changing contract across generations*. New York: Aldine de Gruyter.

Silverstein, M., & Bengtson, V. L. (1997). Intergenerational solidarity and the structure of adult child-parent relationships in American families. *American Journal of Sociology*. 103. 429–460.

Silverstein, M., & Litwak, E. (1993). A task-specific typology of intergenerational family structure in later life. *Gerontologist*, 33, 256–264.

Silverstein, M., Giarrusso, R., & Bengtson, V. L. (1998). Intergenerational solidarity and the grandparent role. In M. Szinovacz (Ed.), *Handbook on grandparenthood* (pp. 144–158). Westport. CT: Greenwood Press.

Skolnick, A. (1991). *Embattled paradise: The American family in an age of uncertainty*. New York: Basic Books.

Soldo, B. J., & Hill, M. S. (1993). Intergenerational transfers: Economic, demographic, and social perspectives. *Annual Review of Gerontology and Geriatrics*, 13, 187–216.

Stacey, J. (1990). *Brave new families: Stories of domestic upheaval in late twentieth century America.* New York: Basic Books.

Stacey, J. (1993). Is the sky falling? *Journal of Marriage and the Family*, 55, 555–559.

Stacey, J. (1996). *In the name of the family: Rethinking family values in the postmodern age.* Boston: Beacon Press.

Stacey, J., & Biblarz, T. (in press). How does the sexual orientation of parents matter? *American Sociological Review.*

Stack, C. (1974). *All our kin: Strategies for survival in a Black community.* New York: Harper and Row.

Szinovacz, M. (1998). *Handbook on grandparenthood.* Westport, CT: Greenwood Press.

Thorne, B., & Yalom, M. (1992). *Rethinking the family: Some feminist questions.* Boston: Northeastern University Press.

Treas, J. (1995a). Commentary: Beanpole or beanstalk? Comments on "The Demography of Changing Intergenerational Relations." In V. L. Bengtson, K. W. Schaie, & L. M. Burton (Eds.), *Adult intergenerational relations* (pp. 26–29). New York: Springer.

Treas, J. (1995b). Older Americans in the 1990s and beyond. *Population Bulletin*, 50, 2–46.

Walker, A. J., & McGraw, L. A. (2000). Who is responsible for responsible fathering? *Journal of Marriage and the Family*, 62, 563–569.

KEY CONCEPTS

divorce rate heterogeneity marriage rate nuclear family

DISCUSSION QUESTIONS

1. How many different types of families can you identify among the people you know? Is your own family structure different than those of your friends and peers? How have these different family forms influenced the way children are raised?
2. Consider the family as a social institution. How does the family influence the individual? If there are multiple generations of people in a family, what benefits does that offer the children of that family? What problems might multigenerational families face?

42

"Why Can't I Have What I Want": Timing, Employment, Marriage, and Motherhood

ROSANNA HERTZ

This piece summarizes findings from interviews with single working mothers. The ideas expressed by the respondents summarize feelings of frustration that "having it all" was not as possible as they once believed. Despite feminist advancements that allow women to work, marry, and have children, for the women in this study, all three were not easily obtained simultaneously. Work and motherhood were chosen over marriage.

Stuck. Virtually every woman I interviewed expressed the feeling. Something conspired to disrupt the trajectory of love to marriage to children. Joy McFadden pointed the finger at her demanding job and a shortage of candidates in the marriage market.[1] She declared herself unwilling to settle for compromises: a marriage arrived at to serve other ends. Claudia D'Angelo acknowledged her tug-of-war between independence and intimacy and the difficulties it caused her in her relationships with men. She worried about marriage transforming her independence into narrowed opportunities, as it had for her mother. And when she did become involved with a man, he didn't share her desire for children.

In some instances, being stuck meant being mired hip deep in a bog of commitments and bereft of energy or time to search for alternatives. For the vast majority of women I interviewed, however, being stuck was a dynamic thing, like Claudia's tug-of-war. It might appear to outsiders as motionlessness, passivity, or even resignation. But, listening to women such as Claudia and Joy, I clearly got the sense that although it may have been enervating, it was rarely passive.

What are the opposing forces that keep women stuck? Middle-class women, I found, are caught between a battered but resilient ideology of marriage-then-motherhood and the experience of independence and self-fulfillment in a

SOURCE: SINGLE BY CHANCE, MOTHERS BY CHOICE by Hertz (2008) Chp. 1 "Why Can't I Have What I Want?" pp. 3–20. By permission of Oxford University Press, USA.

workplace that poses fewer barriers to women than previously. In the late 1970s, when at least half the women I interviewed reached the age of majority, women stopped sporting engagement rings at college graduation and started brandishing their degrees, which galvanized them as agents of change. As they took to heart the expectation for equality in the workplace, middle-class women no longer had to strike a risky bargain with men to achieve economic stability in their adult lives. Marriage receded in importance as women had other options and a greater range of opportunities for defining themselves in the world. While women did not stop seeking marriage, their expectations for the institution were transformed as the need for a man for economic security and social stability fell away, leaving only the idealized image of marriage for love.... Unlike generations of middle-class women before them who believed their fate was either marriage and motherhood or spinsterhood and career, these women always expected they would have it all.

While some scholars suggested that it would be difficult to have children and continue to be employed simultaneously, and some early second-wave feminists argued that family obligations to nurture children would make competing equally with male peers difficult, neither scholars nor activists urged women to give up children entirely. Academics, by contrast, argued that it was possible to have it all, but maybe it would be easier to have baby and career sequentially; marriage and heterosexuality were taken for granted....

Joy and Claudia represent the two-thirds of women in this study who are middle class. These women grew up imagining white picket fences and perfect children. They worked hard in school with the goal of going to college, even though they did not necessarily anticipate lifelong careers. Everyone assumed they would settle down and raise a family. In their families, men would be the providers. However, when they graduated from college in the mid to late 1970s, Joy and Claudia found themselves in a time of enormous flux.... As young women in their early twenties, they were in the midst of a rapid expansion of employment opportunities and an influential women's liberation movement. Most parents, even conservative ones, encouraged their daughters to be "whatever they wanted to be." Family and marriage were put on hold as exciting job opportunities arose and young women started to bring home a paycheck. Joy and Claudia liked making decisions about how to spend the money they earned, and they reveled in the many different ways they could shape their lives as self-sufficient women. The irony, they discovered, was that although they had been raised to follow in the footsteps of their mothers, they were actually imitating their fathers.

By contrast, women with working-class origins usually came from dual-earner families. Working-class moms rarely left the labor force except when their children were babies, and even then some were employed. Even though their parents may have only completed high school, these daughters were likely to have some college education. Unlike the middle-class pattern in which mothers stayed home and raised their families until the youngest child was entering high school, working-class daughters watched their mothers bring home a paycheck, even if the hours they worked made it appear that they were waiting at home for the school bus (Garey 1999).

For example, when Abby Pratt-Evans was growing up, her dad, upon returning home from his construction job, would sit in the living room watching TV, his reward for a long day. Her mom, on the other hand, rushed back from her nursing shift to prepare dinner while Abby and her preteen sister peppered her with questions about carpool arrangements, weekend plans with friends, and math homework. Dinner was always on the table on time, but as Abby grew older she noted that her mom also had worked all day. Abby loved her mother but never could stand how her mother allowed her father to sit and not help every night. Women who came from working-class backgrounds were simultaneously proud of their mothers' employment achievements and sad that it was their mothers who were doubly burdened with keeping family life together. These women worried that unless they redefined marriage, their husband's employment would slowly overshadow their own and they would become their mothers. Women in this study were too committed to pursuing their employment and independence to let that happen. The surge of importance employment took on for the women in this study is not without context. Second-wave feminism, emerging in the 1960s and 1970s, emphasized the struggle for equal opportunity. Focusing on the transformation of social structures, including law, education, and employment, second-wave feminism sought to change and expand all aspects of women's and men's lives (Rosen 2001). A few of the oldest women in this study, who had been the first to achieve in the workplace, spoke of a strong attachment to feminism. Mostly, however, these women were free riders, reaping the benefits of feminist activism without feeling part of the movement themselves. They attended college and established their careers in a time of economic expansion when equality was already mandated.

Although higher education and workplace norms have changed to reflect new legislation, the social revolution initiated by women's entry into the economy scans to have stalled the threshold of the home (Hochschild 1989). Husbands have continued serving as main providers and wives as primary caretakers, even if breadwinning is an increasingly shared endeavor. We once assumed, perhaps naively, that when women became a permanent part of the paid labor force, their husbands would begin to share equally in housework and child care (Hertz 1986). As we now know from over three decades of research on work and family among two-income couples, the division of labor between spouses is not equally shared. Further, the clash between work and family life has become an increasing concern as academics expose the ways American families experience a shortage of time (see also Schor 1991; Hochschild 1989, 1997; Moen 2003; Jacobs and Gerson 2004; Milkie et al. 2004).

The majority of the women I interviewed described themselves as having been strongly committed to work prior to motherhood. They have occupations as diverse as lawyers, managers, consultants, waitresses, and aerobics instructors. Many work in the service sector in feminized occupations (such as nurses, secretaries, social workers, and elementary schoolteachers). Others work in major corporate, university, and nonprofit settings as managers, professors, and lawyers. A smaller group is self-employed, including small-business owners, writers, Web designers, and contract workers for corporations and hospitals. Annette Barker, a senior manager with a local high-tech firm, described with pride her rapid rise in the company's ranks in the early to mid-1980s after completing her MBA:

> I was working for a company that I had been with for six years, that I had grown up with, that I had gone from being an entry-level programmer to a senior programmer, a project leader and a manager, a director. My career developed there and my identity was my work. I kept getting promoted and with each promotion I continued my pace. I reached a pretty lofty position.

Likewise, Abby, who has master's degrees in both educational administration and educational psychology, threw herself into her work as an elementary schoolteacher, winning awards for her innovative style. Abby dreamed of moving up the ladder, too:

> I teach gifted and talented children and I love working with young children, particularly troubled kids. I put in extra time developing a new curriculum. I was always working on making myself as a teacher more child-friendly ... Still ... I wanted to move up. So I went back to get certified to be a principal.

Both Annette and Abby received great personal satisfaction from their workplace accomplishments, but, like Leigh, a journalist groomed for the national stage, they marveled at how they had gotten so ensnared in the rush to status that they lost track of their plans for motherhood. Leigh Newell explained:

> It was an era where we were constantly reading about the first woman lawyer to do this, the first woman to *senior* VP at that. And there was very little discussion about motherhood. Now, when I got into my late thirties and began looking back even then and thinking, "Why didn't I do some of these things? Why didn't I think more about having children?" Then of course you get into that thinking, "Well, was the message right?" But motherhood, just wasn't on my radar screen. It just wasn't.

The belief that it takes a partner to have a child was a cultural mandate that even these successful women were unable to ignore. Most busied themselves with work, hoping the right partner would materialize and start the sequence. As a result of women's diminishing economic need for marriage, they were more fully able to invest in the romanticized vision of the institution, in which the magic of love overshadows more practical considerations.

Claudia, who shared similar experiences with many of the other women in this study, talked about how she couldn't fathom moving out of her college dorm to follow a boyfriend who had entered graduate school. In her words, "the relationship had potential but the timing was off." Timing, I was told repeatedly, was critical to forming intimate and lasting relationships. Timing was not simply a matter of finding the right person. Both people had to be ready to commit, and not just to a relationship but also to a future that included marriage and children. Lily Baker, age thirty-nine and with a one-year-old conceived through anonymous donor insemination, described succinctly what for her turned out to be nearly two decades of bad timing and weak commitments:

> The way I look at it in a nutshell is that in my twenties, I wasn't ready. In my thirties I was, I dated three men in my thirties. The first one wasn't ready to get married; I would have married him, wise or unwise. The second one I would have married, but he said, "You're the most wonderful woman I've ever dated, but I'm not in love with you." And the third one I definitely expected to marry. I was like, "Oh, I'm glad die other two didn't work because this is the One for me." And he wasn't ready.

Like Claudia, Lily felt she had time on her side. She was convinced that a much better prospect was just around the comer.

This is not to say that these women, heterosexual or lesbian, lacked in relationships. The overwhelming majority described long-lasting and, in many instances, very fulfilling commitments. But in every instance, whether the relationship was simply romantic or included cohabitation, marriage, or partnership, circumstances conspired to stop things short of the full package—commitment with children. Women, whether heterosexual or lesbian, wanted ideally to parent with partners. Men's ambivalence about commitment loomed large in women's accounts, and a man's waffling often derailed plans for moving forward. Over coffee, Nadine Margolis gave me the last story of the man she had dated.

> I looked at his drawings for this house he was building. And I said to myself, "This man does not want to have children." And that night I asked him about it, and he said that he did not want to have children. And I said, "Yes, I did want children." We parted quietly from that night. I could just tell from the way the drawing was done, there was no place for a child to get next to a parent. This was definitely not someone who wanted children, or who had thought about it.

Nadine abandoned a secure, richly drawn future plan for a life together and stepped out alone into unmapped territory. Similar to other women in this study her relationship ended when she raised her wish for children. She was not willing to settle for a life without children; he could not commit to her vision of a family life....

Frequently, some women succumbed to moments of self-blame for not finding a marriage partner. The pressure to find suitable partners is not limited to straight women; gay women in this study also felt a sense of failure to find someone with whom they could both spend their life and have a child. At some level, self-doubt and self-criticism insinuated themselves into every woman's story. "Is there something wrong with me? Am I to blame? Did I refuse to compromise? Am I naive?"

The women I interviewed looked toward marriage and found themselves depressed by the dwindling odds of finding love that would lead to children.

As they believed marriage to be slipping further and further out of their reach, motherhood, on the other hand, moved closer, drawn in by their desire for children. As Claudia put it, women were "running two races and losing at both." Faced with the decision to choose one or the other in order to win, women found themselves making a difficult life decision. While social norms dictated throwing the baby out with the bathwater—that is, discarding motherhood because marriage seemed unattainable—women salvaged the baby. Women shed the burden of marriage, determined to win the race to motherhood alone.

In earlier generations, this sense of being stuck most likely would have resulted in spinsterhood—in becoming the "favorite aunt," to use Joy's words. Both Joy and Claudia believed that marriage and children would happen naturally and effortlessly. Joy was caught up in enormous professional demands that limited her social life. Claudia, framed her story around ambivalent relationships, even though she also had a demanding career as a clinical psychologist. Neither woman rejected marriage as a social institution; indeed, both honored it by exhausting virtually every route to marriage before electing to have a child as a single mom. Similarly, the lesbian women in this study embraced the idea of a stable partnership with the same fervor as the straight women; they, too, clung tenaciously to the ideal of motherhood even when the possibility of having a partner was remote.

Stymied in their efforts to give and get commitment, many women have abandoned the belief that marriage is an essential part of the family equation. The women in this study set marriage aside, fully realizing that as they aged motherhood could slip out of their grasp. These particular women refused to be driven to the altar by their desire for children. Reserving marriage only for love, they no longer reserved motherhood for marriage. Lori-Ann Stuart, a forty-one-year-old lesbian with a four-year-old son described how her plans to parent with her partner of three years fell apart when her partner left her: "So when we were involved, we always said, "Yeah, we'll have kids together." She left. She fell in love with someone else and left conveniently right around the time when we had had said, "OK, this is when we're going to start finding a donor, etc." After we broke up, for a couple of years I couldn't deal with the whole idea of having a kid. I just couldn't separate that we were going to have a kid together to what was I going to do now? I couldn't see myself parenting without a partner." Entering the dating market again, Lori-Ann decided that her own biological clock overrode finding a partner. She called back the close friend whom she had approached to become a known donor when she was partnered.

To become a single mother is not an inherently selfish act, because what drives parenthood is neither wholly altruistic nor completely self-absorbed. These women's decisions to become mothers reflect the broader mandates of American culture that tie motherhood to womanhood, parenthood to adulthood. Their decision is akin to that made by their partnered heterosexual peers, although it is not sheltered by social norms. Parenthood in the context of single mothers is regarded with needling suspicion by much of the rest of society, as it is a threat of the unthinkable—families without dads, the ultimate displacement.... No longer constrained by social pressures in the same ways as before, women are still stuck by the fear that two parents (even same-sex parents) are inherently better than one. However, at some point motherhood becomes a more compelling force than fear. As Susan Jaffe, forty-eight with a nine-year-old, described it:

> Time was running out and I began to start wrestling with the idea that I might not meet someone in time. And so I knew that by this time I really had a *strong* desire to have a child. And it didn't abate. It got increasingly larger. I knew that I would be really unhappy if I didn't have a child. I'd made a decision that come hell or high water, I was gonna have a child. But I hadn't decided how, even though my first choice was not to be a single parent. My first choice was to have a partner.

Many women suddenly and overwhelmingly become hungry for the motherhood they have put on the back burner up until this point.... As their desire for a child overtakes them, other considerations such as work, success, and the search for a suitable partner pale. Single motherhood can be the solution to the dilemma for these women.

However, the decision to go it alone is nor made overnight. Joy elected an anonymous donor after deciding against the possibility of complications that might occur with a known donor. Claudia delayed applying for adoption until her age nearly disqualified her. The paths to single motherhood are diverse and involve complex decisions about timing, insemination or adoption, and racial, ethnic, religious, and ideological considerations (Hertz 2006).

NOTE

1. All women quoted in the text are identified by pseudonym, and I have changed certain details for some women to protect their identity, such as sex of child, exact occupation, and community of residence.

REFERENCES

Garey, Anita Ilta. 1999. *Weaving Work and Motherhood.* Philadelphia: Temple University Press.

Hertz, Rosanna. 1986. *More Equal than Others: Women and Men in Dual-Career Marriages.* Berkeley: University of California Press.

———. 2006. *Single by Chance, Mothers by Choice: How Women Are Choosing Parenthood without Marriage and Creating the New American Family.* New York: Oxford University Press.

Hochschild, Arlie, with Anne Machung. 1989. *The Second Shift.* New York: Viking.

———. 1997. *The Time Bind: When Work Becomes Home and Home Becomes Work.* New York: Henry Holt and Co.

Jacobs, Jerry A., and Kathleen Gerson. 2004. *The Time Divide.* Cambridge, MA: Harvard University Press.

Moen, Phyllis, ed. 2003. *It's About Time: Couples and Careers.* Ithaca, NY: Cornell University Press.

Milkie, Melissa, Marybeth Mattingly, Kei Nomaguchi, Suzanne M. Bianchi, and John P. Robinson. 2004. "The Time Squeeze: Parental Statuses and Feelings About Time with Children. *Journal of Marriage and Family* 66 (3): 739–761.

Rosen, Ruth. 2001. *The World Split Open: How the Modern Women's Movement Changed America.* New York: Penguin Books.

Schor, Juliet. 1991. *The Overworked American.* New York: Basic Books.

KEY CONCEPTS

ideology

labor force participation rate

DISCUSSION QUESTIONS

1. What are the most recent statistics on the number of families that are headed by single mothers? How has this changed in your lifetime? What challenges are unique to single mothers?
2. Why is marriage so elusive for these working women? Beyond the personal choice issues, what structural issues can explain career and motherhood *instead of* marriage?

43

Gay Marriage: Why Now? Why At All?

REESE KELLY

This reading outlines the reasons the gay marriage debate has emerged only recently. Previously, the lesbian and gay movement focused on tolerance and acceptance in the public sphere. The AIDS crisis and the aftermath of 9/11 moved the focus to one of the politics of equality, looking for ways to give gays and lesbians access to the same benefits of heterosexuals. Kelly concludes with a brief outline of the three sides to the debate over gay marriage.

Currently, the most polarizing issue in the United States regarding the gay and lesbian community is gay marriage. The dominant agenda in the lesbian and gay (L/G) movement is the fight for legal recognition of same-sex couples. The theme of gay marriage is presently on the rise, with coverage of political demonstrations, weddings and commitment ceremonies, and civil rights court proceedings. The most noteworthy illustration of the gay marriage issue in popular culture is its importance in the 2005 US presidential election. Alongside the September 11 terrorist attacks and the war in Iraq, gay marriage, often disguised as "family values," stood as a key point of contention among most of the candidates. But why now? Gays and lesbians have been challenging society and the courts for the right to marry since the early 1970s. What is so unique about the current social climate that marriage is the leading gay issue?

TOLERATED BUT NOT EQUAL

Let's start with the emergence of the L/G movement in the late 1960s and early 1970s. During this time people were fired from their jobs as teachers, hospital workers, and state employees because of their actual or suspected homosexuality. The representation of gays and lesbians in the media was negative, if at all. Bars and clubs were raided nightly, limiting the available social space for people to be open about their homosexuality. Consequently, the gay and lesbian movement spawned

SOURCE: Gay Marriage: Why Now? Why at All? By Reese Kelly. Reprinted with permission of the author.

in the 1970s focused on visibility, self-affirming identities, and ending discrimination based on sexuality, mirroring other civil rights movements of the time.

Lesbians and gays who did fight for the right to marry were but a small minority within the gay and lesbian movement. The movement was driven by a young generation of gay and lesbian activists who believed that marriage was a patriarchal heterosexual institution that upheld the oppression of gays and lesbians. To them, gay marriage was the antithesis to their agenda of freedom from society's dominant institutions. Both national and local gay and lesbian organizations ignored the issue of marriage, saw it as a hopeless cause, or more commonly, had higher priorities. Accordingly, the 1970s was a decade of fighting for a self-affirming positive gay or lesbian identity and fighting off social stigmas not for adult concerns, such as gay marriage.

Beginning in the 1980s and 1990s, marriage became part of the larger fight for gay and lesbian equality due to three major social factors: the impact of AIDS, the growing visibility and integration of gays and lesbians in society, and a growing middle-aged L/G population. All of these factors contributed to a change in the L/G movement from tolerance to a politics of equality. AIDS had considerable influence on the direction of the gay and lesbian agenda during the 1980s and early 1990s. First recognized in 1981, HIV, the virus that causes AIDS, surfaced as the "gay plague," leading to hundreds of thousands of deaths among gay men by the early 1990s. The fear, ignorance, and complacency on the part of the medical community as well as of the general public forced the issue to the front of the gay and lesbian political agenda. The AIDS crisis became a platform for pinpointing specific legal and social acts of discrimination towards gays and lesbians. The origin of many AIDS-related issues was that gays and lesbians had no legal partner status. More specifically, health coverage, visitation rights, inheritance and residency rights were a necessity for those infected as well as their partners.

Also during this time, the residual effects of the gay rights movement could be seen in the media's representation of gays and lesbians. In the late 1980s and increasingly through the 1990s, mainstream television and film began to have gay characters and present issues unique to gay culture. The most notable movie, *Philadelphia*, confronts issues of workplace discrimination and AIDS. A few years later, Ellen DeGeneres became the first openly gay character (and actor) in a leading role on US prime-time television, and Pedro Zamora became the first openly gay and HIV–positive reality television star on MTV's Real World, San Francisco. Gay people, more than ever before, were becoming part of the cultural landscape. Unfortunately, depictions of gays and lesbians were heavily based on stereotypes, while television shows that offered nuanced portrayals were often canceled. The overall gay presence in the media reflected a growing tolerance, if not quite acceptance, of gays, a widespread decriminalization of homosexual behavior, and more so, the increasing number of Americans who were "coming out" about their gay, lesbian, and bisexual lifestyles. The heightened visibility of gays and lesbians reminded all Americans of the existence of gay culture and its demand for equal rights.

In the late 1980s and early 1990s we also see an increasingly middle-aged and coupled gay population. Of those who came out of the closet during the 1960s and 1970s, many sought long-term relationships, contrary to the stereotype of the promiscuous gay man. Many lesbians were also coupling off and starting families with children from previous marriages, through adoption, and through *in vitro* fertilization. These gay families expressed the need for rights to which only married heterosexuals had access, such as second-parent adoption, healthcare benefits, and hospital visitation rights. In regard to the law, only the biological parent had rights to the child. In the case of a separation or death, the non-biological parent had no legal recourse to gain custody, even partial, of shared children. An even more common occurrence in child custody discrimination was that gays and lesbians who had children from previous heterosexual marriages often lost custody in divorce settlements. Even if it was in the best interest of the child, emotionally or economically, to remain with the gay parent, courts often ruled in favor of the heterosexual parent for fear of raising the child in an "unsuitable household." The need for parental rights became part of the agenda of gay rights organizations.

By the mid-1990s, then, gays and lesbians faced a complicated and often contradictory combination of social acceptance and discrimination. The L/G movement had shifted from a politics of tolerance to a politics of equality, including employment rights and struggles in academia over knowledge. They were becoming progressively more visible in culture, and national gay and lesbian organizations were on the rise and gaining economic and political strength. Former President Bill Clinton even claimed that gay people were a part of his "vision" of America. Nevertheless, there were still no laws protecting them from workplace housing, and family discrimination in most states. Moreover, in contradiction to Clinton's campaign promise to include gays in America's "vision," he enacted two federal laws endorsing the national discrimination against gays and lesbians. The first law, in 1993, was the "Don't ask, don't tell" military policy banning gays from being open about their sexuality or engaging in homosexual acts while in military service. Furthermore, Clinton signed the Defense of Marriage Act in September 1996, which allows each state to recognize or deny any marriage-like same-sex relationship which has been recognized in another state. It also defines marriage as a legal union between one man and one woman only. Accordingly, gays and lesbians, although accepted by the dominant culture, were accepted on the condition of being second-class citizens, restricted from access to two of America's principal social institutions, the military and marriage.

9/11, AND THE POLITICS OF FEAR

Through the late 1990s and into the new millennium, the United States, as a whole, experienced considerable contradictions, with a rise in social acceptance and certain legal protections for gays, but also a rise in political conservatism. With the exportation of jobs to other countries and the response to the 9/11 attacks, there is a budding culture of anger and hostility towards anyone who is seen as an outsider. The "other" or "outsider" is seen as a threat to the lives of the everyday American. Although this fear is mostly directed at Americans with

Middle-Eastern, Mexican, or Asian heritage, it has also spread to include anyone who is considered to be "unpatriotic." In regards to sexuality, those who are perceived as "unpatriotic" are predominantly single mothers and gays. To be "patriotic" by the current dominant standards one must be heterosexual, married, and family-centered. Religious activist and dominant spokesperson for the Christian Right Jerry Falwell went so far as to claim that gays and lesbians were part of the reason for the 9/11 attacks. Led by President George W. Bush, conservative politicians aim to promote "family values" which enact discriminatory legislation against gays and lesbians. The most notable of these campaigns is a constitutional amendment to protect marriage, proposed by President Bush in February of 2004. This amendment would define marriage as being between only a man and a woman, leaving states to define other legal arrangements for same-sex couples....

Although 9/11 helped shape a conservative political climate, gays and lesbians have not retreated into the closet. Instead, gays are moving into the mainstream and the battle for sexual justice grows ever more public. Gay characters can be found in almost every prime-time television show, and in major movie box office hits such as *As Good As It Gets.* Many popular TV programs focus solely on gay culture, such as *Queer Eye for the Straight Guy* and the two Showtime hits *Queer as Folk* and *The L Word.* Gay culture is being disseminated from US urban meccas such as San Francisco and New York City's Chelsea Village to suburban areas across the country, such as Oak Park, Illinois, and Somerville, Massachusetts. Furthermore, gay culture is no longer confined to those who identify as gay, as one can see with the invention of the "metrosexual," a clean-cut, well-dressed, urban-dwelling heterosexual man. Moreover, a more radical faction of the gay and lesbian movement, the queer movement, is gaining popularity among the younger generation, predominantly college students. The queer movement calls into question what we define as man/woman and heterosexual/homosexual. It aims to break down the dichotomous categories of sex, gender, and sexuality by turning assumptions about what is normal and natural on their heads. Thus, as gays become a larger part of the dominant culture, the queer agenda continues to be a driving force for broader social and legal change.

Moreover, despite some serious setbacks to the movement, gays and lesbians have achieved a significant number of civil rights advancements in the past decade. In the private sector, many companies have made domestic partner benefits available and adopted anti-discrimination policies. By 2004, almost half of the United States' Fortune 500 companies offered same-sex partner health insurance and other benefits (Chauncey 2004: 117). On the state level, almost half of the states allow second-parent adoption in some jurisdictions, if not the entire state (Human Rights Campaign 2004). There are also approximately a dozen states which allow protection from workplace discrimination for state employees based on sexual orientation (Human Rights Campaign 2004). The most significant legal advancement, however, was the 2003 Supreme Court decision in the *Lawrence v. Texas* case that sodomy laws are unconstitutional. In other words, the US Supreme Court ruled that private sexual conduct is protected under the constitution, ending legal discrimination against homosexual sex acts. The growing visibility and integration of gay culture, and the struggle against an increasingly conservative national politics, have set the stage for marriage to be the fighting ground for gay and lesbian civil rights at this time.

CHANGING GENDER ROLES AND CHANGES IN MARRIAGE

To understand some of the passion behind the gay marriage debate, we must look at some key social forces that have changed the institution of marriage. In particular, I want to comment on changing gender aspects of marriage.

Consider that the role of motherhood has been somewhat uncoupled from marriage. Many women are today raising children on their own as a result of divorce, teenage pregnancy, or simply choosing not to marry. Workforce participation has allowed many women to opt out of marriage, for they can rely on their own salary instead of seeking out a second income through marriage. Moreover, the stigma attached to being single has lessened over the years due to the overwhelming number of single adult women, with or without children. Although this social trend has prompted conservative rhetoric regarding the need for marriage and the family, these alternative family structures have poked holes in the notion that a nuclear family is the only legitimate option.

More generally, the changing roles of men and women have dramatically altered the social organization of marriage. Only a few decades ago, men and women had distinct roles in marriage. Women were relegated to the home, subservient to their male partners, and were expected to be the primary caregiver in the family. In contrast, men were expected to be dominant heads of households and primary breadwinners, with little obligation for childcare. Nowadays, women are less constrained to the private world of their homes and participate, at least minimally, in the public corporate world. As a result, women generally have more financial independence, giving them leverage in delineating familial duties. Also, the notion of masculinity seems to be loosening up a bit. There is less of a stigma attached to "stay-at-home" husbands or to those men with less earning potential than their wives. Additionally, gay couples and non-nuclear family arrangements provide examples of alternative roles for all family members that are not necessarily gender-specific. The home, and more specifically marriage, is becoming a place of intimacy and commitment between equals, and gender roles are less constraining or defining.

ALTERNATIVES TO MARRIAGE

There are a few alternatives to marriage for gays and lesbians who cannot legally marry and for those who merely choose not to....

The first step towards accruing marriage benefits for gay couples was the advent of "domestic partner" (DP) status, which allows non-married couples to gain marriage–like benefits from private agencies. For instance, health benefits, bereavement and sick leave, and pension benefits offered by employers are increasingly available to "domestic partners" of the employed. Some employers ask for proof of codependence to qualify for DP benefits, while others require registry with the city or town of residence. DP benefits are available mostly to gay couples who cannot legally marry, but in many instances are offered to heterosexual cohabiting couples as well. A substantial number of corporations and professions offer DP benefits, the first being the software company Lotus in 1992. DPs vary in the extent of included benefits, from very few to matching those given through marriage. Moreover, DP benefits are insufficient because they do not involve any federal recognition of an intimate relationship, and therefore DP benefits cannot be carried across states or claim any federal-level benefits such as tax breaks.

The legal status available to gays and lesbians which comes closest to marriage is a civil union.... In short, civil unions provide most, if not all, benefits and rights of marriage on the state level. Healthcare benefits, second-parent adoption, hospital visitation, and state tax benefits can be acquired through civil union status.... Unlike DP benefits, civil unions are only open to those couples who cannot legally marry. Not only are the benefits of civil unions unequal to those of marriage, the mere fact that they exist as a separate status just for gay people symbolizes a powerful inequality in our legal system. Simply put, civil unions are not equal to marriage in practice and principle.

SHOULD LESBIANS AND GAYS BE ALLOWED TO MARRY?

There are three general arguments in the gay marriage debate: the anti-marriage side, the pro-marriage side, and a queer critique of the place of marriage in our society. It is important to keep in mind that people of any sexual orientation or political identification can and do align with each of the sides. Many heterosexuals support gay marriage, just as some gays are passionately against it.

The anti-gay marriage side emphasizes the notion that marriage and heterosexuality are the cornerstones of the Judeo–Christian tradition. Critics of gay marriage appeal to the bible, which claims that the relationship between Adam and Eve is the foundation of civilization. The bible also explicitly states that a man should not lie with a man as he lies with a woman. As a society that rests on Christian foundations, marriage in America must be restricted to heterosexuality in order to preserve its moral and social coherence. Those who are against gay marriage also believe that heterosexual marriage provides the right moral environment for a healthy family. Marriage encourages family values such as monogamy, respect for authority, and the importance of the differences between men and women. Homosexuals are stereotyped as being unable to have stable relationships and dangerous to healthy gender development in children. For many critics, homosexuality is the very antithesis of a healthy family and society.

The pro-gay marriage side argues that, whether or not gays and lesbians want to marry, they should have the right to decide. The argument from the anti-gay marriage side that marriage is a religious matter is misleading. In fact, marriage is a civic or legal relationship between the two individuals and the government. Marital status confers more than 1,400 government-recognized rights and benefits. Individuals may choose to have a religious marriage, but the institution of marriage is a secular legal affair. Moreover, it is important to note that those who argue for gay marriage continue to define marriage as, ideally, a long-term, loving commitment that should be valued and respected. Other than having a same-sex partner, they believe that many gays are already in such long-term relationships or wish to commit to relationships based on shared love and commitment. Also, pro-gay marriage advocates argue that, since the right to marry is viewed as part of being a first-class citizen, denying this right to gays consigns them to a second-class status.

There is a third party to the gay marriage debate, the so called "queer" position. The term "queer" is less a sexual identity than a viewpoint that questions the privileging of certain specific social norms, identities, and institutions as normal and beyond question. Queers are critical of marriage because it is the one intimate arrangement that the state sanctions as natural and preferable. What's wrong with this? Queers argue that, by conferring recognition and rights on marriage, the state renders all other non-marital relationships as inferior. A state-recognized institution of marriage creates a division between marriage, which is respectable, good, even ideal, and other intimate and family arrangements which are less than ideal, if not viewed as deviant and abnormal. Queers also make the compelling point that it's not only gays and lesbians who are disadvantaged by associating marriage with many exclusive benefits, but the poor and many non-white people are disadvantaged because less of them marry. In sum, queers do not advocate the inclusion of gays and lesbians into the institution of marriage, but the extension of full marital benefits to all who need them.

The controversy of gay marriage is complicated and … it affects people of all sexual orientations. Fortunately, whether or not gays are … able to marry, the dispute does not stop us from deciding who we choose to love, in what ways, and for however long.

REFERENCES

Chauncey, George. 2004. *Why marriage?: The history shaping today's debate over gay equality*. New York: Basic Books.

Human Rights Campaign. 2004. "HRC Marriage Center." Available online at http://www.hrc.org (accessed March 23 2006).

Warner, Michael. 1999. *The trouble with normal: Sex, politics and the ethics of queer life*. Cambridge, MA: Harvard University Press.

KEY CONCEPTS

domestic partner | gay marriage | politics of equality

DISCUSSION QUESTIONS

1. Research the language of the Defense of Marriage Act. What language in there supports denying gays and lesbians the right to marry? How is this justified?
2. How will gay marriage change the institution of the family? What are the positive outcomes for marriage among gays and lesbians?

44

The Myth of the Missing Black Father

ROBERTA L. COLES AND CHARLES GREEN

Roberta Coles and Charles Green explore the common belief that black men are largely absent as fathers. They challenge this idea through a careful review of research on black fathers and black families, and they also identify the historical, economic, and demographic trends that are affecting black family life.

The black male. A demographic. A sociological construct. A media caricature. A crime statistic. Aside from rage or lust, he is seldom seen as an emotionally embodied person. Rarely a father. Indeed, if one judged by popular and academic coverage, one might think the term "black fatherhood" an oxymoron. In their parenting role, African American men are viewed as verbs but not nouns; that is, it is frequently assumed that black men *father* children but seldom *are* fathers. Instead, as the law professor Dorothy Roberts (1998) suggests in her article "The Absent Black Father," black men have become the symbol of *fatherlessness*. Consequently, they are rarely depicted as deeply embedded within and essential to their families of procreation. This stereotype is so pervasive that when black men are seen parenting, as Mark Anthony Neal (2005) has personally observed in his memoir, they are virtually offered a Nobel Prize.

But this stereotype did not arise from thin air. In 2000, only 16 percent of African American households were married couples with children, the lowest of all racial groups in America. On the other hand, 19 percent of Black households were female-headed with children, the highest of all racial groups. From the perspective of children's living arrangements, ... over 50 percent of African American children lived in mother-only households in 2004, again the highest of all racial groups. Although African American teens experienced the largest decline in births of all racial groups in the 1990s, still in 2000, 68 percent of all births to African American women were nonmarital, suggesting the pattern of single-mother parenting may be sustained for some time into the future (Martin et al. 2003). This statistic could easily lead observers to assume that the fathers are absent.

While it would be remiss to argue that there are not many absent black fathers, absence is only one slice of the fatherhood pie and a smaller slice than is normally thought. The problem with "absence," as is fairly well established now, is that it's an ill-defined pejorative concept usually denoting nonresidence with the child, and it is sometimes *assumed* in cases where there is no legal marriage to the mother. More importantly, absence connotes invisibility and noninvolvement, which further investigation has proven to be exaggerated (as will be discussed below). Furthermore, statistics on children's living arrangements also indicate that nearly 41 percent of black children live with their fathers, either in a married or cohabiting couple household or with a single dad.

These African American family-structure trends are reflections of large-scale societal trends—historical, economic, and demographic—that have affected all American families over the past centuries. Transformations of the American society from an agricultural to an industrial economy and, more recently, from an industrial to a service economy entailed adjustments in the timing of marriage, family structure, and the dynamics of family life. The transition from an industrial to a service economy has been accompanied by a movement of jobs out of cities; a decline in real wages for men; increased labor-force participation for women; a decline in fertility; postponement of marriage; and increases in divorce, nonmarital births, and single-parent and non-family households.

These historical transformations of American society also led to changes in the expected and idealized roles of family members. According to Lamb (1986), during the agricultural era, fathers were expected to be the "moral teachers"; during industrialization, breadwinners and sex-role models; and during the service economy, nurturers. It is doubtful that these idealized roles were as discrete as implied. In fact, LaRossa's (1997) history of the first half of the 1900s reveals that public calls for nurturing, involved fathers existed before the modern era. It is likely that many men had trouble fulfilling these idealized roles despite the legal buttress of patriarchy, but it was surely difficult for African American men to fulfill these roles in the context of slavery, segregation, and, even today, more modern forms of discrimination. A comparison of the socioeconomic status of black and white fathers illustrates some of the disadvantages black fathers must surmount to fulfill fathering expectations. According to Hernandez and

Brandon (2002), in 1999 only 33.4 percent of black fathers had attained at least a college education, compared to 68.5 percent of white fathers. In 1998, 25.5 percent of black fathers were un- or underemployed, while 17.4 percent of white fathers fell into that category. Nearly 23 percent of black fathers' income was half of the poverty threshold, while 15 percent of white fathers had incomes that low.

The historical transformations were experienced across racial groups but not to the same extent. The family forms of all racial groups in America have become more diverse, or at least recognition of the diversity of family structure has increased, but the proportions of family types vary across racial groups. Because African American employment was more highly concentrated in blue-collar jobs, recent economic restructuring had harsher implications for black communities and families (Nelson 2004). The higher and more concentrated poverty levels and greater income and wealth inequality—both among African Americans and between African Americans and whites—expose African American men, directly and indirectly, to continued lower life expectancy, higher mortality, and, hence, a skewed gender ratio that leaves black women outnumbering black men by the age of eighteen.

All of these societal and family-level trends affect black men's propensity to parent and their styles of parenting in ways we have yet to fully articulate. For instance, Americans in general have responded to these trends by postponing marriage by two to four years over the last few decades, but that trend is quite pronounced among African Americans, to the point that it is estimated that whereas 93 percent of whites born from 1960 through 1964 will eventually marry, only 64 percent of blacks born in the same period ever will (Goldstein and Kenney 2001). Consequently, in 1970 married-couple families accounted for about 68 percent of all black families, but in 2000, after several decades of deindustrialization, only 46 percent were married couples. The downstream effect of marriage decline is that the majority of black children no longer live in married-couple homes.

Certainly, the skewed gender ratio mentioned earlier contributes to this declining marriage trend, but the role of other factors is under debate. Wilson (1987) and others have suggested that black men's underemployment, along with black women's higher educational attainment in relation to black men (and smaller wage gap than between white men and women, according to Roberts 1994), may decrease both men's and women's desire to marry and may hinder some black men's efforts to be involved fathers (Marsiglio and Cohan 2000). However, other research (Lerman 1989; Ellwood and Crane 1990) has found that even college-educated and employed black men have exhibited declines in marriage, and yet additional research points to attitudinal factors (South 1993; Tucker and Mitchell-Kernan 1995; Crissey 2005), with black men desiring marriage less than white and Latino men.

Other parenting trends may also be affected by black men's unique status. Their higher mortality rate and lower life expectancy may affect the timeline of parenting, increasing pressure to reproduce earlier. If married or cohabiting, black women's higher employment rate may increase the amount of time black men spend with their children (Fagan 1998). Higher poverty and collective values also pull extended family members into the mix, diffusing parenting responsibilities, which may lead to more protective or more neglectful styles of parenting.

Because of these society-wide and race-specific changes in family formation and gender roles, academia and popular culture have exhibited an increasing fascination with the diversifying definitions of masculinity and the roles men play in families, particularly as fathers....

In conjunction with this increased amount of research and, in fact, frequently fueling the research, has been a proliferation of public and private programs and grants aimed at creating "responsible fatherhood." While many of the programs have been successful in educating men on how to be qualitatively better fathers, many have aimed primarily either at encouraging fathers to marry the mothers of their children or at securing child support. Marriage and child support are important aspects of family commitment, but marriage is no guarantee of attentive fathering, and garnished child support alone, particularly if it goes to the state and not to the mother and child, is hardly better parenting. Within this policy focus, African American men are most frequently attended to under the rubric of "fragile families" (Hobson 2002; Gavanas 2004; Mincy, Garfinkel, and Nepomnyaschy 2005). Although this classification may be intended to bring attention to structural supports that many families lack, once again it promulgates the idea that black men cannot be strong fathers.

Given the increased focus on fatherhood in scholarly and popular venues, what do we really know about black men and parenting?

So let's start ... with what we know about nonresident or so-called absent fathers. Studies on this ilk of fathers indicate that generally a large portion of nonresident fathers are literally absent from their children's lives or, if in contact, their involvement decreases substantially over time. A number of memoirs by black men and women, sons and daughters of literally absent fathers, attest to the painful experience that this can be for the offspring—both sons and daughter—of these physically or emotionally missing father.

Although ... anguished experiences are too common, they remain only one part, though often the more visible part, of the larger fatherhood picture. An increasing number of quantitative and qualitative studies find that of men who become fathers through nonmarital births, black men are least likely (when compared to white and Hispanic fathers) to marry or cohabit with the mother (Mott 1994; Lerman and Sorensen 2000). But they were found to have the highest rates (estimates range from 20 percent to over 50 percent) of visitation or provision of some caretaking or in-kind support (more than formal child support). For instance, Carlson and McLanahan's (2002) figures indicated that only 37 percent of black nonmarital fathers were cohabiting with the child (compared to 66 percent of white fathers and 59 percent of Hispanic), but of those who weren't cohabiting, 44 percent of unmarried black fathers were visiting the child, compared to only 17 percent of white and 26 percent of Hispanic fathers. These studies also suggested that black nonresident fathers tend to maintain their level of involvement over time longer than do white and Hispanic nonresident fathers (Danziger and Radin 1990; Taylor et al. 1990; Seltzer 1991; Stier and Tienda 1993; Wattenberg 1993; Coley and Chase-Lansdale 1999).

Sometimes social, fictive, or "other" fathers step in for or supplement nonresident biological fathers. Little research has been conducted on social fathers, but it is known they come in a wide variety: relatives, such as grandfathers and uncles; friends, romantic partners and new husbands of the mother, cohabiting or not; and community figures, such as teachers, coaches, or community-center staff. Although virtually impossible to capture clearly in census data, it is known that a high proportion of black men act as social fathers of one sort or another, yet few studies exist on this group of dads. Lora Bex Lempert's 1999 study of black grandmothers as primary parents found that many families rely on grandfathers, other male extended family members, or community members to fill the father's shoes, but unfortunately her study did not explore the experience of these men....

... McLanahan and Sandefur (1994) found that, compared to those who live in single-parent homes, black male teens who lived with stepfathers were significantly less likely to drop out of school and black teen females were significantly less likely to become teen mothers. The authors speculated that the income, supervision, and role models that stepfathers provide may help compensate for communities with few resources and social control. Although they are often pictured as childless men, these social fathers may also be some other child's biological father, sometimes a nonresident father himself. Consequently, it is not easy and is certainly misleading to discuss fathers as if they come in discreet, nonoverlapping categories of biological or social.

A smaller amount of research has been conducted on black fathers in two-parent families, which are more likely to also be middle-class families. Allen (1981), looking at wives' reports, found black wives reported a higher level of father involvement in childrearing than did white wives. McAdoo (1988) and Bowman (1993) also concluded that black fathers are more involved than white fathers in childrearing. However, Roopnarine and Ahmeduzzaman (1993), and Hossain and Roopnarine (1994) find no or insignificant racial differences in the level and quality of married fathers' involvement. Across races, fathers in married-couple families were about equally involved with their children, which in all cases was less than mothers.

In terms of parenting style, studies of black two-parent families have found that African American parenting styles tend to be more authoritarian, with an emphasis on obedience and control or monitoring, than those of white parents. This style difference is frequently explained by lower income and neighborhood rather than by race itself (Garcia-Coll 1990; Hofferth 2003). Bright and Williams (1996) conducted a small qualitative study of seven low- to middle-income black fathers in two-parent families in an urban area. They found these fathers worked collaboratively with their wives to nurture their children and that chief among their concerns were rearing children with high self-esteem, protecting their family members in unsafe environments, securing quality education, and having a close relationship with their children. Marsiglio (1991) also found black fathers to talk more and have positive engagement with their older children.

Finally, and ironically, most *absent* in the literature on black fatherhood have been those fathers who are most *present:* black, single full-time fathers. About 6 percent of black households are male-headed, with no spouse present; about half of those contain children under eighteen years old. These men also may be biological or adoptive fathers, but little is known about them....

In sum, research on black fathers has been limited in quantity and has narrowly focused on nonmarital, nonresident fathers and only secondarily on dads in married-couple households. This oversight is not merely intentional, for black men are only about 6 percent of the U.S. population and obviously a smaller percent are fathers. They are not easy to access, particularly by an academy that remains predominantly white....

... We do not intend to decide which set of dads is better (whether by race or by type of father within races). We are not interested in the good dad–bad dad typology. We make the assumption that good fathering is best for children, but we also assume good fathering can take many forms and styles....

REFERENCES

Allen, W. 1981. "Mom, Dads, and Boys: Race and Sex Differences in the Socialization of Male Children." In *Black Men*, ed. L. Gary, 99–114. Beverly Hills, Calif.: Sage.

Bowman, P. 1993. "The Impact of Economic Marginality on African-American Husbands and Fathers." In *Family Ethnicity*, ed. H. McAdoo, 120–137. Newbury Park, Calif.: Sage.

Bright, J. A., and C. Williams. 1996. "Child-rearing and Education in Urban Environments: Black Fathers' Perspectives." *Urban Education* 31(3): 245–60.

Carlson, M. J., and S. S. McLanahan. 2002. "Fragile Families, Father Involvement, and Public Policy" In *Handbook of Father Involvement: Multidisciplinary Perspectives*, ed. Catherine Tamis-LeMonda and Natasha Cabrera, 461–88. Mahwah, N.J.: Lawrence Erlbaum.

Coley, R. L., and P. L. Chase-Lansdale. 1999. "Stability and Change in Paternal Involvement Among Urban African American Fathers." *Journal of Family Psychology* 13(3): 1–20.

Crissey, S. R. 2005. "Race/Ethnic Differences in the Marital Expectations of Adolescents: The Role of Romantic Relationships." *Journal of Marriage and Family* 67: 697–709.

Danziger, S., and N. Radin. 1990. "Absent Does Not Equal Uninvolved: Predictors of Fathering in Teen Mother Families." *Journal of Marriage and the Family* 52(3): 636–42.

Ellwood, D. T., and J. Crane. 1990. "Family Change Among Black Americans: What Do We Know?" *The Journal of Economic Perspectives* 4(4): 65–84.

Fagan, J. 1998. "Correlates of Low-Income African American and Puerto Rican Fathers' Involvement with Their Children." *Journal of Black Psychology* 24(3): 351–67.

Garcia-Coll, C. 1990. "Developmental Outcome of Minority Infants: A Process-Oriented Look Into our Beginnings. *Child Development* 61: 270–89.

Gavanas, A. 2004. *Fatherhood Politics in the United States: Masculinity, Sexuality, Race, and Marriage*. Chicago: University of Illinois Press.

Goldstein, J. R., and C. T. Kenney. 2001. "Marriage Delayed or Marriage Forgone? New Cohort Forecasts of First Marriage for U.S. Women," *American Sociological Review* 66(4): 506–19.

Hernandez, D. J., and P. D. Brandon. 2002. "Who Are the Fathers of Today?" In *Handbook of Father Involvement*, ed. C. S. Tamis-LeMonda and N. Cabrera, 33–62. Mahwah, N.J.: Lawrence Erlbaum.

Hobson, B. 2002. *Making Men Into Fathers*. Cambridge: Cambridge University Press.

Hofferth, S. 2003. "Race/Ethnic Differences in Father Involvement in Two-Parent Families: Culture, Context, or Economy?" *Journal of Family Issues* 24(2): 185–216.

Hossain, Z., and Roopnarine, J. 1994. "African-American Fathers' Involvement with Infants: Relationship to Their Functioning Style, Support, Education, and Income." *Infant Behavior and Development* 17: 175–84.

Lamb, M. E. 1986. "The Changing Role of Fathers." In *The Father's Role: Applied Perspectives*, ed. M. E. Lamb, 3–27. New York: Wiley.

LaRossa, R. 1997. *The Modernization of Fatherhood: A Social and Political History*. Chicago: University of Chicago Press.

Lempert, L. B. 1999. "Other Fathers: An Alternative Perspective on African American Community Caring." In *The Black Family: Essays and Studies*, ed. R. Staples, 189–201. Belmont, Calif.: Wadsworth.

Lerman, R. L. 1989. "Employment Opportunities of Young Men and Family Formation." *American Economic Review* (May): 62–66.

Lerman, R., and E. Sorensen. 2000. "Father Involvement with Their Nonmarital Children: Patterns, Determinants, and Effects on Their Earnings." *Marriage and Family Review* 29(2/3): 137–58.

Marsiglio, W. 1991. "Paternal Engagement Activities with Minor Children." *Journal of Marriage and the Family* 53: 973–86.

Marsiglio, W., and M. Cohan. 2000. "Contextualizing Father Involvement and Paternal Influence: Sociological and Qualitative Themes." In *Fatherhood: Research, Interventions, and Policies*, ed. H. E. Peters, G. W. Peterson, S. K. Steinmetz, and R. D. Day, 75–95. New York: Haworth.

Martin, J. A., B. E. Hamilton, P. D. Sutton, S. J. Ventura, F. Menacker, and M. L. Munson. 2003. "Births: Final Data for 2002." *National Vital Statistics Reports* 52(10). Washington, D.C.: Government Printing Office.

McAdoo, J. L. 1988. "The Roles of Black Fathers in the Socialization of Black Children." In *Black Families*, ed. H. P. McAdoo, 257–69. Newbury Park, Calif.: Sage.

McLanahan, S., and G. Sandefur. 1994. *Growing Up with a Single Parent: What Hurts, What Helps*. Cambridge, Mass.: Harvard University Press.

Mincy, R., I. Garfinkel, and L. Nepomnyaschy. 2005. "In-Hospital Paternity Establishment and Father Involvement in Fragile Families." *Journal of Marriage and Family* 67: 611–26.

Mott, E. L. 1994. "Sons, Daughters, and Fathers' Absence: Differentials in Father-Leaving Probabilities and in Home Environments." *Journal of Family Issues* 5: 97–128.

Nelson, T. J. 2004. "Low-Income Fathers." *Annual Review of Sociology* 30: 427–51.

Roberts, D. 1998. "The Absent Black Father." In *Lost Fathers: The Politics of Fatherlessness in America*, 144–61. New York: St. Martin's Press.

Roberts, S. 1994. "Black Women Graduates Outpace Male Counterparts." *New York Times*. October 31.

Roopnarine, J. L., and M. Ahmeduzzaman. 1993. "Puerto Rican Fathers' Involvement with Their Preschool-Age Children." *Hispanic Journal of Behavioral Sciences* 15(1): 96–107.

Seltzer, J. A. 1991. "Relationships Between Fathers and Children Who Live Apart: The Father's Role After Separation." *Journal of Marriage and the Family* 53: 79–101.

South, S. J. 1993. "Racial and Ethnic Differences in the Desire to Marry." *Journal of Marriage and Family* 55(2): 357–70.

Stier, H. and M. Tienda. 1993. "Are Men Marginal to the Family? Insights from Chicago's Inner City." In *Men, Work, and Family*, ed. J. C. Hood, 23–44. Newbury Park, Calif.: Sage.

Taylor, R., L. Chatters, M. B. Tucker, and E. Lewis. 1990. "Developments in Research on Black Families: A Decade Review." *Journal of Marriage and the Family* 52: 993–1014.

Tucker, M. B., and Mitchell-Kernan, C. 1995. "Trends in African American Family Formation: A Theoretical and Statistical Overview." In *The Decline in Marriage Among African American: Causes, Consequences, and Policy Implications*, ed. M. B. Tucker and C. Mitchell-Kernan, 3–26. New York: Russell Sage Foundation.

Wattenberg, E. 1993. "Paternity Actions and Young Fathers." In *Young Unwed Fathers: Changing Roles and Emerging Policies*, ed. R. Lerman and T. Ooms, 213–34. Philadelphia: Temple University Press.

Wilson, W. J. 1987. *The Truly Disadvantaged: The Inner City, the Underclass, and Public Policy*. Chicago: University of Chicago Press.

KEY CONCEPTS

fatherhood

fictive fatherhood

industrialization

myth of the absent father

DISCUSSION QUESTIONS

1. What are the myths about black fathers that Coles and Green identify in their article, and how are they challenged by sociological research?
2. What are the social and historical trends that have shaped the experiences of black men as fathers? How are these both similar to and different from the experiences of fathers in other racial or ethnic groups?

Applying Sociological Knowledge: An Exercise for Students

Design a qualitative research project where you propose interviewing individuals who come from all different family structures. Consider including people who grew up in multigenerational households, those who were raised by a single parent, and those from blended families. What questions would you ask of all of them? If you had to craft a definition of family, what would it be? What about gay and lesbian couples, with or without children? How does your definition of family include them?

PART XIV

Religion

45

The Protestant Ethic and the Spirit of Capitalism

MAX WEBER

Max Weber's classic analysis of the Protestant ethic and the spirit of capitalism shows how cultural belief systems, such as a religious ethic, can support the development of specific economic institutions. His multidimensional analysis shows how capitalism became morally defined as something more than pursuing monetary interests and, instead, has been culturally defined as a moral calling because of its consistency with Protestant values.

The impulse to acquisition, pursuit of gain, of money, of the greatest possible amount of money, has in itself nothing to do with capitalism. This impulse exists and has existed among waiters, physicians, coachmen, artists, prostitutes, dishonest officials, soldiers, nobles, crusaders, gamblers, and beggars. One may say that it has been common to all sorts and conditions of men at all times and in all countries of the earth, wherever the objective possibility of it is or has been given. It should be taught in the kindergarten of cultural history that this naïve idea of capitalism must be given up once and for all. Unlimited greed for gain is not in the least identical with capitalism, and is still less in its spirit. Capitalism may even be identical with the restraint, or at least a rational tempering, of this irrational impulse. But capitalism is identical with the pursuit of profit, and forever renewed profit, by means of continuous, rational, capitalistic enterprise....

If any inner relationship between certain expressions of the old Protestant spirit and modern capitalistic culture is to be found, we must attempt to find it, for better or worse, not in its alleged more or less materialistic or at least antiascetic joy of living, but in its purely religious characteristics....

In the title of this study is used the somewhat pretentious phrase, the *spirit* of capitalism. What is to be understood by it? The attempt to give anything like a definition of it brings out certain difficulties which are in the very nature of this type of investigation.

If any object can be found to which this term can be applied with any understandable meaning, it can only be an historical individual, i.e., a complex of elements associated in historical reality which we unite into a conceptual whole from the standpoint of their cultural significance....

Remember, that *time* is money. He that can earn ten shillings a day by his labour, and goes abroad, or sits idle, one half of that day, though he spends but sixpence during his diversion or idleness, ought not to reckon *that* the only expense; he has really spent, or rather thrown away, five shillings besides.

Remember, that *credit* is money. If a man lets his money lie in my hands after it is due, he gives me the interest, or so much as I can make of it during that time. This amounts to a considerable sum where a man has good and large credit, and makes good use of it....

The most trifling actions that affect a man's credit are to be regarded. The sound of your hammer at five in the morning, or eight at night, heard by a creditor, makes him easy six months longer; but if he sees you at a billiard-table, or hears your voice at a tavern, when you should be at work, he sends for his money the next day; demands it, before he can receive it, in a lump....

Truly what is here preached is not simply a means of making one's way in the world, but a peculiar ethic. The infraction of its rules is treated not as foolishness but as forgetfulness of duty. That is the essence of the matter. It is not mere business astuteness, that sort of thing is common enough, it is an ethos. *This* is the quality which interests us.

When Jacob Fugger, in speaking to a business associate who had retired and who wanted to persuade him to do the same, since he had made enough money and should let others have a chance, rejected that as pusillanimity and answered that "he (Fugger) thought otherwise, he wanted to make money as long as he could," the spirit of his statement is evidently quite different from that of Franklin.[1] What in the former case was an expression of commercial daring and a personal inclination morally neutral, in the latter takes on the character of an ethically colored maxim for the conduct of life. The concept spirit of capitalism is here used in this specific sense, it is the spirit of modern capitalism. For that we are here dealing only with Western European and American capitalism is obvious from the way in which the problem was stated. Capitalism existed in China, India, Babylon, in the classic world, and in the Middle Ages. But in all these cases, as we shall see, this particular ethos was lacking....

And in truth this peculiar idea, so familiar to us today, but in reality so little a matter of course, of one's duty in a calling, is what is most characteristic of the social ethic of capitalistic culture, and is in a sense the fundamental basis of it. It is an obligation which the individual is supposed to feel and does feel toward the content of his professional activity, no matter in what it consists, in particular no matter whether it appears on the surface as a utilization of his personal powers, or only of his material possessions (as capital).

Rationalism is an historical concept which covers a whole world of different things. It will be our task to find out whose intellectual child the particular concrete form of rational thought was, from which the idea of a calling and the devotion to labor in the calling has grown, which is, as we have seen, so irrational from the standpoint of purely eudæmonistic self-interest, but which has been and still is one of the most characteristic elements of our capitalistic culture. We are here particularly interested in the origin of precisely the irrational element which lies in this, as in every conception of a calling....

... Like the meaning of the word, the idea is new, a product of the Reformation. This may be assumed as generally known. It is true that certain suggestions of the positive valuation of routine activity in the world, which is contained in this conception of the calling, had already existed in the Middle Ages, and even in late Hellenistic antiquity. We shall speak of that later. But at least one thing was unquestionably new: the valuation of the fulfillment of duty in worldly affairs as the highest form which the moral activity of the individual could assume. This it was which inevitably gave everyday worldly activity a religious significance, and which first created the conception of a calling in this sense.... Late Scholasticism, is, from a capitalistic viewpoint, definitely backward. Especially, of course, the doctrine of the sterility of money which Anthony of Florence had already refuted.

... For, above all, the consequences of the conception of the calling in the religious sense for worldly conduct were susceptible to quite different interpretations. The effect of the Reformation as such was only that, as compared with the Catholic attitude, the moral emphasis on and the religious sanction of, organized worldly labor in a calling was mightily increased....

The real moral objection is to relaxation in the security of possession, the enjoyment of wealth with the consequence of idleness, and the temptations of the flesh, above all of distraction from the pursuit of a righteous life. In fact, it is only because possession involves this danger of relaxation that it is objectionable at all. For the saints' everlasting rest is in the next world; on earth man must, to be certain of his state of grace, "do the works of him who sent him, as long as it is yet day." Not leisure and enjoyment, but only activity serves to increase the glory of God, according to the definite manifestations of His will.

Waste of time is thus the first and in principle the deadliest of sins. The span of human life is infinitely short and precious to make sure of one's own election. Loss of time through sociability, idle talk, luxury, even more sleep than is necessary for health, six to at most eight hours, is worthy of absolute moral condemnation. It does not yet hold, with Franklin, that time is money, but the proposition is true in a certain spiritual sense. It is infinitely valuable because every hour lost is lost to labor for the glory of God. Thus inactive contemplation is also valueless, or even directly reprehensible if it is at the expense of one's daily work....

It is true that the usefulness of a calling, and thus its favor in the sight of God, is measured primarily in moral terms, and thus in terms of the importance of the goods produced in it for the community. But a further, and, above all, in practice the most important, criterion is found in private profitableness. For if that God, whose hand the Puritan sees in all the occurrences of life, shows one of His elect a chance of profit, he must do it with a purpose. Hence the faithful Christian must follow the call by taking advantage of the opportunity. "If God shows you a way in which you may lawfully get more than in another way (without wrong to your soul or to any other), if you refuse this, and choose the less gainful way, you cross one of the ends of your calling, and you refuse to be God's steward, and to accept His gifts and use them for Him when He requireth it: you may labour to be rich for God, though not for the flesh and sin."

Wealth is thus bad ethically only in so far as it is a temptation to idleness and sinful enjoyment of life, and its acquisition is bad only when it is with the purpose of later living merrily and without care. But as a performance of duty in a calling it is not only morally permissible, but actually enjoined....

Let us now try to clarify the points in which the Puritan idea of the calling and the premium it placed upon ascetic conduct was bound directly to influence the development of a capitalistic way of life. As we have seen, this asceticism turned with all its force against one thing: the spontaneous enjoyment of life and all it had to offer....

On the side of the production of private wealth, asceticism condemned both dishonesty and impulsive avarice. What was condemned as covetousness, Mammomsm etc., was the pursuit of riches for their own sake. For wealth in itself was a temptation. But here asceticism was the power "which ever seeks the good but ever creates evil"; what was evil in its sense was possession and its temptations. For, in conformity with the Old Testament and in analogy to the ethical valuation of good works, asceticism looked upon the pursuit of wealth as an end in itself as highly reprehensible; but the attainment of it as a fruit of labor in a calling was a sign of God's blessing. And even more important: the religious valuation of restless, continuous, systematic work in a worldly calling, as the highest means to asceticism, and at the same time the surest and most evident proof of rebirth and genuine faith, must have been the most powerful conceivable lever for the expansion of that attitude toward life which we have here called the spirit of capitalism.

When the limitation of consumption is combined with this release of acquisitive activity, the inevitable practical result is obvious: accumulation of capital through ascetic compulsion to save. The restraints which were imposed upon the consumption of wealth naturally served to increase it by making possible the productive investment of capital....

One of the fundamental elements of the spirit of modern capitalism, and not only of that but of all modern culture: rational conduct on the basis of the idea of the calling, was born—that is what this discussion has sought to demonstrate—from the spirit of Christian asceticism....

The Puritan wanted to work in a calling; we are forced to do so. For when asceticism was carried out of monastic cells into everyday life, and began to dominate worldly morality, it did its part in building the tremendous cosmos of the modern economic order. This order is now bound to the technical and economic conditions of machine production which today determine the lives of all the individuals who are born into this mechanism, not only those directly concerned with economic acquisition, with irresistible force....

Since asceticism undertook to remodel the world and to work out its ideals in the world, material goods have gained an increasing and finally an inexorable power over the lives of men as at no previous period in history. Today the spirit of religious asceticism—whether finally, who knows?—has escaped from the cage. But victorious capitalism, since it rests on mechanical foundations, needs its support no longer. The rosy blush of its laughing heir, the Enlightenment, seems also to be irretrievably fading, and the idea of duty in one's calling prowls about in our lives like the ghost of dead religious beliefs. Where the fulfillment of the calling cannot directly be related to the highest spiritual and cultural values, or when, on the other hand, it need not be felt simply as economic compulsion, the individual generally abandons the attempt to justify it at all. In the field of its highest development, in the United States, the pursuit of wealth, stripped of its religious and ethical meaning, tends to become associated with purely mundane passions, which often actually give it the character of sport.

No one knows who will live in this cage in the future, or whether at the end of this tremendous development entirely new prophets will arise, or there will be a great rebirth of old ideas and ideals, or, if neither, mechanized petrification, embellished with a sort of convulsive self-importance. For of the last stage of this cultural development, it might well be truly said: "Specialists without spirit, sensualists without heart; this nullity imagines that it has attained a level of civilization never before achieved."

The modern man is in general, even with the best will, unable to give religious ideas a significance for culture and national character which they deserve. But it is, of course, not my aim to substitute for a one-sided materialistic an equally one-sided spiritualistic causal interpretation of culture and of history. Each is equally possible, but each, if it does not serve as the preparation, but as the conclusion of an investigation, accomplishes equally little in the interest of historical truth.

NOTE

1. The quotations are attributed to Benjamin Franklin.

KEY CONCEPTS

capitalism

Protestant ethic

DISCUSSION QUESTIONS

1. Weber is known for developing a multidimensional view of human society. What role does he see the Protestant ethic playing in the development of capitalism?

2. Weber's analysis sees Western capitalists as not pursuing money just for the sake of money, but because of the moral calling invoked by the Protestant ethic. Given the place of consumerism in contemporary society, how do you think Weber might modify his argument were he writing now? In other words, are there still remnants of the Protestant ethic in our beliefs about stratification? If so, how do they fit with contemporary capitalist values?

46

Muslims in America

JEN'NAN GHAZAL READ

Jen'nan Ghazal Read's research challenges many of the stereotypes about Muslim Americans that have been particularly strong in the aftermath of 9/11.

Seven years after the terrorist attacks on U.S. soil catapulted Muslims into the American spotlight, concerns and fears over their presence and assimilation remain at an all-time high.

Recent national polls find that four in 10 Americans have an unfavorable view of Islam, five in 10 believe Islam is more likely than other religions to encourage violence, and six in 10 believe Islam is very different from their own religion. All this despite the fact that seven in 10 admit they know very little about Islam. And yet Americans rank Muslims second only to atheists as a group that doesn't share their vision of American society.

These fears have had consequences. In 2001, the U.S. Department of Justice recorded a 1,600 percent increase in anti-Muslim hate crimes from the prior year, and these numbers rose 10 percent between 2005 and 2006. The Council on American-Islamic Relations processed 2,647 civil rights complaints in 2006, a 25 percent increase from the prior year and a 600 percent increase since 2000. The largest category involved complaints against U.S. government agencies (37 percent).

Clearly, many Americans are convinced Muslim Americans pose some kind of threat to American society.

SOURCE: Jen'nan Ghazal Read, Muslims in America, *Contexts* 2008, 7: 39 by SAGE Publications. Reprinted by permission of SAGE Publications.

Two widespread assumptions fuel these fears. First, that there's only one kind of Islam and one kind of Muslim, both characterized by violence and anti-democratic tendencies. Second, that being a Muslim is the most salient identity for Muslim Americans when it comes to their political attitudes and behaviors, that it trumps their social class position, national origin, racial/ethnic group membership, or gender—or worse, that it trumps their commitment to a secular democracy.

Research on Muslim Americans themselves supports neither of these assumptions. Interviews with 3,627 Muslim Americans in 2001 and 2004 by the Georgetown University Muslims in the American Public Square (MAPS) project and 1,050 Muslim Americans in 2007 by the Pew Research Center show that Muslim Americans are diverse, well-integrated, and largely mainstream in their attitudes, values, and behaviors.

The data also show that being a Muslim is less important for politics than how Muslim you are, how much money you make, whether you're an African-American Muslim or an Arab-American Muslim, and whether you're a man or a woman.

The notion that Muslims privilege their Muslim identity over their other interests and affiliations has been projected onto the group rather than emerged from the beliefs and practices of the group itself. It's what sociologists call a social construction, and it's one that has implications for how these Americans are included in the national dialog.

SOME BASIC DEMOGRAPHICS

Let's start with who Muslim Americans really are. While size estimates of the population range anywhere from 2 million to 8 million, there is a general agreement on the social and demographic characteristics of the community.

Muslim Americans are the most ethnically diverse Muslim population in the world, originating from more than 80 countries on four continents. Contrary to popular belief, most are not Arab. Nearly one-third are South Asian, one-third are Arab, one-fifth are U.S.-born black Muslims (mainly converts), and a small but growing number are U.S.-born Anglo and Hispanic converts. Roughly two-thirds are immigrants to the United States, but an increasing segment is second- and third-generation U.S.-born Americans. The vast majority of immigrants have lived in the United States for 10 or more years.

Muslim Americans also tend to be highly educated, politically conscious, and fluent in English, all of which reflects the restrictive immigration policies that limit who gains admission into the United States. On average, in fact, Muslim Americans share similar socioeconomic characteristics with the general U.S. population: one-fourth has a bachelor's degree or higher, one-fourth lives in households with incomes of $75,000 per year or more, and the majority are employed. However, some Muslims do live in poverty and have poor English language skills and few resources to improve their situations.

SOCIOECONOMIC CHARACTERISTICS OF AMERICAN MUSLIMS

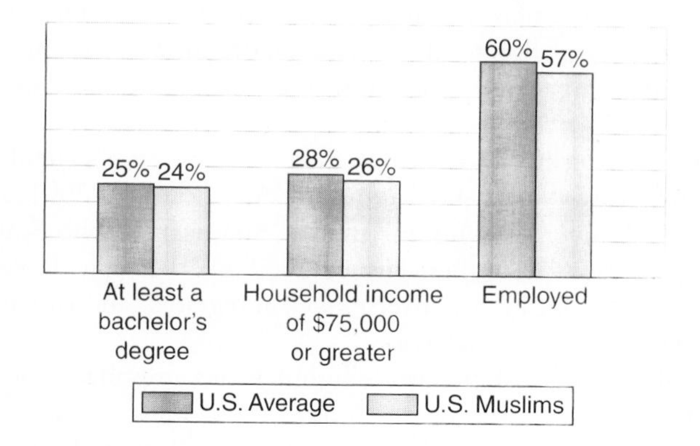

One of the most important and overlooked facts about Muslim Americans is that they are not uniformly religious and devout. Some are religiously devout, some are religiously moderate, and some are non-practicing and secular, basically Muslim in name only, similar to a good proportion of U.S. Christians and Jews. Some attend a mosque on a weekly basis and pray every day, and others don't engage in either practice. Even among the more religiously devout, there is a sharp distinction between being a good Muslim and being an Islamic extremist.

None of this should be surprising. Many Muslim Americans emigrated from countries in the Middle East (now targeted in the war on terror) in order to practice—or not practice—their religion and politics more freely in the United States. And their religion is diverse. There is no monolithic Islam that all Muslims adhere to. Just as Christianity has many different theologies, denominations, and sects, so does Islam. And just like Christianity, these theologies, denominations, and sects are often in conflict and disagreement over how to interpret and practice the faith tradition. This diversity mimics other ethnic and immigrant groups in the United States.

AMERICAN MUSLIMS MOSQUE ATTENDANCE

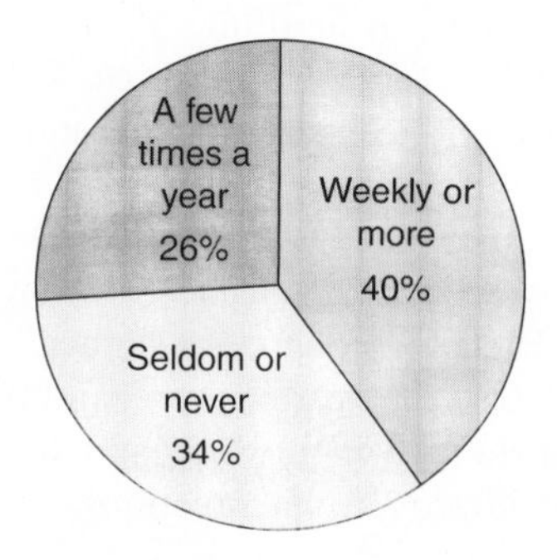

Evidence from the MAPS project, Pew Research Center, and General Social Survey demonstrates that Muslim Americans are much more politically integrated than the common stereotypes imply. Consider some common indicators of political involvement, such as party affiliation, voter registration, and contact with politicians. Compared to the general public, Muslim Americans are slightly less likely to be registered to vote, reflecting the immigrant composition and voter eligibility of this group (63 percent compared to 76 percent of the general population), slightly more likely to have contacted a politician (51 percent compared to 44 percent of the general population), and slightly more likely to affiliate with the Democratic Party, which falls in line with other racial and ethnic minorities (63 percent compared to 51 percent of the general population).

POLITICAL INVOLVEMENT OF AMERICAN MUSLIMS

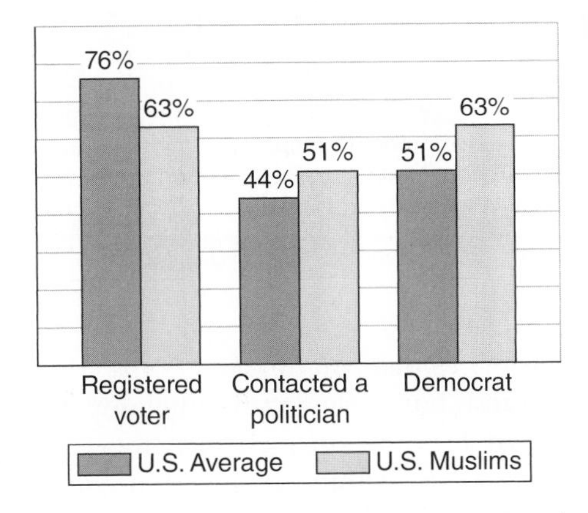

All these data demonstrate that, contrary to fears that Muslim Americans comprise a monolithic minority ill-suited to participation in American democracy, Muslim Americans are actually highly diverse and already politically integrated. They are also in step with the rest of the American public on today's most divisive political issues.

ATTITUDES, VALUES, AND VARIATION

The majority of both Muslim Americans (69 percent) and the general public (76 percent) oppose gay marriage, favor increased federal government spending to help the needy (73 percent and 63 percent, respectively), and disapprove of President George W. Bush's job performance (67 percent and 59 percent). Muslim Americans are slightly more conservative than the general public when it comes to abortion (56 percent oppose it, compared to 46 percent) as well as the federal government doing more to protect morality in society (59 percent compared to 37 percent).

The one area in which American Muslims are not entirely in step with the general public is foreign policy, especially having to do with the Middle East. In 2007, for example, the general public was nearly four times as likely to say the war in Iraq was the "right decision" and twice as likely to provide the same response to the war in Afghanistan (61 percent compared to 35 percent of Muslim Americans).

VIEWS ON SOCIAL AND DOMESTIC ISSUES

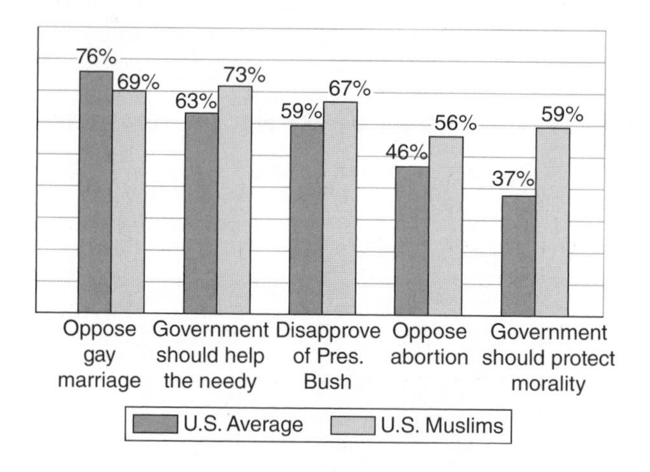

In short, these numbers tell us that Muslim Americans lean to the right on social issues (like most Americans), but to the left on foreign policy. But these generalizations don't tell the whole story—in particular, these averages don't demonstrate the diversity that exists within the Muslim population by racial and ethnic group membership, national origin, socioeconomic status, degree of religiosity, or nativity and citizenship status.

Consider, for example, Muslim Americans' levels of satisfaction and feelings of inclusion (or exclusion) in American society—major building blocks of a liberal democracy. In examining how these perceptions vary by racial and ethnic group membership within the group, we see that African-American Muslims express more dissatisfaction and feel more excluded from American society than Arab or South Asian Muslims. They're more likely to feel that the United States is fighting a war against Islam, to believe Americans are intolerant of Islam and Muslims, and to have experienced discrimination in the past year (whether racial, religious, or both is unclear). South Asians feel the least marginalized, and Arab Muslims fall in between. These racial and ethnic differences reflect a host of factors, including the immigrant composition and higher socioeconomic status of the South Asian and Arab populations and the long-standing racialized and marginalized position of African Americans. Indeed, many (though not all) African Americans converted to Islam seeking a form of religious inclusion they felt lacking in the largely white Judeo-Christian traditions.

DIVERSITY AMONG AMERICAN MUSLIMS

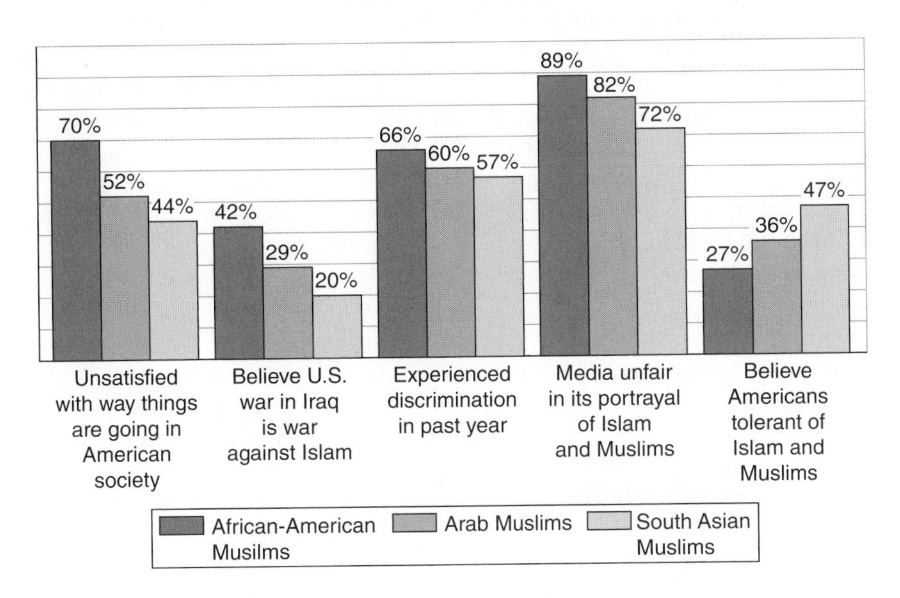

(Incidentally, most African-American Muslims adhere to mainstream Islam [Sunni or Shi'a], similar to South Asian and Arab Muslim populations. They should not be confused with the Nation of Islam, a group that became popular during the civil rights era by providing a cultural identity that separated black Americans from mainstream Christianity. Indigenous Muslims have historically distanced themselves from the Nation of Islam in order to establish organizations that focus more on cultural and religious [rather than racial] oppression.)

Before we can determine whether religion is the driving force behind all Muslims' political opinions and behaviors—whether Islam, as is popularly assumed, trumps Muslim Americans' other commitments and relationships to nationality, ethnicity, race, and even democracy—let's step back and place Muslim Americans in a broader historical context of religion and American politics.

WHEN RELIGION MATTERS, AND DOESN'T

Muslim Americans aren't the first religious or ethnic group considered a threat to America's religious and cultural unity. At the turn of the 20th century, Jewish and Italian immigrants were vilified in the mainstream as racially inferior to other Americans. Of course today those same fears have been projected onto Hispanic, Asian, and Middle Eastern immigrants. The Muslim American case shares with these other immigrant experiences the fact that with a religion different from the mainstream comes the fear that it will dilute, possibly even sabotage, America's thriving religious landscape.

Yes, thriving. By all accounts, the United States is considerably more religious than any of its economically developed Western counterparts. In 2000,

93 percent of Americans said they believed in God or a universal spirit, 86 percent claimed affiliation with a specific religious denomination, and 67 percent reported membership in a church or synagogue. The vast majority of American adults identify themselves as Christian (56 percent Protestant and 25 percent Catholic), with Judaism claiming the second largest group of adherents (2 percent), giving America a decidedly Judeo-Christian face. There are an infinite number of denominations within these broad categories, ranging from the ultra-conservative to the ultra-liberal. And there is extensive diversity among individuals in their levels of religiosity within any given denomination, again ranging from those who are devout, practicing believers to those who are secular and non-practicing.

This diversity has sparked extensive debates among academics, policymakers, and pundits over whether American politics is characterized by "culture wars," best summarized as the belief that Americans are polarized into two camps, one conservative and one liberal, on moral and ethical issues such as abortion and gay rights. Nowhere has the debate played out more vividly than the arena of religion and politics, where religiously based mobilization efforts by the Christian right helped defeat liberal-leaning candidates and secure President Bush's reelection in 2004. Electoral victories, however, haven't usually translated into policy victories, as evidenced by the continued legality of abortion and increasing protection of gay rights. So when does religion matter for politics and when doesn't it?

Here we come back to the Muslim American case. Like Muslim Americans, Americans generally have multiple, competing identities that shape their political attitudes and behaviors—93 percent of Americans may believe in God or a universal spirit but 93 percent of Americans don't base their politics on that belief alone. In other words, just because most Americans are religiously affiliated doesn't mean most Americans base their politics on religion. To put it somewhat differently, the same factors that influence other Americans' attitudes and behaviors influence Muslim Americans' attitudes and behaviors. Those who are more educated, have higher incomes, higher levels of group consciousness, and who feel more marginalized from mainstream society are more politically active than those without these characteristics. Similar to other Americans, these are individuals who feel they have more at stake in political outcomes, and thus are more motivated to try to influence such outcomes.

Muslims, on average, look like other Americans on social and domestic policies because, on average, they share the same social standing as other Americans, and on average, they are about as religious as other Americans. Consider two common indicators of religiosity, frequency of prayer and frequency of church attendance, and compare Muslim Americans to Christian Americans. Both groups are quite religious, with the majority praying every day (70 percent of Christians compared to 61 percent of Muslims) and a sizeable proportion attending services once a week or more (45 percent compared to 40 percent). And they look similar with respect to attitudes on gay rights and abortion, in part because Christian and Muslim theology take similar stances on procreation and gender roles.

Again, these numbers tell only part of the story. What's missing is that religion's relationship to politics is multidimensional. In more complex analyses it has become clear that the more personal dimensions of religious identity—or being a devout Muslim who prays every day—have little influence on political attitudes or behaviors, which runs counter to stereotypes that link Islamic devotion to political fanaticism.

AMERICAN MUSLIMS COMPARED TO CHRISTIANS

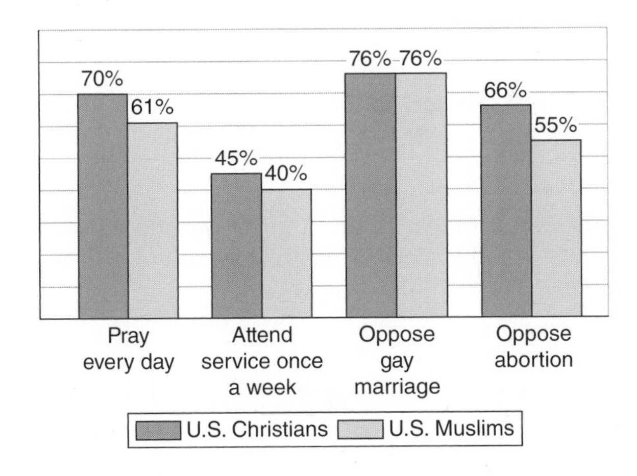

In contrast, the more organized dimensions of Muslim identity, namely frequent mosque attendance, provide a collective identity that stimulates political activity. This is similar to what we know about the role of the church and synagogue for U.S. Christians and Jews. Congregations provide a collective environment that heightens group consciousness and awareness of issues that need to be addressed through political mobilization. Thus, it is somewhat ironic that one of the staunchest defenders of the war on terror—the Christian right—may be overlooking a potential ally in the culture wars—devout Muslim Americans.

AN EXCEPTIONAL EXPERIENCE

In many ways these findings track closely with what we know about the religion–politics connection among other U.S. ethnic and religious groups, be they Evangelical Christians or African Americans. They also suggest that the Muslim experience may be less distinct than popular beliefs imply. In fact, Muslim Americans share much in common with earlier immigrant groups who were considered inassimilable even though they held mainstream American values (think Italian, Irish, and Polish immigrants).

At the same time, though, we can't deny that the Muslim American experience, particularly since 9/11, has been "exceptional" in a country marked by a declining salience of religious boundaries and increasing acceptance of religious difference. Muslim Americans have largely been excluded from this ecumenical trend. If we're going to face our nation's challenges in a truly democratic way, we need to move past the fear that Muslim Americans are un-American so we can bring them into the national dialogue....

KEY CONCEPTS

ethnic group religion religiosity

DISCUSSION QUESTIONS

1. In what specific ways does Read's research debunk some of the popularly held beliefs about Muslims?
2. In what other ethnic and/or religious groups would you find diversity similar to that in Read's research? What does this teach you about social stereotypes of religious/ethnic groups?

47

All Creatures Great and Small: Megachurches in Context

MARK CHAVES

Mark Chaves notes the growth in very large churches since the 1970s and reviews several explanations for this trend, offering, in the end, his own additional explanation of this transformation in the social organization of religion.

Megachurches—by which I mean very large Protestant churches—are increasingly difficult to ignore. By the latest count there are approximately 1,200 Protestant churches in the country with weekly attendance of at least

SOURCE: Mark Chaves, "All Creatures Great and Small: Megachurches in Context," *Review of Religious Research* 47 (June 2006): 329–346. Reprinted with permission of Religious Research Association.

2,000 people (Thumma 2005), and by every account these very large churches have proliferated in recent decades. Journalists and scholars have by now paid a lot of attention to these churches, and as a result we know a lot about them....

There are many questions one might ask about the megachurch phenomenon. Why do some churches, and not others, grow very large? What kinds of people are attracted to megachurches? How do these churches operate internally? How variable or similar are they to each other in content and style? How influential are they on American religious culture? How politically active and influential are they?

These are all interesting and important questions, but I'm not going to address any of them. Instead, I want to tackle the question of why these churches have become an increasingly visible part of the religious landscape in recent decades....

Incidentally, this way of approaching the subject provides a principled sociological reason for understanding large Protestant churches as qualitatively different than large Catholic churches: they are produced by qualitatively different underlying processes. The Catholic size distribution is skewed like the Protestant distributions, but Protestant churches in the United States are, of course, not as strongly tied to their geographical location through a parish system, and they operate in systems with weaker central control of church starts, closings, and responses to rapid growth. Some of the consequences of large size might be similar in very large churches within both Protestantism and Catholicism, but their origins are different. Very large Protestant churches represent a qualitatively different phenomenon than very large Catholic churches because the size distributions in which they reside are produced by qualitatively different underlying systems. The question I am addressing, then, is why the Protestant system has produced so many very large churches in recent decades....

... How has the church size distribution changed over time?...

Results

... I will emphasize three patterns. First, across the Protestant spectrum, there are more very big churches. There were 145 Southern Baptist churches with more than 2,000 members in 1972; in 2002 there were 458. The number of Assemblies of God churches of this size increased from 60 in 1981 to 149 in 2003. This increase might not be surprising for these two denominations, since they have grown in membership over this period, but the same trend is evident in denominations that have declined in membership over this period. The number of Episcopalian churches of this size increased from 65 in 1990 to 82 in 2002. The number of Evangelical Lutheran Church in America churches of this size went from 199 in 1987 to 230 in 2002. The number of United Methodist churches of this size went from 245 in 1974 to 287 in 2002. The bottom line is that the number of very large Protestant churches has increased in almost every denomination on which we have data, and it does not matter whether the denomination is big or small, liberal or conservative, growing or declining.

The rate of increase in the number of very large churches seems to pick up after 1970, but the trend toward more very big churches did not begin in the 1970s. It is a longer-term trend. The number of Presbyterian churches with more than 2,000 members, for example, increased from 5 in 1900 to 74 in 1983. The number of Episcopalian churches of that size increased from 7 in 1930 to 33 in 1960; for Missouri Synod Lutherans, the number goes from 2 in 1900 to 23 in 1967. Conventional wisdom on this subject says that the number of very large churches has increased *recently,* and the results bear that out, but conventional wisdom does not recognize that this is not entirely a post-1970 trend.

The long-term nature of this trend suggests that simple population growth and increased population density are partly responsible for megachurch proliferation. If you need a town or city or suburb of a certain size in order to support a 2,000-person church, then the more communities of that size there are, the more 2,000-person churches there will be. So simple population growth probably is part of the story of the increasing number of very large churches.

A second pattern I want to emphasize is that the very biggest churches are getting bigger. It is not news that this has occurred in recent decades, but once again this trend did not begin in the 1970s. It too is a longer-term trend, although again it seems that the rate at which the largest churches are getting bigger has accelerated since the 1970s....

To me, the most interesting development is a third trend: *people are increasingly concentrated in the very largest churches*.... This, to me, is an extraordinarily interesting, even astonishing, picture because *every* denomination shows the same pattern of steadily, in some cases rapidly, increasing concentration from 1970 to the present, with no end in sight to this trend. Denominations vary in how concentrated their people are in the very largest churches, but all of them show the same trend towards increasing concentration since about 1970....

This increased concentration may be changing American religion's social and political significance. For one thing, increased concentration makes religion more visible, since one 2,000-person church is more visible, if only because of the size of its building, than ten 200-person churches. Increased concentration also probably increases religion's potential for social and political influence, since one 2,000-person church is easier to mobilize for social or political action than ten 200-person churches; a politician is more likely to address one 2,000-person church than ten 200-person churches; and the pastor of one 2,000-person church probably gets an appointment with the mayor more easily than the pastors of ten 200-person churches. Increasing concentration seems likely also to have repercussions for intrade-nominational politics and the development and diffusion of worship practices. And increased concentration also can fool observers into thinking that there is a religious revival occurring when really there is a change in the social organization of religion. These consequences of increasing religious concentration make trying to understand what is behind it a worthy agenda.

WHY THE INCREASING CONCENTRATION?

Explaining this concentration trend is much more difficult than establishing it, and I am going to be more tentative in what follows than I have been so far. Not too tentative, though. I am going to argue that the usual ways of explaining the rise of megachurches do not quite work as explanations of this increased concentration, at least not in their simple versions, and I am going to suggest a new kind of explanation for the phenomenon.

It sometimes is said that the secret to megachurch success is that they have figured out how to attract the unchurched, and they thereby bring people into churches who were not previously involved. Willow Creek is famous, for example, for their profile of "Unchurched Harry" as their ideal-typical target recruit. If new churchgoers were going disproportionately to the very largest churches, this would indeed increase religious concentration.

One problem with this explanation, however, is that church attendance overall is not increasing. Reasonable people disagree about whether church attendance is stable or declining, but it clearly has not increased in recent decades. Perhaps more telling, the only study I know of that compares very large churches with smaller churches with respect to the percentage of new members who were not previously involved in a church finds no difference (Thumma and Petersen 2003). So the increasing concentration of people in the very largest churches is *not* a consequence of megachurches tapping into a previously uninvolved population. Increased concentration is occurring mainly because people are shifting from smaller to larger churches, not because people are shifting from uninvolvement to involvement in big churches.

Neither can we explain this trend by reference to some constant advantage of size. It surely is true that there are Durkheimian, collective effervescence, attractions to worshiping as part of a big group. It also surely is true that big groups enjoy a kind of status advantage that comes from being perceived as where the action is, or where something interring is happening....

So there are constant advantages to size, but these cannot explain the concentration trend we have observed. Constant advantages of bigness imply a trend toward concentration that should have begun long ago with the appearance of the first big churches, not just in 1970....

A third, and somewhat more subtle explanation, invokes suburbanization, the proliferation of automobiles and, more generally, decreasing travel costs, perhaps in interaction with the eternal advantages of bigness. Maybe people always did prefer big churches ... but perhaps they could not act on those preferences because it was too difficult to get to a big church. Perhaps increases in population density outside of central cities made it easier for religious entrepreneurs to build very large buildings on affordable tracts of land at the edge of the developed zone. And once these big churches are built and everyone has a car and access to paved roads on which to drive quickly and cheaply to the very big church at the outer edge of the suburbs just like we drive to the mall, perhaps only then were people able to easily attend the big churches they always would have preferred....

... [T]he problem with the suburbanization and travel cost explanation of increasing concentration is that the trend lines do not line up properly. American society has become steadily suburbanized throughout the twentieth century, with the fastest rate of population shift to suburbs occurring between 1945 and 1970 (Fischer and Hout 2006...). There was no increase in the suburbanization rate in the 1970s; indeed, the rate at which population shifted to the suburbs seems to have *declined* after 1970. The most rapid population increase in the rings around cities seems to have been accompanied by *de*centralization of religion more than by its centralization....

The upshot here is that the suburbanization and travel-cost explanation implies that religious concentration should have started to increase decades sooner than it actually did. Megachurches often are compared with shopping malls, but the cars and suburbanization story works much better for malls than it does for churches. Large regional shopping malls start to proliferate immediately after World War II, directly on the heels of rapidly increasing suburbanization (Hanchett 1996, Jackson 1996). Religious concentration and the fast increase in very large churches smarts 25 or 30 years later....

My view is that suburbanization, the ubiquity of cars, and decreasing travel costs explains something about the *location* of megachurches—why the very largest churches used to be in central cities but are now mainly in suburbs—but I do not see how this urban geography story can explain why the size distribution itself has changed so dramatically since 1970....

A fourth possible explanation is that megachurches are a new, innovative organizational form designed by religious entrepreneurs, perhaps those associated with the church growth movement, who were particularly attuned to post-1970 society and culture. Several early megachurches—both Willow Creek and Saddleback, for example—began only after a religious entrepreneur walked neighborhoods and surveyed people about what kind of church they would want to attend, and only then designing the model that proved to be wildly successful and therefore mimicked by churches across the country. This kind of explanation rests on the creative action of these innovators in developing a new type of church.

The main problem with this sort of explanation is that, in key ways, today's megachurches are *not* a new organizational form....

Note that characteristics typical of today's megachurches—rapid growth under a gifted leader, high quality music, multi-function buildings, many and varied small groups and activities—are evident in ... descriptions from the 1920s....

... In short, several key megachurch characteristics that often are described as innovative developments designed to appeal specifically to late 20th century Americans—bigness itself, auditorium-style worship spaces with stages instead of pulpits and little or no Christian symbolism, lots of small groups and varieties of activities for all subgroups, multi-purpose buildings, high quality preaching and music, extravagant theatrical display—are not at all new....

Perhaps, however—and this is a fifth potential explanation for the increasing concentration of people in the very largest churches—recent cultural change has given a new advantage to an organizational form that has existed for a long time. Perhaps people today are *more* comfortable with bigness, *more* attracted to spectacle, or *more* drawn to a church in which they can choose to be anonymous, or in which they can choose between anonymity in a big crowd and intimacy in a small group. It is difficult to rule out some version of this argument, but it also is difficult to rule it in....

I think there is indeed a cultural affinity between what megachurches offer and what contemporary churchgoers want, and I think this affinity helps explain which churches grow and which ones do not grow. But I do not think that cultural change since the 1970s gives us the complete story behind the concentration trend that we have observed....

A NEW HYPOTHESIS ABOUT INCREASING CONCENTRATION IN RELIGION: RISING COSTS

I think the most productive approach to explaining the increasing concentration of people in the very largest churches will bear a family resemblance to the cultural change approach in positing that something changed in the 1970s to give new advantages to very large churches. But I want to propose that something other than, or maybe in addition to, the *culture* changed. I'm going to propose that the increased concentration of people in the very largest churches is caused in part by rising costs that make it more and more difficult to run a church at a customary level of programming and quality. Let me explain....

In an initial effort to track the relationship between rising revenue and rising costs, I used the median salary of full-time male clergy as a proxy for cost.... This is not a perfect measure. Paying the minister is the bulk of most churches' budget, but it is not the only part of the budget. Probably, though, using the median clergy salary as a measure of what it costs to run a church is a conservative estimate of real cost increases after 1970. Benefit cost increases, for example, start to outpace wage increases at about that time....

... [T]he qualitative picture is that, from 1940 to 1960 real increases in revenue for the average congregation far outpaced real increases in clergy salaries, and *that gap narrows* considerably from 1960 to 1980, with the lines even crossing between 1990 and 2000, indicating that real median clergy salary increased at a higher rate than real donations between 1990 and 2000. Again, I do not want to put too much weight on these preliminary results, but they do suggest that churches faced a qualitatively different, more stressful, revenue-cost situation after 1960 or 1970 than they faced before 1960.

When cost increases outpace revenue increases, churches will cut corners and reduce quality by deferring maintenance, declining to replace the youth minister who graduated from seminary and moved on, replacing the recently retired full-time minister with a half-time person, and so on. In short, churches will find it difficult to maintain the same level of programming and quality they had before. And this will be true even if the church loses no members. If costs rise faster than revenues, a 200-person church will be unable to produce the same level of programming and quality it produced before *even if it stays a 200-person church.* Moreover, the minimum size at which a church can be economically viable will increase. The result is that, when cost increases outpace revenue increases, people will be pushed out of smaller churches that no longer meet their minimum standards and into larger churches that still do....

I am not claiming that the cost-disease mechanism is the only thing driving increased concentration in religion. Cultural change, or shifts in economic and urban geography, may well be part of the story. Technological advances in audio-visual equipment, telecommunications, and computers also may have helped mitigate some of the ways in which very large size might otherwise lead to reduced quality in church services, pastoral care, and programming. A complete explanation of increasing religious concentration surely will not be monocausal.

CONCLUSION

Among other things, I have called attention to forces that are pushing people out of smaller churches as well as pulling them into bigger churches. Scholars and journalists who have written about megachurches have focused almost exclusively on the pull factors, but shifting attention from megachurches themselves to the size distribution as a whole leads us to pay attention to the underlying system operating here, and that means attending to push factors as well as pull factors.

Even if the cost-disease explanation leaves you cold, I hope you are as amazed as I am at the discovery that, in every Protestant denomination on which we have data—large or small, liberal or conservative, growing or declining—people are increasingly concentrated in the very largest churches. This increasing concentration is, I think, a significant change in the social organization of American religion. We do not yet fully grasp its causes and consequences, but I hope you will help me try to understand it better....

REFERENCES

Fischer, Claude S. and Michael Hout. 2006. *Century of Difference*. New York: Russell Sage Foundation.

Hanchett, Thomas W. 1996. "U.S. Tax Policy and the Shopping-Center Boom of the 1950s and 1960s." *American Historical Review* 101:1082–1110.

Jackson, Kenneth T. 1996. "All the World's a Mall: Reflections on the Social and Economic Consequences of the American Shopping Center." *American Historical Review* 101:1111–1121.

Thumma, Scott. 2005. "United States Has More Megachurches Than Previously Thought." Web Report. Hartford: Hartford Seminary. Downloaded from www.hartsem.edu/events/news_mega.htm.

Thumma, Scott and Jim Petersen. 2003. "Goliaths in Our Midst: Megachurches in the ELCA." Pp. 102–124 in *Lutherans Today: American Lutheran Identity in the Twenty-First Century*, ed. by Richard Cimino. Grand Rapids, MI: William B. Eerdmans Publishing Co.

KEY CONCEPTS

megachurch

DISCUSSION QUESTIONS

1. What are the organizational characteristics of a megachurch, and how are these related to other patterns of social change in society?
2. What are the major explanations for the growth in megachurches that Chaves identifies? How does his explanation differ from those he reviews.

Applying Sociological Knowledge: An Exercise for Students

Go to the Gallup website (www.gallup.com) and search for recent reports on religion. Summarize the results with particular focus on race and gender differences. How does this information match or contradict what you expected or believed about religion in the United States?

PART XV

Education

48

A School in a Garden

MITCHELL STEVENS

Mitchell Stevens uses the concept of social reproduction to explain how admissions in elite colleges contribute to social inequality in the United States.

Set at a high elevation overlooking farmland, sleepy towns, and hardwood forests, the College enjoys a geographical prominence commensurate with its stunning campus. Lovely old buildings from the early campaigns resemble pieces of a giant chess set, carefully positioned around shady quadrangles. Slate roofs and mullioned windows convey a sense of history. A few of the facades are illuminated in the evenings, making them visible for miles into the surrounding valleys. The most impressive route of arrival carries drivers through a sweeping lawn dotted with perennial beds and specimen trees. Lovingly tended, the trees are a special point of pride. Many employees can name a favorite. Each trunk gets an annual skirting of fresh mulch. The sycamores near the chapel receive special medications.

The campus is an important constant in the College's history. Like many private schools throughout the northeastern United States, this one was built by Protestant churchmen at what was once a cutting edge of American frontier. Hilltops were school builders' preferred sites for hygienic as well as symbolic reasons. Higher elevations were presumed to enjoy cleaner air, a notable advantage in a coal-burning industrial society, and also encouraged flattering allusions to Athens and Zion. The virtues of this particular hill have long been routed by College boosters. An information pamphlet for prospective students published in 1917 promises tidy walks crisscrossing under "fine old trees, which form the backdrop for the brown-grey buildings." "In a situation so beautiful and naturally healthful," explains another passage, "the College is further safeguarded by a modern sanitation system and its own water supply from spring fed reservoirs." Later literature describes the physical plant in other terms but continues to praise its beauty. A 1973 viewbook quotes a student's enthusiastic description: "This is a beautiful campus. In the fall especially, it's the most gorgeous place I ever hope

SOURCE: Reprinted by permission of the publisher from "A School in a Garden" in CREATING A CLASS: COLLEGE ADMISSIONS AND THE EDUCATION OF ELITES by Mitchell L. Stevens, pp. 5–10, 10–12, 12–13, 14–16, 17–19, Cambridge, Mass.: Harvard University Press,

to see. The air is clean and you are just totally removed from all the things that are making it so hard to live in cities these days." ... Technological advances in color photography and the luminous capacities of computer screens would give subsequent advocates ever more vivid tools for disseminating their news. Surveys of admitted students throughout the College's history would confirm the campus as a prominent factor in many matriculation decisions.

Schools like this one—private, lush, residential, and with selective undergraduate admissions—constitute only a tiny fraction of the colleges and universities in the United States, yet they enjoy historical and cultural influence in great disproportion to their number. They are among the nation's most enduring and most emulated organizations. Early Americans built schools to train religious leaders of many different faiths, to gain an edge over neighboring towns and denominations, and to put particular towns and cities on the map. A school on a hill could be a light in the darkness, a glimmer of intellectual sophistication, a sign that a community was going places, making progress, looking up. As the frontier moved westward, the older institutions became models for school founders in every corner of the country. Colleges in the northeastern United States became benchmarks of excellence in virtually all things: curriculum, faculty, athletics, admissions, aesthetics. Even today, with the thousands of U.S. colleges and universities, degrees conferred by a relative handful of private, highly selective, affluent colleges and universities "back East" bear a subtle but unmistakable cachet....

For eighteen months in 2000 and 2001, I lived and worked at one of these schools as a researcher. I resided in an apartment on its campus, ate often in its cafeterias, borrowed books from its library, and took my exercise on its wooded trails. I spent most of my working hours in the College's Office of Admissions and Financial Aid, where I tried to get as close as I could to the people who made decisions.

I was not alone in my interest. Selective admissions policies have been the object of increasing public fascination and debate in recent years. Courts, legislatures, and college presidents argue over the appropriate criteria selective schools should use when figuring out who they will admit. Magazines rank "the best" institutions by how many applicants they turn away. Growing numbers of private consultants make their livings off of the anxieties of people facing the elite college search....

Despite all of the attention being paid to selective admissions, however, we know remarkably little about how admissions officers go about making decisions about real applicants in real time. I wanted to know how the decisions got made, and with what consequence for those who hoped to someday attend schools like the College. There are many excellent historical studies, and quite a few workplace memoirs by admissions officers themselves, but almost no reports based on critical scholarly observation.... Also remarkable is that, despite all the hype about selective college access, apparently no scholar in any field has taken a stab at explaining the hype itself. Many parents, especially those of the affluent upper middle class, worry ever more and ever earlier about their children's fate in the selective college admissions game, but it is not clear why. Why, in a society where a decent college education has become almost as accessible as a good cup of coffee, when virtually every state in the union underwrites at least one good research university, has admission to a handful of very expensive, often geographically remote private schools grown ever more competitive in recent years? What, if anything, has changed that makes attendance at particular institutions, and not just any college, seem so important to so many? I suggest an answer to these questions by looking out on the landscape of contemporary America through the front door of a highly selective private college.

I went to this place with a long-standing interest in two features of our national culture that are as influential as they are contradictory. On the one hand, Americans place very high value on the appraisal of people as individuals. Whether in schools, workplaces, or department stores, we believe that individualized consideration is better than standardized care. We like personalized attention, first names, and custom made. On the other hand, we put great faith in the fairness of universal standards. In our schools, workplaces, and courts of law, we tend to believe that everyone should be evaluated on the same terms. We tend to be suspicious when institutions make exceptions to their officially universal rules, using terms like *special preferences* and *discrimination* to call foul on the deviations. We might in theory settle the contradiction between universalism and individualism by making a clear choice between them when we build our institutions, creating systems for the management of human beings in which either the rules apply to absolutely everyone, or in which there are no hard-and-fast rules at all. But we don't choose. Instead, and despite the contradictions, we tend to create institutions that mix the two ideals together.

Nowhere is the commingling of individualism and universalism more apparent than in schools. On the one hand, we tend to view personalized instruction as the sine qua non of educational excellence. We sing the praises of small classrooms and "individualized education programs." We are understanding when people demand choices about where their children will go to school. Many parents and teachers alike cry to the heavens when school officials ask that standard curricula be taught in standardized ways. On the other hand Americans are zealous educational universalists. On the political left, progressive reformers have long and quite successfully championed a dream of universal schooling—initially to the point of literacy, next to the completion of high school and, in recent years, to college degrees. The 1954 *Brown v. Board of Education* U.S. Supreme Court decision, considered by many to be a sacred event in our national history, preaches a gospel of educational universalism, making explicit the notion that public schooling should be apportioned equally to all citizens. On the political right, reformers have recently, and also quite successfully, pressed for universal measures of students' academic accomplishment and school performance. The centerpiece of the Bush administration's No Child Left Behind Act, for example, is the obligation that schools receiving federal funding demonstrate the progress of their students through standardized tests. It is difficult to imagine a more universal measure of individual performance than machine-graded, multiple-choice exams backed by the authority of the national government. Rather than making a choice between individualism and universalism in our schools, then, we pursue the virtues of both ideals at the same time.

Highly selective liberal arts schools like the College also embody the commingling of individualism and universalism. On the one hand, their signature organizational characteristics are their intimate size and their mission of service to students as whole persons. On the other hand, the competition for admission means that these schools also are beholden to powerful cultural expectations that they evaluate every applicant according to universal standards of merit.

At their admissions front doors, elite liberal arts schools are expected to be individualistic and universalistic simultaneously. This is why it seemed to me that an admissions office would be a good site for examining what happens when these two ideals are brought together routinely, with what advantages and costs to applicants and schools.

An additional thing that had long intrigued me about liberal arts colleges is that they are quintessentially American institutions. The liberal arts organizational ideal—of a small, residential campus, geared primarily, if not exclusively, to highly individualized undergraduate instruction—was invented and nurtured in the United States, and in stark contrast to our model of research universities, it has not traveled beyond national borders. One looks almost in vain for schools built on the liberal arts model anywhere else in the world. I began my inquiry suspecting that the national peculiarity of the liberal arts form might hold some larger lessons about culture, schooling, and social class in America....

COLLEGE AND CLASS

College educations are now crucial components of our national class structure. Most people presume that a college degree is a prerequisite for a financially comfortable adulthood, and a large corpus of sociological research on the relationship between educational attainment and life chances largely confirms the conventional wisdom.... Attainment of the relatively secure, well-compensated jobs held by the affluent upper middle class virtually requires a college education. Those without college degrees increasingly are relegated to less lucrative and less stable work. But even though there is wide agreement about the economic importance of college, there has been enduring controversy on the question of why educational attainment has come to play its now-pivotal role in the American class system.

One answer, often called the *reproduction* thesis, holds that variation in educational attainment essentially is a coating for preexisting class inequalities. The reproduction thesis was built from Karl Marx's insights about how powerful groups inevitably create social and cultural systems that legitimate their own class advantage. From this perspective college degrees, and the classroom time and school-work they represent, provide palatable justification for the tendency of privileged families to hand privilege down to their children. Adherents of the reproduction thesis support their argument by pointing out the obdurate correlation between parents' socioeconomic status and their offspring's school completion in general.... And of the Horatio Algers who do not fit this general pattern—the high academic achievers who graduate from prestigious colleges and go on to positions of wealth and influence, despite the odds—reproduction theorists explain that the exceptions are important in giving the education system its veneer of class neutrality. It is important for public acceptance of the whole enterprise that at least some of the less advantaged can make schooling work for them.

A second answer, which we might call the *transformation* thesis, makes different sense of the very same correlation between family privilege and educational attainment. This thesis argues that the replacement of traditional social hierarchies with educational ones is a definitive chapter in every society's progress toward modernity. German sociologist Max Weber, the first proponent of the transformation thesis, famously argued that as societies modernized, inequalities of family, caste, and tribe gradually give way to hierarchies predicated on individual achievement. In modern times individuals accumulate status and power as they move through the elaborate bureaucracies that characterize all industrial societies: large corporations, centralized governments, highly bureaucratized religious organizations, and schools. These forms of organization tend to distribute rewards on the basis of demonstrated individual accomplishment, not inherited privilege.... The transformation thesis would have us see that the ultimate value of college degrees lies in their capacity to confer advantages independently of their recipients' social backgrounds. If the correlation between parents' privilege and children's educational attainment were exact—if accomplishment in school neatly paralleled class origins—then schooling would not be so coveted by people from humble backgrounds. As it is, education is broadly perceived by people from all social classes as an effective mechanism of social mobility, because it *is* capable of moving people up, and down, the class hierarchy....

In a series of influential writings in the 1970s, sociologist Randall Collins deftly integrated the two theses, creating a term so pithy and evocative that it has shaped public and scholarly conversations about college ever since. Collins argued that the reproduction theorists were correct: the terms of social privilege are deeply contested in every modern society, and the haves perennially seek to translate their advantages into forms that render them legitimate in the eyes of have-nots. But he added that the transformation thesis also is true: privileged groups create educational institutions that have considerable independence from the people who pay for them. Schools function as quasi-autonomous third parties between haves who support them and have-nots. The academically accomplished kids who attend Harvard or Stanford on full scholarships, and the tuition-paying rich kids who flunk out of the same schools, are living embodiments of this institutional autonomy. Collins described this system of educational legitimation as *credentialism*, and the educationally stratified world it engendered *the credential society*.[1]

During the same decades that social scientists were developing this line of inquiry, the U.S. federal and state governments were actively building the largest higher education infrastructure in world history. Part of the justification for this expansion had to do with the more optimistic of social scientists' findings on education and individual life outcomes. If people's employment and earnings prospects were measurably improved through postsecondary schooling, the policy reasoning went, then a virtuous government would be right to expand opportunities for college attendance. In the decades following World War II, the U.S. state and federal governments did precisely that. Between 1945 and 1980 they dramatically grew the size and mission of public research universities, provided many millions of dollars in student grant and loan programs, and elaborately subsidized a whole tier of institutions—community colleges—to provide truly mass higher education opportunity....

Worries about credential inflation notwithstanding, policy makers and the general public have conceived of college so optimistically for so long that pointing out the very limited extent to which expanded college access has changed the distribution of privilege in this country remains an unpopular thing to do. Nevertheless it is true: higher education has not been the great American equalizer. To be sure, there are proportionally more college graduates in this country than in any previous era, but, with only a few exceptions, the overall distribution of educational attainment remains stubbornly correlated with socioeconomic background....

This does not mean, however, that the expansion of higher education has been without consequence for the character of the national class system. My research suggests that one profound result of higher education's expansion has been the entrenchment of a complicated, publicly palatable, and elaborately costly machinery through which wealthy parents hand privilege down to their children.

The pursuit of college credentials is the widest and most dependable path to the good life that American society currently provides, and the terms of college admission have become the instructions families use when figuring out how to ensure their own children's future prosperity. The rise of the credential society has been accompanied by a value system in which the terms of college admission are also the goals of ideal child rearing and the standards of youthful accomplishment in American popular culture. These goals and standards are most explicitly depicted in the attributes elite colleges say they are looking for in applicants: measurable academic and athletic ability, demonstrated artistic accomplishment, and formally recognized philanthropic service.

Affluent families have a big advantage in meeting these goals and standards because they have relatively more resources to invest in doing so. Keenly aware of the terms of elite college admission, privileged parents do everything in their power to make their children into ideal applicants. They pay for academically excellent high schools. They shower their children with books and field trips and lots of adult attention. They nurture athletic talent through myriad youth sports programs. They encourage and fund early glimmers of artistic interest. They channel kids with empathic hearts toward exotic and traceable forms of humanitarian service. In the process of doing all of this, affluent families fashion an entire way of life organized around the production of measurable virtue in children.

On this line of thinking, the ever more frenzied activity surrounding selective admissions in the nation's most comfortable neighborhoods and school districts is essentially ceremonial. By the time upper-middle-class seventeen-year-olds sit down to write their applications, most of the race to top colleges has already been run and they already enjoy comfortable leads. For these kids the big question is not whether they will be admitted to an elite institution, but which particular schools will offer them spots. Nevertheless the intense final lap of the admissions race has profound importance as a ritual of just deserts. The simple fact that precise outcomes remain uncertain for everyone up to the very end serves to assure us that admission prizes are never won without persistence, steady wits, and hard work....

Physical Education

One of the many revealing documents in the College's archive is a survey report from the late 1930s titled "A Study of the Reasons Given by 145 Members of the Freshman Class for Their Coming to the College." It is impossible to know the degree of rigor with which this survey was carried out, but its figures tell an evocative story. The survey appears to have given respondents a choice of some sixty factors that may have influenced their decision to attend the school. The list ranges widely, from "Academic reputation" to "Fraternity connections" and "Infirmary care." The single most frequently cited reason for attending is an item under the category "Physical Aspects": *Attraction of the campus* garners 67 mentions, a virtual tie with *General advantages of small college* and *Academic reputation*. Little wonder, then, that a document titled "Tentative Publicity Program," field alongside the survey results, includes *Beauty of the campus* high on its list of recommended emphases.

This beauty is an asset that the College carefully maintains and actively promotes. Many of the facilities put up in recent years pay homage to the structures surrounding the oldest central quadrangles. As if in defiance of cost, stone facades and slate roofs adorn even some of the newest and largest buildings. Tidy footpaths, immunized to mud by an intricate terra cotta drainage system, lace through terraced gardens so beguiling that they are favored sites for wedding photographers. Otherwise quiet summer afternoons rumble with the din of motorized maintenance as physical plant workers aerate, mow, and fertilize many acres of lawns. In a custom shared by many of its similarly spectacular peer institutions, the College annually produces a full-color calendar of the most favored campus views and distributes it free to the institution's many alumni and friends.

Yet despite the great care and pride with which colleges attend to their physical appearance, sociologists of education have almost entirely ignored campus aesthetics. It is as if we have presumed that the job of conferring credentials is the most, or even the only, important work elite schools do. This blindness to aesthetics is part of a larger myopia in the sociology of education, and in the scholarly literature on stratification more broadly, about the sensual aspects of class. While we have become ever more sophisticated in our appreciation of how educational credentialing works, we have given ever less attention to the myriad ways in which schools produce a whole range of social values: intellectual, physical, aesthetic, and emotional.

Insights of the French sociologist Pierre Bourdieu provide a useful corrective to American sociologists' narrow focus on credentials as the primary produce of schools. Bourdieu argued that social class is about much more than where people fall in a society's distribution of wealth. It also entails particular patterns of aesthetic production, consumption, and sensual experience. What a society calls beautiful, for example, and what it makes beautiful in turn, are every bit as important to marking class distinctions as wealth and credentials are. On this line of thinking, it is no accident that in the schools to which they send their children, as much as in the neighborhoods they live in and the museums they patronize, the upper classes in every society go to great lengths to define what is beautiful and then surround themselves with the material embodiments of those definitions....

On this line of thinking about class distinction, the physical appearance of human bodies matters as much as that of the physical worlds those bodies inhabit. Aspects of our corporeal bodies—how we carry them through space and attend to their shape, adornment, and longevity—also are important ways through which we mark class differences. Because bodies are such visible and consequential embodiments of class, parents go to great lengths to maintain and improve their children's physical health and appearance: through clothing, diet, and personal hygiene, and, significantly for purposes here, through sport and exercise. This ... is how the institutional status interests supporting college athletics and the class interests of families come together....

REFERENCES

Collins, Randall. 1979. *The Credential Society: A Historical Sociology of Education and Stratification*. New York: Academic Press.

KEY CONCEPTS

credentialism

social reproduction thesis

DISCUSSION QUESTIONS

1. Where does your college sit in the hierarchy of social prestige among colleges and universities in the United States? How will your school's place in this hierarchy shape the opportunities of students who graduate?
2. Explain *social reproduction theory,* and illustrate with your own college experience.

49

From the Achievement Gap to the Education Debt: Understanding Achievement in U.S. Schools

GLORIA LADSON-BILLINGS

This article is a speech given in 2006 by the then president of the American Educational Research Association, Gloria Ladson-Billings. In her presidential address, she uses the analogy of the national debt and the national deficit to explain what has happened in American education. She argues that instead of focusing so much attention on the achievement gap between minority disadvantaged students and white privileged students, educational research and policy should focus on the education debt. The problem with the American system of education, according to Ladson-Billings, is that we are accumulating more debt in that all students are suffering from a poor system. The article calls for action, policy, and research that will help reverse this trend.

The questions that plague me about education research are not new ones. I am concerned about the meaning of our work for the larger public—for real students, teachers, administrators, parents, policymakers, and communities in real school settings. I know these are not new concerns; they have been raised by others, people like the late Kenneth B. Clark, who, in the 1950s, was one of the first social scientists to bring research to the public in a meaningful way. His work with his wife and colleague Mamie formed the basis for the landmark *Brown v. Board of Education* (1954) case that reversed legal segregation in public schools and other public accommodations. However, in his classic volume *Dark Ghetto: Dilemmas of Social Power,* first published in 1965, Clark took social scientists to task for their failure to fully engage and understand the plight of the poor:

> To my knowledge, there is at present nothing in the vast literature of social science treatises and textbooks and nothing in the practical and field training of graduate students in social science to prepare them for

SOURCE: Gloria Ladson-Billings, from the Achievement Gap to the Education Debt, 2006 from *Educational Researcher* 35: 3–12 by SAGE Publications. Reprinted by permission of SAGE Publications.

> the realities and complexities of this type of involvement in a real, dynamic, turbulent, and at times seemingly chaotic community. And what is more, nothing anywhere in the training of social scientists, teachers, or social workers now prepares them to understand, to cope with, or to change the normal chaos of ghetto communities. These are grave lacks which must be remedied soon if these disciplines are to become *relevant* [emphasis added] to the stability and survival of our society (p. xxix).

Clark's concern remains some 40 years later. However, the paradox is that education research has devoted a significant amount of its enterprise toward the investigation of poor, African American, Latina/o, American Indian, and Asian immigrant students, who represent an increasing number of the students in major metropolitan school districts. We seem to study them but rarely provide the kind of remedies that help them to solve their problems.

To be fair, education researchers must have the freedom to pursue basic research, just as their colleagues in other social sciences do. They must be able to ask questions and pursue inquiries "just because." However, because education is an applied field, a field that local states manage and declare must be available to the entire public, *most* of the questions that education researchers ask need to address the significant question that challenge and confound the public: Why don't children learn to read? What accounts for the high levels of school dropout among urban students? How can we explain the declining performance in mathematics and science at the same time that science and mathematics knowledge is exploding? Why do factors like race and class continue to be strong predictors of achievement when gender disparities have shrunk?

THE PREVALENCE OF THE ACHIEVEMENT GAP

One of the most common phrases in today's education literature is "the achievement gap." The term produces more than 11 million citations on Google. "Achievement gap," much like certain popular culture music stars, has become a crossover hit. It has made its way into common parlance and everyday usage. The term is invoked by people on both ends of the political spectrum, and few argue over its meaning or its import. According to the National Governors' Association, the achievement gap is "a matter of race and class. Across the U.S., a gap in academic achievement persists between minority and disadvantaged students and their white counterparts." It further states: "This is one of the most pressing education-policy challenges that states currently face" (2005). The story of the achievement gap is a familiar one. The numbers speak for themselves. In the 2005 National Assessment of Educational Progress results, the gap between Black and Latina/o fourth graders and their White counterparts in reading scaled scores was more than 26 points. In fourth-grade mathematics, the gap was more than 20 points (Education Commission of the States, 2005). In eighth-grade reading, the gap was more than 23 points, and in eighth-grade mathematics the gap was more than 26 points. We can also see that these gaps persist over time (Education Commission of the States).

Even when we compare African Americans and Latina/os with incomes comparable to those of Whites, there is still an achievement gap as measured by standardized testing (National Center for Education Statistics, 2001). While I have focused primarily on showing this gap by means of standardized test scores, it also exists when we compare dropout rates and relative numbers of students who take advanced placement examinations; enroll in honors, advanced placement, and "gifted" classes; and are admitted to colleges and graduate and professional programs.

Scholars have offered a variety of explanations for the existence of the gap. In the 1960s, scholars identified cultural deficit theories to suggest that children of color were victims of pathological lifestyles that hindered their ability to benefit from schooling (Hess & Shipman, 1965; Bereiter & Engleman, 1966; Deutsch, 1963). The 1966 Coleman Report, *Equality of Educational Opportunity* (Coleman et al. 1966), touted the importance of placing students in racially integrated classrooms. Some scholars took that report to further endorse the cultural deficit theories and to suggest that there was not much that could be done by schools to improve the achievement of African American children. But Coleman et al. were subtler than that. They argued that, more than material resources alone, a combination of factors was heavily correlated with academic achievement. Their work indicated that the composition of a school (who attends it), the students' sense of control of the environments and their futures, the teachers' verbal skills, and their students' family background all contribute to student achievement. Unfortunately, it was the last factor—family background—that became the primary point of interest for many school and social policies.

But I want to use this opportunity to call into question the wisdom of focusing on the achievement gap as a way of explaining and understanding the persistent inequality that exists (and has always existed) in our nation's schools. I want to argue that this all-out focus on the "Achievement Gap" moves us toward short-term solutions that are unlikely to address the long-term underlying problem.

NATIONAL DEBT VERSUS NATIONAL DEFICIT

Most people hear or read news of the economy every day and rarely give it a second thought. We hear that the Federal Reserve Bank is raising interest rates, or that the unemployment numbers look good. Our ears may perk up when we hear the latest gasoline prices or that we can get a good rate on a mortgage refinance loan. But busy professionals rarely have time to delve deeply into all things economic. Two economic terms—"national deficit" and "national debt"—seem to befuddle us. A deficit is the amount by which a government's, company's, or individual's spending exceeds income over a particular period of time. Thus, for each budget cycle, the government must determine whether it has a balanced budget, a budget surplus, or a deficit. The debt, however is the sum of all previously incurred annual federal deficits. Since the deficits are financed by government borrowing, national debt is equal to all government debt.

Most fiscal conservatives warn against deficit budgets and urge the government to decrease spending to balance the budget. Fiscal liberals do not necessarily embrace deficits but would rather see the budget balanced by increasing tax revenues from those most able to pay. The debt is a sum that has been accumulating since 1791, when the U.S. Treasury recorded it as $75,463,476.52 (Gordon, 1998).

But the debt has not merely been going up. Between 1823 and 1835 the debt steadily decreased, from a high of almost $91 million to a low of $33,733.05. The nation's debt hit the $1 billion mark in 1863 and the $1 trillion mark in 1981. Today, the national debt sits at more than $8 trillion. This level of debt means that the United States pays about $132,844,701,219.88 in interest each year. This makes our debt interest the third-largest expenditure in the federal budget after defense and combined entitlement programs such as Social Security and Medicare (Christensen, 2004).

Even in those years when the United States has had a balanced budget, that is, no deficits, the national debt continued to grow. It may have grown at a slower rate, but it did continue to grow. President Clinton bragged about presenting a balanced budget—one without deficits—and not growing the debt (King, J., 2000). However, the debt was already at a frighteningly high level, and his budget policies failed to make a dent in the debt.

THE DEBT AND EDUCATION DISPARITY

What does a discussion about national deficits and national debt have to do with education, education research, and continued education disparities? It is here where I began to see some metaphorical concurrences between our national fiscal situation and our education situation. I am arguing that our focus on the achievement gap is akin to a focus on the budget deficit, but what is actually happening to African American and Latina/o students is really more like the national debt. We do not have an achievement gap; we have an education debt.

... I have taken a somewhat different tack on this notion of the education debt. The yearly fluctuations in the achievement gap give us a short-range picture of how students perform on a particular set of achievement measures. Looking at the gap from year to year is a misleading exercise. Lee's (2002) look at the trend lines shows us that there was a narrowing of the gap in the 1980s both between Black and White students and between the Latina/o and White students, and a subsequent expansion of those gaps in the 1990s. The expansion of the disparities occurred even though the income differences narrowed during the 1990s. We do not have good answers as to why the gap narrows or widens. Some research suggests that even the combination of socioeconomic and family conditions, youth culture and student behaviors, and schooling conditions and practices do not fully explain changes in the achievement gap (Lee).

However, when we begin looking at the construction and compilation of what I have termed the education debt, we can better understand why an achievement gap is a logical outcome. I am arguing that the historical, economic, sociopolitical, and moral decisions and policies that characterize our society have created an education debt. So, at this point, I want to briefly describe each of those aspects of the debt.

THE HISTORICAL DEBT

Scholars in the history of education … have documented the legacy of educational inequities in the United States. These inequities initially were formed around race, class, and gender. Gradually, some of the inequities began to recede, but clearly they persist in the realm of race. In the case of African Americans, education was initially forbidden during the period of enslavement. After emancipation we saw the development of freedmen's schools whose purpose was the maintenance of a servant class. During the long period of legal apartheid, African Americans attended schools where they received cast-off textbooks and materials from White schools. In the South, the need for farm labor meant that the typical school year for rural Black students was about 4 months long. Indeed, Black students in the South did not experience universal secondary schooling until 1968 (Anderson, 2002). Why, then, would we not expect there to be an achievement gap?

The history of American Indian education is equally egregious. It began with mission schools to convert and use Indian labor to further the cause of the church. Later, boarding schools were developed as General George Pratt asserted the need "to kill the Indian in order to save the man." This strategy of deliberate and forced assimilation created a group of people, according to Pulitzer Prize writer N. Scott Momaday, who belonged nowhere (Lesiak, 1991). The assimilated Indian could not fit comfortably into reservation life or the stratified mainstream. No predominately White colleges welcomed the few Indians who successfully completed the early boarding schools. Only historically Black colleges, such as Hampton Institute, opened their doors to them. There, the Indians studied vocational and trade curricula.

Latina/o students also experienced huge disparities in their education. In Ferg-Cadima's report *Black, White, and Brown: Latino School Desegregation Efforts in the Pre- and Post-*Brown. v. Board of Education *Era* (2004), we discover the longstanding practice of denial experienced by Latina/os dating back to 1848. Historic desegregation cases such as *Mendez v. Westminster* (1946) and the Lemon Grove Incident detail the ways that Brown children were (and continue to be) excluded from equitable and high-quality education.

It is important to point out that the historical debt was not merely imposed by ignorant masses that were xenophobic and virulently racist. The major leaders of the nation endorsed ideas about the inferiority of Black, Latina/o, and Native peoples. Thomas Jefferson (1816), who advocated for the education of the American citizen, simultaneously decried the notion that Blacks were capable of education. George Washington, while deeply conflicted about slavery, maintained a substantial number of slaves on his Mount Vernon Plantation and gave no thought to educating enslaved children.

A brief perusal of some of the history of public schooling in the United States documents the way that we have accumulated an education debt overtime. In 1827 Massachusetts passed a law making all grades of public school open to all pupils free of charge. At about the same time, most Southern states already had laws forbidding the teaching of enslaved Africans to read. By 1837, when Horace Mann had become head of the newly formed Massachusetts State Board of Education, Edmund Dwight, a wealthy Boston industrialist, felt that the state board was crucial to factory owners and offered to supplement the state salary with his own money. What is omitted from this history is that the major raw material of those textile factories, which drove the economy of the East, was cotton—the crop that depended primarily on the labor of enslaved Africans (Farrow, Lang, & Frank, 2005). Thus one of the ironies of the historical debt is that while African Americans were enslaved and prohibited from schooling, the product of their labor was used to profit Northern industrialists who already had the benefits of education.

This pattern of debt affected other groups as well. In 1864 the U.S. Congress made it illegal for Native Americans to be taught in their native languages. After the Civil War, African Americans worked with Republicans to rewrite state constitutions to guarantee free public education for all students. Unfortunately, their efforts benefited White children more than Black children. The landmark *Plessy v. Ferguson* (1896) decision meant that the segregation that the South had been practicing was officially recognized as legal by the federal government.

Although the historical debt is a heavy one, it is important not to overlook the ways that communities of color always have worked to educate themselves. Between 1865 and 1877, African Americans mobilized to bring public education to the South for the first time. Carter G. Woodson (1933/1972) was a primary critic of the kind of education that African Americans received, and he challenged African Americans to develop schools and curricula that met the unique needs of a population only a few generations out of chattel slavery.

THE ECONOMIC DEBT

As is often true social research, the numbers present a startling picture of reality. The economics of the education debt are sobering. The funding disparities that currently exist between schools serving White students and those serving students of color are not recent phenomena. Separate schooling allows for differential funding. In present-day dollars, the funding disparities between urban schools and their suburban counterparts present a telling story about the value we place on the education of different groups of students.

The Chicago public schools spend about $8,482 annually per pupil, while nearby Highland Park spends $17,291 per pupil. The Chicago public schools have an 87% Black and Latina/o population, while Highland Park has a 90% White population. Per pupil expenditures in Philadelphia are $9,299 per pupil for the city's 79% Black and Latina/o population, while across City Line Avenue in Lower Merion, the per pupil expenditure is $17,261 for a 91% White population. The New York City public schools spend $11,627 per pupil for a student population that is 72% Black and Latina/o, while suburban Manhasset spends $22,311 for a student population that is 91% White (figures from Kozol, 2005).

One of the earliest things one learns in statistics is that correlation does not prove causation, but we must ask ourselves why the funding inequities map so neatly and regularly onto the racial and ethnic realities of our schools. Even if we cannot prove that schools are poorly funded *because* Black and Latina/o students attend them, we can demonstrate that the amount of funding rises with the rise in White students. This pattern of inequitable funding has occurred over centuries. For many of these populations, schooling was nonexistent during the early history of the nation; and, clearly, Whites were not prepared to invest their fiscal resources in these strange "others."

Another important part of the economic component of the education debt is the earning ratios related to years of schooling. The empirical data suggest that more schooling is associated with higher earnings; that is, high school graduates earn more money than high school dropouts, and college graduates earn more than high school graduates.

THE SOCIOPOLITICAL DEBT

The sociopolitical debt reflects the degree to which communities of color are excluded from the civic process. Black, Latina/o, and Native communities had little or no access to the franchise, so they had no true legislative representation. According to the Civil Rights Division of the U.S. Department of Justice, African Americans and other persons of color were substantially disenfranchised in many Southern states despite the enactment of the Fifteenth Amendment in 1870 (U.S. Department of Justice, Civil Rights Division, 2006).

The Voting Rights Act of 1965 is touted as the most successful piece of civil rights legislation ever adopted by the U.S. Congress (Grofman, Handley, & Niemi 1992). This act represents a proactive attempt to eradicate the sociopolitical debt that had been accumulating since the founding of the nation.

Table 1 shows the sharp contrasts between voter registration rates before the Voting Rights Act of 1965 and after it. The dramatic changes in voter registration are a result of Congress's bold action. In upholding the constitutionality of the act, the Supreme Court ruled as follows:

> Congress has found that case-by-case litigation was inadequate to combat wide-spread and persistent discrimination in voting, because of the inordinate amount of time and energy required to overcome the obstructionist tactics invariably encountered in these lawsuits. After enduring nearly a century of systematic resistance to the Fifteenth Amendment, Congress might well decide to shift the advantage of time and inertia from the perpetrators of the evil to its victims. (*South Carolina v. Katzenbach*, 1966; U.S. Department of Justice, Civil Rights Division, 2006)

It is hard to imagine such a similarly drastic action on behalf of African American, Latina/o, and Native American children in schools. For example, imagine that an examination of the achievement performance of children of color provoked an immediate reassignment of the nation's best teachers to the schools serving the most needy students. Imagine that those same students were guaranteed places in state and regional colleges and universities. Imagine that within one generation we lift those students out of poverty.

The closest example that we have of such a dramatic policy move is that of affirmative action. Rather than wait for students of color to meet predetermined standards, the society decided to recognize that historically denied groups should be given a preference in admission to schools and colleges. Ultimately, the major beneficiaries of this policy were White women. However, Bowen and Bok (1999) found that in the case of African Americans this proactive policy helped create what we now know as the Black middle class.

As a result of the sociopolitical component of the education debt, families of color have regularly been excluded from the decision-making mechanisms that should ensure that their children receive quality education. The parent–teacher organizations, school site councils, and other possibilities for democratic participation have not been available for many of these families. However, for a brief moment in 1968, Black parents in the Ocean Hill–Brownsville section of New York exercised community control over the public schools (Podair, 2003). African American, Latina/o, Native American, and Asian American parents have often advocated for improvements in schooling, but their advocacy often has been muted and marginalized. This quest for control of schools was powerfully captured in the voice of an African American mother during the fight for school desegregation in Boston. She declared: "When we fight about schools, we're fighting for our lives" (Hampton, 1986).

Indeed, a major aspect of the modern civil rights movement was the quest for quality schooling. From the activism of Benjamin Rushing in 1849 to the struggles of parents in rural South Carolina in 1999, families of color have been fighting for quality education for their children (Ladson-Billings, 2004). Their more limited access to lawyers and legislators has kept them from accumulating the kinds of political capital that their White, middle-class counterparts have.

THE MORAL DEBT

A final component of the education debt is what I term the "moral debt." I find this concept difficult to explain because social science rarely talks in these terms.

What I did find in the literature was the concept of "moral panics" (Cohen, 1972; Goode & Ben-Yehuda, 1994a, 1994b; Hall, Critcher, Jefferson, Clarke, & Roberts, 1978) that was popularized in British sociology. People in moral panics attempt to describe other people, groups of individuals, or events that become defined as threats throughout a society. However, in such a panic the magnitude of the supposed threat overshadows the real threat posed.

TABLE 1 Black and White Voter Registration Rates (%) in Selected U.S. States, 1965 and 1988

	March 1965			November 1988		
State	Black	White	Gap	Black	White	Gap
Alabama	19.3	69.2	49.9	68.4	75.0	6.6
Georgia	27.4	62.6	35.2	56.8	63.9	7.1
Louisiana	31.6	80.5	48.9	77.1	75.1	−2.0
Mississippl	6.7	69.9	63.2	74.2	80.5	6.3
North Carolina	46.8	96.8	50.0	58.2	65.6	7.4
South Carolina	37.3	75.7	38.4	56.7	61.8	5.1
Virginia	38.3	61.1	22.8	63.8	68.5	4.7

Note: From the website of the U.S. Department of Justice, Civil Rights Division, Voting Rights Section (http://www.usdoj.gov/crt/voting/intro/intro_c.html), "introduction to Federal Voting Rights Laws."

... [A] moral debt reflects the disparity between what we know is right and what we actually do. Saint Thomas Aquinas saw the moral debt as what human beings owe to each other in the giving of, or failure to give, honor to another when honor is due. This honor comes as a result of people's excellence or because of what they have done for another. We have no trouble recognizing that we have a moral debt to Rosa Parks, Martin Luther King, Cesar Chavez, Elie Wiesel, or Mahatma Gandhi. But how do we recognize the moral debt that we owe to entire groups of people? How do we calculate such a debt?

... What is it that we might owe to citizens who historically have been excluded from social benefits and opportunities? Randall Robinson (2000) states:

> No nation can enslave a race of people for hundreds of years, set them free bedraggled and penniless, pit them, without assistance in a hostile environment, against privileged victimizers, and then reasonably expect the gap between the heirs of the two groups to narrow. Lines, begun parallel and left alone, can never touch up. (p. 74)

Robinson's sentiments were not unlike those of President Lyndon B. Johnson, who stated in a 1965 address at Howard University: "You cannot take a man who has been in chains for 300 years, remove the chains, take him to the starting line and tell him to run the race, and think that you are being fair" (Miller, 2005).

Taken together, the historic, economic, sociopolitical, and moral debt that we have amassed toward Black, Brown, Yellow, and Red children seems insurmountable, and attempts at addressing it seem futile. Indeed, it appears like a task for Sisyphus. But as legal scholar Derrick Bell (1994) indicated, just because something is impossible does not mean it is not worth doing.

WHY WE MUST ADDRESS THE DEBT

On the face of it, we must address it because it is the equitable and just thing to do. As Americans we pride ourselves on maintaining those ideal qualities as hallmarks of our democracy. That represents the highest motivation for paying this debt. But we do not always work from our highest motivations.

Most of us live in the world of the pragmatic and practical. So we must address the education debt because it has implications for the kinds of lives we can live and the kind of education the society can expect for most of its children. I want to suggest that there are three primary reasons for addressing the debt—(a) the impact the debt has on present education progress, (b) the value of understanding the debt in relation to past education research findings, and (c) the potential for forging a better educational future.

The Impact of the Debt on Present Education Progress

As I was attempting to make sense of the deficit/debt metaphor, educational economist Doug Harris (personal communication, November 19, 2005) reminded me that when nations operate with a large debt, some part of their current budget goes to service that debt. I mentioned earlier that interest payments on our national debt represent the third largest expenditure of our national budget. In the case of education, each effort we make toward improving education is counterbalanced by the ongoing and mounting debt that we have accumulated. That debt service manifests itself in the distrust and suspicion about what schools can and will do in communities serving the poor and children of color. Bryk and Schneider (2002) identified "relational trust" as a key component in school reform. I argue that the magnitude of the education debt erodes that trust and represents a portion of the debt service that teachers and administrators pay each year against what they might rightfully invest in helping students advance academically.

The Value of Understanding the Debt in Relation to Past Research Findings

The second reason that we must address the debt is somewhat selfish from an education research perspective. Much of our scholarly effort has gone into looking at educational inequality and how we might mitigate it. Despite how hard we try, there are two interventions that have never received full and sustained hypothesis testing—school desegregation and funding equity. Orfield and Lee (2006) point out that not only has school segregation persisted, but it has been transformed by the changing demographics of the nation. They also point out that "there has not been a serious discussion of the costs of segregation or the advantages of integration for our most segregated population, white students" (p. 5). So, although we may have recently celebrated the 50th anniversary of the *Brown* decision, we can point to little evidence that we really gave *Brown* a chance. According to Frankenberg, Lee, and Orfield (2003) and Orfield and Lee (2004), America's public schools are more than a decade into a process of resegregation. Almost three-fourths of Black and Latina/o students attend schools that are predominately non-White. More than 2 million Black and Latina/o students—a quarter of the Black students in the Northeast and Midwest—attend what the researchers call apartheid schools. The four most segregated states for Black students are New York, Michigan, Illinois, and California.

The funding equity problem … also has been intractable. In its report entitled *The Funding Gap 2005*, the Education Trust tells us that "in 27 of the 49 states studied, the highest-poverty school districts receive fewer resources than the lowest-poverty districts. … Even more states shortchange their highest minority districts. In 30 states, high minority districts receive less money for each child than low minority districts" (p. 2). If we are unwilling to desegregate our schools *and* unwilling to fund them equitably, we find ourselves not only backing away from the promise of the *Brown* decision but literally refusing even to take *Plessy* seriously. At least a serious consideration of *Plessy* would make us look at funding inequities.

In one of the most graphic examples of funding inequity, new teacher Sara Sentilles (2005) described the southern California school where she was teaching:

> At Garvey Elementary School, I taught over thirty second graders in a so-called temporary building. Most of these "temporary" buildings have been on campuses in Compton for years. The one I taught in was old. Because the wooden beams across the ceiling were being eaten by termites, a fine layer of wood dust covered the students' desks every morning. Maggots crawled in a cracked and collapsing area of the floor near my desk. One day after school I went to sit in my chair, and it was completely covered in maggots. I was nearly sick. Mice raced behind cupboards and bookcases. I trapped six in terrible traps called "glue lounges" given to me by the custodians. The blue metal window coverings on the outsides of the windows were shut permanently, blocking all sunlight. Someone had lost the tool needed to open them, and no one could find another. … (p. 72)

Rothstein and Wilder (2005) move beyond the documentation of the inequalities and inadequacies to their *consequences*. In the language that I am using in this discussion, they move from focusing on the gap to tallying the debt. Although they focus on Black–White disparities, they are clear that similar disparities exist between Latina/os and Whites and Native Americans and Whites. Contrary to conventional wisdom, Rothstein and Wilder argue that addressing the achievement gap is not the most important inequality to attend to. Rather, they contend that inequalities in health, early childhood experiences, out-of-school experiences, and economic security are also contributory and cumulative and make it near impossible for us to reify the achievement gap as *the* source and cause of social inequality.

The Potential for Forging a Better Educational Future

Finally, we need to address what implications this mounting debt has for our future. In one scenario, we might determine that our debt is so high that the only thing we can do is declare bankruptcy. Perhaps, like our airline industry, we could use the protection of the bankruptcy laws to reorganize and design more streamlined, more efficient schooling options. Or perhaps we could be like developing nations that owe huge sums to the IMF and apply for 100% debt relief. But what would such a catastrophic collapse of our education system look like? Where could we go to begin from the ground up to build the kind of education system that would aggressively address the debt? Might we find a setting where a catastrophic occurrence, perhaps a natural disaster—a hurricane—has completely obliterated the schools? Of course, it would need to be a place where the schools weren't very good to begin with. It would have to be a place where our Institutional Review Board and human subject concerns would not keep us from proposing aggressive and cutting-edge research. It would have to be a place where people were so desperate for the expertise of education researchers that we could conduct multiple projects using multiple approaches. It would be a place so hungry for solutions that it would not matter if some projects were quantitative and others were qualitative. It would not matter if some were large-scale and some were small-scale. It would not matter if some paradigms were psychological, some were social, some were economic, and some were cultural. The only thing that would matter in an environment like this would be that education researchers were bringing their expertise to bear on education problems that spoke to pressing concerns of the public. I wonder where we might find such a place?

REFERENCES

Anderson, J. D. (2002, February 28). Historical perspectives on Black academic achievement. Paper presented for the Visiting Minority Scholars Series Lecture. Wisconsin Center for Educational Research, University of Wisconsin, Madison.

Bell, D. (1994). *Confronting authority: Reflections of an ardent protester.* Boston, MA: Beacon Press.

Bereiter, C., & Engleman, S. (1966). *Teaching disadvantaged children in preschool.* Englewood Cliffs, NJ: Prentice Hall.

Bowen, W., & Bok, D. (1999). *The shape of the river.* Princeton, NJ: Princeton University Press.

Brown v. Board of Education 347 U.S. 483(1954).

Bryk, A., & Schneider, S. (2002). *Trust in schools: A core resource for improvement.* New York: Russell Sage Foundation.

Christensen, J. R. (Ed.). (2004). *The national debt: A primer.* Haupauge, NY: Nova Science Publishers.

Clark, K. B. (1965). *Dark ghetto: Dilemmas of social power.* Hanover, NH: Wesleyan University Press.

Cohen, S. (1972). *Folk devils and moral panics: The creation of mods and rockers*. London: McGibbon and Kee.

Coleman, J., Campbell, E., Hobson, C., McPartland, J., Mood, A., Weinfeld, F. D., et al. (1966). *Equality of educational opportunity*. Washington, DC: Department of Health, Education and Welfare.

Deutsch, M. (1963). The disadvantaged child and the learning process. In A. H. Passow (Ed.), *Education in depressed areas* (pp. 163–179). New York: New York Bureau of Publications, Teachers College, Columbia University.

Education Commission of the States. (2005). *The nation's report card*. Retrieved January 2, 2006, from http://nces.ed.gov/nationsreportcard

Farrow, A., Lang, J., & Frank, J. (2005). *Complicity: How the North promoted, prolonged and profited from slavery*. New York: Ballantine Books.

Ferg-Cadima, J. (2004, May). *Black, White, and Brown: Latino school desegregation efforts in the pre- and post-*Brown v. Board of Education *era*. Washington, DC: Mexican-American Legal Defense and Education Fund.

Frankenberg, E., Lee, C., & Orfield, G. (2003, January). *A multiracial society with segregated schools: Are we losing the dream?* Cambridge, MA: The Civil Rights Project, Harvard University.

Goode, E., & Ben-Yehuda, N. (1994a). Moral panics: Culture, politics, and social construction. *Annual Review of Sociology*, 20, 149–171.

Goode, E., & Ben-Yehuda, N. (1994b). *Moral panics: The social construction of deviance*. Oxford: Blackwell.

Gordon, J. S. (1998). *Hamilton's blessing: The extraordinary life and times of our national debt*. New York: Penguin Books.

Grofman, B., Handley, L., & Niemi, R. G. (1992). *Minority representation and the quest for voting equality*. New York: Cambridge University Press.

Hall, S., Critcher, C., Jefferson, T., Clarke, J., & Roberts, B. (1978). *Policing the crisis: Mugging, the state, and law and order*. London: Macmillan.

Hampton, H. (Director). (1986). *Eyes on the prize* [Television video series]. Blackside Productions (Producer). New York: Public Broadcasting Service.

Hess, R. D., & Shipman, V. C. (1965). Early experience and socialization of cognitive modes in children. *Child Development*, 36, 869–886.

Jefferson, T. (1816, July 21). Letter to William Plumer. The Thomas Jefferson Paper Series. 1. General correspondence, 1651–1827. Retrieved September 11, 2006, from http://rs6.loc.gov/cgi-bin/ampage

King, J. (2000, May 1). Clinton announces record payment on national debt. Retrieved February 7, 2006, from http://archives.cnn.com/2000/ALLPOLITICS/stories/05/01/Clinton.debt

Kozol, J. (2005). *The shame of the nation: The restoration of apartheid schooling in America*. New York: Crown Publishing.

Ladson-Billings, G. (2004). Landing on the wrong note: The price we paid for *Brown*. *Educational Researcher*, 33(7), 3–13.

Lee, J. (2002). Racial and achievement gap trends: Reversing the progress toward equity. *Educational Researcher*, 31(1), 3–12.

Lesiak, C. (Director). (1992). *In the White man's image* [Television broadcast]. New York: Public Broadcasting Corporation.

Mendez v. Westminster 64F. Supp. 544 (1946).

Miller, J. (2005, September 22). New Orleans unmasks apartheid American style [Electronic version]. *Black Commentator, 151.* Retrieved September 11, 2006, from http://www.blackcommentator.com/151/151_miller_new_orleans.html

National Center for Education Statistics. (2001). *Education achievement and Black–White inequality.* Washington, DC: Department of Education.

National Governors' Association. (2005). *Closing the achievement gap.* Retrieved October 27, 2005, from http://www.subnet.nga.org/educlear/achievment/

Orfield, G., & Lee, C. (2004, January). *Brown at 50: King's dream or Plessy's nightmare?* Cambridge, MA: The Civil Rights Project, Harvard University.

Orfield, G., & Lee, C. (2006, January). *Racial transformation and the changing nature of segregation.* Cambridge, MA: The Civil Rights Project, Harvard University.

Plessy v. Ferguson 163 U.S. 537 (1896).

Podair, Jerald E. 2003. *The strike that changed New York: Blacks, whites, and the Ocean Hill-Brownsville Crises.* New Haven, CT: Yale University.

Robinson, R. (2000). *The debt: What America owes to Blacks.* New York: Dutton Books.

Rothstein, R., & Wilder, T. (2005, October 24). *The many dimensions of racial inequality.* Paper presented at the Social Costs of Inadequate Education Symposium, Teachers College, Columbia University, New York.

Sentilles, S. (2005). *Taught by America: A story of struggle and hope in Compton.* Boston, MA: Beacon Press.

South Carolina v. Katzenbach 383 U.S. 301, 327–328 (1966).

U.S. Department of Justice, Civil Rights Division. (2006, September 7). *Introduction to federal voting rights laws.* Retrieved September 11, 2006, from http://www.usdoj.gov/crt/voting/intro/intro.htm

Woodson, C. G. (1972). *The mis-education of the Negro.* Trenton, NJ: Africa World Press. (Original work published 1933)

KEY CONCEPTS

achievement gap

education debt

DISCUSSION QUESTIONS

1. What did this article tell us about the achievement gap in education? What has typically been the pattern between disadvantaged students and privileged students with regard to school performance?
2. Why is the achievement gap bad for everyone? In what ways does inequality in education hurt the United States as a whole?

50

Education: Inclusive Disability Studies

DAN GOODLEY

Dan Goodley discusses how an inclusive education, including for students with disabilities, can give all students the education needed to function in a democratic society. But he also argues that the move to neoliberal education thwarts inclusion and recognition of the diverse needs of students.

INTRODUCTION

To what extent are disabled children fully included in their schools? How do different approaches of special and inclusive education understand disabled children? How can teachers respond in enabling ways? Contemporaneously, more and more disabled children have entered their local mainstream schools. Disability studies and social justice meet at the crossroads of inclusive education. But this approach can only be understood in relation to its antithesis: special education. Mass public education was never designed with disabled learners in mind.... Apple (1982) argues that compulsory schooling in the Global North is bound to the demands of capitalism but school curricula refer to far more than what is taught in schools. The "hidden curricula" of schools are where the values, rituals and routines of wider society are acculturated within students. Corbett and Slee (2002: 134) urge inclusive educators to ask questions about the deep cultures of schools to find alternative curricula that are more in tune with needs of diverse learners. ... [W]e will explore the meaning of inclusive/ special education but contextualise this in relation to the influence of marketisation....

SOURCE: Dan Goodley, Inclusive Disability Studies, from Disability Studies, by SAGE Publications. Reprinted by permission of SAGE Publications.

INCLUSIVE EDUCATION

... [I]nclusive education refers to the increasing participation of learners in the culture, curricula, and communities of their neighbourhood centres of learning. ... [A]chievements of schools mean nothing if school communities fail to enhance the spirit of all teachers and pupils. Clough and Corbett (2000) sketch out an historical overview of educational responses to disability. The 1950s was typified by a psychomedical view of the disabled child requiring specialist intervention. These were the halcyon days for special educators and psychologists and their knowledge production around the disabled learner. The 1960s saw a sociological response that viewed special educational needs and learning disabilities as the direct product of exclusionary schooling. The 1970s was characterised by curricular approaches that met, or failed to meet, the learning needs of pupils. The 1980s saw attention turn to school improvement via comprehensive schooling and the systemic reorganisation of schools. The focus was on communities of schools rather than the achievements of individuals. The 1990s onwards has seen interventions from disability studies. At the heart of this critique is a call for inclusive education.

WHAT IS INCLUSION (AND WHAT ABOUT SPECIAL)?

In order to understand inclusive education we need to briefly probe special education. The latter has been described by Armstrong (2002) as a "wild profusion of entangled ideas," including charity, medicalisation and psychologisation. Special education places the disabled child in a specialist setting supported by specially trained professionals who intervene to improve the child. Proponents argue that this provides a more suitable context for the needs of disabled children to be met. But children who experience special education often have a narrow education, achieve low levels of academic attainment and leave schools only to enter other segregated arenas of work and education (Lipsky and Gartner, 1996). Special schools transplant the failings of mass education into the minds and bodies of disabled children. The special child is viewed through the lens of functionalism...; a learner who fails to fit in and learn (Thomas and Loxley, 2001:5), marked by the processes of psychologisation. Special schools collude in the failure of all schools to accommodate difference....

But special education does not just appear in segregated schools. It can refer to special procedures and systems, that are also to be found in mainstream schools. Special educators, according to Isaacs (1996: 41–42), succeed in "promoting limited ontologies of personhood, exaggerated legitimations of the normal, based on an ever-growing and insidious idea that the failings of children can be explained in terms of a medical deficit-model of the disabled child." Special education has therefore been described as a "segregating, insulated, self-protecting, racially biased philosophy and array of practices, a product of ...

misguided scientific positivism, or merely as an ineffective, over-blown problem to easily solvable school problems" (Gerber, 1996: 159).

... [T]he first special education classes in North America included the urban poor, Native, Hispanic and African Americans. Black and working-class children have always been over-represented in special schools, leading Mercer (1973) to argue that measures of intelligence, social incompetence and maladaptive functioning say more about the racialised biases of professional assessment than they do of the deficits of "mentally handicapped children." Whether or not one accepts these criticisms, it is safe to say that the drive for special schools and the growing professionalism of special educators are intimately connected (Meekosha and Jakubowicz, 1996: 81). When the problems of educational achievement are understood as residing within the "special child," then schools and teachers "remain outside of this diagnostic glare" (Slee, 1996: 105). And, special schools breed "unusual practices enacted by specialist teachers to instruct or manage the exotic and homogenised groups of special needs children" (Gerber, 1996: 157). According to Slee (2004: 47), special education remains bound to an "inherent technicism, rooted in deficit-bound psycho-medical paradigms of individual pathological defects." These assumptions continue to influence inclusive education (Ware, 2004b).

Practices of integration emerged in the 1960s in reaction to special education (Vislie, 2003). Integration promoted the right to education for disabled children in their local schools. In Britain, the Warnock Committee (1978) report, *Special Educational Needs*, was hugely influential in unsettling special school categories of impairment. Warnock proposed the use of "Special Educational Needs" or "SEN" as a catchall term for identifying those children who required extra support in mainstream schools.

For Rioux (1994b), integration was doomed to fail disabled children because it smuggled in unproblematic ideas of disability, meritocracy, and self-reliance that ignored wider structural inequalities. She argued that integrated students still remained the focus of special educators even when they were in mainstream schools. According to Christensen (1996), the development of techniques such as the Individual Education Plan (IEP)—an artefact of special schools and widely used in schools in WENA and as an accompanying practice to the identification of SEN—allocate resources to a child based on the identification of individual requirements (often understood as learning problems and educational deficits).... Special Educational Needs Co-ordinators (SEN-COs) in British schools (trained teachers with specialist interests in the support of disabled pupils) were established to provide professional support to those with this new label. But IEPs and SENCOs still addressed individual children rather than the school culture. Too often SENCOs were the sole agitators in schools and they, like their SEN children, were marginalised by the wider school culture (Booth, 2002).

Slee (1997: 411) observes that simply cleaning up exclusionary contexts do not make them more inclusionary spaces if the values of regular schooling remain. Following Tomlinson (1982) and Troyna and Vincent (1996), integration is a reformist agenda that is both exclusive (in, for example, only allowing the least disruptive to be included in schools) and inadequate (failing to address the structural inequalities within education)....

In the mid-1990s, a body of inclusive education literature emerged to challenge functionalist deficit "within child" thinking of special education and the conservative compromises of integration.... Armed with damning evidence on the failure of special schools, inclusive educators called for a change to the disabling ethos and philosophies of *all* mainstream schools.

Sociological theory was a key resource and a host of materialist, radical humanist, and social constructionists theories were employed to critique existing educational arrangements and foster more inclusive alternatives (e.g. Skritc, 1995). Inclusive education sought to broaden the options available to a variety of learners, develop the skills and confidence of all teachers, and develop a general political project that opened up the complex realities of educational settings (Slee, 1997: 410), while exposing the deleterious impact of sexist, disablist and racist education (Slee, 2004: 55).... [W]e can see that accommodation, assimilation and integration fit with a *consensus* model of society: disabled children are expected to fit into existing schooling arrangements. In contrast, inclusive education approaches seek changes in line with *conflict* approaches to sociology, demanding educators to rethink education and disability....

Inclusion is therefore a response to special education and integration in which children with SEN are understood as comprising a group constituted by a "bureaucratic device for dealing with the complications arising from clashes between narrow waspish curricula and disabled students" (Slee, 1997: 412). Inclusive education:

- is a process by which a school attempts to respond to all pupils as individuals;
- records inclusion and exclusion as connected processes, schools developing more inclusive practices may need to consider both;
- emphasises overall school effectiveness;
- is of relevance to all phases and types of school, possibly including special schools, since within any educational provision teachers face groups of students with diverse needs and are required to respond to this diversity. (Vislie, 2003: 21)...

... While more and more disabled students are entering post-compulsory education and universities, questions still remain about the extent to which they belong. The narrowness of curricula means exclusion for many non-normative children....

NEOLIBERAL EDUCATION

... Neoliberalism refers to monetary and trade policies of a pro-corporate free market economy that has dominated ... economic and cultural politics and global markets since the early 1980s (Richardson, 2005). The state is "rolled back," unproductive welfare spending is reduced, and public services and social provisions are increasingly taken over by, or aligned with the principles of, business. Public entitlements, such as welfare and education, have become dismantled through an alliance with market freedom and the essence of the Washington consensus, the driving force of global progression and the deregulation of the market. It is linked to a context of austerity; a change of economic context in which countries have to compete internationally and economic rationality overtakes social welfare reform (Rizvi and Lingard, 1996: 13)....

Neoliberal education, at its most seemingly benign, calls upon common standards, assessment, and accountability of schools and teachers to pupils and parents. At its most damaging, these values are characterised by shrinking resources, and schools are pulled into the competitive marketplace where productivity and accountability (to consumers and government assessment bodies) are paramount. We see, too, the application of increasingly more stringent academic criteria and higher standards of education, a narrowing of the curriculum and an increase in educational testing and assessment (Jung, 2002).... Thus "McDonaldisation of education" (surveillance, testing, targeting, performativity and marketisation, according to Gabel and Danforth, 2008b) or "human capital paradigm approach to education" (a focus on work-related competencies and skills, Peters et al., 2008) has led to a splintering of teachers unions and an erosion of morality. In the context of Neoliberalism, services are no longer regarded as a civic or human right but as commodities for consumers. The "common good" equates with free competition, consumer power and profitability (Rizvi and Lingard, 1996: 14–15).

Schools are high-pressure places; subjected to league tables, children to endless SATS, teachers to inspection. Curricula are nationalised, allowing comparison between schools, teachers and pupils, and focused on science, maths and literacy—key requirements of capitalist economies (McLaren, 2009)....

Neoliberal schools are stressful places. Teachers are held to greater accountability, more assessment, and a loss of autonomy.... Teachers are implementers of the decisions of others—like policy makers—and are assessed in terms of how well they implement these decisions....

The problem for *Every Child Matters* and *No Child Left Behind* is their reliance on normative understandings of the child.... The (preferred) child, the child that is meaningfully included, is the neoliberal subject of a neoliberal capitalist society. While the majority world might view the child in terms of community contribution, economic capital and lineage (Ghai, 2006:157), in minority world contexts like Britain, the neoliberal child is a vessel (though a knowing and autonomous one) for all of contemporary society's contradictions. The child is innocent/responsible, player/worker, achiever/learner. This criss-crossing of discourses raise questions about the values we place on (disabled) children and the futures we envisage for them. The marketisation of education creates a paradox: the development of new norms and the promotion of new forms of deviancy....

REFERENCES

Apple, M. (1982). *Education and Power*. Boston: Routledge & Kegan Paul.

Armstrong, F. (2002). The historical development of special education: humanitarian rationality or 'wild profusion of entangled events'?. *History of Education*, 31 (5), 437–450.

Bogdan, R. and Taylor, S.J. (1982). *Inside Out: The Social Meaning of Mental Retardation*. Toronto: University of Toronto Press.

Booth, T. (2002). Inclusion and exclusion policy in England: who controls the agenda? In F. Armstrong, D. Armstrong and L. Barton (eds), *Inclusive Education: Policy, Contexts and Comparative Perspectives*. (pp. 78–98). London: David Fulton.

Christensen, C. (1996). Disabled, handicapped or disordered: what's in a name? In C. Christensen, and F. Rizvi (eds), *Disability and the Dilemmas of Education and Justice*. (pp. 63–78). Buckingham: Open University Press.

Clough, P. and Corbett, J. (2000). *Theories of Inclusive Education: A Students' Guide*. London: Paul Chapman.

Corbett, J. and Slee, R. (2002). An international conversation on inclusive education. In F. Armstrong, D. Armstrong and L. Barton (eds), *Inclusive Education: Policy, Contexts and Comparative Perspectives*. (pp. 133–146). London: David Fulton.

Gerber, M. (1996). Reforming special education: beyond 'inclusion'. In C. Christensen and F. Rizvi (eds), *Disability and the Dilemmas of Education and Justice*. (pp. 156–174). Buckingham: Open University Press.

Ghai, A. (2006). *(Dis)embodied Form: Issues of Disabled Women*. Delhi: Shakti Books.

Isaacs, P. (1996). Disability and the education of persons. In C. Christensen and F. Rizvi (eds), *Disability and the Dilemmas of Education and Justice*. (pp. 27–45). Buckingham: Open University Press.

Jung, K.E. (2002). Chronic illness and educational equity: the politics of visibility. *NWSA Journal*, 14(3), 178–200.

Lipsky, D. and Gartner, A. (1996). Equity requires inclusion: the future for all students with disabilities. In C. Christensen and F. Rizvi (eds), *Disability and the Dilemmas of Education and Justice*. (pp. 145–155). Buckingham: Open University Press.

McLaren, P. (2009). Critical pedagogy: a look at major concepts. In A. Darder, M.P. Baltodano and R.D. Torres (eds), *The Critical Pedagogy Reader* (2nd edition). (pp. 69–96). New York: Routledge.

Meekosha, H. and Jakubowicz, A. (1996). Disability, participation, representation and social justice. In C. Christensen and F. Rizvi (eds), *Disability and the Dilemmas of Education and Justice*. (pp. 79–95). Buckingham: Open University Press.

Mercer, J.R. (1973). *Labelling the Mentally Retarded: Clinical and Social System Perspectives on Mental Retardation*. Los Angeles: University of California Press.

Peters, S., Wolbers, K. and Dimling, L. (2008). Reframing global education from a disability rights movement perspective. In S. Gabel and S. Danforth (eds), *Disability and the International Politics of Education*. (pp. 291–310). New York: Peter Lang.

Richardson, D. (2005). Desiring sameness? The rise of neoliberal politics of normalization. *Antipode*, 37 (3), 515–535.

Rioux, M. (1994b). Towards a concept of equality of well being: overcoming the social and legal construction of inequality. In M. Rioux and M. Bach (eds), *Disability Is Not Measles: New Directions in Disability*. (pp. 67–108). Ontario: L'Institut Roeher.

Rizvi, F. and Lingard, B. (1996). Disability, education and the discourses of justice. In C. Christensen and F. Rizvi (eds), *Disability and the Dilemmas of Education and Justice*. (pp. 9–26). Buckingham: Open University Press.

Skrtic, T.M. (ed.) (1995). *Disability and Democracy: Reconstructing (Special) Education for Postmodernity*. New York: Teachers College Press.

Slee, R. (1996). Disability, class and poverty: school structures and policising identities. In C. Christensen and E. Rizvi (eds), *Disabiltiy and the Dilemmas of Education and Justice*. (pp. 96–188). Buckingham: Open University Press.

Slee, R. (1997). Imported or important theory? Sociological interrogations of disablement and special education. *British Journal of Sociology of Education*, 18 (3), 407–419.

Slee, R. (2004). Meaning in the service of power. In L. Ware (ed.), *Ideology and the Politics of (In)exclusion*. New York: Peter Lang.

Thomas, G. and Loxley, A. (2001). *Deconstructing Special Education and Constructing Inclusion*. Buckingham: Open University Press.

Tomlinson, S. (1982). *A Sociology of Special Education*. London: Routledge.

Troyna, B. and Vincent, C. (1996). 'The ideology of expertism': the framing of special education and racial equality policies in the local state. In C. Christensen and F. Rizvi (eds), *Disability and the Dilemmas of Education and Justice*. (pp. 131–144). Buckingham: Open University Press.

Vislie. (2003). From integration to inclusion: focusing on global trends and changes in the Western European societies. *European Journal of Special Needs Education*, 18, 17–35.

Ware, L. (2004b). Introduction. In L. Ware (ed.), *Ideology and the Politics of (In)exclusion*. (pp. 1–12). New York: Peter Lang.

Warnock Committee (1978). *Special Educational Needs*. The Warnock Report. London: Department of Schools and Education.

KEY CONCEPTS

hidden curriculum meritocracy neoliberalism

DISCUSSION QUESTIONS

1. What is meant by neoliberal education, and how does Goodley see this as affecting students with disabilities?
2. In what ways was your school experience shaped by neoliberal ideals?

Applying Sociological Knowledge: An Exercise for Students

Research the news for a recent article on urban schools. What are the key issues in educational reform for inner cities? What are the two or more sides to the debate regarding funding for schools? Consider your own educational experience compared to that described in the news report you read. What advantages or disadvantages did you have in your schooling? What about the trend for online schooling? Who benefits most from online learning? What are the problems with online education?

PART XVI

Economy and Work

51

Children of the Great Recession

A Tour of the Generational Landscape, from Struggles to Successes, Coast to Coast

RONALD BROWNSTEIN

This reading summarizes the problems facing young people today as they try to gain employment after school. The recession created a problem of too many workers and too few jobs. Brownstein finds that young people are not necessarily discouraged. They turn, instead, to finding education bargains, starting their own companies, or deciding on public service work.

The April sun shines brightly over the career development center at California State University (Los Angeles). Yet, despite the perfect weather, the center's conference room is crammed with students taking notes intently. Judy Narcisse, a counselor at the center, is running a workshop on résumé writing.

Gesturing toward a PowerPoint presentation on a large screen, Narcisse marches her charges through the basics. Her tone is casual, and her humor deadpan, but her advice is hard-headed. "If you're going to leave your cellphone number, don't have rap music on your voice mail," she explains. "You can't say, 'Hey, this is Naomi! Give me a holla!'"

As Narcisse speaks, heads bob and pens scribble. Her students at this heavily Hispanic and Asian, working-class university of about 20,000 located east of downtown Los Angeles are attentive to a fault. "What if I don't have enough experience to fill a piece of paper?" one young woman asks. Print your résumé with wider margins or a larger font size, Narcisse replies. She is full of such practical suggestions: Find internships, she recommends, attend career fairs, volunteer, proofread your résumé. At times, Narcisse sounds like a coach delivering a half-time pep talk. 'We're in a tough economic market," she says bluntly. "You're going to have to hustle."

SOURCE: Reprinted with permission from *National Journal*, May 8, 2010.

The message is not lost on these children of the Great Recession, or on the massive Millennial Generation they embody. The Millennials, best defined as people born between 1981 and 2002, are now the largest generation in the United States, with nearly 92 million members according to the latest Census Bureau figures. They are the most diverse generation in American history (more than two-fifths are nonwhite, according to the census), and they are en route to possibly becoming the best-educated: About half of Millennial men and 60 percent of Millennial women older than 18 have at least some college education. They are frequently described as a civic generation, more oriented toward public service and volunteerism and more instinctively optimistic than any cohort of young people since the fabled GI Generation that fought World War II and built postwar America.

And now their great expectations are colliding with diminished circumstances. Millions in the generation are navigating the transition to adulthood through the howling gale of the worst economic downturn since the Depression. Federal statistics tell a bracing story. The unemployment rate among 20-to-24-year-olds stood at nearly 16 percent in March, more than double the level of three years earlier, before the recession began. For those 25 to 29, unemployment stood at 11.5 percent, also more than double the level of three years ago. For teenagers, 20-something African-Americans and Hispanics, and young people without a college degree, the unemployment rates spike as high as 28 percent—debilitating levels of the kind that Americans haven't seen in decades.

EMPLOYMENT

Adult Millennials have been hit hard by the recession, with their rate of unemployment half again as high as that of older workers. But the situation looks still worse when you factor in those who have stopped looking for work or opted to remain in school. The percentage of Millennials with jobs—called the employment–population ratio—has dropped much more sharply than that for other age groups.

Even in this dire climate, some Millennials are finding new ways to weather the storm, displaying the instinct for pragmatic problem-solving that generational theorists consider one of their defining characteristics. They are flocking into service-oriented alternatives such as Teach for America or partnering with peers to launch innovative entrepreneurial ventures.

But for many others, the working world has become an inscrutable maze of part-time jobs, temporary gigs, and full-time positions that abruptly dissolve into layoffs and start the entire disorienting cycle again. Such widespread uncertainty could impose lasting costs, measured not only in diminished opportunity and earnings for the Millennials themselves, but lost productivity and multiplying social challenges for American society at large. "If the jobless recovery … drags on for three or four years, then I think we will face some large problems that we haven't faced since the Great Depression," says Robert Reich, who served as Labor secretary in the Clinton administration. "Young people are just not going to form the habits and attachments to work that their older siblings or parents have had, and that could conceivably create a whole variety of social problems."

The Broken Escalator

When the job market works well, it functions like an escalator. Young people finish their schooling, whether high school or higher education, and move onto the escalator's bottom steps with entry-level jobs. As they acquire more skills and experience, they rise into better-paying and more-senior positions, clearing space on the first steps for the next crop of new graduates. Meanwhile, on the top steps, older workers step off into retirement, creating room that allows everyone below them to ascend through their own careers.

But now the escalator is jammed at every level. With jobs scarce, many young people are stuck at the bottom, unable to take that first step. Those who have been lucky or skillful enough to get on the escalator in the past few years are often not rising smoothly. They might gain a job, lose it, and fall back several steps or off the escalator altogether. There they must jostle with each successive class of graduates trying to squeeze on at the bottom. Meanwhile, at the other end, with the stock market's collapse decimating 401(k) plans, fewer older workers are moving briskly off the escalator into retirement. (Federal statistics show that while participation in the labor force is declining slightly for middle-aged workers, and sharply for younger workers, it has actually increased since 2007 among workers ages 55 to 64.) The continued presence of this older generation on the escalator makes it more difficult for everyone behind them to ascend, intensifying the pileup at the lower levels. Compounding the pressure, more middle-aged workers are being toppled from the upper steps by layoffs, which force them to compete for space lower down that junior colleagues might once have occupied. Rather than advancing in smooth procession, everyone is stepping on everybody else.

The pileup is being felt keenly this spring on college campuses. Last year, the bottom fell out of the job market for the class of 2009. The Collegiate Employment Research Institute at Michigan State University, which tracks these trends nationwide, found that last spring large employers hired 42 percent fewer graduating students than they had originally targeted when the school year began. It was a "rout," the group reported in its latest annual survey. This year, the job market for graduates appears to have stabilized and may even be rebounding slightly—albeit from that sharply reduced new starting point—says Phil Gardner, the institute's director. "We see sporadic little shoots grow up here and there," he explains, "but … no surge."

Career counselors at a wide variety of colleges and universities echo Gardner's equivocal verdict. "Things are starting to thaw," says Paula Klempay, the director of M.B.A. career services at the University of Washington's Foster School of Business. "We're not as robust as we were two years ago, but we're better than last year." Jaime Velasquez, assistant director of career services at the University of Illinois (Chicago), says that the number of employers at the school's job fair jumped from 50 last year to 62 this spring, "and they all had positions available." Still, he notes, "five years ago, we had 110 [firms]."

School Daze

The climate has improved enough—modestly but perceptibly—that success stories on campus are again growing more common. Kesha McLaren, a senior at the University of Missouri's College of Agriculture, struck out when she tried submitting her résumé directly to employers, but she was hired by an agricultural finance company in St. Louis after a professor steered a recruiter in her direction. "It was the only position I was offered, but that doesn't bother me, because it was kind of my dream job," she says. Halfway across the country, the undergraduate business school at Howard University, a historically black college in Washington, D.C., this year has placed 90 percent of its honors students in graduate school or full-time work. Ann Jackson, a 21-year-old graduating this spring, is moving to Portland, Ore., after winning a job at Intel's human relations department. "I got lucky, honestly," she says. "I'd been in a lot of interviews, and there were quite a few I didn't get."

Despite these flickers of improvement, the challenges remain substantial, even for graduates of elite institutions. In 2009, the unemployment rate among 20-to-24-year-olds with four-year college degrees stood at 8.8 percent, twice what it was just two years earlier. Many college officials concur with John Noble, the director of career counseling at Williams College in Massachusetts, who says that even with some recent revival, "this is the worst market I've seen since I started in this line of work in 1982."

Complicating matters for the graduating class of 2010 are the substantial number of students from previous classes who are still competing for entry-level positions. Sean Sposito, for instance, graduated from the University of Missouri's journalism school in 2009 but is looking for work again after completing a one-year internship with the Newark, N.J., *Star-Ledger*—which he supplemented with jobs as a bar bouncer. "To me this is regular," he says of the uncertainty he faces. "I haven't known a professional life outside of this."

Economists say that young people who enter the workforce during recessions never entirely recover from the delays they face at the foot of the escalator. Over the course of their working lives, their wages remain lower than those of young people who enter the workforce when times are flush. The social cost for these detours and reversals is high too. Gardner says the economy will be dealing for years with a mismatch between the skills that new graduates have acquired and the work that is available for them. "When this all gets said and done ... we are going to have a pile of highly skilled people underutilized in the economy, frustrated because they don't know how to get out of these positions, and maybe just giving up," he says.

The challenges facing Millennials only rise as their education levels decline. In 2009, 20 percent of 20-to-24-year-olds with only a high school degree were unemployed, according to federal statistics. For those in the age group who did not finish high school, the unemployment rate soared to almost 28 percent. Carlotta Workman feels the weight of those numbers from her position as guidance counselor at the high school in blue-collar Zanesville, Ohio. "Kids are thinking, 'My mom just got laid off, so how am I going to get a job?'" she says.

Somewhat surprisingly, the federal data show that the unemployment rate was no greater last year for graduates of two-year community colleges than for graduates with four-year degrees. But that statistic says nothing about the quality of jobs available to the community college graduates. Roseanne Buckley, director of career services at Burlington County College in New Jersey, finds graduates still working in pizza parlors or holding down the other unskilled jobs that they used to help pay for school in the first place. "Before," she says, "you just expected that once you graduated and got your degree, you'd be able to land a job in that field." Instead, her graduates are often jostling for positions with older workers who have been laid off from more-lucrative positions. "They have no experience," Buckley says of her graduates. "How are they supposed to compete with somebody that's a little bit older, that has a lot of work experience?"

Coping Mechanisms

Social analysts say that as a "civic generation," Millennials are more inclined to light candles than to curse the darkness. They tend to be pragmatic, team-oriented, resilient, and optimistic. In a recent Pew Research Center survey, for instance, only about one-third of Millennials said they were earning enough money now to live the kind of life they wanted. But among the two-thirds who said they weren't earning enough money now, nearly nine in 10 said they expected to earn enough someday. Morley Winograd, co-author of the 2008 book *Millennial Makeover: My Space, YouTube, and the Future of American Politics,* says that although this generation is as likely as its elders to believe that big institutions such as government and business are broken, its reaction to that conclusion is different. "It isn't like the Baby Boomers who said, 'It's broken, let's tear down everything,'" says Winograd, a fellow at the Democratic advocacy group NDN. "In this generation, they say, 'These institutions are broken, we can fix them.'"

It is in that spirit that many Millennials are blazing innovative paths through the Great Recession. Alexander Develov, a senior finishing this spring at business-oriented Babson College in Wellesley, Mass., for instance, has already started a business that provides online video marketing services. "The recession is fantastic for somebody trying to start a business," he says. "You can't raise as much money, but you can get the best people."

Winograd notes that polls have found that Millennials list among their top economic priorities the opportunity to collaborate with peers and the chance to have an impact on the institutions where they work. Such attitudes could lead them to develop new ways of organizing their work lives, especially if their opportunities for stable long-term employment at a single company continue to narrow.

Other Millennials are choosing to ride out the labor market's squalls by pursuing more education. Community college applications have soared 17 percent in the past two years—although the increase is driven in part by older people seeking new skills after layoffs. Demand at community colleges is so high that

Bunker Hill Community College in Boston has been forced to add classes on a graveyard shift that begins at 11 p.m. Many young people with four-year degrees are seeking further education as well, although this reliance on graduate school as a harbor could prove a mixed blessing: While some experts say that acquiring more skills almost always pays off, others worry that students will simply accumulate more debt without obtaining the experience they need to begin moving up the workplace escalator. "Graduate school never solved a labor market problem," says Gardner of Michigan State.

But arguably this generation's most distinctive response to hard times is not entrepreneurship or extended education but rather a pursuit of public service, in all of its variations—a trend that was evident even before the recession struck. Edna Medford, who has spent more than two decades teaching history at Howard, sees the impulse toward service as a defining characteristic of this generation. "In the last couple of years, there is a return to that whole idea of being willing to commit to the larger group," she says. "It's truly refreshing."

Closed doors in the private workforce have only reinforced this inclination. Applications are surging at virtually every form of service-oriented institution. AmeriCorps, the national service program created by President Clinton, received nearly 250,000 applications last year—more than two and a half times the number it received in 2008. And so far this year, applications are running almost 60 percent higher than even that peak. Applications for the Peace Corps jumped almost 20 percent last year, to their highest level in the program's nearly 50-year history. At Teach for America, which recruits young college graduates to teach for two years in high-need urban and rural public schools, applications have nearly doubled since 2008, to more than 46,000. "I think the silver lining of the economic downturn is that it's caused people to re-examine what's truly important to them, and that's led folks to choose careers where they can make a difference," says Kerci Marcello Stroud, the communications director of Teach for America.

A similar wave is washing over the military, which is now able to choose from a growing pool of better-prepared applicants. The share of Army recruits with a high school degree jumped to 95 percent in 2009, from 79 percent as recently as 2007, and the Marines have a backlog of young people trying to join. "Continued high unemployment is one reason for interest in the military," acknowledges Curtis Gilroy, director of accession policy in the Office of the Defense Undersecretary for Personnel and Readiness. "But another is the patriotism exhibited by young people today." Howard University senior Frank Bonner, who enlisted to fly aircraft in the Navy after he graduates this spring, exemplifies these intertwined motivations. "A lot of it comes down to job security," he says, "and dedication to my country."

The Big Fixes

In many respects, the economic challenges facing young people are simply a more acute version of the threats confronting all Americans in this sharp downturn. And, as a result, many of the questions about how to create more opportunity for young Americans inevitably become tangled in the larger debates between the political parties about how to promote more prosperity for society overall—that is, whether the economy is more likely to respond to an agenda focused mostly on tax cuts and deregulation or one centered on government investment in areas such as education and alternative energy.

Yet in some ways the problems confronting the Millennial Generation are unique. Although unemployment has increased among all age groups, younger workers are falling out of the workforce much more rapidly than older workers—and more rapidly than they did in previous recessions. In fact, federal statistics show that the share of young people either working or actively looking for work was declining through the first part of the past decade, even before the recession hit. Such data suggest a growing problem in connecting young people to the world of work—in effect, a breakdown in the process of getting them to step onto the escalator. "They are just simply delaying their careers and their labor force connections," laments Reich, who is now a public policy professor at the University of California (Berkeley). "It's a social problem if young people, fresh out of school ... can't get into the labor force and have to sit around for a year or two or more."

What can be done to ease the pathway into work for more young people? Obviously, nothing will help more than an overall acceleration in job creation. But smaller, targeted steps could make a difference as well. Among them:

Expanding access to community college. These two-year institutions are a powerful means of providing young people with skills that improve their marketability. Last year, President Obama set a goal of graduating an additional 5 million people from community colleges by 2020 and to that end has sought $12 billion in increased funding; Congress recently approved $2 billion in grants as a down payment on that figure. Experts caution, however, that for community colleges to fulfill expectations, they must ensure that their programs are tied to hiring needs in the local workforce—and they must also improve their record on guiding students through to completion. Federal data show that fewer than three in 10 community college students who enrolled in 2004 had received degrees by the summer of 2007. "We believe [community colleges] are one of the smartest investments we can make" says Melody Barnes, the chief White House domestic policy adviser. "But in doing so, we also feel that it is important to make sure that they are matching their programs to the needs of students today."

Helping students manage the burden of debt. With two-thirds of students now graduating with debts, Obama recently won congressional approval for a significant expansion in federal Pell Grants, as well as reforms in student loan programs that will limit repayments to 10 percent of a student's discretionary income after 2014. But policy makers also need to explore more careful ways to break the link between expanded student aid and rising tuition—such as proposals from congressional Republicans to limit federal subsidies for schools that raise costs too rapidly.

Broadening internship opportunities. College placement officials agree that employers increasingly expect graduating students to have acquired experience through internships. But because most internships offer little or no pay, low-income students who must earn money either to help meet their family's bills or fund their educations often can't afford to pursue them, which leaves these students disadvantaged in the job hunt after graduation. In a recent study, the liberal think-tank Demos argued that to even the playing field, Washington should provide grants to subsidize internships for low-income students. House Education and Labor Committee Chairman George Miller, D-Calif., recently proposed $500 million in government-funded apprenticeships that could serve a similar function, although with federal deficits gaping, the idea's prospects are uncertain.

Providing more chances to serve. Michigan State's Gardner says that his strongest advice to young people unable to find work is "to find an issue and get involved in it" through volunteering. Nationally, he says, one of the best ways to respond to youth unemployment would be "to expand opportunities for service." To this end, last year Obama proposed and Congress approved legislation that will more than triple the size of AmeriCorps by 2017. Other pending proposals would enlarge public service alternatives such as the Peace Corps.

Updating the safety net for freelance workers. The American social safety net delivers most of its benefits, such as pensions and health care, through the workplace. But that model may be increasingly obsolete for Millennials, given a job environment where so many move back and forth between contract and conventional employment, notes Sara Horowitz, founder of the Freelancers Union, which advocates for freelance workers. Health care reform moved the nation toward accommodating this new reality by creating exchanges and federal subsidies that will make it easier for the self-employed to obtain coverage. But Washington has only just begun rethinking how to provide security to a more fluid workforce.

Perhaps the one sure thing is that the Millennials themselves, as they assume growing responsibility in public and private institutions, will increasingly shape the debate over how to provide opportunity and security for America's rising generations. Few of them seem daunted by that prospect; indeed, many in this enormous, civically oriented, politically engaged generation appear impatient to place their stamp on society. The most farsighted and ambitious among them see the recession that has raged across the economy as something akin to a forest fire that leaves terrible damage in its wake but cultivates new growth by clearing away obstructions. "Recessions are moments of realignment where everyone realizes the way people have been producing work and making money has broken down somehow and it's time to retool," says Julienne Alexander, another partner in Washington's Steadfast Associates. "We want to be a part of that retooling by giving smart, creative youngsters a place to do both commercial and creative work." With luck, this generation's efforts may someday be seen as seedlings on a charred forest floor—the first step toward regeneration after a season of fearsome loss.

KEY CONCEPTS

labor force participation   millennial generation   recession

DISCUSSION QUESTIONS

1. Based on your reading of this article, do you think the value of a college education has gone down? Is the cost of your higher education going to pay off with better work and better pay?
2. What consequences of the recession have you observed or experienced? How are your plans for the future influenced by the economic situation?

52

Harder Times: Undocumented Workers and the U.S. Informal Economy

RICHARD D. VOGEL

In this article Vogel describes the informal economy that employs people "off the book." He argues that these "underground" workers experience conditions much worse than those employed in the formal economic sector, including low pay and lack of benefits. He also argues that workers within the informal economy experience greater gender, ethnic, and racial discrimination. Undocumented workers also cost economies money in that the lack of taxes from these workers further burdens the public sector. He outlines the case of Los Angeles and also looks at other states' data on undocumented workers. His conclusion is that the persistence of a system that feels the need for undocumented workers further worsens conditions for the working class in America. For both native and immigrant workers in the informal economy, the historical patterns of exploitation continue.

Many of the informal economies operating in the world today are the offspring of globalization and need to be understood as such. The economic and social prospects for people engaged in informal employment—sometimes referred to as "precarious" and "off-the-books employment"—as well as their families and communities, are substantially inferior to those associated with formal employment, and the current boom of informal economic activity bodes ill for all working people.

SOURCE: "Harder Times: Undocumented Workers and the U.S. Informal Economy", by Richard D. Vogel from *Monthly Review* 2006, Volume 58, Issue 03 (July–August). Reprinted by permission.

Referred to variously as "underground," "shadow," "invisible," and "black" (as in "black market") economies, many informal economies have developed around the economic survival activities of workers who have been excluded from the formal economy of their nation or region and exploited by entrepreneurs willing to take advantage of their desperation. Initially considered phenomena of the third world and developing nations, informal economies are now expanding rapidly in the free market nations of the western world, including the United States.

Work in the informal economy contrasts sharply with formal employment: wages and working conditions are substandard; there are no guaranteed minimum or maximum hours of work, paid holidays, or vacations; gender, ethnic, and racial discrimination are uncontrolled and systematically exploited; no mandatory health or safety regulations protect workers from injury or death on the job; and workers are denied the traditional benefits of employment in the formal economy: workman's compensation; health, unemployment and life insurance; and pensions. Needless to say, few workers join the informal labor force voluntarily—the vast majority are recruited primarily through economic desperation unmitigated by even a minimal social safety net.

Worker participation in the informal economy is a complex phenomenon. While immigrant workers play a significant role in the informal U.S. economy, native workers also participate. These workers in the informal sector include widely divergent groups from professionals who do unreported jobs on the side, and craft workers who exchange work in kind, to marginalized native workers who, because of cutbacks in welfare programs, must accept any work they can find.

The informal sector of the U.S. economy is growing as the working population expands and employment opportunities in the formal economy do not keep pace. Not only are workers being displaced as companies move their operations offshore in search of lower labor costs, but an increasing number of U.S. corporate start-ups are overseas ventures. These current trends contribute to the exclusion of both new and veteran workers from the formal economy.

LOS ANGELES: A DEFINITIVE CASE STUDY

Although no comprehensive national studies of the informal U.S. economy have been published to date, a study of work in the informal economy of the City of Los Angeles by the Economic Roundtable, a nonprofit, public policy research organization, offers an in-depth look at a local informal economy. Since the recently reported case study of Los Angeles's informal economy presents the most refined estimates of informal employment and its impact available for any specific area of the country, it deserves special attention.

This [research] reveals a widening gap between worker and employer reported employment signaling an ongoing informalization of the Los Angeles County economy that has continued unabated through boom and bust for the last eighteen years.

The most remarkable feature ... is in the period following the economic recession of 2001. The data indicates that as late as 2004 economic recovery was still out of sight. In spite of that, the informal economy held relatively steady during this period while the formal economy continued in serious decline. The 2005 data, not included in the Economic Roundtable study, indicates a continuation of the same trend. The conclusion of the Economic Roundtable researchers, that the economic stagnation of southern California that was triggered by the recession of 2001 would have been worse without the ameliorating effects of the informal economy, appears to be fully warranted.

In addition to documenting the general trends of informal economic activity, the Economic Roundtable study also offers credible estimates of the actual size of the informal economy in Los Angeles County. These researchers calculated low-range, mid-range, and high-range estimates for the number of county workers in the informal labor force and, after careful consideration, settled on the mid-range estimate of 679,000 workers in 2004. This number represented a substantial 15 percent of Los Angeles County's labor force in that same year.

The Economic Roundtable researchers used this number to determine the cost of informal employment to the public social safety net in Los Angeles County. Determining that the average annual wage for off-the-books jobs was slightly over $12,000 in 2004, they calculated that the informal economy produced an $8.1 billion payroll. This unreported and untaxed payroll shortchanged the public sector in Los Angeles County by the following amounts:

- $1 billion in Social Security taxes (paid by both employers and workers)
- $236 million in Medicare taxes (paid by both employers and workers)
- $96 million in California State Disability Insurance payments (paid by workers)
- $220 million in Unemployment Insurance payments (paid by employers)
- $513 million in Workers Compensation Insurance payments (paid by employers)

These losses added up to over $2 billion in unpaid payroll benefits and insurance that were needed to fund a minimal social safety net for workers. The payroll tax shortfall is continuing, and resistance on the part of California taxpayers to underwrite public relief measures has resulted in the widespread deterioration of social services for informal workers and their families across the state.

Though the Economic Roundtable study is focused primarily on work in the informal economy, researchers point out that informal workers in Los Angeles County spend an estimated $4.1 billion per year that should generate $440 million in sales tax revenue. However these workers purchase many goods and services from informal retailers and service providers who do not collect sales taxes and submit them to the state, further eroding support for the public sector.

The social and political crises of the region are being fueled by the fact that the expanding informal economy in southern California is based on the widespread exploitation of undocumented Mexican and Central American workers. This state of affairs has elevated the issue of illegal immigration in California (and, of course, the nation) to center stage.

THE EXPLOITATION OF UNDOCUMENTED WORKERS

The issue of the exploitation of undocumented workers is fundamental to understanding the informal economy in Los Angeles. While it is true that many native workers and legal immigrants participate in Los Angeles's informal economy, undocumented immigrants represent the majority of the off-the-books workers. At the same time, while many individuals, both natives and immigrants, participate in both economies, these dual roles should not be allowed to obscure the dominant role of undocumented immigrant workers in Los Angeles's informal economy.

The Economic Roundtable study substantiates the dominant role of undocumented workers in Los Angeles County when it addresses the question of how many workers in the informal economy are undocumented immigrants and which industries employ them. Based on U.S. Immigration and Naturalization Service (INS) and Census 2000 data, the Economic Roundtable estimates that undocumented immigrant workers make up 61 percent of the informal labor force in Los Angeles County and 65 percent in the city. Census 2000 data also establishes that while there are sizable numbers of Asian and Middle Eastern immigrants in California, the vast majority of the immigrants residing and working in southern California are from Latin America.

While the proportion of undocumented workers in the informal economy of Los Angeles is truly remarkable, where these workers are employed is also surprising. The picture of the informal labor force … contrasts sharply with popular perceptions of immigrant workers that are based on glimpses of day laborers soliciting on street corners, domestic workers waiting at bus stops, and landscape crews packed into pickup trucks. Derived from INS and Census 2000 data, this [research] indicates that a wide array of industries in Los Angeles County systematically exploit undocumented immigrant labor. The Economic Roundtable study reveals that, in fact, the majority of the informal workers labor regularly and out of public view in factories, mills, restaurants, warehouses, workshops, nursing homes, office buildings, and private homes. In actual numbers, the ERT researchers estimate that these top twenty industries in Los Angeles County alone employ over 180,000 undocumented workers.

Most of these undocumented workers are obviously not casual day laborers. The nature of the majority of the jobs … entails stable relationships between employers and workers and requires extensive community networks to recruit and support those workers.

Although the earnings of immigrant workers in the informal economy are inferior to that of native workers across the board, they do not represent the rock bottom of exploitation—the Economic Roundtable's breakdown of earnings by gender reveals the severity of the super-exploitation of undocumented women workers in the Los Angeles informal economy.

THE SUPER-EXPLOITATION OF UNDOCUMENTED IMMIGRANT WOMEN

Since this super-exploitation of undocumented women workers takes place almost exclusively out of sight, it all but escapes public attention. However, the Economic Roundtable documents significant employment of undocumented women workers in the top twenty industries that employ the most informal workers. Generally, women tend to work in homes, personal service jobs, and light manufacturing, traditionally low-paying jobs, while men hold the majority of the jobs in transportation, construction, and other relatively higher-paying blue-collar jobs.

Economic Roundtable researchers highlight the super-exploitation of undocumented women workers in the Los Angeles informal economy when they compare wages by gender in specific job categories for the year 1999. For example, in the category of services to buildings, where the gender shares of jobs were exactly equal, men averaged $13,308 while women made an average of $6,869 (51 percent of the earnings of men). Even in private households, beauty salons, department stores, and health care services, jobs clearly dominated by women workers, men were paid more. While women held 75 percent of the jobs in these categories, they earned only 66 percent of the wages of men doing the same work. Overall, the average wage for undocumented women workers in the informal economy of Los Angeles County was $7,630 compared to $16,553 for men. That amounts to only 46 percent of undocumented men's annual earnings.

The economic basis of Los Angeles's informal economy can be summed up succinctly—undocumented women workers make less than one-half (46 percent) of what documented men workers make, and the average wage of all undocumented workers is less than one-half (41 percent) that of native workers in the same job categories in the formal economy. Even these glaring inequalities underestimate the super-exploitation of informal laborers because they do not account for the other employment benefits that workers in the formal economy receive, which range from 27 to 30 percent of their total employment compensation.

The Economic Roundtable study indicates that the informal sector of the Los Angeles economy, which depends on the exploitation of undocumented workers, has become an integral part of the area's economy. The implications of this extraordinary case study clearly suggest that the trends of the national informal economy deserve careful reconsideration.

NATIONAL TRENDS

Clearly the informal economies of Los Angeles and Los Angeles County are unique. The proximity of southern California to Mexico and the ready access to Mexican and Central American labor have historically shaped the economy, social structure, and culture of the area. The fact that Los Angeles remains the primary destination of Mexican immigrants, both legal and undocumented, is a matter of public record.

The Immigration and Naturalization Service statistics verify the concentration of undocumented immigrants in California but also reveal that Texas, New York, and Illinois (primarily New York City and Chicago) have considerable concentrations. All of these areas also have sizable established informal economies.

The big surprises in the INS study are the recent trends of worker immigration from Mexico and Central America to other destinations in the United States. This [research] shows that six of the top ten states that experienced the highest growth rate of undocumented immigrants during the last decade of the twentieth century are in the southeastern United States, with three states in mid-America not far behind. The phenomenal increase in immigration to the southeast is to the booming metropolitan areas where new construction and service-oriented businesses have created a huge demand for low-cost labor. During the period 1990–2000, the number of undocumented immigrants doubled from 3.5 to 7 million for the United States overall. Current estimates indicate that those immigration trends have accelerated and the nation's total is now between 11.5 and 12 million (http://www.pewhispanic.org).

The present patterns of illegal undocumented worker immigration in the United States signal a rapid expansion of the national informal economy. The practice of basing the informal economy on the exploitation of undocumented workers, well established in California and Texas, is rapidly spreading across the nation. The Pew Hispanic Center estimates that in March 2005, 7.2 million of the 12 million undocumented immigrants in the United States were working, making up about 4.9 percent of the total work force. Applying the same correction factor to this Pew assessment that the Economic Roundtable researchers applied to employment data in the Current Population Survey for L.A. County, to estimate total employment in the local informal economy, suggests that 8–10 percent of all workers might be employed in the informal economy nationwide.

This emerging informal economy signals perennial hard times for the undocumented workers and a continuing war of attrition against the U.S. working class at large.

HARDER TIMES, U.S. STYLE

Global economic theory discusses informal economies in terms of "restructuring." This neoclassical and neoliberal theory maintains that all the workers of the world are competing with each other in a zero-sum game. Informal economies arise when the workers in one economic region of the world lose out to the workers in another region, and the formal economy of the loser region contracts or collapses. According to this theory, informal economies arise to fill the resulting economic vacuum, and the mechanism of the free market will determine the outcome.

Neoliberal economists argue that displaced immigrant workers, primarily from Mexico and Central America, migrate to the United States and compete with native workers for scarce jobs in a post-industrial economy, as dictated by the "invisible hand" of the "free market." The ideological cover story for public consumption, meanwhile, is that immigrant workers take the jobs that native workers refuse.

But it is *political economics*, not free markets, that shape global, national, and local economies. The movement of industrial and service production offshore to take advantage of cheap labor requires the compliance, or outright cooperation, of both home and foreign governments and is therefore a quintessentially political act. Also political is the widespread subversion of immigration and labor law for the sake of profits. These political machinations—not some mythical "invisible hand"—are the engines of the informal economy in the United States.

The burgeoning informal economy in the United States is introducing new elements into familiar historical patterns of exploitation. While the U.S. economy has traditionally been fueled by immigrant labor, the current dependence on undocumented labor is unprecedented. Although the masses of Western and Eastern Europeans who immigrated to do America's dirty work in the past were rewarded with citizenship for their sacrifices, there is no indication that the current Mexican and Central American workers can expect the same. Naturalization is not part of the deal. Dirty and poorly paid work in the contemporary informal economy is not an initiation into the mainstream U.S. working class. The current wave of undocumented immigrants who work in the informal economy today are more likely than not to stay in the informal U.S. economy—as long as they are needed.

The U.S. economy is indebted to undocumented immigrant workers, but there is no indication that the debt is going to be paid. The proposed "guest worker" programs currently being debated in Congress are nothing more than reruns of the Bracero Program that lasted from 1942 to 1964, and the debts owed to workers from that program are still outstanding. The contemporary strategy is to use Mexican and Central American workers displaced by NAFTA as long as possible.

If, or when, they are no longer needed, they can be repatriated. It happened to their compatriots who were uprooted from their homes and communities and driven across the border at the onset of the Great Depression and again during the paramilitary repatriation campaign designated "Operation Wetback" in the recession that followed the Second World War.

The current boom in the informal economy bodes no better for native workers in both the informal and formal sectors of the U.S. economy. Real wages, benefits, and standards of living continue to decline for all workers, and the labor movement is stalled. Realizing the American Dream through hard work in a promising job is becoming a remote possibility rather than an accessible opportunity. And there's nothing like the naked specter of wage slavery in the informal economy and the ghettoization of the poor to keep the expectations of all workers in check. The revolution of rising expectations that gripped workers and minorities in the 1960s has been overshadowed by the prospect of pauperization.

Until the labor movement develops the solidarity necessary to confront the current assault of capitalism, the expanding informal economies of the world will continue to impoverish the lives of all working people. The call to all workers of the world to unite has never been more urgent than it is today.

KEY CONCEPTS

informal economy labor force undocumented worker

DISCUSSION QUESTIONS

1. Why do companies decide to employ undocumented workers? Outline the ways undocumented workers are exploited.
2. In your own work experience, have you been employed in the informal economy? Did you see this as an advantage? Can you understand how this is not ideal for immigrants and working-class people?

53

Working on People

ROBIN LEIDNER

Robin Leidner examines the relationship between workers and employers in service work where the routinization of work routines shapes how workers do their work—and resist the routinization that now characterizes much service work.

The logic of work routinization is simple, elegant, and compelling. Adam Smith's famous discussion of pin manufacture laid the groundwork: instead of paying high-priced, skilled workers to do a job from start to finish, employers

SOURCE: Fast Food, Fast Talk: Service Work and the Routinization of Everyday Life, by Robin Leidner, © 1994 by the Regents of the University of California. Published by the University of California Press.

could split the job into its constituent parts and assign each task to minimally qualified workers, thus greatly reducing costs and increasing output (Smith 1776). Frederick Taylor's system of scientific management, aimed at removing decision making from all jobs except managers' and engineers', pushed the division of labor still further. Routinizing work processes along these lines offers several benefits to managers. First, designing jobs so that each worker repeatedly performs a limited number of tasks according to instructions provided by management increases efficiency, through both greater speed and lower costs. Second, it results in products of uniform quality. And its final and most enticing promise is that of giving management increased control over the enterprise.

These systems for dividing and routinizing work were conceived for manual labor and were first widely implemented in manufacturing; therefore most sociological work on routinization has been based on studies of this sector (e.g., Blauner 1964; Burawoy 1979; Sabel 1982). As the clerical work force grew in importance, the principles of routinization were applied to what had been considered "brain work," transforming some clerical functions into rote manual tasks and reducing the scope, variety, and opportunity for decision making in many jobs (see Braverman 1974; Garson 1975; Glenn and Feldberg 1979a; Mills 1951). In recent years the service sector has come increasingly to dominate the U.S. economy (Mills 1986; Noyelle 1987; Smith 1984), and the routinization of jobs that require workers to interact directly with customers or clients—what I call "interactive service work"—has been expanding as well, challenging employers to find ways to rationalize ... workers' self-presentation and feelings as well as their behavior. As routinization spills over the boundaries of organizations to include customers as well as employees, employers' strategies for controlling work affect not only workers but the culture at large.

Although many of the rationales, techniques, and outcomes associated with routinizing other kinds of work are applicable to interactive service work, the presence of customers or clients complicates the routinization process considerably. When people are the raw materials of the work process, it is difficult to guarantee the predictability of conditions necessary for routinization. Organizations attempting routinization must try to standardize the behavior of non-employees to some extent in order to make such routinization possible (see Mills 1986). It is part of the labor process of many interactive service workers to induce customers and clients to behave in ways that will not interfere with the smooth operation of work routines.

In addition to the issues of deskilling, autonomy, and power already associated with work routinization, the routinization of interactive service work raises questions about personal identity and authenticity. In this sort of work there are no clear distinctions among the product being sold, the work process, and the worker. Employers of interactive service workers may therefore claim authority over many more aspects of workers' lives than other kinds of employers do, seeking to regulate workers' appearance, moods, demeanors, and attitudes. But even when management makes relatively little effort to reshape the workers, interactive service workers and their customers or clients must negotiate interactions in which elements of manipulation, ritual, and genuine social exchange are subtly mixed.

Routinized interactive service work thus involves working on people in two senses. The workers work on the people who are their raw materials, including themselves, and the organizations work on their employees.... These processes can be problematic, both practically and morally. This [article] explores these practical difficulties and moral and psychological stresses through case studies of two organizations that have taken the routinization of interactive service work to extremes, McDonald's and Combined Insurance.

ROUTINES AND RESISTANCE

The literature on work routinization is centrally concerned with the relative power of employers and workers. Braverman's influential account of work routinization as a process of deskilling (1974) presents the struggle for control in its starkest terms. When workers rather than employers are most knowledgeable about how best to accomplish tasks, management does not have absolute control over the work process: how long tasks take; how many people are required to do a job; how much work can reasonably be completed in one shift. When management determines exactly how every task is to be done, it loses much of its dependence on the cooperation and good faith of workers and can impose its own rules about pace, output, quality, and technique. Moreover, since the removal of skill from jobs makes workers increasingly interchangeable, management's power to dictate wages, hours, and working conditions is expanded.

Among the most serious criticisms of Braverman's account is that it overstates the capacity of management to get its own way (Friedman 1977; Littler 1990; Thompson 1989). Historical and sociological research has documented the capacity of workers acting individually and collectively to resist managerial plans in ways that affect how routinization is implemented or even undermine it altogether (e.g., Edwards 1979; Halle 1984; Montgomery 1979; Paules 1991). The amended picture depicts workplaces in which the interests of workers and employers are, almost invariably, directly at odds, with outcomes determined by the resources each group has at its disposal in the struggle for control.

This revised account is still inadequate, however, for understanding interactive service work. There, the basic dynamic of labor and management struggling for control over the work process is complicated by the direct involvement of service-recipients: customers or clients, respondents or patients, prospects or passengers.... All three parties are trying to arrange the interactions to their own advantage, whether they want to maximize speed, convenience, pleasantness, efficiency, customization of service, degree of exertion, or any other outcome they feel to be beneficial. And in many instances the interests of workers are at odds with those of management. Telephone companies want each operator to handle as many calls as possible, while operators prefer a less harried pace; employers insist that the customer is always right, while workers who deal with the public want to defend their own dignity when faced with customers who behave outrageously. In other circumstances, however, workers' and management's common goals, especially their mutual interest in exerting control over service-recipients, are at least as salient as their differences.... For example, commission salespeople are as eager to maximize sales as their bosses are, and nurses and other medical workers want patients to cooperate with hospital regimens. Whether workers ally themselves primarily with service-recipients or with management in a given situation depends on the type of service being rendered, the design of the service system, and the preferences of those involved.

This variability suggests that we need a more nuanced understanding of the meanings of routines for workers than the sociological literature provides. There is a high degree of consensus among analysts on how routinization affects workers: it oppresses them. It robs their work of interest and variety, it eliminates the opportunity for workers to use and develop their capacities by solving problems and making decisions, and it prevents them from deriving self-worth, meaning, or deep satisfaction from work by detaching it from workers' wills (see, e.g., Braverman 1974; Garson 1975, 1988). Routinizing interactive service work seems to extend this oppression to deeper levels. Hochschild (1983) has shown in the case of flight attendants how regulating what she calls "emotion work" legitimizes employers' intervention in the very thought processes and emotional reactions of workers, alienating workers from their feelings, their faces, and their moods....

Rather than assuming that workers who do not resist routines are either miserable or duped, it would seem more fruitful to consider whether there are circumstances in which routines, even imposed routines, can be useful to workers. In the case of interactive service workers, this question is closely tied to how the interests of the three parties to the service interactions—workers, management, service-recipients—are aligned. Workers have good reason to embrace any imposed routines they see as expanding their ability to control service interactions. The insurance agents I studied, for example, believed that their routines gave them power in sales encounters and helped maximize their sales commissions. When workers do not see the routines as enhancing control over service-recipients, they are more likely to resist standardization, as did many of the fast-food workers I studied. Some, however, accepted and even welcomed tight scripting, for, in addition to sparing them some kinds of effort and clarifying the standards for good work, the routines could act as shields against the insults and indignities these workers were asked to accept from the public.

Although the presence of service-recipients can provide a motivation for workers to cooperate with routinization, managers of interactive service workers are not necessarily more successful than other types of managers in seeing that the work is done according to their designs. Even when workers do their best to implement routines as management would like, there is no guarantee that service-recipients will do the same. A major feature distinguishing routinized interactive service work from other kinds of routinized work is the requirement that the behavior of nonemployees be standardized to some extent. In some circumstances service-recipients do their best to fit smoothly into routines, perhaps because they believe that doing so is the surest way for them to get the service they want, perhaps because they feel they have little choice. Some service-recipients, however, resist routines.

Resistance is especially likely when the service-recipients have not chosen to become involved in the interaction (as in many types of sales), but it is also common when the service-recipients do not believe that compliance will get them the outcome they want, either because they think the routine is badly designed or ill-suited to the circumstances, or because they want customized service. Even cooperative service-recipients can inadvertently upset routines if they are unable to understand how the routine is supposed to work, if they are unable to play their part correctly, or if the routine is ill-suited to their situation. The workers of both McDonald's and Combined Insurance dealt with some service-recipients who were cooperative and some who were not, but their typical experiences differed sharply. The counter workers dealt exclusively with people who had already decided to do business with McDonald's, most of whom were very familiar with the service routine and wanted to play their parts smoothly. The insurance agents called on "prospects," many of whom would have preferred to be left alone and some of whom made clear that they did not want the service interaction to proceed at all.

In any setting, whether service-recipients are more eager than workers to see that the routine is followed correctly or are motivated to keep the workers from carrying out the routine at all, their behavior is never entirely predictable. Organizations have a variety of means of standardizing service-recipients' actions. Cues in the setting, printed directions, and taken-for-granted norms may provide sufficient order, since tailoring behavior to meet the needs of anonymous, bureaucratic organizations is a common enough experience for most adults in the United States. When the task is more difficult or the service-recipients more recalcitrant, it is likely to be up to interactive service workers to guide or pressure them to behave in organizationally convenient ways....

Interactive service workers are located on the boundaries of organizations, where they must mediate between the organization and outsiders. They have to try, simultaneously, to meet their employer's requirements, to satisfy the demands of service-recipients, and to minimize their own discomfort (see Glenn and Feldberg 1979b; Lipsky 1980; Prottas 1979). Corporate spokespeople, employee trainers, and other professional cheerleaders frequently make the case that the interests of the three groups are harmonious, since all benefit from efficient and amicable service interactions. In fact, though, there is a great deal of variation in how closely the preferences of the three parties are aligned. The degree of congruity of interests determines much of the character of an interactive service job: its degree of difficulty, the nature of jobholders' incentives and disincentives, and the quality of relations between workers and service-recipients.

The rules set down by management often require that interactive service workers displease service-recipients, either because the routine does not allow workers the flexibility to meet customers' requests, because the service-recipient does not want to participate in the interaction at all, or because the logic of the routine runs counter to interactive norms. Service-recipients can become frustrated by their inability to get what they want, and they may perceive the workers as pushy, unresponsive, stupid, or robot-like. In any case, even if the dissatisfaction felt by the service-recipient is attributable to the rules set down by the employer, it is likely to be experienced by the service-recipient as the fault of the worker, and it is the worker who has to deal with the problem. Interactive service workers thus serve as buffers, absorbing the hostilities consumers feel when organizational routines do not meet their needs or expectations (Glenn and Feldberg 1979b: 12–13).

In interactive service work, as in other kinds of work, understanding employers' designs for routinization is only the first step in understanding how the routines function in practice. We must also examine how employers try to persuade or coerce the relevant actors (in this case, service-recipients as well as workers) to cooperate and how the behavior of these actors, and other contingencies, alter the routines in action. In interactive service work, the patterns of shared and opposed preferences among the three parties to the interactions are crucial determinants of the outcomes, including the degree to which workers and service-recipients resist organizational attempts to standardize their behavior....

When the service-recipient has any doubt about the spontaneity or uniqueness of the interaction, choosing a response can be tricky. If one assumes sincerity and responds accordingly, there is the danger of being a trusting fool, allowing oneself to be manipulated. Suspicion and aloofness, on the other hand, cut one off from potentially gratifying human contact. Responding in kind to surface cues despite recognition that the situation is contrived, an interim strategy, is problematic as well. It signals acceptance of the terms of the interaction although one might in fact resent them.

And what about the actors themselves? What does participation in routinized interactions do to their sense of self? Employers who manage the self-presentations and emotions of employees are aware that they must pay attention to the relation between the workers' self-conceptions and their behavior on the job. While airlines try to get their flight attendants to throw themselves fully into the cheerful, solicitous, unresentful persona assigned to them (Hochschild 1983), the employers of the phone-sex operators are apparently conscious that role distance (Goffman 1961b) can provide protection from any demeaning implications of fulfilling others' fantasies. "Remember," they counsel in the "Professionalism" section of the employee manual, "you are not your character on the phone" (*Harper's*, December 1990: 27).

Sociologists need not believe in the existence of unmediated emotional responses or of an inviolate individual with a stable core of self in order to recognize that participation in routinized interactions presents dilemmas of identity. When such participation is a condition of employment, as it is for many interactive service workers, these dilemmas are everyday concerns.

First, there is the problem of managing the disjunction between actions and feelings. Hochschild (1983: 7) argues that workers whose emotions are managed by their employers become alienated from their feelings in a process parallel to that described by Marx of the alienation of proletarians from the actions of their bodies and the products of their labor. These workers thus have difficulty experiencing themselves as authentic even off the job, for they lose track of which feelings are their own. As Wharton and Erickson (1990: 21) argue, "An essential facet of experiencing one's self as real is the congruence between action and self-feeling regarding that action." How able are workers to maintain a sense of self when their actions do not reflect their feelings, or when their feelings are manipulated to produce the desired effect?

Second, there is the question of the content of the routinized behavior, the ways the worker is required to act. Some workers' routines call upon them to take on characteristics they may value, such as patience, good cheer, or self-confidence, but in other cases the required behavior is more problematic. Because we read character from actions, workers who must behave in ways they ordinarily would not somehow have to reconcile the contradictions between their self-conception and their behavior. Can they manipulate people without thinking of themselves as underhanded and disrespectful? Can they stifle their ordinary responses without thinking of themselves as doormats? The problem of detaching oneself from the implications of work behavior is not unique to interactive service workers whose jobs have been routinized. All interactive service workers—salespeople, clergy, teachers, doctors, waitresses, and so on—face similar tensions, whether or not their jobs are scripted and standardized. But routinization calls attention to the inauthenticity, thus heightening the discordance between workers' self-identities and the identities they enact at work.

Problems of identity are raised by how interactive service workers are treated as well as by how they behave, and routinization can shape the interactive style of service-recipients. Sociologists do not generally hold that identity can be created and maintained autonomously. As Cooley's idea of the "looking glass self" (1902) and Mead's elaboration of the sources of self in interaction (1934) convey, people's self-conceptions are shaped by the treatment they are accorded. A sense of self is formed in childhood, but throughout life it remains vulnerable to messages from others. If car salespeople are treated with suspicion (Lawson 1991), phone solicitors with rudeness, restaurant servers with condescension or familiarity, domestic servants as non-people (Rollins 1985), how will they react? Certainly they need not accept the implied judgment of the service-recipients. But they must construct some means of defending themselves from demeaning treatment or some account that allows them to accept such treatment without thinking of themselves as demeaned. Hochschild (1983) stresses the difficulties of protecting oneself from the implications of treatment on the job, but Hughes (1984 [1970]) and Rollins (1985) show how "the humble" can salvage a sense of control and self-esteem or even construe themselves as superior to "the proud." My account acknowledges both workers' agency in interpreting their own behavior and the treatment they receive from others, and the constraints that the conditions of employment set on the possible range of workers' interpretations and responses.

Routines may actually offer interactive service workers some protection from assaults on their selves. Where workers are often exposed to rude, insulting, or depersonalizing treatment from service-recipients, routines can make that treatment easier to bear. Workers who follow scripts need not feel that they are being personally attacked when they are subjected to slammed phones, nasty comments, or tirades about poor service. Employers often provide training in how to think about and respond to poor treatment. Philadelphia's parking authority workers, for example, receive instruction on how to react to being spit on by people angry at receiving a ticket (Matza 1990: 19).

Whether or not workers appreciate all or some aspects of routinization, they have available a range of possible responses to it. They may reshape their self-concept to fit what they see as a positive new identity (see Biggart 1989 on direct sales workers); hold on to their own identity but set it aside at work, where they assume a new "situational identit[y]" (Van Maanen and Kunda 1989: 68); try to alter or vary the routine to reflect their own will and personality; distance themselves from the work role by willing "emotional numbness" (Van Maanen and Kunda 1989: 69); or interpret their own behavior and the treatment they receive in ways that are positive or at least neutral. Which of these responses are used depends not only on the workers' personal characteristics but also on the nature of the service interactions designed by their employers. Almost inevitably, however, their jobs require them to take an instrumental approach to their own identity and to relations with others.

THE RESEARCH

I did not choose to study McDonald's and Combined Insurance because I thought they were typical interactive service organizations. Rather, both companies took routinization to an extreme, and I was interested in examining how far working on people could be pushed. I believed that extreme cases would best illuminate the kinds of tensions and problems that routinizing human interactions creates, and that a study of long-established and successful companies would provide clues to the attractions of routinization and the means by which managers can overcome the difficulties of standardizing workers and service-recipients.

Although these two companies both pushed routinization almost to its logical limits, their public-contact employees did very different kinds of work and had different kinds of relations with service-recipients. The companies therefore adopted dissimilar approaches to routinization, with distinctive ramifications for workers and service-recipients. McDonald's took a classic approach to routinization, making virtually all decisions about how work would be conducted in advance and imposing them on workers. Since Combined's sales agents worked on their own and faced a broader range of responses than McDonald's workers did, their company had to allow them some decision-making scope. It did create routines that were as detailed as possible, but it pushed the limits of routinization in another direction, extending it to many aspects of workers' selves both on and off the job. It aimed to make a permanent, overall change in the ways its agents thought about themselves and went about their lives.

At each company, I collected information through interviewing and participant-observation. I examined routinization at two levels. First, I learned as much as I could about the company's goals and strategies for routinization by attending corporate training programs and interviewing executives. Next,

I explored how the routines worked out in practice by doing or observing the work and by interviewing interactive service workers.

At both McDonald's and Combined Insurance I received official permission to do my research. The managers I interviewed, the trainees I accompanied through classes, and the workers I interviewed, observed, and worked with were all aware that I was conducting a study and that my project had been approved by higher levels of management.[1] The customers I served and observed were the only participants who did not know that I was collecting data.

I carried out my research at McDonald's in the spring through fall of 1986. I began by attending management training classes at corporate headquarters. Next, managers at the regional office arranged my placement at a franchise that was apparently chosen largely on the basis of its convenience for me. The manager of the franchise arranged for me to be trained to serve customers; once trained, I worked without pay for half a dozen shifts, or a total of about twenty-eight hours of work. I then conducted structured interviews with workers who served customers, obtaining twenty-three complete interviews, as well as several that were incomplete for a variety of reasons. I also spent long hours hanging around in the crew room, where I talked informally with workers, including those not in customer-service jobs, and listened as workers talked with each other about their job experiences and their reactions to those experiences. I supplemented the participant-observation and worker interviews with interviews or informal talks with most of the franchise managers and its owner, with management trainers, with a former McDonald's executive who had been in charge of employee research, and with numerous former employees and current customers of McDonald's.

Most of my research at Combined Insurance was conducted in the winter and spring of 1987. I received permission to study the work of its life-insurance agents and was assigned a place in a training class. When the two-week training period ended I was put in touch with a regional sales manager who arranged for me to work with a sales team.... Because one must be licensed to sell insurance, I did not actually do the agent's job, but I spent a week and a half in the field with the sales team, going out on sales calls with each of its members in turn. Since these daily rounds included a lot of time spent driving from one prospective customer to another, I had ample opportunity for informal interviews with the agents. I also conducted a formal interview with the sales manager, and I attended an evening team training meeting and two morning team meetings, one of which was attended by the district manager. In addition, I interviewed several Combined Insurance executives, including a vice president in charge of introducing a new sales presentation, a regional sales manager, the director of market research and one of his assistants, and a manager responsible for developing sales routines and materials.

In the summer of 1989, I conducted follow-up interviews at Combined Insurance to learn about major changes the company had made in its life insurance products and sales approach. I reinterviewed the director of market research and the man who had been the regional sales manager but who now was responsible for overseeing one-half of the company's life-insurance agents. I also interviewed two regional sales managers, one the man who now managed the region where the sales team I had studied was located, and one located in a part of the country where the new sales system had been introduced more recently. The follow-up research allowed me to learn about the subsequent careers of the agents I had known and, more important, to find out how the agents' jobs and the company's results had changed under the new sales system.

I supplemented my research at these two companies by gathering information on other kinds of interactive service work through interviews, site visits, and examination of employee scripts. I visited a Burger King training facility and interviewed a training manager there; I interviewed an executive from Hyatt Hotels and attended an orientation session for Hyatt workers; and I interviewed people who worked, or had worked, as an AT&T customer service representative, a waitress, a door-to-door canvasser, and a psychologist.

OVERVIEW

One obstacle [to routinizing interactive service work] arises from the difficulty of defining quality in the provision of services. Routinization is intended to ensure consistency in the quality of work, but, as Garson (1988: 60) asks, "Is a more uniform conversation a positive good?" Can consistency be equated with high quality? Since workers cannot be separated from products in interactive service work, the question next arises: how much intervention in workers' fives is justifiable in order to ensure a uniform appearance, attitude, and work process?

A separate problem is that the raw materials of interactive service work, people, are never entirely predictable, and routinization makes sense only if stable conditions of work can be guaranteed. Given the intrinsic fluidity, negotiability, and indexicality of interaction (see Prus 1989a), how can prepared scripts replace flexible individual response? A service interaction must work on two levels: as a work process with a specific goal, and as a human interaction. The danger in routinizing human interactions is that if workers do not, or are not allowed to, respond flexibly, form may come to supersede meaning, thus increasing the likelihood of failure on both levels. In other words, rigid routines strictly enforced can actually prevent workers from doing an adequate job, harming both customer satisfaction and employee morale (Koepp 1987b; Normann 1984).

Employers have responded to these dilemmas in several ways.... First, they employ a variety of strategies intended to reduce the unpredictability of work specifications by standardizing the behavior of service-recipients. Second, they try to overcome resistance to mass-produced service by finding ways to personalize routines, or to seem to personalize them. Finally, in circumstances where too much unpredictability remains to make it possible to dispense with worker flexibility, employers may use a strategy I call "routinization by transformation," which is intended to change workers into the kinds of people who will make decisions that management would approve....

The processes of routinization at McDonald's and Combined Insurance were similar in many ways, despite the dissimilarity of their products and organizations.

Both companies were faced with the two basic challenges of routinizing interactive service work: to standardize the behavior of employees, and to control the behavior of customers. Both companies paid close attention to how their workers looked, spoke, and felt, rather than limiting standardization to the performance of physical tasks. Emotion work was an integral part of serving McDonald's food and selling Combined's insurance, and smiling was a job requirement shared by the food-service workers and the insurance agents. Both groups were expected to behave deferentially and to try to please their customers.

Nevertheless, the jobs were dissimilar enough to allow a comparison of differing approaches to routinization practicable in different circumstances. Some of the major distinctions between the two cases were the extent of efforts to affect workers' personalities, the gender of the workers, and the degree of supervision. Moreover, the companies' routines varied greatly in scope and content, since interactions at McDonald's are briefer and less dependent on workers' initiative than insurance sales calls. Many of the differences between the two jobs stemmed from differences in the predictability of service-recipients' behavior and demands. This variable influenced the complexity of the work routines and the nature of the relations between workers and service-recipients. Where the behavior of service-recipients is relatively unpredictable, interactive service workers may be responsible for getting them to cooperate, as was the case at Combined Insurance.

Mills regards customers as subordinates of service providers (1986: 121), but the extent to which they were experienced as such varied greatly in these cases. Both the food-service workers and the insurance agents had to sell and to serve, but their jobs differed in the emphasis placed on each. Service-recipients therefore played very different roles in the interactions. At McDonald's the work was done for them, while at Combined Insurance the work seemed to be done to them. Variation in the nature of the worker–customer relation helps explain several important differences between the two cases, including differences in the types of attitudes workers were expected to cultivate, the range of personal qualities subject to standardization, the scope of decision making left in the job, and workers' reactions to routinization.

... [T]he two cases also differed markedly in how the interests of managers, workers, and service-recipients were aligned. These variations were determined in part by managers' decisions about how to set up routines, how to compensate workers, and how to balance customer satisfaction with efficiency and profitability. Both companies tried to ensure that the workers would take the customers' interests as their own, while defining the customers' interests in ways that were organizationally convenient, but the degree to which the three parties actually shared interests varied. At McDonald's, customers and managers generally had similar preferences about how the interactions were to proceed, which limited the workers' range of action. Combined's agents and management, by contrast, shared an interest in controlling prospective customers in order to maximize sales. For this reason, Combined's agents' routines enhanced their power over customers, while McDonald's workers' limited theirs. In each setting, however, there were many specific situations in which the interests of the three parties parted from these general patterns, and these shifting stakes helped determine the outcomes of the carefully planned routines imposed by management.

... [T]he relevant contrasts are not only those between these two jobs, but also those between routinized interactive service work and other kinds of routinized work, and those between routinized service interactions and nonroutinized interactions.... I discuss the dilemmas raised for workers' sense of themselves and the range of workers' responses to routinization. These responses varied both between the two companies and, especially at McDonald's, within the company. In each setting, some workers resisted in various ways the identities imposed on them, and others embraced the routines and willingly took on the persona prepared for them by the company.

The degree to which workers can benefit from embracing routines depends largely on the kinds of relations among workers, managers, and service-recipients that routines set up. How particular individuals weigh the benefits and drawbacks of the routines and how they respond to them is more complicated. The meanings workers assign to their own behavior at work and to that of the managers and service-recipients they work with are to some degree fluid and indeterminate. Workers interpret actions, their own and others', and the interpretations they construct, often with the guidance and influence of their employers and their peers, can heighten or mitigate dissatisfaction, allow for some satisfactions and close off others, and lead workers to evade or to embrace their imposed routines.

To underline the importance of interpretation in shaping workers' responses to their jobs, the construction of gender meanings in the two jobs [are analyzed]. I show that when jobs are segregated by gender, workers can interpret similar job characteristics as suitable either for men or for women, as they try to make sense of the existing pattern of employment or to bolster their sense of entitlement to the job. Despite the scripting of their words and actions, both they and the public can interpret workers' behavior as expressing preexisting gender differences that account for the gendered division of labor.

[This research] draws out the implications of routinizing service interactions, arguing that employers who standardize these labor processes exert a cultural influence that extends beyond the workplace. Their organizational control strategies reach deeply into the lives of workers, encouraging them to take an instrumental stance toward their own personalities and toward other people. Moreover, service routines subject people who are not employees of service organizations to organizational efforts to standardize their behavior. Through participation in encounters that violate the norms of social interaction even as they exploit those norms, service-recipients learn not to take for granted that the ground rules that govern interaction will be respected. As they adjust their expectations and behavior accordingly, the ceremonial forms that bolster individual identity (Goffman 1956) and social cohesion (Durkheim 1965 [1915]) tend to be treated as indulgences rather than obligations.

My analysis of interactive service work addresses several shortcomings of the literature on work and routinization. Sociologists of work have not kept pace with the rapid shift of jobs out of manufacturing. Despite some excellent studies of service jobs (e.g., Biggart 1989; Hochschild 1983), generalizations about work under capitalism have not been recast to encompass interactive service work. The enormous increase in the proportion of the labor force engaged in service occupations is widely acknowledged, but models derived from industrial work continue to guide the thinking and research of theorists of the labor process. In these models, workers and capitalists (or managers) are the relevant parties, the creation and sale of a product is the focus of analysis, and workers' attitudes and emotions are relevant only insofar as they affect willingness to accept the conditions of work. When these models are applied to interactive service work, we find that some of the basic analytic distinctions fit awkwardly at best and that important areas of inquiry are overlooked. The role of service-recipients in the labor process is not analogous to that of either workers or managers, for example, and the manipulation of participants' selves cannot be satisfactorily analyzed as a form of deskilling.[2]

My approach to analyzing routinization takes from the labor-process tradition its emphases on concrete work practices and on issues of control and autonomy. Early work in that tradition has been criticized for overstating employers' success in imposing work routines and for assuming that these routines necessarily serve the goals management intended (see Littler 1990). Such oversimplifications are untenable when applied to interactive service work, where not only workers but service-recipients vie with management in a three-way contest for control and satisfaction. My analysis highlights the various means by which each party tries to exert control and the shifting patterns of alliance among them, allowing for a relatively fine-grained consideration of outcomes. In sharp contrast to the traditional picture of routinization, my analysis shows that, in some situations, service routines provide workers and service-recipients with benefits that help account for their frequent acquiescence in managerial designs.

A major shortcoming of labor-process theory is its unsatisfactory treatment of workers' subjectivity. Braverman's work (1974) excluded this topic altogether, and subsequent theorists have too often limited it to class-consciousness and consent or resistance to oppressive structures of work (Burawoy 1979; Thompson 1989).[3] As Hochschild (1983) showed, questions of subjectivity are not separable from the analysis of actual work practices in interactive service work, since workers' identities are actively managed by employers. Hochschild emphasized the distress felt by workers subjected to organizational exploitation of their feelings and personalities, but I find that not all workers resist the extension of standardization to their inner selves. Rather, many attempt to construct interpretations of their work roles that do not damage their conceptions of themselves (cf. Hughes 1984 [1951]; 1984 [1970]).

I do not take the position that the routinization of service work and the standardization of human personality that it can entail are in fact benign, benefiting workers, customers, and companies in a happy congruence of interests. Often, these manipulations are invasive, infantilizing, exasperating, demeaning. But if we are to understand how routinization works in service jobs and what outcomes it produces, we need an account of the relations among routinization, skill, control, interaction, and self that is both more precise and more subtle than we have yet been given. These outcomes of routinization are felt throughout society as participants adjust their assumptions about the norms guiding social interaction. The study of interactive service work thus provides a bridge between the sociology of work and broader examinations of culture and society.

HOW CAN WORK ON PEOPLE BE ROUTINIZED?

Employers routinize work both to assure a uniform outcome and to make the organization less dependent on the skills of individual workers. When satisfactory outcomes can be obtained by unskilled or semi-skilled workers, the organization benefits through lower labor costs and increased managerial control (Braverman 1974). The related goals of establishing quality control and lowering the skill levels necessary to complete tasks are achieved by determining in advance exactly how each task is to be done and then instituting a system of control that persuades or coerces workers to follow the predetermined routines.

Routinization appeals to employers of interactive service workers for the same reasons that it attracts employers of manufacturing and clerical workers. The routinization of interactive service work is not a new phenomenon—the work of telephone operators, for example, was routinized by the first years of the twentieth century (Norwood 1990: 33)—but it is becoming increasingly widespread. With the rapid expansion of the service sector in the United States in the postwar period (see Albrecht and Zemke 1985; Mills 1986; Smith 1984), the cost and quality of interactive work have assumed greater significance in many organizations, making it more attractive to try to apply the principles of routinization developed for use in manufacturing work to human interactions.

In interactive service work, however, routinization cannot so readily accomplish its intended goals. When routinization is applied to human interactions, difficulties arise in creating products of uniform quality and in decreasing organizational dependence on workers' skills.... [I]n jobs where the outcome or product of the work is not easily separable from the quality of the interaction, quality control is not a matter of standardizing products but of standardizing the workers themselves. This involves extending organizational control to aspects of the workers' selves that are usually considered matters of personal choice or judgment. Furthermore, the goal of uniformity is itself problematic in interactive work. Since good service is often equated with "personal" service, standardization may undercut quality in human interactions.

Employers become less dependent on workers' skills when they are able to prespecify exactly how the work is to be done. In interactive service work, the presence of customers or clients who are integrally involved in the work process introduces a degree of unpredictability that makes routinization difficult. Standardizing the work may be impossible unless the behavior of these nonemployees can be standardized as well.

For routinization to succeed, then, interactive service organizations must work on people—both their workers and their customers or clients—to bring their attitudes and behavior into line with organizational needs. This ... [research] elaborates the problems inherent in such a project and describes some of the strategies interactive service organizations use to make routinization work.

QUALITY CONTROL IN INTERACTIVE SERVICE WORK

In manufacturing work, quality control means assuring that all of the products created meet a minimum standard. The work process and the personal characteristics of the workers are clearly distinguishable from the products they make, and quality is a property of the products. Routinizing manufacturing work involves standardizing the workers' movements so that they will most efficiently produce items that meet the organization's standards.

Almost all interactive service jobs have both an interactive and a noninteractive component. For example, fast-food servers greet customers, take orders, and collect payment (interactive work), and place orders, gather the requested items, and put them in a bag or on a tray (noninteractive work). Even work that is almost entirely interactive, such as providing psychotherapy, involves offstage noninteractive tasks such as completing forms. The process of routinizing the noninteractive components of the work differs little from the process of routinizing manufacturing or clerical jobs. Frederick Taylor's principles can be applied to assembling hamburgers as easily as to other kinds of tasks, and the quality of the products can be evaluated apart from the qualities of the workers who created them.

For the interactive parts of the work, routinization and quality control are more complicated, because the distinctions among the product or outcome of the work, the work process, and the worker may be blurred or nonexistent. In many kinds of interactive work, employers cannot introduce specifications for the products or instructions for the work processes without trying to standardize the workers themselves. If the successful completion of the work depends on how workers look, feel, and speak as well as on what they do, then employers' intervention in these matters can be justified as a legitimate interest in quality control. Chase (1978: 140–41), for example, writing in the *Harvard Business Review*, encouraged managers to think of workers' personal attributes in that light:

> Managers have long recognized the desirability of having "attractive" personnel greet the public in such job classifications as receptionist, restaurant hostess, and stewardess.... Any interaction with the customer makes the direct worker in fact part of the product, and therefore his attitude can affect the customer's point of view of the service provided.

There are three types of interactive service work in which the success of the work depends on the quality of the interaction, and therefore on the personal attributes of the worker. In one type, the interaction is inseparable from the product being sold or delivered—for instance, in psychotherapy, prostitution, or teaching. In the second type, a product exists apart from the interaction, but a particular type of experience is an important part of the service. For example, patrons of Playboy Clubs expected titillation and deference as well as food and drink (see Miller 1984; Steinem 1983), and airline passengers, who buy tickets primarily to get from one place to another, are promised friendly service on their journey (Hochschild 1983). Finally, in some jobs the interaction is a crucial part of the work process even though it is not part of a product being sold or provided. The success of salespeople, fund-raisers, bill collectors, and survey interviewers depends on the workers' ability to construct particular kinds of interactions.

In all of these kinds of jobs, workers' looks, attitudes, demeanors, speech, and ways of thinking can be integral to the work process and outcome. The jobs all require what Hochschild (1983) calls "emotion work," that is, the work of creating a particular emotional state in others, often by manipulating one's own feelings. Such work may be necessary because it is part of the service being sold—for example, a flight attendant is not doing her job properly if she is not smiling—or because the worker's task cannot go forward without it—a family-planning counselor may not get the information she needs unless she is perceived as nonjudgmental (Joffe 1986). To routinize the interactive parts of service jobs, including emotion work, organizations can use uniforms or other controls on personal appearance, scripts, rules about demeanor and about proper ways to treat customers or clients, and even far-reaching attempts at psychological reorientation.

Some employers routinize service delivery through rigid scripting, but leave the management of the emotional texture of the interactions to their workers.... Others go far beyond scripting, shaping workers' personalities, attitudes, and styles. Three factors determine whether employers choose to routinize emotion work. First, they must believe that the quality of the interaction is important to the success of the enterprise. If workers can accomplish their tasks acceptably regardless of the quality of the interactions (as can cab drivers, unlike telephone solicitors) or if the organization is assured of a steady demand for its services regardless of the quality of the interactions (as are government agencies, unlike restaurants), employers are not likely to be much concerned with establishing a consistent quality of service interaction.

Second, employers will routinize service interactions only if they think workers will be unable or unwilling to conduct the interactions appropriately on their own. If employers do not believe emotion work requires any special skills, presumably they will not routinize that part of the work. If they believe that particular kinds of personal attributes are necessary to do the work well—good looks, empathy, assertiveness—they can try to hire people who have those qualities, perhaps using such characteristics as gender and social class to judge whether proper attitudes and skills can be taken for granted. If the interactive task is more difficult or if workers with the desired attributes are unavailable or too expensive, employers may choose routinization.

Finally, the nature of the service dictates whether routinization in fact can ensure quality control. When the tasks are too complex or context-dependent to be standardized, organizations must depend on the skills of more expensive workers....

NOTES

1. Pseudonyms are used throughout the text, except for top-level officials of McDonald's and Combined Insurance.
2. Although sociologists have been slow to give full weight to the distinctive aspects of labor processes in service jobs, the management literature on service work has boomed. Some early work on managing services stressed the possibilities for extending industrial methods of production and assessment to service provision (e.g., Levitt 1972). Soon, however, writers advising managers came to focus on the peculiar attributes of service work, examining such questions as how to define quality in a service, how to manage customers' expectations, and how to inspire front-line workers (Albrecht 1988; Cziepel, Solomon, and Suprenant 1985; Heskett, Sasser, and Hart 1990; Lovelock 1988; Mills 1986; Normann 1984). This trend suggests that controlling service work poses problems that methods applicable to other kinds of work do not address. However, the management literature, focusing on technique, treats a relatively narrow range of issues.
3. Knights (1990) and Willmott (1990) take on the challenge of integrating subjectivity into labor-process theory.

REFERENCES

Albrecht, Karl and Ron Zemke. 1985. *Service America! Doing Business in the New Economy*. Homewood, Ill.: Dow Jones-Irwin.

Biggart, Nicole Woolsey. 1983 "Rationality, Meaning, and Self-Management: Success Manuals, 1950–1980." *Social Problems* 30: 298–311.

Biggart, Nicole Woolsey. 1989. *Charismatic Capitalism: Direct Selling Organizations in America*. Chicago: University of Chicago Press.

Blauner, Robert. 1964. *Alienation and Freedom*. Chicago: University of Chicago Press.

Braverman, Harry. 1974. *Labor and Monopoly Capital: The Degradation of Work in the Twentieth Century*. New York: Monthly Review Press.

Burawoy, Michael. 1979. *Manufacturing Consent: Changes in the Labor Process Under Capitalism*. Chicago: University of Chicago Press.

Chase, Richard B. 1978 "Where Does the Customer Fit in a Service Operation?" *Harvard Business Review* 56 (Nov.–Dec.): 137–42.

Cooley, Charles Horton. 1902. *Human Nature and the Social Order*. New York: Charles Scribner's Sons.

Durkheim, Emile. 1965 [1915]. *The Elementary Forms of the Religious Life*. Trans. Joseph Ward Swain. New York: Free Press.

Edwards, Richard. 1979. *Contested Terrain: The Transformation of the Workplace in the Twentieth Century*. New York: Basic Books.

Friedman, Andrew. 1977. *Industry and Labour: Class Struggle at Work and Monopoly Capitalism*. London: Macmillan.

Garson, Barbara. 1975. *All the Livelong Day: The Meaning and Demeaning of Routine Work*. New York: Doubleday.

Garson, Barbara. 1988. *The Electronic Sweatshop: How Computers Are Transforming the Office of the Future into the Factory of the Past*. New York: Simon and Schuster.

Glenn, Evelyn Nakano and Roslyn L. Feldberg. 1979a. "Proletarianizing Clerical Work: Technology and Control." Pp. 51–72 in Andrew Zimbalist, ed., *Case Studies on the Labor Process*. New York: Monthly Review Press.

Glenn, Evelyn Nakano and Roslyn L. Feldberg. 1979b. "Women as Mediators in the Labor Process." Paper presented at the annual meeting of the American Sociological Association, Boston, Mass.

Goffman, Erving. 1956. "The Nature of Deference and Demeanor." *American Anthropologist* 58: 473–502.

Goffman, Erving. 1961b. "Role Distance." Pp. 83–152 in *Encounters: Two Studies in the Sociology of Interaction*. Indianapolis, Ind.: Bobbs-Merrill.

Halle, David. 1984. *America's Working Man: Work, Home, and Politics Among Blue Collar Property Owners*. Chicago: University of Chicago Press.

Hochschild, Arlie Russell. 1983. *The Managed Heart: Commercialization of Human Feeling*. Berkeley: University of California Press.

Hughes, Everett C. 1984 [1970]. "The Humble and the Proud: The Comparative Study of Occupations." Pp. 417–27 in *The Sociological Eye*. New Brunswick, N.J.: Transaction Books.

Joffe, Carole. 1986. *The Regulation of Sexuality: Experiences of Family Planning Workers*. Philadelphia: Temple University Press.

Knights, David and Hugh Willmott, eds. 1990. *Labour Process Theory*. London: Macmillan.

Koepp, Stephen. 1987a. "Big Mac Strikes Back." *Time* (April 13): 58–60.

Koepp, Stephen. 1987b. "Pul-eeze! Will Somebody Help Me?" *Time* (February 2) 48–55.

Lawson, Helene. 1991. "Learning to Suspect and Betray: The Transformation of Car Sales Women." Paper presented at the annual meeting of the Midwest Sociological Society, Des Moines, Iowa.

Lipsky, Michael. 1980. *Street Level Bureaucracies: Dilemmas of the Individual in Public Service*. New York: Russell Sage.

Littler, Craig R. 1990. "The Labour Process Debate: A Theoretical Overview, 1974–1988." Pp. 46–94 in David Knights and Hugh Willmott, eds., *Labour Process Theory*. London: Macmillan.

Littler, Craig R. and Graham Salaman. 1982. "Bravermania and Beyond: Recent Theories of the Labour Process." *Sociology* 16: 251–69.

Matza, Michael. 1990. "Write 'em Up, Move 'em Out." *Philadelphia Inquirer Magazine* (May 13): 14–20.

Mead, George Herbert. 1934. *Mind, Self, and Society*, ed. Charles W. Morris. Chicago: University of Chicago Press.

Miller, Russell. 1984. *Bunny: The Real Story of Playboy*. London: Michael Joseph.

Mills, C. Wright. 1951. *White Collar: The American Middle Class*. New York: Oxford University Press.

Mills, Peter K. 1986. *Managing Service Industries: Organizational Practices in a Post-industrial Economy*. Cambridge, Mass.: Ballinger.

Montgomery, David. 1979. *Workers' Control in America: Studies in the History of Work, Technology, and Labor Struggles*. Cambridge: Cambridge University Press.

Normann, Richard. 1984. *Service Management: Strategy and Leadership in Service Businesses*. Chichester: John Wiley and Sons.

Norwood, Stephen H. 1990. *Labor's Flaming Youth: Telephone Operators and Worker Militancy*. Urbana, Ill.: University of Illinois Press.

Noyelle, Thierry J. 1987. *Beyond Industrial Dualism: Market and Job Segmentation in the New Economy*. Boulder, Colo.: Westview Press.

Paules, Greta Foff. 1991. *Dishing It Out: Power and Resistance Among Waitresses in a New Jersey Restaurant*. Philadelphia: Temple University Press.

Prottas, Jeffrey. 1979. *People Processing*. Lexington, Mass.: Lexington Books.

Prus, Robert. 1989a. *Making Sales: Influence as Interpersonal Accomplishment*. Newbury Park, Calif.: Sage Publications.

Rollins, Judith. 1985. *Between Women: Domestics and Their Employers*. Philadelphia: Temple University Press.

Roman, Murray. 1979. *Telephone Marketing Techniques*. New York: AMACOM (American Management Association).

Sabel, Charles F. 1982. *Work and Politics: The Division of Labor in Industry*. Cambridge: Cambridge University Press.

Smith, Adam. 1982 [1776]. *The Wealth of Nations*. New York: Penguin.

Smith, Joan. 1984. "The Paradox of Women's Poverty: Wage-earning Women and Economic Transformation." *Signs* 10: 291–310.

Steinem, Gloria. 1983. "I Was a Playboy Bunny." Pp. 29–69 in Gloria Steinem, *Outrageous Acts and Everyday Rebellions*. New York: Holt, Rinehart and Winston.

Thompson, Paul. 1989. *The Nature of Work: An Introduction to Debates on the Labour Process*, second edition. London: Macmillan.

Wharton, Amy S. and Rebecca J. Erickson. 1990. "The Emotional Division of Labor: Women's Work and Family Life." Paper presented at the annual meeting of the American Sociological Association, San Francisco, Calif.

KEY CONCEPTS

alienation routinization service sector work

DISCUSSION QUESTIONS

1. Explain what Leidner means by "routinization," and give an example.
2. Identify one type of service work (perhaps a job you have held). Does routinization characterize how this work is done? What impact does this have on workers?

Applying Sociological Knowledge: An Exercise for Students

Look online for job advertisements or collect local newspapers that include job postings. How many of the jobs would you be qualified for? What types of jobs are posted online and in newspapers? Consider the implications of finding work for people of color, those without education beyond high school, and immigrants.

PART XVII

Government and Politics

54

The Power Elite

C. WRIGHT MILLS

C. Wright Mills's classic book, The Power Elite, *first published in 1956, remains an important analysis of the system of power in the United States. He argues that national power is located in three particular institutions: the economy, politics, and the military. An important point in his article is that the power of elites is derived from their institutional location, not their individual attributes.*

The powers of ordinary men are circumscribed by the everyday worlds in which they live, yet even in these rounds of job, family, and neighborhood they often seem driven by forces they can neither understand nor govern. "Great changes" are beyond their control, but affect their conduct and outlook nonetheless. The very framework of modern society confines them to projects not their own, but from every side, such changes now press upon the men and women of the mass society, who accordingly feel that they are without purpose in an epoch in which they are without power.

But not all men are in this sense ordinary. As the means of information and of power are centralized, some men come to occupy positions in American society from which they can look down upon, so to speak, and by their decisions mightily affect, the everyday worlds of ordinary men and women. They are not made by their jobs; they set up and break down jobs for thousands of others; they are not confined by simple family responsibilities; they can escape. They may live in many hotels and houses, but they are bound by no one community. They need not merely "meet the demands of the day and hour"; in some part, they create these demands, and cause others to meet them. Whether or not they profess their power, their technical and political experience of it far transcends that of the underlying population. What Jacob Burckhardt said of "great men," most Americans might well say of their elite: "They are all that we are not."

The power elite is composed of men whose positions enable them to transcend the ordinary environments of ordinary men and women; they are in positions to make decisions having major consequences. Whether they do or do not make such decisions is less important than the fact that they do occupy such pivotal positions: their failure to act, their failure to make decisions, is itself an

SOURCE: THE POWER ELITE by Wright Mills (1956) 2511w from pp. 3–29, 269–297.

act that is often of greater consequence than the decisions they do make. For they are in command of the major hierarchies and organizations of modern society. They rule the big corporations. They run the machinery of the state and claim its prerogatives. They direct the military establishment. They occupy the strategic command posts of the social structure, in which are now centered the effective means of the power and the wealth and the celebrity which they enjoy.

The power elite are not solitary rulers. Advisers and consultants, spokesmen and opinion-makers are often the captains of their higher thought and decision. Immediately below the elite are the professional politicians of the middle levels of power, in the Congress and in the pressure groups, as well as among the new and old upper classes of town and city and region. Mingling with them, in curious ways which we shall explore, are those professional celebrities who live by being continually displayed but are never, so long as they remain celebrities, displayed enough. If such celebrities are not at the head of any dominating hierarchy, they do often have the power to distract the attention of the public or afford sensations to the masses, or, more directly, to gain the ear of those who do occupy positions of direct power. More or less unattached, as critics of morality and technicians of power, as spokesmen of God and creators of mass sensibility, such celebrities and consultants are part of the immediate scene in which the drama of the elite is enacted. But that drama itself is centered in the command posts of the major institutional hierarchies.

The truth about the nature and the power of the elite is not some secret which men of affairs know but will not tell. Such men hold quite various theories about their own roles in the sequence of event and decision. Often they are uncertain about their roles, and even more often they allow their fears and their hopes to affect their assessment of their own power. No matter how great their actual power, they tend to be less acutely aware of it than of the resistances of others to its use. Moreover, most American men of affairs have learned well the rhetoric of public relations, in some cases even to the point of using it when they are alone, and thus coming to believe it. The personal awareness of the actors is only one of the several sources one must examine in order to understand the higher circles. Yet many who believe that there is no elite, or at any rate none of any consequence, rest their argument upon what men of affairs believe about themselves, or at least assert in public.

There is, however, another view: those who feel, even if vaguely, that a compact and powerful elite of great importance does now prevail in America often base that feeling upon the historical trend of our time. They have felt, for example, the domination of the military event, and from this they infer that generals and admirals, as well as other men of decision influenced by them, must be enormously powerful. They hear that the Congress has again abdicated to a handful of men decisions clearly related to the issue of war or peace. They know that the bomb was dropped over Japan in the name of the United States of America, although they were at no time consulted about the matter. They feel that they live in a time of big decisions; they know that they are not making any.

Accordingly, as they consider the present as history, they infer that at its center, making decisions or failing to make them, there must be an elite of power.

On the one hand, those who share this feeling about big historical events assume that there is an elite and that its power is great. On the other hand, those who listen carefully to the reports of men apparently involved in the great decisions often do not believe that there is an elite whose powers are of decisive consequence.

Both views must be taken into account, but neither is adequate. The way to understand the power of the American elite lies neither solely in recognizing the historic scale of events nor in accepting the personal awareness reported by men of apparent decision. Behind such men and behind the events of history, linking the two, are the major institutions of modern society. These hierarchies of state and corporation and army constitute the means of power; as such they are now of a consequence not before equaled in human history—and at their summits, there are now those command posts of modern society which offer us the sociological key to an understanding of the role of the higher circles in America.

Within American society, major national power now resides in the economic, the political, and the military domains. Other institutions seem off to the side of modern history, and, on occasion, duly subordinated to these. No family is as directly powerful in national affairs as any major corporation; no church is as directly powerful in the external biographies of young men in America today as the military establishment; no college is as powerful in the shaping of momentous events as the National Security Council. Religious, educational, and family institutions are not autonomous centers of national power; on the contrary, these decentralized areas are increasingly shaped by the big three, in which developments of decisive and immediate consequence now occur.

Families and churches and schools adapt to modern life; governments and armies and corporations shape it; and, as they do so, they turn these lesser institutions into means for their ends. Religious institutions provide chaplains to the armed forces where they are used as a means of increasing the effectiveness of its morale to kill. Schools select and train men for their jobs in corporations and their specialized tasks in the armed forces. The extended family has, of course, long been broken up by the industrial revolution, and now the son and the father are removed from the family, by compulsion if need be, whenever the army of the state sends out the call. And the symbols of all these lesser institutions are used to legitimate the power and the decisions of the big three.

The life-fate of the modern individual depends not only upon the family into which he was born or which he enters by marriage, but increasingly upon the corporation in which he spends the most alert hours of his best years; not only upon the school where he is educated as a child and adolescent, but also upon the state which touches him throughout his life; not only upon the church in which on occasion he hears the word of God, but also upon the army in which he is disciplined.

If the centralized state could not rely upon the inculcation of nationalist loyalties in public and private schools, its leaders would promptly seek to modify the decentralized educational system. If the bankruptcy rate among the top five hundred corporations were as high as the general divorce rate among the thirty-seven million married couples, there would be economic catastrophe on an international scale. If members of armies gave to them no more of their lives than do believers to the churches to which they belong, there would be a military crisis.

Within each of the big three, the typical institutional unit has become enlarged, has become administrative, and, in the power of its decisions, has become centralized. Behind these developments there is a fabulous technology, for as institutions, they have incorporated this technology and guide it, even as it shapes and paces their developments.

The economy—once a great scatter of small productive units in autonomous balance—has become dominated by two or three hundred giant corporations, administratively and politically interrelated, which together hold the keys to economic decisions.

The political order, once a decentralized set of several dozen states with a weak spinal cord, has become a centralized, executive establishment which has taken up into itself many powers previously scattered, and now enters into each and every cranny of the social structure.

The military order, once a slim establishment in a context of distrust fed by state militia, has become the largest and most expensive feature of government, and, although well versed in smiling public relations, now has all the grim and clumsy efficiency of a sprawling bureaucratic domain.

In each of these institutional areas, the means of power at the disposal of decision makers have increased enormously; their central executive powers have been enhanced; within each of them modern administrative routines have been elaborated and tightened up.

As each of these domains becomes enlarged and centralized, the consequences of its activities become greater, and its traffic with the others increases. The decisions of a handful of corporations bear upon military and political as well as upon economic developments around the world. The decisions of the military establishment rest upon and grievously affect political life as well as the very level of economic activity. The decisions made within the political domain determine economic activities and military programs. There is no longer, on the one hand, an economy, and, on the other hand, a political order containing a military establishment unimportant to politics and to moneymaking. There is a political economy linked, in a thousand ways, with military institutions and decisions. On each side of the world-split running through central Europe and around the Asiatic rimlands, there is an ever-increasing interlocking of economic, military, and political structures. If there is government intervention in the corporate economy, so is there corporate intervention in the governmental process. In the structural sense, this triangle of power is the source of the *interlocking directorate* that is most important for the historical structure of the present....

At the pinnacle of each of the three enlarged and centralized domains, there have arisen those higher circles which make up the economic, the political, and the military elites. At the top of the economy, among the corporate rich, there are the chief executives; at the top of the political order, the members of the political directorate; at the top of the military establishment, the elite of soldier-statesmen clustered in and around the Joint Chiefs of Staff and the upper echelon. As each of these domains has coincided with the others, as decisions tend to become total in their consequence, the leading men in each of the three domains of power—the warlords, the corporation chieftains, the political directorate—tend to come together, to form the power elite of America....

By the powerful we mean, of course, those who are able to realize their will, even if others resist it. No one, accordingly, can be truly powerful unless he has access to the command of major institutions, for it is over these institutional means of power that the truly powerful are, in the first instance, powerful. Higher politicians and key officials of government command such institutional power; so do admirals and generals, and so do the major owners and executives of the larger corporations. Not all power, it is true, is anchored in and exercised by means of such institutions, but only within and through them can power be more or less continuous and important.

Wealth also is acquired and held in and through institutions. The pyramid of wealth cannot be understood merely in terms of the very rich; for the great inheriting families, as we shall see, are now supplemented by the corporate institutions of modern society: every one of the very rich families has been and is closely connected—always legally and frequently managerially as well—with one of the multi-million dollar corporations.

The modern corporation is the prime source of wealth, but, in latter-day capitalism, the political apparatus also opens and closes many avenues to wealth. The amount as well as the source of income, the power over consumer's goods as well as over productive capital, are determined by position within the political economy. If our interest in the very rich goes beyond their lavish or their miserly consumption, we must examine their relations to modern forms of corporate property as well as to the state; for such relations now determine the chances of men to secure big property and to receive high income....

If we took the one hundred most powerful men in America, the one hundred wealthiest, and the one hundred most celebrated away from the institutional positions they now occupy, away from their resources of men and women and money, away from the media of mass communication that are now focused upon them—then they would be powerless and poor and uncelebrated. For power is not of a man. Wealth does not center in the person of the wealthy. Celebrity is not inherent in any personality. To be celebrated, to be wealthy, to have power requires access to major institutions, for the institutional positions men occupy determine in large part their chances to have and to hold these valued experiences....

KEY CONCEPTS

interlocking directorate

power

power elite model

DISCUSSION QUESTIONS

1. What evidence do you see of the presence of the power elite in today's economic, political, and military institutions? Suppose that Mills were writing his book today; what might he change about his essay?
2. Mills argues that the power elite use institutions such as religion, education, and the family as the means to their ends. Find an example of this from the daily news, and explain how Mills would see this institution as being shaped by the power elite.

55

The Rise of the New Global Elite

CHRYSTIA FREELAND

Chrystia Freeland sees the new global economy as creating a plutocracy, wherein a small number of extremely wealthy individuals hold increasing world power. This represents a shift, she thinks, in traditional models of political power and authority.

If you happened to be watching NBC on the first Sunday morning in August last summer, you would have seen something curious. There, on the set of *Meet the Press*, the host, David Gregory, was interviewing a guest who made a forceful case that the U.S. economy had become "very distorted." In the wake of the recession, this guest explained, high-income individuals, large banks, and major corporations had experienced a "significant recovery"; the rest of the economy, by contrast—including small businesses and "a very significant amount of the labor force"—was stuck and still struggling. What we were seeing, he argued, was not a single economy at all, but rather "fundamentally two separate types of economy," increasingly distinct and divergent.

This diagnosis, though alarming, was hardly unique: drawing attention to the divide between the wealthy and everyone else has long been standard fare on the left.... What made the argument striking in this instance was that it was being offered by none other than the former five-term Federal Reserve Chairman Alan Greenspan: iconic libertarian, preeminent defender of the free market, and (at least until recently) the nation's foremost devotee of Ayn Rand. When the high priest of capitalism himself is declaring the growth in economic inequality a national crisis, something has gone very, very wrong.

This widening gap between the rich and non-rich has been evident for years. In a 2005 report to investors, for instance, three analysts at Citigroup advised that "the World is dividing into two blocs—the Plutonomy and the rest":

> In a plutonomy there is no such animal as "the U.S. consumer" or "the UK consumer," or indeed the "Russian consumer." There are rich consumers, few in number, but disproportionate in the gigantic slice of income and consumption they take. There are the rest, the "non-rich," the multitudinous many, but only accounting for surprisingly small bites of the national pie.

Before the recession, it was relatively easy to ignore this concentration of wealth among an elite few. The wondrous inventions of the modern economy—Google, Amazon, the iPhone—broadly improved the lives of middle-class consumers, even as they made a tiny subset of entrepreneurs hugely wealthy. And the less-wondrous inventions—particularly the explosion of subprime credit—helped mask the rise of income inequality for many of those whose earnings were stagnant.

But the financial crisis and its long, dismal aftermath have changed all that. A multibillion-dollar bailout and Wall Street's swift, subsequent reinstatement of gargantuan bonuses have inspired a narrative of parasitic bankers and other elites rigging the game for their own benefit. And this, in turn, has led to wider—and not unreasonable—fears that we are living in not merely a plutonomy, but a plutocracy, in which the rich display outsize political influence, narrowly self-interested motives, and a casual indifference to anyone outside their own rarefied economic bubble....

... [T]he rich of today are also different from the rich of yesterday. Our light-speed, globally connected economy has led to the rise of a new super-elite that consists, to a notable degree, of first- and second-generation wealth. Its members are hardworking, highly educated, jet-setting meritocrats who feel they are the deserving winners of a tough, worldwide economic competition—and many of them, as a result, have an ambivalent attitude toward those of us who didn't succeed so spectacularly. Perhaps most noteworthy, they are becoming a transglobal community of peers who have more in common with one another than with their countrymen back home. Whether they maintain primary residences in New York or Hong Kong, Moscow or Mumbai, today's super-rich are increasingly a nation unto themselves.

THE WINNER-TAKE-MOST ECONOMY

The rise of the new plutocracy is inextricably connected to two phenomena: the revolution in information technology and the liberalization of global trade. Individual nations have offered their own contributions to income inequality—financial deregulation and upper-bracket tax cuts in the United States; insider privatization in Russia; rent-seeking in regulated industries in India and Mexico. But the shared narrative is that, thanks to globalization and technological innovation, people, money, and ideas travel more freely today than ever before....

From a global perspective, the impact of these developments has been overwhelmingly positive, particularly in the poorer parts of the world. Take India and China, for example: between 1820 and 1950, nearly a century and a half, per capita income in those two countries was basically flat. Between 1950 and 1973, it increased by 68 percent. Then, between 1973 and 2002, it grew by 245 percent, and continues to grow strongly despite the global financial crisis.

But within nations, the fruits of this global transformation have been shared unevenly. Though China's middle class has grown exponentially and tens of millions have been lifted out of poverty, the super-elite in Shanghai and other east-coast cities have steadily pulled away. Income inequality has also increased in developing markets such as India and Russia, and across much of the industrialized West, from the relatively laissez-faire United States to the comfy social democracies of Canada and Scandinavia. Thomas Friedman is right that in many ways the world has become flatter; but in others it has grown spikier.

One reason for the spikes is that the global market and its associated technologies have enabled the creation of a class of international business megastars. As companies become bigger, the global environment more competitive, and the rate of disruptive technological innovation ever faster, the value to shareholders of attracting the best possible CEO increases correspondingly. Executive pay has skyrocketed for many reasons—including the prevalence of overly cozy boards and changing cultural norms about pay—but increasing scale, competition, and innovation have all played major roles.

Many corporations have profited from this economic upheaval. Expanded global access to labor (skilled and unskilled alike), customers, and capital has lowered traditional barriers to entry and increased the value of an ahead-of-the-curve insight or innovation. Facebook, whose founder, Mark Zuckerberg, dropped out of college ..., is already challenging Google, itself hardly an old-school corporation. But the biggest winners have been individuals, not institutions....

Meanwhile, the vast majority of U.S. workers, however devoted and skilled at their jobs, have missed out on the windfalls of this winner-take-most economy—or worse, found their savings, employers, or professions ravaged by the same forces that have enriched the plutocratic elite. The result of these divergent trends is a jaw-dropping surge in U.S. income inequality. According to the economists Emmanuel Saez of Berkeley and Thomas Piketty of the Paris School of Economics, between 2002 and 2007, 65 percent of all income growth in the United States went to the top 1 percent of the population. The financial crisis interrupted this trend temporarily, as incomes for the top 1 percent fell more than those of the rest of the population in 2008. But recent evidence suggests that, in the wake of the crisis, incomes at the summit are rebounding more quickly than those below. One example: after a down year in 2008, the top 25 hedge-fund managers were paid, on average, more than $1 billion each in 2009, quickly eclipsing the record they had set in pre-recession 2007.

PLUTOCRACY NOW

... As with the aristocracies of bygone days, ... vast wealth has created a gulf between the plutocrats and other people, one reinforced by their withdrawal into gated estates, exclusive academies, and private planes. We are mesmerized by such extravagances as Microsoft co-founder Paul Allen's 414-foot yacht, the *Octopus,* which is home to two helicopters, a submarine, and a swimming pool.

But while their excesses seem familiar, even archaic, today's plutocrats represent a new phenomenon. The wealthy of F. Scott Fitzgerald's era were shaped, he wrote, by the fact that they had been "born rich." They knew what it was to "possess and enjoy early."

That's not the case for much of today's super-elite. "Fat cats who owe it to their grandfathers are not getting all of the gains," Peter Lindert told me. "A lot of it is going to innovators this time around. There is more meritocracy in Bill Gates being at the top than the Duke of Bedford." Even Emmanuel Saez, who is deeply worried about the social and political consequences of rising income inequality, concurs that a defining quality of the current crop of plutocrats is that they are the "working rich." He has found that in 1916, the richest 1 percent of Americans received only one-fifth of their income from paid work; in 2004, that figure had risen threefold, to 60 percent....

... [M]any American plutocrats suggest ... that the trials faced by the working and middle classes are generally their own fault. When I asked one of Wall Street's most successful investment-bank CEOs if he felt guilty for his firm's role in creating the financial crisis, he told me with evident sincerity that he did not. The real culprit, he explained, was his feckless cousin, who owned three cars and a home he could not afford. One of America's top hedge-fund managers made a near-identical case to me—though this time the offenders were his in-laws and their subprime mortgage. And a private-equity baron who divides his time between New York and Palm Beach pinned blame for the collapse on a favorite golf caddy in Arizona, who had bought three condos as investment properties at the height of the bubble.

It is this not-our-fault mentality that accounts for the plutocrats' profound sense of victimization in the Obama era. You might expect that American elites—and particularly those in the financial sector—would be feeling pretty good, and more than a little grateful, right now. Thanks to a $700 billion TARP bailout and hundreds of billions of dollars lent nearly free of charge by the Federal Reserve (a policy Soros himself told me was a "hidden gift" to the banks), Wall Street has surged back to pre-crisis levels of compensation even as Main Street continues to struggle. Yet many of America's financial giants consider themselves under siege from the Obama administration—in some cases almost literally....

You might say that the American plutocracy is experiencing its John Galt moment. Libertarians (and run-of-the-mill high-school nerds) will recall that Galt is the plutocratic hero of Ayn Rand's 1957 novel, *Atlas Shrugged.* Tired of being dragged down by the parasitic, envious, and less talented lower classes, Galt and his fellow capitalists revolted, retreating to "Galt's Gulch," a refuge in the Rocky Mountains. There, they passed their days in secluded natural splendor, while the rest of the world, bereft of their genius and hard work, collapsed....

This plutocratic fantasy is, of course, just that: no matter how smart and innovative and industrious the super-elite may be, they can't exist without the wider community. Even setting aside the financial bailouts recently supplied by the governments of the world, the rich need the rest of us as workers, clients, and consumers. Yet, as a metaphor, Galt's Gulch has an ominous ring at a time when the business elite view themselves increasingly as a global community, distinguished by their unique talents and above such parochial concerns as national identity, or devoting "their" taxes to paying down "our" budget deficit. They may not be isolating themselves geographically, as Rand fantasized. But they appear to be isolating themselves ideologically, which in the end may be of greater consequence....

... America really does need many of its plutocrats. We benefit from the goods they produce and the jobs they create. And even if a growing portion of those jobs are overseas, it is better to be the home of these innovators—native and immigrant alike—than not. In today's hypercompetitive global environment, we need a creative, dynamic super-elite more than ever.

There is also the simple fact that someone will have to pay for the improved public education and social safety net the American middle class will need in order to navigate the wrenching transformations of the global economy. (That's not to mention the small matter of the budget deficit.) Inevitably, a lot of that money will have to come from the wealthy—after all, as the bank robbers say, that's where the money is.

It is not much of a surprise that the plutocrats themselves oppose such analysis and consider themselves singled out, unfairly maligned, or even punished for their success. Self-interest, after all, is the mother of rationalization, and—as we have seen—many of the plutocracy's rationalizations have more than a bit of truth to them: as a class, they are generally more hardworking and meritocratic than their forebears; their philanthropic efforts are innovative and important; and the recent losses of the American middle class have in many cases entailed gains for the rest of the world.

But if the plutocrats' opposition to increases in their taxes and tighter regulation of their economic activities is understandable, it is also a mistake. The real threat facing the super-elite, at home and abroad, isn't modestly higher taxes, but rather the possibility that inchoate public rage could cohere into a more concrete populist agenda—that, for instance, middle-class Americans could conclude that the world economy isn't working for them and decide that protectionism or truly punitive taxation is preferable to incremental measures such as the eventual repeal of the upper-bracket Bush tax cuts....

The lesson of history is that, in the long run, super-elites have two ways to survive: by suppressing dissent or by sharing their wealth. It is obvious which of these would be the better outcome for America, and the world. Let us hope the plutocrats aren't already too isolated to recognize this. Because, in the end, there can never be a place like Galt's Gulch....

KEY CONCEPTS

global elite philanthropy plutocracy

DISCUSSION QUESTIONS

1. What evidence does Freeland provide for the development of a new plutocracy?
2. Study the national and international news for several days in a row and list every news item that you think illustrates Freeland's idea that a new plutocracy is shaping global politics. How does the emergence of this global class help explain some of what you observe in contemporary politics?

56

Cultures of the Tea Party

ANDREW J. PERRIN, STEVEN J. TEPPER, NEAL CAREN, AND SALLY MORRIS

This team of sociologists has conducted a telephone survey of those who are affiliated with the Tea Party. They discuss the cultural beliefs held by Tea Party supporters, thus helping to explain the rise of this populist movement.

The Tea Party Movement (TPM) was *the* story of the 2010 midterm elections.

On February 19, 2009, CNBC commentator Rick Santelli complained from the floor of the Chicago Mercantile Exchange about the proposed mortgage relief program: "How many people want to pay for your neighbor's mortgages that has [sic] an extra bathroom and can't pay their bills? Raise their hand!" He called for "another Tea Party" to protest the policies. In the wake of the 2008 election, in the midst of the economic meltdown and the health care reform debate, local protests loosely organized under the banner of the Tea Party

SOURCE: Cultures of the Tea Party, Andrew J. Perrin, Steven J. Tepper, Neal Caren, and Sally Morris, from *Contexts* 2011, 10: 74 by SAGE Publications. Reprinted by permission of SAGE Publications.

began to emerge nationwide. In December 2009, 41 percent of Americans said they viewed the movement positively.

It's just that no one can really pin down what the TPM is. We know it fuses populist anger with limited-government politics, conservative social concerns, support for free markets, and a nostalgic loyalty to a vision of Revolutionary America. These various political strands are interwoven with powerful cultural motifs drawn from history and a theatrical use of images, language, and stories.

We set out to determine if general public support for the movement represents a new political and cultural phenomenon, or if it's simply realignment within the Republican Party. We needed to know the political and cultural dispositions of Tea Party supporters.

Based upon two telephone polls of registered voters in North Carolina and Tennessee and a set of interviews and observations at a TPM rally in Washington, North Carolina, we determined that there are four primary cultural dispositions among people in those states who feel positively toward the TPM: authoritarianism, ontological insecurity, libertarianism, and nativism.

First, TPM supporters (not necessarily members) tend to hold more authoritarian views than others. For example, 81 percent of those who approve of the TPM agree that it's more important that a child obey his parents than be responsible for his own actions. Only 65 percent of non-TPM supporters agree. Second, they experience "ontological insecurity," or fear of change. In our poll, 51 percent of people who are very concerned about "changes taking place in American society these days" were TPM supporters, compared to just 21 percent of those who were only somewhat or not at all concerned.

Tea Party supporters were more likely to be libertarian, believing that there shouldn't be regulations on expressions such as clothing, television shows, or musical lyrics. In our poll, 24 percent of TPM supporters believe fewer rules should regulate what can be posted on the Internet and who can read it; in contrast, 16 percent of non-TPM supporters hold this view. Finally, supporters are more likely to share nativist sentiments: 18 percent feel very negatively toward immigrants compared to 12 percent of non-TPM supporters.

The TPM is best understood as a new cultural expression of the late-20th century Republican Party, reinforcing preexisting strands of nativism and ontological insecurity, while highlighting libertarian and authoritarian strands.

The TPM's activation of cultural imagery, metaphor, and history gives this movement its powerful symbolic resonance, animates its activists, and dominates media coverage of the movement. Culture gives people the tools to understand and relate new events to old ones; it gives the past meaning for the present. Culture also provides collective accounts of individual experiences, frustrations, and aspirations.

The TPM's cultural work begins with the name itself—a nostalgic connection to the American Revolution's protest against "taxation without representation"—and extends to the recurring cultural theme of a return to the ideals of the Constitution. According to one Tea Party volunteer, "We don't want the big government that's taking over everything we worked so hard for … we want to take back what our Constitution said. You read the Constitution. Those values—that's what we stand for." Such statements are rooted in ontological insecurity and consistent with expressions of earlier right wing movements based in status defense and organized around traditionalism and desire for a simpler, purer past.

In our follow-up poll, 84 percent of those who felt positively toward the TPM said the Constitution should be interpreted "as the Founders intended," compared to only 34 percent of other respondents. But this support is not absolute. The Tea Partiers were twice as likely to favor a constitutional amendment banning flag burning; many also support efforts to overturn citizenship as defined by the Fourteenth Amendment. That they simultaneously want to honor the founders' Constitution and alter that same document highlights the political flexibility of the cultural symbols they draw on.

The TPM supporters' inconsistent views suggest they are animated more by a network of Constitutional cultural associations than a commitment to the original text. In fact, such inconsistencies around policy, whether on the right or left, underscore what many sociologists see as the growing importance of culture in political life. The Constitution—and Tea Party more generally—take on heightened symbolic value, coming to represent a "way of life" or a "world view" rather than a specific set of laws or policy positions.

The memory of the Boston Tea Party itself, emphasized in press coverage of the movement, is heavily linked to the libertarian cultural disposition that favors limited government. The implied claim is that the tax revolt of the Revolution matches today's call for lower taxes. One Tea Partier stated, "You have a government which can always tax and tax and tax ... and this is just one concept that the Tea Party recognizes as a problem: the greed of centralized government ... which takes away the responsibility of individuals to do anything."

Like the Constitution, then, the Boston Tea Party is a cultural image only loosely connected to the historical record. While the original revolt was about taxation *without representation*, current complaints are directed at a democratically-elected leaders. Interestingly, the deployment of Tea Party imagery in politics has historically been flexible; in the 1970s, activists drew upon the revolt of 1773 to argue not for lower taxes but for a more progressive tax system. These are slippery symbols.

While the libertarian strand is strong, the wistful nostalgia of TPM supporters is primarily composed of their blend of ontological insecurity and nativist dispositions. One rally organizer summed up the sentiment at the rally: "And why do you think they [illegal immigrants] wanted to come? They wanted to come because it [the U.S.] is successful. But after they get here, what do they want to do? They want to change it. And see, you can't change it and be successful. It just won't work..." This mix of nativism and fear of change is also expressed by the distance TPM supporters feel from President Obama. While 41 percent of TPM supporters felt that former president Clinton was "not at all" like them, 81 percent feel that way about President Obama.

It is a truism of American politics that the first midterm election in a presidential administration results in losses for the President's party. Recognizing that the Tea Party Movement is one such backlash is key to understanding what's new here (and what's not). The coalition of views that makes up the TPM is largely the same set of dispositions that makes up the Republican Party. What's new about the TPM is its syncretic cultural work melding 21st century discontent to the symbolic memory of 18th century America. One TPM supporter says it all: "We want to save America. We want to see this country go back to the original success formula that made us what we are today."

KEY CONCEPTS

ontology

populism

DISCUSSION QUESTIONS

1. What four cultural dispositions do the authors identify among people who feel positively toward the Tea Party?
2. What social conditions do you think have led to the upsurge in populist anger?

Applying Sociological Knowledge: An Exercise for Students

Go to the official websites for the Democratic National Committee (www.democrats.org) and the Republican National Committee (www.gop.com). Look through the statements from each of the major political parties and identify key differences. What about similarities? Mills's work focuses on the economy and the military as two key issues; how does each political party address these two topics?

PART XVIII

Health Care

57

The Social Meanings of Illness

ROSE WEITZ

In this article, Rose Weitz outlines the sociological study of health and illness, summarizes the medical model of illness, and compares the two models. She explains that sociologists largely see illness as socially constructed and that race, class, and gender inequalities play a part in how we define health and sickness.

All Marco Oriti has ever wanted, ever imagined, is to be taller. At his fifth birthday party at a McDonald's in Los Angeles, he became sullen and withdrawn because he had not suddenly grown as big as his friends who were already five: in his simple child's calculus, age equaled height, and Marco had awakened that morning still small. In the six years since then, he has grown, but slowly, achingly, unlike other children. "Everybody at school calls me shrimp and stuff like that," he says.

"They think they're so rad. I feel like a loser. I feel like I'm nothing." At age 11, Marco stands 4 feet 1 inch—4 inches below average—and weighs 49 pounds. And he dreams, as all aggrieved kids do, of a sudden, miraculous turnaround: "One day I want to, like, surprise them. Just come in and be taller than them."

Marco, a serious student and standout soccer player, more than imagines redress. Every night but Sunday, after a dinner he seldom has any appetite for, his mother injects him with a hormone known to stimulate bone growth. The drug, a synthetic form of naturally occurring human growth hormone (HGH) produced by the pituitary, has been credited with adding up to 18 inches to the predicted adult height of children who produce insufficient quantities of the hormone on their own—pituitary dwarfs. But there is no clinical proof that it works for children like Marco, with no such deficiency. Marco's rate of growth has improved since he began taking the drug, but his doctor has no way of knowing if his adult height will be affected. Without HGH, Marco's predicted height was 5 feet 4 inches, about the same as the Nobel Prize-winning economist Milton Friedman and ... Masters golf champion, Ian Woosnam, and an inch taller than the basketball guard Muggsy Bogues of the Charlotte Hornets. Marco has been taking the shots for six years, at a cost to his family and their insurance company of more than $15,000 a year [$21,000 in 2005 dollars]....

SOURCE: From WEITZ, The Sociology of Health, Illness, and Health Care, 4E.

A Cleveland Browns cap splays Marco Oriti's ears and shadows his sparrowish face. Like many boys his age, Marco imagines himself someday in the NFL. He also says he'd like to be a jockey—making a painful incongruity that mirrors the wild uncertainty over his eventual size. But he is unequivocal about his shots, which his mother rotates nightly between his thighs and upper arms. "I hate them," he says.

He hates being short far more. Concord, the small Northern California city where the Oriti family now lives, is a high-achievement community where competition begins early. So Luisa Oriti and her husband, Anthony, a bank vice president, rationalize the harshness of his treatment. "You want to give your child that edge no matter what," she says, "I think you'd do just about anything." (Werth, 1991)

Does Marco have an illness? According to his doctors, who have recommended that he take an extremely expensive, essentially experimental, and potentially dangerous drug, it would seem that he does. To most people, however, Marco simply seems short....

MODELS OF ILLNESS

The Medical and Sociological Models of Illness

What do we mean when we say something is an illness? As Marco's story suggests, the answer is far from obvious. Most Americans are fairly confident that someone who has a cold or cancer is ill. But what about the many postmenopausal women whose bones have become brittle with age, and the many older men who have bald spots, enlarged prostates, and urinary problems? Or the many young boys who have trouble learning, drink excessively, or enjoy fighting? Depending on who you ask, these conditions may be defined as normal human variations, as illnesses, or as evidence of bad character. As these questions suggest, defining what is and is not an illness is far from a simple task. In this section we explore the medical model of illness: what doctors typically mean when they say something is an illness. This medical model is not accepted in its entirety by all doctors—those in public health, pediatrics, and family practice are especially likely to question it—and is not rejected by all sociologists, but it is the dominant conception of illness in the medical world. The sociological model of illness summarizes critical sociologists' retort to the medical model of illness. This sociological model reflects sociologists' view of how the world currently operates, not how it ideally should operate....

The medical model of illness begins with the assumption that illness is an *objective* label given to anything that deviates from normal biological functioning (Mishler, 1981). Most doctors, if asked, would explain that polio is caused by a virus that disrupts the normal functioning of the neurological system, that menopause is a "hormone deficiency disease" that, among other things, impairs the body's normal ability to regenerate bone, and that men develop urinary problems when their prostates grow excessively large and unnaturally compress the urinary tract. Doctors might further explain that, because of scientific progress, all educated doctors can now recognize these problems as illnesses, even though they were not considered as such in earlier eras.

In contrast, the sociological model of illness begins with the statement that illness (as the term is actually used) is a *subjective* label, which reflects personal and social ideas about what is normal as much as scientific reasoning (Weitz, 1991). Sociologists point out that ideas about normality differ widely across both individuals and social groups. A height of 4 feet 6 inches would be normal for a Pygmy man but not for an American man. Drinking three glasses of wine a day is normal for Italian women but could lead to a diagnosis of alcoholism in American medical circles. In defining normality, therefore, we need to look not only at individual bodies but also at the broader social context. Moreover, even within a given group, "normality" is a range and not an absolute. The median height of American men, for example, is 5 feet 9 inches, but most people would consider someone several inches taller or shorter than that as still normal. Similarly, individual Italians routinely and without social difficulties drink more or less alcohol than the average Italian. Yet medical authorities routinely make decisions about what is normal and what is illness based not on absolute, objective markers of health and illness but on arbitrary, statistical cutoff points—deciding, for example, that anyone in the fifth percentile for height or the fiftieth percentile for cholesterol level is ill. Culture, too, plays a role: Whereas [in America] ... breast enlargement for small breasts [is recommended], ... in Brazil large breasts are denigrated as a sign of African heritage and breast *reduction* is the most popular cosmetic surgery (Gilman, 1999).

Because the medical model assumes illness is an objective, scientifically determined category, it also assumes there is no *moral* element in labeling a condition or behavior as an illness. Sociologists, on the other hand, argue that illness is inherently a moral category, for deciding what is illness always means deciding what is good or bad. When, for example, doctors label menopause a "hormonal deficiency disease," they label it an undesirable deviation from normal. In contrast, many women consider menopause both normal and desirable and enjoy the freedom from fear of pregnancy that menopause brings (E. Martin, 1987). In the same manner, when we define cancer, polio, or diabetes as illnesses, we judge the bodily changes these conditions produce to be both abnormal and undesirable, rather than simply normal variations in functioning, abilities, and life expectancies. (Conversely, when we define a condition as healthy, we judge it normal and desirable.)

Similarly, when we label an individual as ill, we also suggest that there is something undesirable about that *person*. By definition, an ill person is one whose actions, ability, or appearance do not meet social norms or expectations within a given culture regarding proper behavior or appearance. Such a person will typically be considered less whole and less socially worthy than those deemed healthy. Illness, then, like virginity or laziness, is a moral status: a social condition that we believe indicates the goodness or badness, worthiness or unworthiness, of a person.

From a sociological stand, illness is not only a moral status but (like crime or sin) a form of deviance (Parsons, 1951). To sociologists, labeling something deviant does not necessarily mean that it is immoral. Rather, deviance refers to behaviors or conditions that socially powerful persons within a given culture *perceive*, whether accurately or inaccurately, as immoral or as violating social norms. We can tell whether behavior violates norms (and, therefore, whether it is deviant) by seeing if it results in *negative social sanctions*. This term refers to any punishment, from ridicule to execution. (Conversely, positive social sanctions refers to rewards, ranging from token gifts to knighthood.) These social sanctions are enforced by social control agents including parents, police, teachers, peers, and doctors. Later in this [article] we will look at some of the negative social sanctions imposed against those who are ill.

For the same reasons that the medical model does not recognize the *moral* aspects of illness labeling, it does not recognize the *political* aspects of that process. Although some doctors at some times are deeply immersed in these political processes—arguing, for example, that insurance companies should cover treatment for newly labeled conditions such as fibromyalgia or multiple chemical sensitivity—they rarely consider the ways that politics underlie the illness-labeling process in general. In contrast, sociologists point out that any time a condition or behavior is labeled as an illness, some groups will benefit more than others, and some groups will have more power than others to enforce the definitions that benefit them. As a result, there are often open political struggles over illness definitions (a topic we will return to later in this [article]). For example, vermiculite miners and their families who were constantly exposed to asbestos dust and who now have strikingly high rates of cancer have fought with insurance companies and doctors, in clinics, hospitals, and the courts, to have "asbestosis" labeled an illness; meanwhile, the mining companies and the doctors they employed have argued that there is no such disease and that the high rates of health problems in mining communities are merely coincidences (A. Schneider and McCumber, 2004).

In sum, from the sociological perspective, illness is a *social construction*, something that exists in the world not as an objective condition but *because we have defined it as existing*. This does not mean that the virus causing measles does not exist, or that it does not cause a fever and rash. It does mean, though, that when we talk about measles as an illness, we have organized our ideas about that virus, fever, and rash in only one of the many possible ways. In another place or time, people might conceptualize those same conditions as manifestations of witchcraft, as a healthy response to the presence of microbes, or as some other illness altogether. To sociologists, then, *illness*, like *crime* or *sin*, refers to biological, psychological, or social conditions subjectively defined as undesirable by those within a given culture who have the power to enforce such definitions.

In contrast, and as we have seen, the medical model of illness assumes that illness is an objective category. Based on this assumption, the medical model of health care assumes that each illness has specific features, universally recognizable in all populations by all trained doctors, that differentiate it both from other illnesses and from health (Dubos, 1961; Mishler, 1981). The medical model thus assumes that diagnosis is an objective, scientific process.

Sociologists, on the other hand, argue that diagnosis is a subjective process. The subjective nature of diagnosis expresses itself in three ways. First, patients with the same symptoms may receive different diagnoses depending on various social factors. Women who seek medical care for chronic pain, for example, are more likely to receive psychiatric diagnoses than are men who report the same symptoms. Similarly, African Americans (whether male or female) are more likely than whites are to have their chest pain diagnosed as indigestion rather than as heart disease (Hoffman and Tarzian, 2001; Nelson, Smedley, and Stith, 2002). Second, patients with the same underlying illness may experience different symptoms, resulting in different diagnoses. For example, the polio virus typically causes paralysis in adults but only flu-like symptoms in very young children, who often go undiagnosed. Third, different cultures identify a different range of symptoms and categorize those symptoms into different illnesses. For example, U.S. doctors assign the label of attention deficit disorder (ADD) to children who in Europe would be considered lazy troublemakers. And French doctors often attribute headaches to liver problems, whereas U.S. doctors seek psychiatric or neurological explanations (Payer, 1996). In practice, the American medical model of illness assumes that illnesses manifest themselves in other cultures in the same way as in American culture and, by extension, that American doctors can readily transfer their knowledge of illness to the treatment and prevention of illness elsewhere.

Finally, the medical model of illness assumes that each illness has not only unique symptoms but also a unique *etiology*, or cause (Mishler, 1981). Modern medicine assumes, for example, that *tuberculosis*, polio, HIV disease, and so on, are each caused by a unique microorganism. Similarly, doctors continue to search for limited and unique causes of heart disease and cancer, such as high-cholesterol diets and exposure to asbestos. Yet even though illness-causing microorganisms exist everywhere and environmental health dangers are common, relatively few people become ill as a result. By the same token, although cholesterol levels and heart disease are strongly correlated among middle-aged men, many men eat high-cholesterol diets without developing heart disease, and others eat low-cholesterol diets but die of heart disease anyway. The doctrine of unique ecology discourages medical researchers from asking why individuals respond in such different ways to the same health risks and encourages researchers to search for magic bullets—a term first used by Paul Ehrlich, discoverer of the first effective treatment for syphilis, in referring to drugs that almost miraculously prevent or cure illness by attacking one specific etiological factor....

MEDICINE AS SOCIAL CONTROL

Creating Illness: Medicalization

The process through which a condition or behavior becomes defined as a medical problem requiring a medical solution is known as medicalization (Conrad and Schneider, 1992; Conrad, 2005). For example, as social conditions have changed, activities formerly considered sin or crime, such as masturbation, homosexual activity, or heavy drinking, have become defined as illnesses. The same has happened to various natural conditions and processes such as uncircumcised penises, limited sexual desire, aging, pregnancy, and menopause (e.g., F. Armstrong, 2000; Barker, 1998; Figert, 1996; Rosenfeld and Faircloth, 2005). The term *medicalization* also refers to the process through which the definition of an illness is *broadened*. For example, when the World Health Organization (WHO) in 1999 lowered the blood sugar level required for diagnosis with diabetes, the number of persons eligible for this diagnosis increased in some populations by as much as 30 percent (Shaw, de Courten, Boyko, and Zimmet, 1999).

For medicalization to occur, one or more organized social groups must have both a vested interest in it and sufficient power to convince others (including doctors, the public, and insurance companies) to accept their new definition of the situation. Not surprisingly, doctors often play a major role in medicalization, for medicalization can increase their power, the scope of their practices, and their incomes. For example, during the first half of the twentieth century, improvements in the standard of living coupled with the adoption of numerous public health measures substantially reduced the number of seriously ill children. As a result, the market for pediatricians declined, and their focus shifted from treating serious illnesses to treating minor childhood illnesses and offering well-baby care. Pediatrics thus became less well-paid, interesting, and prestigious. To increase their market while obtaining more satisfying and prestigious work, some pediatricians have expanded their practices to include children whose behavior concerns their parents or teachers and who are now defined as having medical conditions such as attention deficit disorder or antisocial personality disorder (Halpern, 1990). Doctors have played similar roles in medicalizing premenstrual syndrome (Figert, 1996), drinking during pregnancy (E. Armstrong, 1998), impotence (Loe, 2004; Tiefer, 1994), and numerous other conditions....

The Consequences of Medicalization In some circumstances, medicalization can be a boon, leading to social awareness of a problem, sympathy toward its sufferers, and development of beneficial therapies. Persons with epilepsy, for example, lead far happier and more productive lives now that drugs usually can control their seizures, and few people view epilepsy as a sign of demonic possession. But defining a condition as an illness does not necessarily improve the social status of those who have that condition. Those who use alcohol excessively, for example, continue to experience social rejection even when alcoholism is labeled a disease. Moreover, medicalization also can lead to new problems, known by sociologists as unintended negative consequences (Conrad and Schneider, 1992; Zola, 1972)....

The Rise of Demedicalization The dangers of medicalization have fostered a counter movement of demedicalization (R. Fox, 1977). A quick look at medical textbooks from the late 1800s reveals many "diseases" that no longer exist. For example, nineteenth-century medical textbooks often included several pages on the health risks of masturbation. One popular textbook from the late nineteenth century asserted that masturbation caused "extreme emaciation, sallow or blotched skin, sunken eyes, ... general weakness, dullness, weak back, stupidity, laziness, ... wandering and illy defined pains," as well as infertility, impotence, consumption, epilepsy, heart disease, blindness, paralysis, and insanity (Kellogg, 1880: 365). Today, however, medical textbooks describe masturbation as a normal part of human sexuality.

Like medicalization, demedicalization often begins with lobbying by consumer groups. For example, medical ideology now defines childbirth as an inherently dangerous process, requiring intensive technological, medical assistance. Since the 1940s, however, growing numbers of American women have attempted to redefine childbirth as a generally safe, simple, and natural process and have promoted alternatives ranging from natural childbirth classes, to hospital birthing centers, to home births assisted only by midwives (Sullivan and Weitz, 1988). Similarly ... gay and lesbian activists have at least partially succeeded in redefining homosexuality from a pathological condition to a normal human variation. More broadly, in recent years, books, magazines, television shows, and popular organizations devoted to teaching people to care for their own health rather than relying on medical care have proliferated. For example, in the early 1970s, the Boston Women's Health Book Collective published a 35-cent mimeographed booklet on women's health. From this, they have built a virtual publishing empire that has sold to consumers worldwide millions of books (including the best-selling *Our Bodies, Ourselves*) on the topics of childhood, adolescence, aging, and women's health....

Social Control and the Sick Role

... Medicine also can work as an institution of social control by pressuring individuals to *abandon* sickness, a process first recognized by Talcott Parsons (1951).

Parsons was one of the first and most influential sociologists to recognize that illness is deviance. From his perspective, when people are ill, they cannot perform the social tasks normally expected of them. Workers stay home, homemakers tell their children to make their own meals, students ask to be excused from exams. Because of this, either consciously or unconsciously, people can use illness to evade their social responsibilities. To Parsons, therefore, illness threatened social stability.

Parsons also recognized, however, that allowing some illness can *increase* social stability. Imagine a world in which no one could ever "call in sick." Over time, production levels would fall as individuals, denied needed recuperation time, succumbed to physical ailments. Morale, too, would fall while resentment would rise among those forced to perform their social duties day after day without relief. Illness, then, acts as a kind of pressure valve for society—something we recognize when we speak of taking time off work for "mental health days."

From Parsons's perspective, then, the important question was how did society control illness so that it would increase rather than decrease social stability? The author's emphasis on social stability reflected his belief in the broad social perspective known as functionalism. Underlying functionalism is an image of society as a smoothly working, integrated whole, much like the biological concept of the human body as a homeostatic environment. In this model, social order is maintained because individuals learn to accept society's norms and because society's needs and individuals' needs match closely, making rebellion unnecessary. Within this model, deviance—including illness—is usually considered dysfunctional because it threatens to undermine social stability.

Defining the Sick Role Parsons's interest in how society manages to allow illness while minimizing its impact led him to develop the concept of the sick role. The sick role refers to social expectations regarding how society should view sick people and how sick people should behave. According to Parsons, the sick role as it currently exists in Western society has four parts. First, the sick person is considered to have a legitimate reason for not fulfilling his or her normal social role. For this reason, we allow people to take time off from work when sick rather than firing them for malingering. Second, sickness is considered beyond individual control, something for which the individual is not held responsible. This is why, according to Parsons, we bring chicken soup to people who have colds rather than jailing them for stupidly exposing themselves to germs. Third, the sick person must recognize that sickness is undesirable and work to get well. So, for example, we sympathize with people who obviously hate being ill and strive to get well and question the motives of those who seem to revel in the attention their illness brings. Finally, the sick person should seek and follow medical advice. Typically, we expect sick people to follow their doctors' recommendations regarding drugs and surgery, and we question the wisdom of those who do not.

Parsons's analysis of the sick role moved the study of illness forward by highlighting the social dimensions of illness, including identifying illness as deviance and doctors as agents of social control. It remains important partly because it was the first truly sociological theory of illness. Parsons's research also has proved important because it stimulated later research on interactions between ill people and others. In turn, however, that research has illuminated the analytical weaknesses of the sick role model.

Critiquing the Sick Role Model Many recent sociological writings on illness—including this [article]—have adopted a conflict perspective rather than a functionalist perspective. Whereas functionalists envision society as a harmonious whole held together largely by socialization, mutual consent, and mutual interests, those who hold a conflict perspective argue that society is held together largely by power and coercion, as dominant groups impose their will on others. Consequently, whereas functionalists view deviance as a dysfunctional element to be controlled, conflict theorists view deviance as a necessary force for social change and as the conscious or unconscious expression of individuals who refuse to conform to an oppressive society. Conflict theorists therefore have stressed the need to study social control agents as well as, if not more than, the need to study deviants....

In sum, the sick role model is based on a series of assumptions about both the nature of society and the nature of illness. In addition, the sick role model confuses the experience of *patienthood* with the experience of *illness* (Conrad,

1987). The sick role model focuses on the interaction between the ill person and the mainstream health care system. Yet interactions with the medical world form only a small part of the experience of living with illness or disability. For these among other reasons, research on the sick role has declined precipitously; whereas *Sociological Abstracts* listed 71 articles on the sick role between 1970 and 1979, it listed only 7 articles between 1990 and 1999, even though overall far more academic articles were published during the 1990s than during the 1970s.

CONCLUSION

The language of illness and disease permeates our everyday lives. We routinely talk about living in a "sick" society or about the "disease" of violence infecting our world, offhandedly labeling anyone who behaves in a way we don't understand or don't condone as "sick."

This metaphoric use of language reveals the true nature of illness: behaviors, conditions, or situations that powerful groups find disturbing and believe stem from internal biological or psychological roots. In other times or places, the same behaviors, conditions, or situations might have been ignored, condemned as sin, or labeled crime. In other words, illness is both a social construction and a moral status.

In many instances, using the language of medicine and placing control in the hands of doctors offers a more humanistic option than the alternatives. Yet, as this [article] has demonstrated, medical social control also carries a price. The same surgical skills and technology for cesarean sections that have saved the lives of so many women and children now endanger the lives of those who have cesarean sections unnecessarily. At the same time, forcing cesarean sections on women potentially threatens women's legal and social status. Similarly, the development of tools for genetic testing has saved many individuals from the anguish of rearing children doomed to die young and painfully, but has cost others their jobs or health insurance.

In the same way, then, that automobiles have increased our personal mobility in exchange for higher rates of accidental death and disability, adopting the language of illness and increasing medical social control bring both benefits and costs. These benefits and costs will need to be weighed carefully as medicine's technological abilities grow.

REFERENCES

Armstrong, Elizabeth M. 1998 "Diagnosing moral disorder: The discovery and evolution of fetal alcohol syndrome." *Social Science and Medicine* 47: 2025–2042.

———. 2000. "Lessons in control: Prenatal education in the hospital." *Social Problems* 47: 583–605.

Barker, K. K. 1998. "A ship upon a stormy sea: The medicalization of pregnancy." *Social Science and Medicine* 47: 1067–1076.

Conrad, Peter. 1987. "The experience of illness: Recent and new directions." *Research in the Sociology of Health Care* 6: 1–32.

———. 2005. "The shifting engines of medicalization." *Journal of Health and Social Behavior* 46: 3–14.

Conrad, Peter, and Joseph W. Schneider. 1992. *Deviance and Medicalization: From Badness to Sickness*. Philadelphia: Temple University Press.

Dubos, Rene. 1961. *Mirage of Health*. New York: Anchor.

Figert, Anne E. 1996. *Women and the Ownership of PMS: The Structuring of a Psychiatric Disorder*. New York: Aldine DeGruyter.

Fox, Renee. 1977. "The medicalization and demedicalization of American society." *Daedalus* 106: 9–22.

Gilman, Sander L. 1999. *Making the Body Beautiful: A Cultural History Of Aesthetic Surgery*. Princeton, NJ: Princeton University Press.

Halpern, S. A. 1990. "Medicalization as a professional process: Postwar trends in pediatrics." *Journal of Health and Social Behavior* 31: 28–42.

Hoffman, Diane E., and Anita J. Tarzian. 2001. "The girls who cried pain: A bias against women in the treatment of pain." *Journal of Law, Medicine, and Ethics* 29: 13–27.

Kellogg, J. H. 1880. *Plain Facts for Old and Young*. Burlington, IA: Segner and Condit.

Loe, Meika. 2004. *The Rise of Viagra: How the Little Blue Pill Changed Sex in America*. New York: New York University Press.

Martin, Emily. 1987. *The Woman in the Body*. Boston: Beacon.

Mishler, Elliot G. 1981. "Viewpoint: Critical perspectives on the biomedical model." Pp. 1–23 in *Social Contexts of Health, Illness, and Patient Care*, edited by Elliot G. Mishler. Cambridge, UK: Cambridge University Press.

Nelson, Alan R., Brian D. Smedley, and Adrienne Y. Stith. 2002. *Unequal Treatment: Confronting Racial and Ethnic Disparities in Health Care*. Washington, DC: Institute of Medicine, National Academy Press.

Parsons, Talcott. 1951. *The Social System*. New York: Free Press.

Payer, Lynn. 1996. *Medicine and Culture*. Rev. ed. New York: Holt.

Rosenfeld, Dana, and Christopher A. Faircloth (eds.). 2005. *Medicalized Masculinities*. Philadelphia: Temple University Press.

Schneider, Andrew, and David McCumber. 2004. *An Air That Kills: How the Asbestos Poisoning of Libby, Montana Uncovered a National Scandal*. New York: Putnam's Sons.

Shaw, Jonathan E., Maximilian de Courten, Edward J. Boyko, and Paul Z. Zimmet. 1999. "Impact of new diagnostic criteria for diabetes on different populations." *Diabetes Care* 22: 762–766.

Sullivan, Deborah A., and Rose Weitz. 1988. *Labor Pains: Modern Midwives and Home Birth*. New Haven, CT: Yale University Press.

Tiefer, Leonore. 1994. "The medicalization of impotence: Normalizing phallocentrism." *Gender & Society* 8: 363–377.

Weitz, Rose. 1991. *Life with AIDS*. New Brunswick, NJ: Rutgers University Press.

Werth, Barry. 1991. "How short is too short? Marketing human growth hormone." *New York Times Magazine* June 16: 14+.

Zola, Irving K. 1972. "Medicine as an institution of social control." *Sociological Review* 20: 487–504.

KEY CONCEPTS

conflict theory
demedicalization
functionalism
medicalization
sick role

DISCUSSION QUESTIONS

1. Compare and contrast the sociological and the medical models of illness. What are the pros and cons of each model?
2. What does the author mean by saying that "illness is a moral status"?

58

Beyond the Affordable Care Act: Achieving Real Improvements in Americans' Health

DAVID R. WILLIAMS, MARK B. McCLELLAN, AND ALICE M. RIVLIN

The authors identify the patterns shaping poor health among Americans, even when we have one of the most expensive and sophisticated health care systems in the world.

With the enactment of the Patient Protection and Affordable Care Act of 2010, estimates suggest that thirty-two million Americans will acquire health insurance coverage by 2018. This would reduce the proportion of uninsured people to about 6 percent of the U.S. population. But expanded coverage alone will not make this a healthy nation. Although improved access to health care is essential, especially when people are already sick, reducing the risk of getting sick in the first place is even more important to improving the nation's health....

SOURCE: Copyrighted and published by Project HOPE/Health Affairs as Williams D R, McClellan M B, Rivlin A M. Beyond the affordable care act: achieving real improvements in Americans' health. Health Aff (Millwood). 2010; 29(8): 1481-1488. The published article is archived and available online at www.healthaffairs.org.

AMERICA'S UNREALIZED HEALTH POTENTIAL

Cross-National Comparisons

Americans are far less healthy than they could be. Compared with other developed countries, the United States ranks low on many health indicators, and its relative position is slipping. In some respects—obesity is a prime example—Americans are becoming less healthy over time. Some experts predict a reversal of long-standing trends: Many of the next generation of Americans are likely to live shorter, sicker lives than their parents did.[1]

The health of many Americans is declining despite the fact that other countries spend far less on health care than the United States does. Indeed, the data indicate that our worsening health is a driver of higher medical costs.[2]

Relative Importance of Medical Care

There is no question that medical care is essential for relieving suffering and curing illness. But medical care prevents only 10–15 percent of premature deaths.[3] Research shows that social factors such as education, income, and the quality of neighborhood environments play a much more important role in shaping our health than medical care does.[4]

For example, college graduates can expect to live five years longer than those who have not completed high school. And scientific evidence indicates that although poor health can reduce a person's economic resources, it is equally true that income is an important driver of health.[5]

Other Contributors to Disparities

Differences in income and education contribute to well-documented disparities in health among racial and ethnic groups. Moreover, although immigrants from the major racial and ethnic groups initially enjoy better health than their native-born counterparts, this health advantage declines as the length of their stay in the United States increases and with the passage of generations.[6]

Health gaps between the United States and other developed countries are not limited to poor Americans and those with little education. The gaps shrink but persist with each increase in income and education level. Even the best-off Americans are behind citizens of other countries on key measures of health.

Analyses conducted for the commission found that American children and adults at all levels of education and income, and in every racial and ethnic group, fell below an attainable national benchmark of good health.[7] These shortfalls were evident even in the states that fared best in overall health.

These findings demonstrate a tremendous unrealized potential. If all Americans are enabled to become as healthy as those in the healthiest groups, major improvements could be accomplished.

Effects on U.S. Productivity

Deficiencies in health not only cause unnecessary suffering and premature death but also affect our nation's productivity. When people are sick, they can't function as well at school, home, or work as when they are healthy. Nearly one in three low-income American adults is functionally limited by chronic illness....

Especially now, as the nation struggles to regain its economic footing, it is important to consider the health shortfalls in our workforce. An analysis commissioned by the Robert Wood Johnson Foundation [RWJF] estimates that if all Americans enjoyed the same level of good health as college graduates, the national economy would achieve annual average savings of $1 trillion in health care expenses....

Longer and healthier lives would result in higher worker productivity, reduced spending on social programs, and increases in tax revenues. Another recent report placed the annual price tag of racial and ethnic disparities in health at $309 billion.[8]

Example: Obesity

The costs of our nation's obesity epidemic illustrate this economic burden. According to the Partnership to Fight Chronic Disease, the doubling of obesity in the United States since 1987 accounts for nearly 30 percent of the increase in health care spending.[9] Furthermore, obesity-related costs—such as those stemming from the increased incidence of diabetes and heart disease—are expected to continue rising. According to a recent study, if current trends continue, within twenty-five years there will be more than forty-four million Americans with diabetes, and the annual cost of caring for them will triple to $336 billion, in 2007 dollars....

CLOSING THE GAPS IN HEALTH

Closing America's health gaps requires a commitment to a broad view of the needed improvements. The recent health reform debate highlighted the importance of preventing illness, and the resulting Patient Protection and Affordable Care Act places increased emphasis on both preventive and primary health care.

However, the Commission to Build a Healthier America concluded that the most important prevention activities occur outside the traditional medical care setting, in the places where we live, learn, work, play, and worship. What most influences population health are the consequences of the actions we take and the choices we make in the course of our daily lives, such as how we raise our children, what food we eat, how physically active we are, whether we smoke, and the safety and health of the neighborhoods we live in.[10]

Social factors begin influencing our health very early in life. Brain, cognitive, and behavioral development in early childhood are strongly linked to an array of important health outcomes later in life, including cardiovascular disease, diabetes, obesity, drug use, and depression. Children who do not receive high-quality supportive care and education begin life with a distinct disadvantage and have a higher risk of becoming unhealthy adults.[11]

Importance of Education

It has been well established that education plays a key role in economic well-being, but the profound effects that education has on health are less widely recognized.... People with more education not only live longer, but they also lead substantially healthier lives than people who have less education....

Education is associated with opportunities that steer people toward better health. More education typically means higher-paying employment with better benefits, including health insurance. The more education parents have, the better able they are to purchase nutritious food; give their children secure child care; provide leisure activities that can reduce the negative health effects of stress; and buy homes in safer, healthier neighborhoods with parks and sidewalks....

Conversely, limited education is linked to lower-paying jobs, inadequate health insurance, higher levels of stress that can affect health, and fewer resources for making such healthy choices as engaging in regular physical activity and eating nutritious food. Everyday life can be an all-consuming struggle for the less educated, leaving little or no time, money, or energy to adopt a healthier lifestyle, and even reducing personal motivation for doing so....

The opportunities created by education are lifelong and often cross generations. Better-educated parents are better equipped to raise healthy youngsters, and better-educated children are better prepared to achieve and maintain good health for themselves. The same is true of income. Parents with higher incomes pass on to their children certain health advantages and opportunities that parents with lower incomes frequently cannot....

The impacts of education and income show that good health is not achieved primarily in hospitals and doctors' offices. Rather, better education and income are associated with differences in how we live our daily lives, which in turn profoundly influence health.

Culture of Health

Living healthier lives also requires the creation of a culture of health. In much the same way that America is going "green" to combat global climate change, the nation must better incorporate health into its homes, schools, neighborhoods, and workplaces. This requires that we as a nation embrace a broad vision of health and take responsibility for decisions that support good health. It also requires commitment from leaders in the private and public sectors to support appropriate decisions, not only in the area of health care delivery, but also in the spheres of education, business, and community planning.

The commission reached two main conclusions, which became the long-term goals set forth in its report. First, individuals can and must take responsibility for improvements in their own health. Second, society must create conditions that will help people make healthy choices. Many people live and work in circumstances and places that make healthy living extremely difficult. Our society's leaders and major institutions must create incentives and lower barriers, to bring better health within everyone's reach....

Conclusion

A great deal of hard work lies ahead. Although improving access to and rethinking health care is necessary, we must do more than that. We need to modify where and how Americans live, learn, work, and play.

We have strong evidence that major strides can be made outside of medical care, and clear opportunities exist to implement, refine, and expand proven strategies. With informed choices on the part of individuals and policy makers, all Americans can have a healthier future.

NOTES

1. Olshansky SJ, Passaro DJ, Hershow RC, Lyden J, Carnes BA, Brody J, et al. A potential decline in life expectancy in the United States in the 21st century. N Engl J Med. 2005;352(11):1138–45.
2. Huang ES, Basu A, O'Grady M, Capretta JC. Projecting the future diabetes population size and related costs for the U.S. Diabetes Care. 2009;32(12):2225–9.
3. McGinnis JM, Williams-Russo P, Knickman JR. The case for more active policy attention to health promotion. Health Aff (Millwood). 2002;21(2):78–93.
4. Smedley B, Syme SL. Promoting health. Washington (DC): National Academies Press; 2000.
5. Braveman P, Egerter S. Overcoming obstacles to health: report from the Robert Wood Johnson Foundation to the Commission to Build a Healthier America [Internet]. Princeton (NJ): Robert Wood Johnson Foundation; 2008 Feb [cited 2010 Apr 21]. Available from: http://www.commissiononhealth.org/PDF/ObstaclesToHealth-Report.pdf
6. Williams DR, Mohammed SA, Leavell J, Collins C. Race, socioeconomic status, and health: complexities, ongoing challenges, and research opportunities. Ann N Y Acad Sci. 2010 Feb;1186:69–101.
7. Egerter S, Braveman P, Pamuk E, Cubbin C, Dekker M, Pedregon V, et al. America's health starts with healthy children: how do states compare? [Internet]. Washington (DC): Robert Wood Johnson Foundation Commission to Build a Healthier America; 2008 Oct [cited 2010 Apr 21]. Available from: http://www.commissiononhealth.org/PDF/819a3435-8bbb-4549-94db-7758248075cf/ChildrensHealth_Chartbook.pdf
8. LaVeist TA, Gaskin DJ, Richard P. The economic burden of health inequalities in the United States. Washington (DC): Joint Center for Political and Economic Studies; 2009.
9. Partnership to Fight Chronic Disease. The 2009 almanac of chronic disease [Internet]. Washington (DC): The Partnership; 2009 [cited 2010 Apr 21]. Available from: http://www.fightchronicdisease.org/pdfs/2009AlmanacofChronicDisease_updated81009.pdf
10. Goodarz D, Ding EL, Mozaffarian D, Taylor B, Rehm J, Murray CJL, et al. The preventable causes of death in the United States: comparative risk assessment of dietary, lifestyle, and metabolic risk factors. PloS Med [serial on the Internet]. 2009

Apr 28 [cited 2010 Apr 22]; 6(4):e1000058. Available from: http://www.plosmedicine.org/article/info%3Adoi%2F10.1371%2Fjournal.pmed.1000058

11. Shonkoff JP, Boyce WT, McEwen BS. Neuroscience, molecular biology, and the childhood roots of health disparities: building a new framework for health promotion and disease prevention. JAMA. 2009;301(21):2252–9.

KEY CONCEPTS

Affordable Care Act (2010) obesity

DISCUSSION QUESTIONS

1. What social factors do the authors identify as producing the overall poor health of Americans?
2. What policy suggestions are implied by the authors' argument?

59

Beyond Caring: The Routinization of Disaster

DANIEL F. CHAMBLISS

Daniel Chambliss analyzes how nurses manage the work of nursing in formal organizations (hospitals), especially as they are required to do work in the context of human disaster and tragedy, such as death.

The moral system of the nurse's world, the hospital, is quite different from that of the lay world. In the hospital it is the good people, not the bad, who take knives and cut people open; here the good stick others with needles and push

SOURCE: Beyond Caring: The Routinization of Disaster, by Daniel Chambliss,

fingers into rectums and vaginas, tubes into urethras, needles into the scalp of a baby; here the good, doing good, peel dead skin from a screaming burn victim's body and tell strangers to take off their clothes. Here, in the words of an old joke, the healthy take money from the sick, and the most skilled cut up old ladies and get paid for it. The layperson's horrible fantasies here become the professional's stock in trade.... Some people believe, reasonably, that nurses' work must, therefore, be terribly stressful. "How do they cope with it?" is a question I am frequently asked. But the layperson has not seen the routinization of activities and the parallel flattening of emotion that takes place as one becomes a nurse. Nursing is stressful, to be sure, but not in the way the layperson imagines. Intravenous lines started, meds delivered, baths given, trays passed out, vital signs taken, a nonstop round of filling out forms, of writing reports, sending off blood samples—this fills the nurse's day, routine tasks done dozens of times. The moral ambiguity of any one task becomes utterly lost in the pile of repeated events; the routine blurs the moral difficulties.

Problems arise against this background of routine. In the large medical center, deaths occur every day. Only a few of them become "ethics problems" for the staff: a patient wants to die, and the staff won't let her; when the staff does let her die, then the family doesn't understand why. Tests and procedures cause pain, but few tests and procedures are challenged on moral grounds. Privacy is violated in the most egregious ways, all the time; but usually the violation is unhappily accepted by the suffering patient, who understands that the professional is just doing her job....

THE HOSPITAL SETTING

As a formal organization, the hospital can be readily described, in objective terms, as a professional bureaucracy. For readers unfamiliar with general hospitals, a brief description of the organization, and some key terms describing its structure, will be useful.

Consider Northern General, one of the hospitals where this research was conducted. Northern General is a major medical care center for its region. It provides services to walk-in patients and accepts referrals from physicians in the surrounding communities and beyond. Some 70 percent of its patients are from the immediate urban area, 28 percent from the rest of the state in which it is located, and the remainder from the greater United States and even outside the country. They are drawn to Northern by the reputation of the hospital and its staff.

The hospital provides an impressive array of health care services. Its Emergency Room (ER) alone treats over 100,000 patients a year. Next door to the ER is a Primary Care Center (PCC) where regular patients come to see their assigned physicians for routine checkups and minor procedures. In the building rising over the ER and the PCC are housed many clinics for outpatient services—Eye Clinic, Radiation Clinic, Dental Clinic, and others. In the basement below, there is the Poison Control Center, a Rape Crisis Center, and a Cancer Hotline. The hospital also has a special staff which deals with child and spouse abuse. And these are only outpatient services, provided to people who are not staying overnight in the hospital. The inpatient floors rising beyond the clinics, in several buildings, house nearly 1,000 patients and the nurses who support their basic medical care in surgical, pediatric, internal medicine, obstetrics and gynecology, psychiatric, and other specialties. The hospital, of course, owns CAT scanners, MRI devices, a linear accelerator, and what are now the entire array of technologies. Pediatricians here treat relatively rare diseases such as Reye's syndrome and Cooley's anemia, and the hospital boasts one of the nation's oldest and most renowned Newborn Intensive Care Units.

With such magnificent facilities, it is no wonder that Northern General is a major training center for physicians, nurses, technicians, and other health care workers. It is one of the 900 or so "teaching hospitals" in the nation directly affiliated with a major medical school.... There is a steep hierarchy of physicians, typical of such hospitals. The medical staff includes, first, senior physicians who hold positions as professors in the university's medical school as well as physicians from the community. These doctors are the "attending" physicians of the hospital and hold the top rank in the medical hierarchy. Just below the attendings are "Fellows," doctors at an advanced stage of training in specialized fields such as renal disease or endocrinology. Below these come the "house staff." They are employed by the hospital, not by the patients, as attendings are, and they work long hours (thirty-six-hour shifts perhaps twice a week, twelve hours other days). House staff are physicians, M.D.s, who are still in training; they are also called "residents." In their first year of training residents are usually called "interns." Most of the day-to-day medical decision making in the hospital is in the hands of the house staff; they are the physicians who are in the hospital handling the emergencies that come up at any time of the day or night.

The nursing administration is larger but a bit simpler. All nurses are employed by the hospital, and the nursing hierarchy performs a large part of the hospital's administrative work. At the top of the Northern General nursing staff is a vice president for nursing of the hospital; below her are a half-dozen directors of nursing for different medical services of the hospital, for example, pediatrics, internal medicine, surgical nursing, etc. The next level is of supervisors who oversee a number of discrete floors or units. Each "floor" consists of a number (probably twenty to thirty) of patient beds and is headed by a head nurse. The head nurse, like a foreman of a work crew, directs the work of the bedside, or "staff," nurses. There are perhaps twenty to forty staff nurses under each head nurse, divided into three shifts (or two twelve-hour shifts in some ICUs). There are some twenty-five floors and a half-dozen units making up the hospital. The head nurse has the immediate daily responsibility for the floor or unit (the generic term for Intensive Care Units). Finally, at the bottom there are nursing students from the university's nursing school, who study for four years to receive a Bachelor of Science degree in nursing. Students typically will work in the hospital for one or two days a week, devoting the rest of their time to classroom studies. While working in the hospital, they usually care for a smaller number of patients than the full-time staff nurse, but the student nurse may well have many of the same responsibilities and tasks that a full-time R.N. does.

Physically, hospital floors look much like those on television soap operas: long hallways with rooms on either side, with the halls intersecting at a nurses' station, a configuration of boothlike desks behind which are clerks and nurses filling out the forms, answering the telephones, checking computer monitors, and taking care of the endless administrative tasks which consume as much of the nurse's time as does patient care. There is also a vast array of supplies and materials. On the desks in the nurse's station are polyurethane stackable boxes holding dozens and dozens (fifty or more) of multicopy forms for ordering all sorts of tests and procedures, billing forms, permission forms, order forms. On supply shelves, stacked to the ceiling, or jamming closets, or filling the walls of a pantry-like room in the hall, are adhesive tape, packaged needles, boxes of syringes, from 1 cubic centimeter (tiny), to 30, 40, 50 cc (a huge thing, the size of a baby's arm, used to deliver food in tube feedings), stacks of sterile isolation gowns in plastic wrappers, rubber bands, blood pressure cuffs, plastic jugs of microbial soap, bottles of saline solution, swabs, scales (for weighing patients' stools), large red plastic disposal jugs for used needles, plastic garbage bags, packaged kits of needles, sterile paper, syringes, and medications for doing lumbar puncture procedures, Foley urinary catheter insertion kits, intravenous kits, arterial catheter lines, and on, and on. The range and volume of visible, usable, disposable equipment is astonishing.

In the Intensive Care Units the equipment is multiplied with red "crash carts" carrying defibrillating machines and the wide range of drugs used for cardiac resuscitation; bottles of drugs like atropine, sodium bicarbonate, and epinephrine wedged into boxes on the fronts of patients' doors, ready to be used in an instant to juice up a failing heart; ring binders, several for each patient, with records of everything that is done to or for him or her. The equipment, the forms, the variety of supplies in themselves reflect the complexity of the organizations in which they are used, which provide and use them. Hospitals are complex, hierarchial organizations, and in that sense are like many other organizations in our society.

HOW THE HOSPITAL IS DIFFERENT

Much here is the same as in other organizations: the daily round of paper processing, answering the phone, making staffing decisions, collecting bills, ordering supplies, stocking equipment rooms; there are fights between departments, arguments with the boss, workers going home tired or satisfied. And medical sociology has made much of these similarities, using its research to create broader theories of, for instance, deviance or of the structure of professions.

But in one crucial respect the hospital remains dramatically different from other organizations: *in hospitals, as a normal part of the routine, people suffer and die.* This is unusual.... Only combat military forces share this feature. To be complete, theories of hospital life need to acknowledge this crucial difference, since adapting themselves to pain and death is for hospital workers the most distinctive feature of their work. It is that which most separates them from the rest of us. In building theories of organizational life, sociologists must try to see how hospitals resemble other organizations … but we should not make a premature leap to the commonalities before appreciating the unique features of hospitals that make a nurse's task so different from that of a teacher or a businessman or a bureaucrat….

A quick survey of typical patients in one Surgical Intensive Care Unit on one Saturday evening should make the point. The words in brackets are additions to my original field notes:

> *Room 1.* 64-year-old white woman with an aortic valve replacement; five separate IMEDs [intravenous drip-control devices] feeding in nitroglycerine, vasopressors, Versed [a pain killer which also blocks memory]. Chest tube [to drain off fluids]. On ventilator [breathing machine], Foley [catheter in the bladder], a pulse oximeter on her finger, a[rterial monitoring] line. Diabetic. In one 30-second period during the night, her blood pressure dropped from 160/72 to 95/50, then to 53/36, before the nurse was able to control the drop. N[urse]s consider her "basically healthy."
>
> *Room 2.* Man with pulmonary atresia, pulmonary valvotomy [heart surgery].
>
> *Room 3.* Woman with CABG [coronary artery bypass graft; a "bypass operation"]. Bleeding out [i.e., hemorrhaging] badly at one point during the night, they sent her back to the OR [Operating Room]. On heavy vasopressors [to keep blood pressure up].
>
> *Room 4.* Older woman with tumor from her neck up to her temple. In OR from 7 A.M. until 2 A.M. the next morning having it removed. Infarct [dead tissue] in the brain.
>
> *Room 5.* 23-year-old woman, MVA [motor vehicle accident]. ICP [intracranial pressure—a measure of brain swelling] measured—terrible. Maybe organ donor. [Patient died next day.]
>
> *Room 6.* Don't know.
>
> *Room 7.* Abdominal sepsis, possibly from surgery. DNR [Do Not Resuscitate] today.
>
> *Room 8.* Big belly guy [an old man with a horribly distended abdomen, uncontrollable. Staff says it's from poor sterile technique in surgery by Doctor M., who is notoriously sloppy. This patient died within the week.] [Field Notes]

This is a typical patient load for an Intensive Care Unit. Eight beds, three patients dead in a matter of days. "Patients and their visitors often find the ICU to be a disturbing, even terrifying place. Constant artificial light, ceaseless activity, frequent emergencies, and the ever-present threat of death create an atmosphere that can unnerve even the most phlegmatic of patients. Some are so sick that they are unaware of their surroundings or simply forget the experience, but for others the ICU is a nightmare remembered all too well."[1] On floors—the larger, less critical care wards of the hospital—fatalities are less common, and patients are not so sick; even so, one-third of the patients may have AIDS, another one-third have cancer, and the rest suffer a variety of serious if not immediately lethal diseases. The ICUs just get patients whose deaths are imminent.

It is interesting that this density of disease presents one of the positive attractions of nursing. People don't become nurses to avoid seeing suffering or to have a quiet day. Every day nurses respond to and share the most intense emotions with total strangers. "People you don't know are going through the most horrible things, and you are supposed to help them. That's intense," says one nurse. And another enthuses about coming home as the sun is coming up; the rest of the world thinks things are just starting, and here you're coming off a big emergency that lasted half the night: "[T]here's a real adrenaline kick in all this stuff. If you deny that, you're denying a big part of [nursing]."...

... On a medical floor, with perhaps two-thirds of the patients suffering eventually fatal diseases, I say to a nurse, "What's happening?" and she replies, walking on down the hall, "Same ol' same ol'." Nothing new, nothing exciting. Or in an Intensive Care Unit the same hospital, "What's going on?" The resident replies, with a little shrug of the shoulders, "People are living, people are dying." Again, no surprises, nothing new. The routine goes on.

As other writers have noted, the professional treats routinely what for the patient is obviously not routine. For the health worker, medical procedures happen to patients every day, and the hospital setting is quite comfortable: "The staff nurse ... belongs to a world of relative health, youth, and bustling activity. She may not yet have experienced hospitalization herself for more than the removal of tonsils or the repair of a minor injury. Although she works in an environment of continuous sickness, she has been so conditioned to its external aspects that she often expresses surprise when someone suggests that the environment must be anxiety evoking."[2] Everett Hughes's formulation of this divergence of experience is classic: "In many occupations, the workers or practitioners ... deal routinely with what are emergencies to the people who receive their services. This is a source of chronic tension between the two." Or, more precisely, "[O]ne man's routine of work is made up of the emergencies of other people."[3]

To the patient, though, the hospital world is special, frightening, a jarring break from the everyday world. For the nurse, it's just the "same ol' same ol'." How extreme the gap is was observed in an ICU one evening:

> Three residents were attempting an LP [lumbar puncture—a "spinal tap" in which a long needle is inserted into the spinal column to draw out spinal fluid]. This is a very painful procedure and is difficult to perform. The television over the foot of the patient's bed was turned on, and "LA Law" was playing. While the resident was inserting the needle, she kept glancing up at the television, trying to simultaneously watch the show and do the LP. The patient, curled into the fetal position to separate the vertebrae, was unaware of this. The other two residents as well were glancing back and forth from procedure to

> television. The resident tried for several minutes and was unable to get the needle in properly, sometimes drawing out blood instead of fluid. Eventually, she called the head resident, who came in and successfully finished the LP. [Field Notes]

This illustrates how casual staff can become, to the point of malfeasance.

How do staff, nurses in particular, routinize the abnormal? Or more fundamentally, what do we even mean by routinization?...

Routinization in the OR or elsewhere in the hospital seems to mean several things: that actions are repeated, that they violate normal taboos, and that routine is embedded in behavior. Consider each in turn, drawing on further examples from other settings in the hospital:

1. *Repeatability*. Each operation is not the first of its kind; most in fact are done several times each day and hundreds of times each year, even by a single team of surgeons, nurses, and technical aides. What the team sees, they have seen many times. Gallbladder removals, hernia repairs and shoulder operations on athletes—these are all very common procedures in the major medical center....

And those repeated procedures take place against the even less dramatic background of the repeated daily events of the nurse's work: starting intravenous lines, taking blood pressures four times a day on every patient on the floor, drawing blood samples, charting vital signs, writing nurses' progress notes, passing food trays, helping patients on and off the bedpan. Both trivial and consequential activities are repeated over and over until each one becomes much like the next; indeed, as Hughes says, the professional's "very competence comes from having dealt with a thousand cases of what the client likes to consider his unique experience."... Says one nurse, "You get to the point where you don't really care for the patients anymore, and one GI [gastrointestinal] bleeder gets to be the same as the next GI bleeder."...

Death becomes a routinized part of daily life, incorporated into the flow. "Mr. Smith died last night," says one nurse to another. "Oh, that's too bad. He was such a nice man"; a casual exchange. One day is like another, if not for Mr. Smith, then at least for the rest of us. For the nurses Mr. Smith will be replaced by another man, a Mr. Jones, with similar ailments and a similar end. Like the fictional prep school teacher Mr. Chips, for whom the students "never get older," nurses see the same patients over and over, even if the names change....

The repeatability of the events—the sense that the same things happen over and over—is part of what is meant by routinization.

2. *Profanation*. Normally, we experience our bodies and the bodies of others as sacred, as areas to be approached with reverence or even with awe. To the healthy person outside the hospital, the body is special, a thing distinct from other things in the world, and must be treated as different. Physical contact with other bodies is emotionally provocative, in ways good and bad. A touch, a hug, a kiss arouse some sensitivity; a slap, however light, provokes

humiliation or perhaps rage. But for patients in the hospital, their bodies are dramatically profaned. The body is often exposed to strangers, older and younger, male and female, even in groups. Many times a day the patient's body is punctured by injection needles. It is the object of teachers' lectures to their students. It is touched frequently, often without special preliminaries. It is probed with fingers and hands and tools in ways that are sometimes brutal, with little respect for the body as a sacred object. Even when professionals are respectful (and many of them always are), the effect on patients is one of secularization of their own flesh....

3. *Existentiality.* Routinization of the abnormal involves not so much a mental leap as an existential action; creating a routine is not some trick of the conscious mind but rather a whole way of acting that involves physical as well as mental components. It comprises embodied habits. Routinization is not "all in the head"; it is something one does. It is carried out in the way a nurse walks and talks while in the presence of abnormal events. It entails that "matter of factness" with which she inserts a bladder catheter or cleans up feces; it is evidenced in the full range of emotions she shows, laughing with a dying man, chatting and even laughing with colleagues during a code (a resuscitation effort), glancing casually through a nursing journal filled with full-color advertisements for ileostomy appliances and crèmes for cancer lesions. After a middle-aged woman died late one night in room 5 of a medical ICU, her family, loudly crying and hugging each other, came into the room to see their dead mother. Outside the room, three nurses who had tried for thirty minutes to resuscitate the patient sat around a table eating corn chips and gossiping, as if nothing had happened. One of those nurses said to me, "We're pretty dehumanized, huh?" but it wasn't true: if she were dehumanized, no such comment would even be made. She knew what was happening, and eating corn chips, even then, is in fact quite human. In a unit where three patients die each week, to get upset with every death would be humanly unacceptable.

So in the Operating Room and elsewhere events are repeated over and over, in an attitude of secularized treatment of the body, and this repeated attitude is built into and expressed in the very ordinary doings of the staff. Sometimes routinization goes beyond mere commonplace into an attitude of detachment, unconcern, or sheer boredom—one of the more common emotions of the nurse's life, to the surprise of laypersons....

NOTES

1. Arnold S. Relman. "Intensive Care Units: Who Needs Them?" *New England Journal of Medicine* 302 (April 1980), p. 965.
2. Esther Lucile Brown, "Nursing and Patient Care." in Fred Davis, *The Nursing Profession: Five Sociological Essays* (New York: John Wiley & Sons, 1966), p. 202.
3. Hughes. *Men and Their Work.* pp. 54, 88.

KEY CONCEPTS

organization routinization

DISCUSSION QUESTIONS

1. What are the three dimensions of routinization that Chambliss notes?
2. Can you think of another occupation where grim work has to be done on a routine basis? How might workers manage this, and how would it be similar to the nurses whom Chambliss studies?

Applying Sociological Knowledge: An Exercise for Students

Identify all of the health care facilities in your community. Who do they serve? Who do they not serve? What does this exercise teach you about various dimensions of inequality in the U.S. health care system.

PART XIX

Environment, Population, and Social Change

60

Poisoning the Planet

The Struggle for Environmental Justice

DAVID NAGUIB PELLOW AND ROBERT J. BRULLE

Toxic waste dumping and other environmental hazards are more common in neighborhoods populated by people of color and poor or working-class people. David Pellow and Robert Brulle discuss how some groups are organizing to challenge this.

One morning in 1987 several African-American activists on Chicago's southeast side gathered to oppose a waste incinerator in their community and, in just a few hours, stopped 57 trucks from entering the area. Eventually arrested, they made a public statement about the problem of pollution in poor communities of color in the United States—a problem known as environmental racism. Hazel Johnson, executive director of the environmental justice group People for Community Recovery (PCR), told this story on several occasions, proud that she and her organization had led the demonstration. Indeed, this was a remarkable mobilization and an impressive act of resistance from a small, economically depressed, and chemically inundated community. This community of 10,000 people, mostly African–American, is surrounded by more than 50 polluting facilities, including landfills, oil refineries, waste lagoons, a sewage treatment plant, cement plants, steel mills, and waste incinerators. Hazel's daughter, Cheryl, who has worked with the organization since its founding, often says, "We call this area the 'Toxic Doughnut' because everywhere you look, 360 degrees around us, we're completely surrounded by toxics on all sides."

THE ENVIRONMENTAL JUSTICE MOVEMENT

People for Community Recovery was at the vanguard of a number of local citizens' groups that formed the movement for environmental justice (EJ). This movement, rooted in community-based politics, has emerged as a significant player at the local, state, national, and, increasingly, global levels. The movement's origins lie in local activism during the late 1970s and early 1980s aimed at combating environmental racism and environmental inequality—the unequal distribution of pollution across the social landscape that unfairly burdens poor neighborhoods and communities of color.

The original aim of the EJ movement was to challenge the disproportionate location of toxic facilities (such as landfills, incinerators, polluting factories, and mines) in or near the borders of economically or politically marginalized communities. Groups like PCR have expanded the movement and, in the process, extended its goals beyond removing existing hazards to include preventing new environmental risks and promoting safe, sustainable, and equitable forms of development. In most cases, these groups contest governmental or industrial practices that threaten human health. The EJ movement has developed a vision for social change centered around the following points:

- All people have the right to protection from environmental harm.
- Environmental threats should be eliminated before there are adverse human health consequences.
- The burden of proof should be shifted from communities, which now need to prove adverse impacts, to corporations, which should prove that a given industrial procedure is safe to humans and the environment.
- Grassroots organizations should challenge environmental inequality through political action.

The movement, which now includes African–American, European–American, Latino, Asian–American/Pacific-Islander, and Native–American communities, is more culturally diverse than both the civil rights and the traditional environmental movements, and combines insights from both causes.

ENVIRONMENTAL INEQUALITIES

Researchers have documented environmental inequalities in the United States since the 1970s, originally emphasizing the connection between income and air pollution. Research in the 1980s extended these early findings, revealing that communities of color were especially likely to be near hazardous waste sites. In 1987, the United Church of Christ Commission on Racial Justice released a groundbreaking national study entitled *Toxic Waste and Race in the United States*, which revealed the intensely unequal distribution of toxic waste sites across the United States. The study boldly concluded that race was the strongest predictor of where such sites were found.

In 1990, sociologist Robert Bullard published *Dumping in Dixie*, the first major study of environmental racism that linked the siting of hazardous facilities to the decades-old practices of spatial segregation in the South. Bullard found that African–American communities were being deliberately selected as sites for the disposal of municipal and hazardous chemical wastes. This was also one of the first studies to examine the social and psychological impacts of environmental pollution in a community of color. For example: across five communities in Alabama, Louisiana, Texas, and West Virginia, Bullard found that the majority of people felt that their community had been singled out for the location of a toxic facility (55 percent); experienced anger at hosting this facility in their community (74 percent); and yet accepted the idea that the facility would remain in the community (77 percent).

Since 1990, social scientists have documented that exposure to environmental risks is strongly associated with race and socioeconomic status. Like Bullard's *Dumping in Dixie*, many studies have concluded that the link between polluting facilities and communities of color results from the deliberate placement of such facilities in these communities rather than from population-migration patterns. Such communities are systematically targeted for the location of polluting industries and other locally unwanted land uses (LULUs), but residents are fighting back to secure a safe, healthy, and sustainable quality of life. What have they accomplished?

LOCAL STRUGGLES

The EJ movement began in 1982, when hundreds of activists and residents came together to oppose the expansion of a chemical landfill in Warren County, North Carolina. Even though that action failed, it spawned a movement that effectively mobilized people in neighborhoods and small towns facing other LULUs. The EJ movement has had its most profound impact at the local level. Its successes include shutting down large waste incinerators and landfills in Los Angeles and Chicago; preventing polluting operations from being built or expanded, like the chemical plant proposed by the Shintech Corporation near a poor African-American community in Louisiana; securing relocations and home buyouts for residents in polluted communities like Love Canal, New York; Times Beach, Missouri; and Norco, Louisiana; and successfully demanding environmental cleanups of LULUs such as the North River Sewage Treatment plant in Harlem.

The EJ movement helped stop plans to construct more than 300 garbage incinerators in the United States between 1985 and 1998. The steady expansion of municipal waste incinerators was abruptly reversed after 1990. While the cost of building and maintaining incinerators was certainly on the rise, the political price of incineration was the main factor that reversed this tide. The decline of medical-waste incinerators is even more dramatic.

Sociologist Andrew Szasz has documented the influence of the EJ movement in several hundred communities throughout the United States, showing that organizations such as Hazel Johnson's People for Community Recovery were instrumental in highlighting the dangers associated with chemical waste incinerators in their neighborhoods. EJ organizations, working in local coalitions, have had a number of successes, including shutting down an incinerator that was once the largest municipal waste burner in the Western Hemisphere. The movement has made it extremely difficult for firms to locate incinerators, landfills, and related LULUs anywhere in the nation, and almost any effort to expand existing polluting facilities now faces controversy.

BUILDING INSTITUTIONS

The EJ movement has built up local organizations and regional networks and forged partnerships with existing institutions such as churches, schools, and neighborhood groups. Given the close association between many EJ activists and environmental sociologists, it is not surprising that the movement has notably influenced the university. Research and training centers run by sociologists at several universities and colleges focus on EJ studies, and numerous institutions of higher education offer EJ courses. Bunyan Bryant and Elaine Hockman, searching the World Wide Web in 2002, got 281,000 hits for the phrase "environmental justice course," and they found such courses at more than 60 of the nation's colleges and universities.

EJ activists have built lasting partnerships with university scholars, especially sociologists. For example, Hazel Johnson's organization has worked with scholars at Northwestern University, the University of Wisconsin, and Clark Atlanta University to conduct health surveys of local residents, study local environmental conditions, serve on policy task forces, and testify at public hearings. Working with activists has provided valuable experience and training to future social and physical scientists.

The EJ movement's greatest challenge is to balance its expertise at mobilizing to oppose hazardous technologies and unsustainable development with a coherent vision and policy program that will move communities toward sustainability and better health. Several EJ groups have taken steps in this direction. Some now own and manage housing units, agricultural firms, job-training facilities, farmers' markets, urban gardens, and restaurants. On Chicago's southeast side. PCR partnered with a local university to win a federal grant, with which they taught lead-abatement techniques to community residents who then found employment in environmental industries. These successes should be acknowledged and praised, although they are limited in their socio-ecological impacts and longevity. Even so, EJ activists, scholars, and practitioners would do well to document these projects' trajectories and seek to replicate and adapt their best practices in other locales.

LEGAL GAINS AND LOSSES

The movement has a mixed record in litigation. Early on, EJ activists and attorneys decided to apply civil rights law (Title VI of the 1964 Civil Rights Act) to the environmental arena. Title VI prohibits all government and industry programs and activities that receive federal funds from discriminating against persons based on race, color, or national origin. Unfortunately, the courts have uniformly refused to prohibit government actions on the basis of Title VI without direct evidence of discriminatory intent. The Environmental Protection Agency (EPA) has been of little assistance. Since 1994, when the EPA began accepting Title VI claims, more than 135 have been filed, but none has been formally resolved. Only one federal agency has cited environmental justice concerns to protect a community in a significant legal case: In May 2001, the Nuclear Regulatory Commission denied a permit for a uranium enrichment plant in Louisiana because environmental justice concerns had not been taken into account.

With regard to legal strategies, EJ activist Hazel Johnson learned early on that, while she could trust committed EJ attorneys like Keith Harley of the Chicago Legal Clinic, the courts were often hostile and unforgiving places to make the case for environmental justice. Like other EJ activists disappointed by the legal system, Johnson and PCR have diversified their tactics. For example, they worked with a coalition of activists, scholars, and scientists to present evidence of toxicity in their community to elected officials and policy makers, while also engaging in disruptive protest that targeted government agencies and corporations.

NATIONAL ENVIRONMENTAL POLICY

The EJ movement has been more successful at lobbying high-level elected officials. Most prominently, in February 1994, President Clinton signed Executive Order 12898 requiring all federal agencies to ensure environmental justice in their practices. Appropriately, Hazel Johnson was at Clinton's side as he signed the order. And the Congressional Black Caucus, among its other accomplishments, has maintained one of the strongest environmental voting records of any group in the U.S. Congress.

But under President Bush, the EPA and the White House did not demonstrate a commitment to environmental justice. Even Clinton's much-vaunted Executive Order on Environmental Justice has had a limited effect. In March 2004 and September 2006, the inspector general of the EPA concluded that the agency was not doing an effective job of enforcing environmental justice policy. Specifically, he noted that the agency had no plans, benchmarks, or instruments to evaluate progress toward achieving the goals of Clinton's Order. While President Clinton deserves some of the blame for this, it should be no surprise that things have not improved under the Bush administration. In response, many activists, including those at PCR, have shifted their focus from the national level back to the neighborhood, where their work has a more tangible influence and where polluters are more easily monitored. But in an era of increasing economic and political globalization, this strategy may be limited.

GLOBALIZATION

As economic globalization—defined as the reduction of economic borders to allow the free passage of goods and money anywhere in the world—proceeds largely unchecked by governments, as the United States and other industrialized nations produce larger volumes of hazardous waste, and as the degree of global social inequality also rises, the frequency and intensity of EJ conflicts can only increase. Nations of the global north continue to export toxic waste to both domestic and global "pollution havens" where the price of doing business is much lower, where environmental laws are comparatively lax, and where citizens hold little formal political power.

Movement leaders are well aware of the effects of economic globalization and the international movement of pollution and wastes along the path of least resistance (namely, southward). Collaboration, resource exchange, networking, and joint action have already emerged between EJ groups in the global north and south. In the last decade EJ activists and delegates have traveled to meet and build alliances with colleagues in places like Beijing, Budapest, Cairo, Durban, The Hague, Istanbul, Johannesburg, Mumbai, and Rio de Janeiro. Activist colleagues outside the United States are often doing battle with the same transnational corporations that U.S. activists may be fighting at home. However, it is unclear if these efforts are well financed or if they are leading to enduring action programs across borders. What is certain is that if the EJ movement fails inside the United States, it is likely to fail against transnational firms on foreign territory in the global south.

Although EJ movements exist in other nations, the U.S. movement has been slow to link up with them. If the U.S. EJ movement is to survive, it must go global. The origins and drivers of environmental inequality are global in their reach and effects. Residents and activists in the global north feel a moral obligation to the nations and peoples of the south, as consumers, firms, state agencies, and military actions within northern nations produce social and ecological havoc in Latin America, the Caribbean, Africa, Central and Eastern Europe, and Asia. Going global does not necessarily require activists to leave the United States and travel abroad, because many of the major sources of global economic decision-making power are located in the north (corporate headquarters, the International Monetary Fund, the World Bank, and the White House). The movement must focus on these critical (and nearby) institutions. And while the movement has much more to do in order to build coalitions across various social and geographic boundaries, there are tactics, strategies, and campaigns that have succeeded in doing just that for many years. From transnational activist campaigns to solidarity networks and letter-writing, the profile of environmental justice is becoming more global each year.

After Hazel Johnson's visit to the Earth Summit in Rio de Janeiro in 1992, PCR became part of a global network of activists and scholars researching and combating environmental inequality in North America, South America, Africa, Europe, and Asia. Today, PCR confronts a daunting task. The area of Chicago in which the organization works still suffers from the highest density of landfills per square mile of any place in the nation, and from the industrial chemicals believed to be partly responsible for the elevated rates of asthma and other respiratory ailments in the surrounding neighborhoods. PCR has managed to train local residents in lead-abatement techniques; it has begun negotiations with one of the Big Three auto makers to make its nearby manufacturing plant more ecologically sustainable and amenable to hiring locals, and it is setting up an environmental science laboratory and education facility in the community through a partnership with a major research university.

What can we conclude about the state of the movement for environmental justice? Our diagnosis gives us both hope and concern. While the movement has accomplished a great deal, the political and social realities facing activists (and all of us, for that matter) are brutal. Industrial production of hazardous wastes continues to increase exponentially; the rate of cancers, reproductive illnesses, and respiratory disorders is increasing in communities of color and poor communities; environmental inequalities in urban and rural areas in the United States have remained steady or increased during the 1990s and 2000s; the income gap between the upper classes and the working classes is greater than it has been in decades; the traditional, middle-class, and mainly white environmental movement has grown weaker; and the union-led labor movement is embroiled in internecine battles as it loses membership and influence over politics, making it likely that ordinary citizens will be more concerned about declining wages than environmental protection. How well EJ leaders analyze and respond to these adverse trends will determine the future health of this movement. Indeed, as denizens of this fragile planet, we all need to be concerned with how the EJ movement fares against the institutions that routinely poison the earth and its people.

KEY CONCEPTS

environmental justice

environmental racism

DISCUSSION QUESTIONS

1. Can you find a newspaper or Internet news story about a community fighting against a polluting industry? What are the key issues facing a neighborhood with an environmental concern?
2. Do you see environmental hazards near your school or your home? In your community, do you see a difference in the racial composition of neighborhoods that are close to environmental hazards versus ones that are not?

61

Zooming In on Diversity

WILLIAM H. FREY

William Frey documents the many changes that are making the U.S. population more diverse than at any other time in history.

America's changing racial and ethnic makeup will profoundly transform the nation's regional landscape for at least the next four decades. Consumer markets, politics and day-to-day personal transactions simply will not go on as they have up to now as this change sweeps the nation. By 2050, only half the population will be non-Hispanic white, the Census Bureau projects. The Hispanic and Asian populations will both triple, the black population will almost double and the white population will barely hold its own.

Yet, what looks to the naked eye like a diversified melting pot at the national level takes on a dramatically different look if you zoom in on specific regions and metropolitan areas. Why? Mostly because people moving to the United States from other countries pick certain places to settle, whereas people who already live within national borders choose others. The lion's share of immigrating Hispanics and Asians tends to cluster in gateway locales, while domestic migration networks of whites and blacks often follow different paths. And there are reasons for this.

Diverging migration patterns have unevenly distributed racial and ethnic diversity into America's regions. Three states—New Mexico, Hawaii and California—already stand out as the nation's first nonwhite majority states. At the same time, there are the 15 states, where minorities account for less than 15 percent of the population. Each minority group, including Hispanics, Asians and blacks, has tended to cluster in geographic patterns that begin to suggest staying power.

Exactly what is the role of race and ethnicity in ingrained and emerging patterns of migration? How rapidly are Hispanics and Asians dispersing away from the traditional gateway regions? And to what extent is their migration converging with the mainstream? New answers to these questions can be drawn from census migration statistics.... [W]hile America may not become a true national melting pot anytime soon, there is a measure of "simmering" going on.

IMMIGRANT MAGNETS, DOMESTIC MIGRANT MAGNETS

Arriving immigrant minorities tend to cluster geographically because destination communities provide them with a comfort zone of familiarity, while others require greater acclimatization for new residents to survive and thrive. Ethnic enclaves in gateway metropolises like Los Angeles and New York contain already established institutions—churches, community centers, stores, neighbors—that make new arrivals feel at home, and give them social and economic support. Immigration laws also foster clustering. Since family reunification is regarded as a priority in legal immigration, family-related migrant "chains" direct many new arrivals to the nation's gateway cities.

If immigrants choose a U.S. destination based on familiarity and cultural support to get started, domestic migrants tend to move for more pragmatic, hardheaded reasons. Most times, it's economic opportunity in the labor market. What's more, whites and blacks are better represented among domestic migrants. The places in which they choose to resettle are less constrained than those of immigrants, for whom a familiar language and the presence of family mean more than the local unemployment rate in selecting a destination.

A distinction between immigration and domestic migration—not a new phenomenon—still holds.... Among the nine leading "magnet metro areas" for immigrants, only one—Dallas—is on the list of the largest metro destinations of domestic migrants. Traditional gateways, New York, Los Angeles, San Francisco, Chicago, Miami and Washington, D.C. attract the greatest number of immigrants. These six areas have led all others since the mid-1960s, when current California, Illinois and New Jersey, show the greatest out-migration of both African Americans and Anglos, mirroring the overall domestic migration flight from these areas.

Both the black and white movement patterns help to reinforce the distinct racial and ethnic structures of their destination regions. The out-migration of whites, in particular, serves to increase the minority profiles of the big immigration states. At the same time, the white movement to the "non-California" West tends to give their states and metros a more suburban feel. And the black migration back to the South tends to reinforce the region's longstanding largely white-black demographic profile.

HISPANICS AND ASIANS

There are two ways that Hispanic and Asian movements affect metro areas and states. As we've said, destinations among recent immigrants tend to focus on traditional gateways. However, there is an increasing tendency for both groups to participate in the domestic migration. This seems to be the primary vehicle for their dispersion beyond the gateways.

The greatest destinations for Hispanic immigrants include the traditional gateways, led by Los Angeles, New York and Miami. Only two non-immigrant magnet metros—Phoenix and Atlanta—rank in the top 10 recent Hispanic immigrant destinations. Domestic migration among Hispanics contrasts with the destinations of recent immigrants.... To a large degree, these Hispanic domestic migration patterns mirror those for whites. Las Vegas and Phoenix represent the top destinations. Similarly, many domestic migrant Hispanics are leaving the traditional immigrant magnet areas of Los Angeles, New York, San Francisco and Chicago. To be sure, the latter metros are gaining many more Hispanics from immigration than are lost through domestic out-migration; but a pattern of dispersal is apparent.

The same pattern is shown at the state level where California, Texas, Florida, New York, Illinois and New Jersey lead all others in gaining Hispanic immigrants.

Yet, four of them are losing Hispanic domestic migrants to a broader swath of states across the country.... This dispersion of Hispanic domestic migrants reflects smaller numbers than those of the more concentrated immigrant flows, but the numbers are trending to increase over time.

The Asian patterns are similar to those of Hispanics, wherein immigrant destinations are losing Asians through domestic migration dispersal. However, compared with Hispanics, the dispersion of Asians away from the gateways is much lighter, reflecting their recency of arrival. The exception to this pattern is San Francisco, which draws Asians from both immigration and domestic migration....

MULTICULTURAL FLIGHT

... [Y]ears ago, I wrote an article called "The New White Flight" (*American Demographics*, April 1994) to highlight the fact that most of the domestic migrants leaving immigrant gateways were non-Hispanic whites. That movement did, and still does, reflect a kind of suburbanization across metropolitan areas and state boundaries toward places that offer affordable housing and improving employment prospects. The difference now is that this domestic out-migration from Los Angeles, New York and other high immigration metros is much more multicultural. There are more Hispanics than whites leaving Los Angeles as a result of domestic migration. In New York, while whites still dominate the exodus, Hispanics and blacks account for a larger share of out-migrants in the late 1990s than 10 years prior. It appears now that blacks and immigrant minorities are responding to the same middle-class crunch as whites, in leaving high immigration metropolitan areas.

The other side of the coin is what is happening in domestic migration magnets. Phoenix and Atlanta provide good examples. In both of these metros, domestic migrating whites and blacks tend to lead the overall gains. But noteworthy in both metros is a phenomenon in the increasing domestic in-migration of Hispanics and, to a lesser degree, Asians. While the overall demographic profiles of these areas do not have nearly the melting pot character of immigrant magnets like Los Angeles or New York, their strong domestic migration flows are creating employment opportunities for new immigrants.

CONCLUSION

Recent race-ethnic migration patterns show that America still has a long way to go before it becomes a coast-to-coast melting pot, where racial and ethnic groups spread evenly across the land. High immigration gateway states and metros continue to stand out as the most racially and ethnically diverse in the nation. They are not, by and large, magnets for white and black domestic migrants who tend to flow to ever-changing constellations of domestic migration magnets. Yet, this analysis also confirms, with migration data from the census, that there exists a true dispersion of Hispanics and Asians as part of a new multicultural flight away from several immigrant gateways. Gradually, these migration patterns will accelerate as they lead to further simmering among our diverse peoples toward a more integrated society.

KEY CONCEPTS

demography social change

DISCUSSION QUESTIONS

1. How are race and ethnicity involved in shaping the migration patterns that are changing the demography of the United States?
2. Would you describe the United States as a melting pot? Why or why not?

62

Climate Denial and the Construction of Innocence: Reproducing Transnational Environmental Privilege in the Face of Climate Change

KARL MARIE NORGAARD

By exploring patterns of denial that climate change is happening, Karl Marie Norgaard explores the idea of environmental privilege.

Global climate change is now altering the biophysical world around which human social systems are organized.... Yet the impacts of changing ecological conditions including unstable weather patterns, sea level rise, intensification of wildfire and increased storm intensity are experienced very differently by communities around the globe. As illustrated by the aftermath of Hurricane Katrina, or the need to relocate disappearing island nations in the South Pacific, the impacts of extreme weather events associated with climate change are disproportionately borne along the lines of gender, race and class, both within and across national boundaries (Parks & Roberts, 2006; Sweetman, 2009; Bullard & Wright, 2009; IPCC, 2007). As a result, the ecological changes that undermine social activities and infrastructure simultaneously reproduce gender, racial and class inequality in complex ways depending on social context (Lane & McNaught, 2009; Carmin & Agyeman, 2011; ... Taylor, 1997; ... Buechler, 2009).

Furthermore, race, class and gender shape people's concern about climate change on the one hand, and their contribution to the problem via carbon emissions on the other. People in nations with higher carbon emissions are not only less likely to be impacted by direct effects of climate change, but they are also less likely to show concern regarding climate change (Dunlap, 1998; Sandvik, 2008). For example, Norwegian scholar Hanno Sandvik reports a negative association between concern and national wealth and carbon dioxide emissions, and describes a "marginally significant" tendency that nations' per capita carbon dioxide emissions are negatively correlated to public concern. Sandvik writes, "the willingness of a nation to contribute to reductions in greenhouse gas emissions decreases with its share of these emissions" (2008:333).

Such trends highlight the fact that in today's globalized risk society, perceptions of near and far, immediate or abstract are politically charged social constructions. How do privileged people with knowledge of climate science recreate a sense of safety in the face of troubling climate events and information? What is the significance of their constructions of risk and concern in reproducing transnational power relations along the lines of race, gender and class? This paper draws upon a larger project that investigates how and why globally privileged people resist information about climate change using ethnographic observation and interviews in a rural Norwegian community I call 'Bygdaby' (see Norgaard, 2011a). In this larger study I describe how people in this community were aware of climate change, but simultaneously re-created a sense that "everything was fine." I then draw upon theory from sociology of emotions and culture to construct a theory of socially organized denial to describe the condition in which people simultaneously "know and don't know" about climate change. In the present paper, I evaluate the material and symbolic significance of both climate change denial and a phenomenon I call "the construction of innocence" in terms of their role in reproducing global material and ideological power relations along the lines of race, class and gender. I reanalyze this existing data with particular attention to the question of transnational environmental privilege. In the community of Bygdaby information about climate change was troubling along two particular dimensions: it disrupted people's sense of security and order, and their sense of innocence. In the course of normalizing a troubling situation through "tools of order" and "tools of innocence," residents simultaneously reproduce transnational privilege.

CLIMATE CHANGE IN THE GLOBAL SOUTH: BEARING THE BRUNT OF A CHANGING CLIMATE

Although the changing climate will eventually impact everyone, it has already begun to precipitate the most extensive and violent impacts to date against the poor, women and people of color of the globe (Sweetman, 2009; World Bank, 2010; Collectif Argos, 2010; Parks & Roberts, 2006). These groups of people contribute least to carbon emissions, making climate change clearly an issue of global environmental justice (Parks & Roberts, 2006, Anguelovski & Roberts, 2011). Unequal effects manifest differently around the world according to geography, as well as the particular intersection of sexism and racism within different social contexts. Unequal impacts can be both acute during crisis events such as flooding or hurricanes (Fothergill & Peek, 2004; Blaikie et al., 1994), or ongoing over long periods of time as is for example the case with extended droughts. Thus on the one hand, women working in subsistence activities face particular vulnerabilities due to long term changing flood and climate regimes (Morton, 2007), while African American women post-Katrina have faced more difficulty securing loans to rebuild businesses and homes than their white counterparts (Bullard & Wright, 2009…).

Not all of those who are most impacted live in the Global South. Race, gender and class radically shape experiences of communities within nations of the Global North as well. For example, indigenous communities in the US and Canadian Arctic are on the front lines of direct impacts from climate change (Arctic Climate Impact Assessment, 2004). Arctic indigenous communities face more direct impacts due to combinations of ecological and social factors including their location in the Arctic where changes are occurring more rapidly, due to more intact cultural ties to land such that people retain subsistence activities and culture and social structure are organized around conditions of the natural world (Trainor et al., 2009). Within indigenous communities, women may face added burdens in the form of increased "reproductive labor" of maintaining family and community cohesion as when entire island communities are displaced by storm surge and relocated to the mainland.… Thus, the particular effects of climate change re-organize gender and racial community relations in unique and often complex ways.

CLIMATE CHANGE IN THE GLOBAL NORTH HIGH EMISSIONS, POLITICAL APATHY AND CLIMATE SKEPTICISM

The high concentrations of carbon dioxide that have so altered our planet's atmosphere have been generated by the nations of the Global North. Twenty percent of total world carbon emissions come from the United States alone. Despite the extreme seriousness of global climate change, there has been meager public response in the way of social movement activity, behavioral changes or public pressure on governments within the wealthy industrialized nations who generate the bulk of carbon emissions and whose economic and political infrastructure affords them the greatest resource for responding (Brechin, 2008; Lorenzoni & Pidgeon, 2006). Yale scholar Anthony Leiserowitz who has conducted extensive work on public opinion on climate change notes that "Large majorities of Americans believe that global warming is real and consider it a serious problem, yet global warming remains a low priority relative to other national and environmental issues and lacks a sense of urgency" (2007)….

Furthermore, within these high emission nations, the little data that exist appear to extend this inverse relationship between concern and carbon footprint across race, class and gender. In a recent study of gender differences in climate change in the United States Aaron McCright finds that, "Women express more concern about climate change than men do. A greater percentage of women than men worry about global warming a great deal, believe global warming will threaten their way of life during their lifetime, and believe the seriousness of global warming is underestimated in the news" (2010:78–79). Although the data are rarely reported, women also tend to have smaller carbon footprints than do men. German policy analyst Meike Spitzer describes differences in the carbon footprint between women and men in Northern countries. Spitzer describes how men use less public transportation than do women, drive personal vehicles longer distances and more often for personal trips than do women. While much attention has pointed to the disproportionate carbon footprint of citizens of the Global North, and to the gendered impacts of climate change (especially for women in the Global South) almost no attention has been paid to the disproportionate carbon footprint between women and men in the global north (Spitzer, 2009).

Meanwhile in their study on race and perceptions of climate change Leiserowitz and Akerlof (2010) find that "Hispanics, African Americans and people of other races and ethnicities were often the strongest supporters of climate and energy policies and were also more likely to support these policies even if they incurred greater costs" (p. 7). Finally, it should be noted that the "climate skeptic" movement (whose followers and key advocates are predominantly male, white and from the United States), outright negates the seriousness of climate change, and thus the claims and experiences of poor women of color in the Global South (Jacques, 2009; McCright, 2010).

While "apathy" in the United States is particularly notable, this gap between the severity of the problem and its lack of public salience is visible in most Western nations (Lorenzoni et al., 2007; Poortinga & Pidgeon, 2003). Especially for urban dwellers in the rich and powerful Northern countries climate change is seen as "no more than background noise" (Brechin, 2008; Lorenzoni & Pidgeon, 2006). Indeed, no nation has a base of public citizens that are sufficiently socially and politically engaged to effect the level of change that predictions of climate science would seem to warrant. Much research has presumed a lack of information to be the limiting factor in this equation. These studies emphasized either the complexity of climate science or political economic corruption as reasons people do not adequately understand what is at stake.... [O]nly two simple facts are essential to understanding climate change: global warming is the result of an increase in the concentration of carbon dioxide in the earth's atmosphere, and the single most important source of carbon dioxide is the combustion of fossil fuels, most notably coal and oil. And while such "information deficit" explanations may explain part of the story, they do not account for the behavior of the significant number of people know about global warming and express concern, yet still fail to take any action. In contrast to existing studies that focus on the public's lack of information or concern about global warming, my own work has illustrated how holding information at a distance is actually an active strategy for people as they negotiate their relationship with climate change. That is to say people had a variety of methods available for normalizing or minimizing disturbing information. These methods can be called strategies of denial (Norgaard, 2011a, 2011b).

CLIMATE DENIAL AND ENVIRONMENTAL PRIVILEGE

This linkage between climate denial and environmental justice draws our attention to the concept of "environmental privilege" (Park & Pellow, 2011; ... Norgaard, 2011a). The intersection of environmental justice and privilege is an extremely important concept that is only now beginning to receive attention. Most scholarly work on privilege examines gender, race or class privilege, while most environmental justice work "studies down," examining the experiences of people who are structurally disadvantaged rather than those with privilege.... Park and Pellow write that "environmental privilege results from the exercise of economic, political and cultural power that some groups enjoy, which enables them exclusive access to coveted environmental amenities such as forests, parks, mountains, rivers, coastal property, open lands and elite neighborhoods. Environmental privilege is embodied in the fact that some groups can access spaces and resources which are protected from the kinds of ecological harm that other groups are forced to contend with everyday" (2011:4).

For people of color living on low lying islands or struggling from flooding or drought, the key questions of the moment may be how to effectively organize to bring attention to their plight and justice to their lives. For middle-class white environmentalists living in wealthy industrialized nations like myself, the key questions related to climate change have to do with the production of social inaction and denial: why are so many of us in the Global North so willing to ignore climate change in our daily lives? How do people manage to ignore disturbing information about climate change and essentially live in denial?

Park and Pellow describe environmental privilege emerging from "the exercise of economic, political and cultural power" (2011:4 [my italics]). Culture is implicated in the reproduction of social relations, but exactly how does this work? How do privileged people create a positive sense of self, normalcy and even sense of their innocence, in the face of knowledge of their high carbon footprints? In fact, societies develop and reinforce a whole repertoire of techniques or "tools" for ignoring disturbing problems. Using Anne Swidler's (1986) model of culture as a tool kit and Jocelyn Hollander and Hava Gordon's (2006) thirteen tools of social construction, I have described how people create a sense of everyday reality using features of everyday life as tools. It is significant that what I describe as "climate denial" felt to people in Bygdaby (and, indeed, to people around the world) like "everyday life." The sense of climate change as a distant problem for which other people are responsible was produced through cultural practices of everyday life. This focus on denial and the construction of innocence looks at the lived experience of privileged people, examining not only the "what" of environmental privilege, but the "why" and the "how."...

Why should we bother to engage with the concept of environmental privilege? Both the complexity of climate change science, and the changing organization of space and time make the global reproduction of race, class and gender inequality increasingly complex and difficult to understand. It is easy to see power operating when key political and economic decision makers negotiate contracts with Shell, British Petroleum, and Exxon and when representatives of nation-states negotiate emissions-trading strategies. Yet the people I spoke with in Bygdaby played a critical role in legitimizing the status quo by not talking about global warming even in the face of late winter snow and a lake that never froze. In normalizing the status quo, the construction of denial and innocence work to silence the needs and voices of women and people of color in the Global South, and thus reproduce global inequality along the lines of gender, race and class. Given that Norwegian economic prosperity and way of life are intimately tied to the production of oil, ignoring or downplaying the issue of climate change serves to maintain Norwegian global economic interests and to perpetuate global environmental injustice.

If we are to really grapple with the how and why of the reproduction of environmental privilege we must take seriously the experiences of privileged social actors themselves. Taking seriously their experience means acknowledging the complexity and difficulty of occupying this social location. While privilege may act like a breeze at one's back, making life easier, people in privileged situations also experience the complex and ambivalent mental and emotional landscape of denial, with all of its feelings of guilt, responsibility and hopelessness. Despite being in positions of power, privileged people may feel very powerless. People occupying privileged social positions encounter "invisible paradoxes," awkward, troubling moments which they seek to avoid, collectively pretend not to have experienced (often as a matter of social tact), and forget as quickly as possible once they have passed.

Global capitalism is currently producing wider divides between the material conditions of the lives of haves and have-nots, and closing gaps in space and time, bringing privileged people ever closer to the worlds of those we exploit through cheap airfares and quality digital internet images. Through the active production of a safe, secure mental world through tools of order and innocence, environmental problems are kept invisible to those with the time, energy, cultural capital and political clout to generate moral outrage in a variety of ways. Wastes and hazards are exported to other nations, other places, other populations. Privileged people are protected from full knowledge of environmental (and many other social problems) by national borders, gated communities, segregated neighborhoods and their own fine-tuned yet unconscious practices of not noticing, looking the other way and normalizing the disturbing information they constantly come across. Ecological collapse and climate change seem fanciful to those in the "safe" and "stable" societies of the North as we buy our fruits and vegetables from South America, our furniture from Southeast Asia and send our wastes into the common atmosphere. In this way social inequality itself works to perpetuate environmental degradation. From this new context emerge new dimensions of invisibility, denial and innocence. Privileged people know their carbon footprints are high, they know many of the reasons why this is so, and some of the global environmental justice implications of their actions.

Ultimately, the phenomenon of climate denial further underscores the need for what Val Plumwood describes as the "ecological self" which is an ethic which extends recognition to all the "others" in the dualistic system. As the impacts of climate change continue privileged people around the world will be faced with more and more opportunities to develop or reformulate their "moral imagination" and "imagine the reality" of what is happening, or to construct their own innocence from the beneficial resources of their culture's particular self-affirming toolkit.

REFERENCES

ACIA. (2004). Impacts of a Warming Arctic. Summary report of the Arctic Climate Impact Assessment. Cambridge University Press, Cambridge, UK.

Anguelovski, I. & Roberts, D. (2011). Spatial Justice and Climate Change. In J. Carmin & J. Agyeman (Eds.), Environmental inequalities beyond borders: Local perspectives on global injustices, pp. 19-44. Cambridge: MIT Press.

Blaikie, P., Cannon, T., Davis, I., & Wisner, B. (1994). At risk: Natural hazards, people's vulnerability, and disasters. Routledge, New York.

Brechin, S.R. (2008). Ostriches and change: A response to "global warming and sociology." Current Sociology, 56:467–474.

Buechler, S. (2009). Gender, water, and climate change in Sonora, Mexico: Implications for policies and programmes on agricultural incomegeneration. Gender and Development, 17(l):51–66.

Bullard, R. & Wright, B. (2009). Race, place, and environmental justice after Hurricane Katrina: Struggles to reclaim, rebuild, and revitalize New Orleans and the Gulf Coast. Boulder, CO: Westview Press.

Carmin, J. & Agyeman, J. (2011). Environmental inequalities beyond borders: Local perspectives on global injustices. Cambridge: MIT Press.

Collectif Argos. (2010). Climate refugees. Cambridge: MIT Press.

Dunlap, R.E. (1998). Lay perceptions of global risk: Public views of global warming in cross-national context. International Sociology, 13:473–498.

Fothergill, A. & Peek, L. (2004). Poverty and disasters in the United States: A review of recent sociological findings. Natural Hazards Journal, 32(1):89–110.

Hollander, J. & Gordon, H. (2006). The process of social construction. Talk Symbolic Interaction, 29(2):183–212.

IPCC (Intergovernmental Panel on Climate Change). (2008). Climate change 2007: Impacts, adaption and vulnerability fourth assessment report. Cambridge: Cambridge University Press for the IPCC.

Jacques, P. (2009). Environmental skepticism: Ecology, power and public life. Ashgate Publishing.

Lane, R. & McNaught, R. (2009). Building gendered approaches to adaptation in the Pacific. Gender and Development, 17(l):77–80.

Leiserowitz, A. (2007). Communicating the risks of global warming: American risk perceptions, affective images and interpretive communities. In S.C. Moser & L. Dilling (Eds.), Creating a climate for change: Communicating climate change and facilitating social change. New York: Cambridge University Press.

Leiserowitz, A. & Akerlof, K. (2010). Race, ethnicity and public responses to climate change. Yale University and George Mason University. New Haven, CT. Yale Project on Climate Change. http://environment.yale.edu/uploads/RaceEthnicity2010.pdf

Lorenzoni, I., Nicholson-Cole S., & Whitmarsh, L. (2007). Barriers perceived to engaging with climate change among the UK public and their policy implications. Global Environmental Change, 17(3-4):445–459.

Lorenzoni, I. & Pidgeon, N.F. (2006). Public views on climate change: European and USA perspectives. Climatic Change, 77:73–95.

McCright, A. (2010). The effects of gender on climate change knowledge and concern in the American public. Population and Environment, 32:66–87.

Morton, J. (2007). The impact of climate change on smallholder and subsistence agriculture. Proceedings of the National Academy of Sciences of the United States of America, 104(50):19680–19685.

Norgaard, K.M. (2011a). Living in denial: Climate change, emotions and everyday life. Cambridge: MIT Press.

Norgaard, K.M. (2011b). The Social organization of climate denial: Emotions, culture and political economy. In J. Dryzek, D. Schlosberg, & R. Norgaard (Eds.), The Oxford handbook of climate change and society, pp. 399–413. Oxford University Press.

Park, L.S-H. & Pellow, D. (2011). The Slums of Aspen: Immigrants versus the environment in America's Eden. New York: New York University Press.

Parks, B. & Roberts, T.J. (2010). Climate change social theory and justice. Theory, Culture & Society, Mar-May, 27(2/3):134–166, 33.

Poortinga, W. & Pidgeon, N. (2003). Public perceptions of risk, science and governance: Main findings of a British survey of five risk cases. University of East Anglia and MORI, Norwich.

Sandvik, H. (2008). Public concern over global warming correlates negatively with national wealth. Climatic Change, 90(3):333–341.

Spitzer, M. (2009). How global warming is gendered. In A. Salleh (Ed.), Ecosufficiency and global justice: Women write political ecology, pp. 218–229. Pluto Press.

Sweetman, C. (2009). Introduction to special issue on climate change and climate justice. Gender and Development, 17(1):1–3.

Swidler, A. (1986). Culture in action. American Sociological Review, 51:273–86.

Taylor, D. (1997). American environmentalism: The role of race, class and gender in shaping activism, 1820–1995. Race, Gender & Class, 5(1):16–62.

Trainor, S., Godduhn, A., Duffy, L.K., Stuart Chapin III, F., Natcher, D.C., Kofinas, G., & Huntington, H.P. (2009). Environmental injustice in the Canadian Far North: Persistent organic pollutants and Arctic climate impacts. In J. Agyeman, R. Haluza-DeLay, P. O'Riley, & P. Cole (Eds.), Speaking for ourselves: Environmental justice in Canada, pp. To be completed. UBC Press.

World Bank. (2010). World development report 2010: Development and climate change international bank for reconstruction and development/The World Bank, Washington D.C.

KEY CONCEPTS

climate change | environmental privilege | transnational

DISCUSSION QUESTIONS

1. What does Norgaard mean by environmental privilege, and how is it illustrated in research findings?
2. What gender differences does Norgaard report with regard to environmental footprints? Does this surprise you and why?

Applying Sociological Knowledge: An Exercise for Students

The BP oil spill in the Gulf of Mexico in the year 2010 is one of the largest environmental disasters in history. Using various sources of evidence, explore the social consequences of this disaster for the different communities and groups of people affected by it. Analyze what you have found by examining such social facts as the race, class, and gender consequences for different communities and groups of this environmental catastrophe.

Glossary

A

achievement gap the differences between groups of students, typically based on socioeconomic status and race/ethnicity, in their academic performance

acquaintance rape rape committed by a person who is known to the victim

adolescence the period of time between childhood and adulthood, often thought of as a period of many adjustments

Affordable Care Act (2010) law passed in 2010 that expands health insurance coverage for many Americans

age cohort group of people born during the same time period

age prejudice negative attitude about an age group that is generalized to all people in that group

alienation feelings of powerlessness and separation from one's group or society

American dream a belief system that the United States is an open society where anyone can be economically and socially successful with hard work

anticipatory socialization the learning of social roles that precedes the actual holding of a given role

assimilation process by which a minority becomes socially, economically, and culturally absorbed within the dominant society

B

bourgeoisie term used to loosely describe either the ruling or middle class in a capitalist society

bureaucracy type of formal organization characterized by an authority hierarchy, with a clear division of labor, explicit rules, and impersonality

C

calculability the idea that the delivery of products in service industries (such as fast food) can be quantified, including portion size, cost, and amount of time it takes for the customer to get the product

capitalism economic system based on the principles of market competition, private property, and the pursuit of profit

capitalist class the group that owns the means of production in a society

caste system system of stratification (characterized by low social mobility) in which one's place in the stratification system is determined and fixed by birth

class see **social class**

class analysis analysis of inequality centered on people's relationship to social class

class consciousness the understanding of one's position in the class system and an awareness of the exploitation of groups because of their class status

climate change the environmental problem associated with global warming

collective consciousness body of beliefs that are common to a community or society and that give people a sense of belonging

color-blind racism belief that racism no longer defines people's status in society, including being oblivious to continuing racial inequities

communism economic system in which the state is the sole owner of the systems of production

community a group of people, often thought to be in rather close proximity, who share a common perception of themselves as linked through social contact

conflict theory perspective that emphasizes the role of power and coercion in producing social order

contingent worker a temporary or nonpermanent worker who works for a fixed period or a specific task or project

conversion process by which a person achieves membership of a religious group

credentialism the practice of requiring increasingly higher educational credentials even when they may not be needed in a given job

cross-cultural pertaining to complex systems of meaning and behavior across more than one group or society; relating to two or more cultures

cultural imperialism forcing all groups in a society to accept the dominant culture as their own

culture complex system of meaning and behavior that defines the way of life for a given group or society

culture of poverty the view that the poor have a different value system thereby causing them to be in poverty

cyberspace interaction sharing by two or more persons of a virtual reality experience via communication and interaction with each other

D

data systematic information that sociologists use to investigate research questions

debunking the sociological process of challenging taken-for-granted assumptions about social reality

deindustrialization structural transformation of a society from a manufacturing-based economy to a service-based economy

democracy system of government based on the principle of representing all people through the right to vote

demography the scientific study of population

demedicalization the process when illness is socially defined as "normal"

deviance behavior that is recognized as violating expected rules and norms

discrimination overt negative and unequal treatment of members of some social group or stratum solely because of their membership in that group or stratum

division of labor systematic interrelation of different tasks that develops in complex societies

divorce rate number of divorces per some number (usually 1,000) in a given year

doing gender analytical framework that interprets gender as an activity accomplished through everyday interaction

domestic partner a person who shares a residence with a sexual partner, usually without a legally recognized union

dominant culture culture of the most powerful group in society

double consciousness realizing that one is being viewed as the "other" in a social situation and simultaneously perceiving and understanding the dominant group

dramaturgical model perspective used to suggest that social interaction has a likeness to dramas presented on a stage

dual labor market theoretical description of the occupational system as divided into two major segments: the primary and secondary labor markets

E

economic crisis the phenomenon beginning around 2009 by which banks failed, people defaulted on their mortgages, and unemployment and general economic malaise was pervasive

education debt the concept that society is creating greater and greater weakness in our educational system by not addressing the needs of all students

educational attainment total years of formal education

elites a class of people enjoying higher intellectual, social, or economic status

environmental justice social movements that challenge the presence of greater pollution and toxic waste dumping in poor and minority communities

environmental privilege immunity from the presence of pollution and toxic waste in one's community environment

environmental racism disproportionate location of sources of toxic pollution in or very near communities of color

environmental sociology an area of study within sociology that examines the interactions between the physical environment and social behavior

epistemology [II] ways of knowing

ethnic group social category of people who share a common culture, such as a common language or dialect, a common religion, and common norms, practices, and customs

ethnography descriptive account of social life and culture in a particular social system based on observations of what people actually do

ethnomethodology a technique for studying human interaction by deliberately disrupting social norms and observing how individuals attempt to restore normalcy

extreme poverty situation in which people live on less than $1 (or its equivalent) per day

F

false negatives a research or test result that erroneously indicates a condition as absent

fatherhood a social status of male parenting that carries with it certain roles, ideals, and responsibilities

feminism a way of thinking and acting that advocates a more just society for women

fictive fatherhood the idea that men can be active parents, even if not the biological parent of a child

functionalism theoretical perspective that interprets each part of society in terms of how it contributes to the stability of the whole society

fundamentalism a form of religion that upholds belief in the strict, literal interpretation of scripture

G

gay marriage the marriage of two people of the same sex

gender socially learned expectations and behaviors associated with members of each sex

gender hierarchy [XI] inequality arrayed along lines of gender

gender identity one's definition of self as a woman or a man

gender role learned expectations associated with being a man or a woman

gender segregation distribution of men and women in different positions in a social system

gender socialization process by which men and women learn the expectations associated with their sex

gendered institution total pattern of gender relationships embedded in social institutions

global assembly line an international division of labor where research and development is conducted in an industrial country and the assembly of goods is done primarily in underdeveloped and poor nations, mostly by women and children

global culture diffusion of a single culture throughout the world

global economy term acknowledging that all dimensions of the economy now cross national borders

global elite a loosely defined group of the richest and most powerful people in the world, working and living transglobally

global stratification the systemic inequalities between and among different groups within nations that result from the differences in wealth, power, and prestige of different societies relative to their position in the international economy

global warming idea that the planet is experiencing a slow and steady change in temperature, resulting in various detrimental environmental consequences

globalization increased economic, political, and social interconnectedness and interdependence among societies in the world

group collection of individuals who interact and communicate, share goals and norms, and who have a subjective awareness as "we"

H

hegemonic masculinity pervasiveness of culturally supported attributes associated with the dominant ideas of masculinity

heterogeneity condition in a group or society in which the members are different from each other

heterosexism institutionalization of heterosexuality as the only socially legitimate sexual orientation

hidden curriculum behaviors or attitudes that are learned at school that are not part of the formal curriculum

homophobia fear and hatred of homosexuality

hooking up casual sexual behavior that does not necessarily result in a committed relationship

hypothesis an "if, then" statement that is tested through rigorous research

I

identity how one defines oneself

ideology shared ideas or beliefs that purport to justify and support the interests of a particular group or organization

immigration migration of people into one society from another society

impression management process by which people control how others perceive them

income the amount of money earned in a given period of time

industrialization sustained economic growth following the application of raw materials and other, more intellectual resources to mechanized production

infant mortality rate the number of deaths per year of infants under age 1 for every 1,000 live births

informal economy (or informal sector) part of the labor force and economic system that is not included in the tax structure and other formal ways of being counted in a review of national economic picture

innumeracy the inability and/or unfamiliarity with mathematical concepts and methods

institution see **social institution**

interlocking directorate organizational linkages created when the same people sit on the boards of directors of a number of different corporations

issues problems that affect large numbers of people and have their origins in the institutional arrangements and history of a society

L

labeling theory a theory that interprets the responses of others as most significant in understanding deviant behavior

labor force all the members of a particular organization or population who are able to work, viewed collectively

labor force participation rate the proportion of a given group employed in the labor market

labor market system in which workers find jobs

life expectancy average number of years individuals in a particular group can expect to live

M

marriage rate number of marriages per some number (usually 1,000) in a given year

masculinity a set of social expectations associated with men's gender roles

McDonaldization process by which increasing numbers of services share the bureaucratic and rationalized processes associated with this food chain

means of production system by which goods are produced and distributed

medicalization identifying a condition or behavior as being a disorder requiring medical attention

medicalization of deviance social process through which a norm-violating behavior is culturally defined as a disease and is treated as a medical condition

megachurch a church with an unusually large congregation

meritocracy a system in which advancement is based on individual ability or achievement

methodology practices and techniques used to gather, process, and interpret theories about social life

middle class an amorphous group generally considered to share a common economic and social status

migration the act of moving from one country or region and settling in another

millennial generation individuals born between 1981 and 2002; the older set of millenials are now in their 20-something years and emerging into adulthood

mortality the quality or state of being a person or thing that is alive and therefore certain to die

mortality inequality inequality in death rate by social characteristics such as gender, race, or social class

mortality rate a measure of the frequency of death in a defined population during a specific interval

multinational corporation corporations that conduct business across national borders

myth of the absent father the misleading idea that African American men are unlikely to assume strong roles as parents

N

natality inequality a preference for male children over girls

neoimperialism dominance of a powerful nation over underdeveloped nations or areas by means of unequal conditions of economic exchange

neoliberalism a modern politico-economic theory favoring free trade, privatization, minimal government intervention in business, and reduced public expenditure on social services

norms specific cultural expectations for how to act in given situations

nuclear family social unit comprised of a man and a woman living together with their children

O

obesity increased body weight caused by excessive accumulation of fat

occupational segregation pattern by which workers are separated into different occupations on the basis of social characteristics such as race and gender

one drop rule the practice of classifying a person with any black ancestry as black

ontology the philosophical study of being or existence

organization large-scale units in society with a complex and hierarchical form

P

participant observation method whereby the sociologist is both a participant in the group being studied and a scientific observer of the group

patriarchy society or group in which men have power over women

philanthropy the practice of charitable aid or donations

plutocracy a government or state in which a wealthy class controls the government

politics of equality egalitarian political theory

popular culture beliefs, practices, and objects that are part of everyday traditions

populism political doctrine that supports the rights and powers of common people

post-racial society the idea that race no longer determines people's social location in society

poverty line figure established by the government to indicate the amount of money needed to support the basic needs of a household

power ability of a person or group to exercise influence and control over others

power elite model theoretical model of power positing a strong link between government and business

predictability process whereby a production process is organized to guarantee uniformity of product with standard outcomes

prejudice negative evaluation of a social group and individuals within that group based on misconceptions about that group

presentation of self Erving Goffman's phrase referring to the way people display themselves during social interaction

prestige subjective value with which different groups or people are judged

proletariat lowest social or economic class of a community; the laboring class, especially the class of industrial workers who lack their own means of production and hence sell their labor to live

Protestant ethic belief that hard work and self-denial lead to salvation and success

Q

queer theory a theoretical perspective that recognizes the socially constructed nature of sexual identity

R

race social category or social construction based on certain characteristics, some biological, that have been assigned social importance in the society

racial inequality the pattern whereby racial groups have differing degrees of advantage and disadvantage

racial formation process by which groups come to be defined as a "race"

racism perception and treatment of a racial or ethnic group or member of that group as intellectually, socially, and culturally inferior to one's own

recession a period of temporary economic decline during which trade and industrial activity are reduced

religion institutionalized system of symbols, beliefs, values, and practices by which a group of people interprets and responds to what they believe is sacred and that provides answers to questions of ultimate meaning

religiosity intensity and consistency of practice of a person's (or group's) faith

representative sample a sample drawn from a larger population to be used in a research project that mirrors very closely the key characteristics of that larger population

research design the plan for a research investigation that is derived from the particular question being asked

resocialization the process by which existing social roles are radically altered or replaced

rite of passage ceremonial events that mark the transition from one status to another; especially common as part of major life events

ritual symbolic activities that express a group's spiritual convictions

role expected behavior associated with a given status in society

role model person (real or fictional) with whom one identifies and patterns oneself after

role strain conflicting expectations within the same role

routinization the reduction of an action to regular, ongoing procedures

S

sample any subset of units from a population that a researcher studies

segregation the spacial and social separation of racial and ethnic groups

self the idea one has about one's own identity

service sector part of the labor market composed of the nonmanual, nonagricultural jobs

sex biological identity as male or female

sexism system of practices and beliefs through which women are controlled and exploited because of the significance given to differences between the sexes

sexual orientation manner in which individuals experience sexual arousal and pleasure

sexual revolution the widespread change in men's and women's roles and a greater public acceptance of sexuality as a normal part of social development

sick role social norms attached to how one who is sick should behave and be viewed by society

social act behavior that takes the perspectives of others into consideration

social change process by which societies are transformed

social class (or class) social structural position that groups hold relative to the economic, social, political, and cultural resources of society

social construction how social phenomena are defined and created by human beings

social control process by which groups and individuals within those groups are brought into conformity with dominant social expectations

social fact social pattern that is external to individuals

social institution established, organized system of social behavior with a recognized purpose

social interaction behavior between two or more people that is given meaning

social mobility movement over time by a person from one social class to another

social process the means by which behavior and social organization evolve

social reproduction thesis theoretical analysis that explains inequality as stemming from groups utilizing privilege to maintain their status over time

social stratification relatively fixed hierarchical arrangement in society by which groups have different access to resources, power, and perceived social worth; a system of structured social inequality

social structure patterns of social relationships and social institutions that comprise society

socialization the process by which one learns the expectations associated with social roles

society system of social interactions that includes both culture and social organization

sociological imagination ability to see the societal patterns that influence individual and group life

sociology study of human behavior in society

stereotype oversimplified set of beliefs about the members of a social group or social stratum that is used to categorize individuals of that group

stigma attribute that is socially devalued and discredited

subculture culture of groups whose values and norms of behavior are somewhat different from those of the dominant culture

survey to query (someone or a group or a sample) in order to collect data for the analysis of some aspect of a group or area

sweatshop a shop or factory in which employees work long hours at low wages under poor conditions

T

theory an explanation that can be tested through research

transnational extending or operating across national boundaries

transsexual a person born with one biological sex, but in the process of, or having completed, the transition to the other sex

troubles privately felt problems that come from events or feelings in one individual's life

U

undocumented worker an employed person who receives pay "off the books" and is not typically counted among full- or part-time workers in national counting of labor force participation rates

unemployment rate the percentage of those not working, but officially defined as looking for work

urban underclass grouping of people, largely minority and poor, who live at the absolute bottom of the socio-economic ladder in urban areas

V

variable Something that can have more than one value or score, used in sociological to measure a social fact

victimization experiences of people who have been victims of crimes or systematic oppression

virtual communities communities that exist via computer interactions

voter turnout percentage of a given population that participates in elections

W

wealth monetary value of what someone owns, calculated by adding all financial assets (stocks, bonds, property, insurance, investments, etc.) and subtracting debts; also called net worth

white privilege system in which white people benefit collectively from the social and economic history of racism

work productive human activity that produces something of value, either goods or services

working class those people who do not own the means of production and therefore must sell their labor in order to earn a living

Name Index

Subject Index